Cell Engineering

Cell Engineering

Volume 5

Editor-in-Chief

Professor Mohamed Al-Rubeai
University College Dublin,
Belfield, Dublin, Ireland

Editorial Board

Dr Hansjorg Hauser
GBF,
Braunschweig, Germany

Professor Michael Betenbaugh
Johns Hopkins University,
Baltimore, U.S.A.

Professor Martin Fussenegger
Swiss Federal Institute of Technology,
Zurich, Switzerland

Dr Nigel Jenkins
Serono Biotech Centre,
Fenil-sur-Corsier, Switzerland

Dr Otto-Wilhelm Merten
A.F.M.-Genethon 11,
Gene Therapy Program,
Evry, France

The titles published in this series are listed at the end of this volume

CELL ENGINEERING
Vol. 5: Systems Biology

Edited by

Mohamed Al-Rubeai

University College Dublin, Ireland

and

Martin Fussenegger

Swiss Federal Institute of Technology,
ETH Hoenggeerberg HCI Institute of Chemical and Bioengineering ICB,
Zurich, Switzerland

 Springer

A C.I.P. Catalogue record for this book is available from the Library of Congress.

ISBN-10 1-4020-5251-0 (HB)
ISBN-13 978-1-4020-5251-4 (HB)
ISBN-10 1-4020-5252-9 (e-book)
ISBN-13 978-1-4020-5252-1 (e-book)

Published by Springer,
P.O. Box 17, 3300 AA Dordrecht, The Netherlands.

www.springer.com

Printed on acid-free paper

CONTENTS

v

LIST OF CONTRIBUTORS

Mohamed B. Al-Fageeh
University of Kent
Canterbury, UK

Mohamed Al-Rubeai
University College Dublin
Dublin, Ireland

Nitin S. Baliga
Institute for Systems Biology
Seattle, WA, USA

Peter Dittrich
Friedrich Schiller University Jena,
Jena, Germany

Zachary L. Fowler
State University of New York at Buffalo
Buffalo, NY, USA

Olaf Heidenreich
Eberhard Karls University Tuebingen
Tuebingen, Germany

Vasiliki Ifandi
Warwick Medical School and University
of Warwick
Coventry, UK

Soo Hean Gary Khoo
University of Birmingham
Birmingham, UK

Thomas R. Kiehl
Rensselaer Polytechnic Institute
Troy, NY, USA

Mattheos Koffas
State University of New York at Buffalo
Buffalo, NY, USA

Effendi Leonard
State University of New York at Buffalo
Buffalo, NY, USA

Kenneth Lundstrom
Flamel Technologies
Vénisieux, France

Natalie Strudwick
University of Durham
Durham, UK

Maria Thomas
Eberhard Karls University Tuebingen
Tuebingen, Germany

Michèle F. Underhill
University of Kent
Canterbury, UK

Rui Zhou
Rensselaer Polytechnic Institute
Troy, NY, USA

Natalia Martínez Soria
Eberhard Karls University Tuebingen,
Tuebingen, Germany

Peter Morin Nissom
Bioprocessing Technology Institute,
Singapore

Robin Philp
Bioprocessing Technology Institute,
Singapore

Sarah J. Scott
University of Kent, Kent, UK

Martin Schröder
University of Durham, Durham, UK

Rosalyn J. Marchant
University of Kent, Kent, UK

C. Mark Smales
University of Kent, Kent, UK

Susan T. Sharfstein
Rensselaer Polytechnic Institute
Troy, NY, USA

D.E. Martens
Wageningen University, Wageningen,
The Netherlands

Amy K. Schmid
Institute for Systems Biology
Seattle, WA, USA

Pietro Speroni di Fenizio
Friedrich Schiller University Jena,
Jena, Germany

CHAPTER 1

STRUCTURAL GENOMICS

A Special Emphasis on Membrane Proteins

KENNETH LUNDSTROM

Flamel Technologies, 33, Avenue du Georges Lévy, 69693 Vénisieux, France

Abstract: Drug discovery based on structural knowledge has proven useful as several structure-based medicines are already on the market. Structural genomics aims at studying a large number of gene products including whole genomes, topologically similar proteins, protein families and protein subtypes in parallel. Particularly, therapeutically relevant targets have been selected for structural genomics initiatives. In this context, integral membrane proteins, which represent 60–70% of the current drug targets, have been of major interest. Paradoxically, membrane proteins present the last frontier to conquer in structural biology as some 100 high resolution structures among the 30,000 entries in public structural databases are available. The modest success rate on membrane proteins relates to the difficulties in their expression, purification and crystallography. To facilitate technology development large networks providing expertise in molecular biology, protein biochemistry and structural biology have been established. The privately funded MePNet program has studied 100 G protein-coupled receptors, which resulted in high level expression of a large number of receptors at structural biology compatible levels. Currently, selected GPCRs have been purified and subjected to crystallization attempts

Keywords: recombinant expression, purification, crystallization, structure-based drug design, structural genomics

1.   INTRODUCTION

Classically the drug discovery process has relied on methods which involve selection of appropriate compounds for biological evaluation to define lead compounds[1] followed by evaluation of structure-activity relationships (SARs). Later, the development of methods for high throughput *in vitro* screening and the synthesis of combinatorial compound libraries have set the stage for drug design based on molecular targets and protein structures. As the requirements for drug safety and

1

M. Al-Rubeai and M. Fussenegger (eds.), Systems Biology, 1–27.
© 2007 *Springer.*

efficacy have become more demanding and complex the costs for drug development have soared and the number of drug candidates even reaching the market has become very rare. The translation of *in vitro* efficacy to a desirable effect in animal models has been of major concern and even finally to obtain a drug which demonstrates the same profile in patients in clinical trials. There have been lengthy discussions related to the problems for pharmaceutical companies to fill their pipelines with novel innovative drug candidates. Cynically, it has been postulated that most major discoveries have already been made in drug development. On a more positive note, the sequencing of the human genome has revealed potential new drug targets. The advent of functional and structural genomics and their combination promises to become important tools in the discovery of novel cellular functions and pathways, which might uncover novel drug functions and open avenues for new medicines.

Structural genomics can be defined as the parallel structural characterization of a large number of gene products. This approach allows one to elaborate on the structure of protein subtypes and families, proteins with similar topology, i.e. membrane proteins, even whole genomes. Structural genomics therefore presents a great opportunity to improve the success in the discovery process of novel drugs. A drawback of studying a large number of protein targets is the need of large resources. For this reason, networks with expertise in various areas such as protein expression, purification and structure determination have been established. Several such consortia are described in this chapter. The MePNet project, which uniquely targets membrane proteins, is described in more detail. Moreover, a brief background in structure based drug design and examples of successful applications are presented. Today, one of the cornerstones for structural biology on therapeutically relevant proteins is recombinant protein production. The essential and most frequently used methods for expression are highlighted. Likewise, the procedures involved in protein purification and the various approaches for structure determination such as X-ray crystallography, nuclear magnetic resonance (NMR) and electron microscopy-based techniques are described. As 60–70% of the current drugs are targeted to membrane proteins a special emphasis is dedicated to this type of proteins.

2.   STRUCTURE-BASED DRUG DESIGN

Structure-based approaches in drug discovery have been frequently applied for lead optimization[2]. In this context, the drug potency and selectivity can be improved based on structural knowledge of proteins and their ligands[3]. The significant impact structural biology has had on lead discovery can be measured by faster definition of drug binding properties and the easier identification of "hit" compounds through screening programs[4]. Recent automated procedures such as AutoSolve® has enabled rapid structure resolution of protein-ligand complexes[5]. For instance, in cocktails of as many as 100 molecules their different shapes can be distinguished by the variation in electron density[6]. Another approach has been to use cocktails of smaller numbers of fragments at very high concentrations[7]. Candidate fragments are ranked automatically, several datasets can be collected at a synchrotron hourly and up to 1000 compounds screened in 2–3 days. In this approach a focused set of fragments from successful drug-like molecules is applied taken into account molecular weight ($<$ 200 Da), the presence of hydrogen bond donors and acceptors and solubility[8]. Also virtual screening can be applied, where large libraries of candidate fragments are systematically docked in the predefined binding site of the target protein in 3D computer models[5]. Additionally, it was demonstrated that NMR-based screening can be used to design small molecule drug candidates to inhibit the aberrant overexpression of c-myc in a variety of tumors by inhibition of the far upstream element (FUSE) binding protein (FBP) expression[9].

An interesting example of structure-based drug design comes from the program of synthesis of novel inhibitors of phosphodiesterase (PDE) to treat hypertension and other cardiovascular indications[3]. Based on the X-ray structure, a rational drug design approach was initiated to develop a series of heterocyclic replacements for the parent ring system in zaprinast to obtain novel inhibitors of PDE. After intermediate steps, the synthesized compound UK 92480 (sildenafil) showed 100-fold increase in PDE5 inhibitory activity compared to zaprinast as well as unprecedented selectivity over other PDE enzymes[10]. Although the pharmacological profile of sildenafil was as expected the clinical performance in patients with coronary heart disease was disappointing. However, further studies demonstrated that sildenafil blocked the PDE5 action which potentiated the natural activity of nitrose oxide and could be used for the treatment of erectile dysfunction (impotence)[11]. Although originally developed for a completely different indication, the globally known blockbuster drug Viagra can therefore be considered as an example of how structural biology at least indirectly could facilitate the drug discovery process. Several other examples exist where structural knowledge has directly influenced drug design as briefly summarized below.

2.1 Examples of Successful Drug Design

Although structure-based drug design is a relatively new approach it has already established itself as a prominent alternative to classical drug screening. The major advantages of applying structural information are the shorter time required and the possibility to develop medicines with improved potency and selectivity. In addition to application in lead discovery many available protein structures have been used directly for the development of drug candidates. Today, 42 structure-based drug compounds have successfully entered clinical trials and at least 7 drugs have been approved on the market[12]. For instance, the AIDS drugs Agenerase® and Viracept® were designed based on the high resolution structure of the human immunodeficiency virus (HIV) proteinase[13]. Similarly, the structure of the influenza virus neuraminidase was the basis for the flu drug Relenza®[14]. The 3D structure of the kinase domain of c-Abl proved also helpful in designing and particularly avoiding resistance to the protein kinase drug Gleevec®[15].

The number of structure-based drugs will certainly increase with time as more structures will be available. The study on kinase inhibitors already applies structure-based approaches and for example the crystal structure of the AKT kinase has presented an attractive target for development of small molecule inhibitors as tumor therapeutics[16]. However, the major boost in structure-based drug discovery is expected when finally structures of the largest topological group of drug targets, membrane proteins, will be available.

3.   RECOMBINANT PROTEIN EXPRESSION

Very few therapeutically interesting drug target proteins are present in native tissues at such high levels that it is possible to isolate them in quantities that allow purification and further structural studies. Among membrane proteins, only the nicotinic acetylcholine receptor isolated from the electronic organ of the *Torpedo marmorata* ray[17] and the bovine rhodopsin from cow retina[18] could be purified in quantities that allowed successful crystallization and structure determination. As a good example of the minute quantities of protein present in native tissue, the equivalent of 1000 pig brains was needed to purify 190 μg of the neuropeptide Y2 receptor for functional studies[19]. Despite efforts of miniaturization, structural studies still require tens to hundreds of milligrams of purified protein. Naturally, in case of human proteins, ethical considerations will prevent any large scale purification efforts from native tissue. It has therefore been essential to establish robust recombinant protein expression systems to provide material for structural biology. In this context, all potential host cells including bacteria, yeast, insect and mammalian cells have been evaluated and expression vectors with a variety of properties engineered. Moreover, cell-free translation systems have been specifically developed for structural biology applications. The major expression systems are briefly presented below.

3.1   Cell-Free Translation

During the last few years the methods for cell-free protein expression have rapidly improved and made this approach a sincere alternative to cell-based recombinant protein expression methods[20]. Cell-free translation systems are currently commercially available based on both *E. coli* and wheat germ lysates. The advantages of cell-free expression are the possibility to directly use PCR fragments omitting any cloning procedures and the expression can be controlled in defined minimal media. Perhaps the most applied approach for cell-free translation is the simple amino-selective or uniform stable isotope labeling, which presents the opportunity of direct sample analysis by NMR[21].

The strongest limitation of cell-free translation systems was for a long time the low expression levels obtained for membrane proteins. However, a cell-free translation system based on a modified *E. coli* S30 extract has provided significant improvement in expression levels also of membrane proteins[22]. For instance, two bacterial membrane proteins, multidrug transporter TehA and YfiK, were expressed at levels of 2.7 mg/mL. Likewise, such GPCRs as the human β2 adrenergic receptor (β2-AR), the human muscarinic acetylcholine M2 receptor (M2R) and the rat neurotensin receptor (NTR) were evaluated for cell-free expression[23]. To obtain functional binding of the GPCRs the β2-AR had to be fused to Gαs[24], the M2R to Gαi[25] and the NTR to maltose binding protein (MBP)[26]. The M2R had additionally the glycosylation sites and the third intracellular loop deleted.

3.2 Prokaryotic Expression

Prokaryotic expression has been the standard procedure for the expression of recombinant proteins. *E. coli*-based expression systems are definitely the most frequently used approach[27]. Alternative bacterial hosts have been verified as summarized below. The obvious advantage of prokaryotic expression is the ease of application and low production costs even at large scale level. The success rate has been very high for bacterial proteins, but also a large number of simple, especially soluble eukaryotic proteins, have been successfully expressed in prokaryotic systems. However, quite a few eukaryotic proteins require special post-translational modifications, which cannot be achieved in bacterial hosts. Whether these functions are necessary depends to a large extent on the further use of the produced material.

3.2.1 E. coli

The popularity of *E. coli* as an expression host stems from the ease to use and scale up of the system, the cheap production and high safety standards. Traditionally, *E. coli*-based expression has been reliable for soluble cytoplasmic proteins especially when applying fusion constructs[28]. Novel high throughput methods for *E. coli* expression including rapid cloning systems such as Gateway vectors[29] are reviewed elsewhere[30]. Recombinant protein secretion has also become feasible from Gram-negative bacteria, where the heterologously expressed protein is targeted to the periplasmic space or secreted into the culture medium[31].

Expression of membrane proteins in *E. coli* has, however, been more difficult. Two approaches have been taken where the recombinant protein is either targeted to the bacterial membrane or accumulated in inclusion bodies as aggregates. In the former case, the success has been relatively good for bacterial membrane transport proteins and receptors[32]. In contrast, expression of eukaryotic membrane proteins has been more difficult and the yields have been disappointingly low due to the toxicity to the host bacteria imposed by the foreign protein inserted in the membrane. However, major engineering involving evaluation of different deletions, fusion partners and tags has resulted in significantly improved expression of GPCRs. In this context, the fusion of the maltose binding protein (MBP) to the N-terminally deleted rat neurotensin receptor (NTR) resulted in milligram quantities of receptor protein[26]. Similarly, when the C-terminally truncated human adenosine A2a receptor was expressed as an MBP fusion protein 10–20 nmol receptor/L was produced in *E. coli* inner membranes[33]. Alternatively, the expression in bacterial inclusion bodies can generate high yields, but as the recombinant protein is present in aggregates refolding is mandatory to restore functional activity. The refolding process has generally been difficult and inefficient for membrane proteins[34]. Recently, technology improvement has provided some promising results for the glucagon-like peptide 1 receptor (GLP-1)[35], the human leukotriene B4 receptor BLT1[36] and the serotonin 5-HT4 receptor[37].

3.2.2   Lactococcus lactis

In search of alternative expression hosts of prokaryotic origin, the Gram-positive bacterium *Lactococcus lactis* has been applied for the expression of both prokaryotic and eukaryotic recombinant proteins[38]. *L. lactis* grows rapidly to high densities and does not require aeration. Isotope labeling is feasible as most strains are auxotrophs for multiple amino acids. Transformation methods have been developed for *L. lactis* and for instance the commonly used nisin NisA promoter with the NisR and NisK regulatory trans-acting factors represents a tightly regulated system[39]. The *L. lactis* system has demonstrated high expression levels of such prokaryotic membrane proteins as ABC transporters and MSF (major facilitator subfamily) efflux pumps[38]. For the yeast mitochondrial carrier proteins CTP1 and AAC3 with a 6 TM topology yields up to 5% of total protein were obtained. However, attempts to express the human KDEL (Lys-Asp-Glu-Leu) receptor resulted only in low levels (< 0.1% of total membrane protein). In another study, 11 yeast mitochondrial transporter proteins were expressed in *L. lactis* in a functional form[40]. The expression levels were 10-fold higher when a lactococcal signal peptide was used or the N-terminus of the transporters was deleted. The expression levels are now compatible with amounts needed for structural biology. Preliminary data for GPCR expression in *L. lactis* also looks promising (Kunji, personal communication).

3.2.3   Other prokaryotes

There is a relatively large number of bacterial species (Table 1), which have been evaluated for homologous and heterologous gene expression. For instance, *Bacillus subtilis* was already in the 1980s considered as a potentially efficient host for recombinant protein production such as interferon[41] and truncated viral membrane proteins[42]. The drawback with this approach was that the recombinant proteins were heavily degraded by bacterial proteases. However, the expression of bacterial proteins such as α-amylase from *Bacillus amyloliquefaciens* resulted in secretion of large quantities of active enzyme[43]. Recently, novel plasmid vectors inducible by IPTG have been engineered for high level intra- and extracellular expression in *B. subtilis*[44].

The characteristic feature of the Archaebacterium *Halobacterium salinarum* is its purple color, which relates to the accumulation of bacterio-opsin protein (Bop) complex formed with the chromophore retinol[45]. Attempts have been made to employ *H. salinarum* for heterologous gene expression. Fusion constructs of C-terminally tagged Bop to *E. coli* aspartate transcarbamylase (AT), human muscarinic M1 receptor, human serotonin 5-HT2 receptor and yeast α mating factor receptor Ste2 have been engineered[46]. The significance of the Bop sequence was evident as introduction of tags in this area reduced both mRNA and protein levels substantially. The Bop-AT fusion protein yields were 7 mg/L. In contrast,

Table 1. Structural Genomics Networks

Network	Expression	Targets
Berkeley Structural Genomics Center (BSGC) www.strgen.org	*E. coli*	*Mycoplasma genitalium* *Mycoplasma pneumoniae*
Center for Eukaryotic Structural Genomics (CESG) www.uwstructuralgenomics.org	*E. coli*	*Arabidopsis thaliana*
European Membrane Proteins (E-MeP) www.e-mep.org	*E. coli* *L. lactis* *S. cerevisiae* *P. pastoris* Baculo, SFV	prokaryotic MPs 100 GPCRs ion channels transporters other eukaryotic MPs
Joint Center for Structural Genomics (JCSG) www.jcsg.org	Cell-free, *E. coli* Baculo Adenovirus	*Thermotoga maritima* mouse genome human GPCRs
Membrane Protein Network (MePNet) www.mepnet.org	E. coli *P. pastoris* SFV	100 GPCRs
Membrane Protein Platform (MPP) www.swegene.org	*E. coli* *S. cerevisiae* *P. pastoris*	bacterial and yeast MPs human GPCRs
Midwest Center for Structural Genomics (MCSG) www.mcsg.anl.gov	*E. coli*	all 3 kingdoms of life
Northeast Structural Genomics Consortium (NSGC) www.nigms.nih.gov/Initiatives/PSI/Centers/NECSG.htm	*E. coli, yeast* insect cells	small proteins from yeast, *C. elegans* and *D. melanogaster*
New York Structural Genomics Research Consortium (NYSGXRC) www.nysgrc.org	*E. coli* yeast	bacterial, yeast and *C. elegans*
Paris-Sud Yeast Structural Genomics (YSG) www.genomics.eu.org	*E. coli*	250 non-membrane yeast proteins
Protein Structure Factory (PSF) www.proteinsturkturfabrik.de	*E. coli* *S. cerevisiae* *P. pastoris*	medically and biotechno-logically relevant proteins
Protein Wide Analysis of Membrane Proteins (ProAmp) www.pst-ag.com	*E. coli*	*S. typhimurium* and *H. pylori* MPs
RIKEN Structural Genomics Initiative (RSGI) www.rsgi.riken,go.jp/rsgi_e/index.html	Cell-free *E. coli*	*A. thaliana, T. thermophilus* and mouse proteins

(*continued*)

Table 1. (Continued)

Network	Expression	Targets
Southeast Collaboratory for Structural Genomics (SECSG) www.secsg.org	*E. coli* Baculo Lentivirus	*Pyrococcus furiousus, C. elegans*, human MPs
Structural Proteomics in Europe (SPINE) www.spineurope.org	*E. coli* Baculo	proteins/protein complexes of relevance to human health and disease
Structural Genomics Consortium (SGC) www.sgc.ox.ac.uk	*E. coli* Baculo	targets related to cancer, diabetes and malaria
Structure 2 Function Project (S2FP) www.s2fp.carb.nist.gov	*E. coli*	Haemophilus influenzae proteins
Swiss National Center for Competence in Research (NCCR) www.structuralbiology.ethz.ch	*E. coli* Baculo	bacterial MPs, transporters and human GPCRs
TB Structural Genomics Consortium (TBSGC) www.mbi-doe.ucla.edu/TB	*E. coli*	*Mycobacterium tuberculosis* proteins

MPs, membrane proteins

the fusions to the human GPCR constructs demonstrated no expression evaluated by immunoblotting. However, the Ste2 receptor showed a positive signal in Western blots, but the expression levels were much lower than for Bop-AT. In another study, the human muscarinic M1 and adrenergic α2B receptors were expressed from Bop fusion vectors[47]. Membranes isolated from cells expressing the Bop-M1 fusion showed no specific signal, whereas membranes from Bop-α2B cells were specifically recognized by a BR polyclonal antibody. Not unexpectedly, no binding activity was detected for Bop-M1. In contrast, specific binding was observed for Bop-α2B, albeit 10 times weaker than obtained in yeast or mammalian cells.

3.3   Expression in Yeast

Yeast represents a good model organism of eukaryotic origin and has frequently been used as host for recombinant protein production. The advantages of yeast are the simple culture and scale-up methods, easy genetic manipulation, the cheap production and yeast cells possessing the machinery for eukaryotic post-translational modifications. Different types of yeast vectors have been engineered, both for episomal expression and chromosomal integration. Yeast vectors have been frequently used for topologically different proteins such as cytoplasmic, membrane and secreted proteins and production of FDA-approved insulin[48] and hepatitis B surface antigen (HBsAg)[49] as pharmaceuticals. Typically, yeast and especially *Saccharomyces cerevisiae* has a strong tendency for hyperglycosylation of heterologous proteins, which could induce immunogenic or allergenic reactions if the product is aimed at therapeutic use. The fission yeast *Schizosaccharomyces pombe*[50] and the methylotropic yeast *Pichia pastoris*[51] have also been developed into efficient expression systems as highlighted below. Among other yeast strains the methylotrophic *Hansenula polymorpha* and the dimorphic *Arxula adeninivorans* and *Yarrowia lipolytica* have not yet been much explored, but might present appropriate alternatives for future recombinant protein expression[52].

3.3.1   Saccharomyces cerevisiae

Saccharomyces cerevisiae is well characterized and its whole genome has been sequenced[53] so it is no surprise that the Baker's yeast has been used frequently for heterologous gene expression[54]. Among the many different proteins expressed from *S. cerevisiae* can be mentioned hepatitis B surface antigen (HBsAg)[55], α1-Antitrypsin[56], human interferon-α[57] and β-Endorphin[58]. A variety of yeast promoters have been used[54], among which two inducible systems apply GAL1 and CUP1[59].

S. *cerevisiae* has also commonly been applied for the expression of membrane proteins. In this context, C-terminal FLAG- and His-tags were engineered to the yeast α-factor Ste2p receptor and expressed at milligram yields[60]. Ste2p receptor activity could be obtained after metal affinity column purification and reconstitution in artificial phospholipid vesicles, but only after addition of solubilized yeast membranes. Also the human dopamine D1A receptor was expressed at high levels in *S. cerevisiae*[61]. Applying metal affinity and immunoaffinity chromatography, the D1A receptor was purified to near homogeneity and showed after reconstitution [^{3}H] SCH23390 antagonist binding. Additionally, large-scale fermentation cultures at the 15 L level resulted in approximately 40 pmol of β2 adrenergic receptor per milligram, which corresponds to yields of 20–30 mg of functionally active receptor[62]. Recently, the biotin-tagged rabbit sarcoplasmic-endoplasmic reticulum Ca^{2+}-ATPase isoform 1a (SERCA1a) was expressed in *S. cerevisiae*[63]. Purification of SERCA1a was performed by avidin agarose affinity chromatography followed by HPLC filtration. The purified protein was successfully crystallized and data collected at a 3.3 Å resolution.

3.3.2   Schizosaccharomyces pombe

Due to certain disadvantages of *S. cerevisiae* such as plasmid instability and hyperglycosylation recombinant protein expression has also been carried out in the fission yeast *Schizosaccharomyces pombe*. For instance, vectors for chromosomal integration to obtain stable expression have been engineered for *S. pombe*[64]. The glycosylation pattern in *S. pombe* is different from *S. cerevisiae*, which could be of advantage for recombinant protein expression[65]. The human blood coagulation factor XVIIIa was expressed in *S. pombe* from an alcohol dehydrogenase (ADH) promoter[66].

Expression vectors based on *S. pombe* have been applied on membrane proteins. In this context, the human dopamine D2 receptor was expressed from a thiamine-repressible *nmt1* promoter, which resulted in 14.6 pmol receptor/mg and localization at the plasma membrane[67]. The expression levels were 5 fold higher compared to *S. cerevisiae*. In contrast, the rat dopamine D2 receptor[68] and the human neurokinin-2 receptor[69] were only expressed at 1 pmol/mg levels.

3.3.3   Pichia pastoris

The metylothrophic yeast strain *Pichia pastoris* has demonstrated high efficacy as an expression host[51]. The expression systems developed for *P. pastoris* are based on integration of the heterologous gene construct in the yeast genome and utilization of strong yeast promoters such as the alcohol oxidase (AOX) promoter[70]. A clear advantage of culturing *P. pastoris* is the possibility to establish extremely high growth density with OD_{600} values up to 500 U/ml. More than 200 recombinant proteins have been expressed in *P. pastoris*, including bacterial, fungal, plant, insect, mammalian and viral proteins. For instance, an endoglucanase from *Streptomyces viridosporus* resulted in yields up to 2.5 g/L[71]. Likewise, human fibrinogen was produced at 100 mg/L[72].

A relatively large number of membrane proteins have been expressed from *P. pastoris* vectors. The mouse serotonin 5HT5 receptor was the first GPCR to be expressed in *P. pastoris*[73]. Engineering of an α-factor signal sequence from *S. cerevisiae* resulted in significantly higher yields (22 pmol/mg). Similarly, when the human dopamine D2 receptor was expressed in *P. pastoris* the receptor density was as low as 1200 receptors/cell, which could be enhanced by 20-fold by the introduction of the α-factor signal sequence[74]. Improvement of expression levels by several folds was observed for the mouse 5-HT5A, the dopamine D2S and the β2 adrenergic receptors by introduction of the biotinylation (biotin-tag) of the transcarboxylase from *Propionibacterium shermanii* at the C-terminal[75]. It is thought that the biotin-tag stabilized the receptor directly against degradation or through prevention of the unfolding protein response. The pharmacological profile of GPCRs expressed in *P. pastoris* was similar to native tissue[74]. However, due to differences in lipid composition the ligand affinities were generally lower in yeast than in mammalian cells. The overexpression of GPCRs resulted in localization of receptors to the endoplasmic reticulum and the Golgi apparatus in contrast to the translocation to the plasma membrane in native mammalian cells. High-level overexpression in mammalian cells has, however, also resulted in prominent retention of receptors in the endoplasmic reticulum[76]. Receptor levels could be significantly enhanced by addition of specific agonists or antagonists or other supplements to the growth medium during methanol induction[77].

Among other membrane proteins, the water channel aquaporins were expressed in *P. pastoris* with C-terminal His- and myc-tags[78]. Recently, the structure of the spinach aquaporin SoPIP2 was solved after purification of the recombinant channel expressed in *P. pastoris*[79]. A rat neuronal voltage-sensitive K^+-channel was overexpressed in *P. pastoris*, which allowed purification and single particle imaging and reconstitution of 2D crystals by cryo-EM[80]. The high resolution structure of another mammalian voltage-dependent K^+ channel from the *Shaker* family was resolved after heterologous expression in *P. pastoris*[81].

3.4   Expression in Insect Cells

Insect cell-based recombinant expression has been applied for mammalian proteins to a large extent because of the similarity of insect and mammalian cells. Insect cells possess many of the post-translational modification mechanisms also characteristic for mammalian proteins. However, differences exist and for instance the N-glycosylation pattern in insect cells is simpler and of high mannose type[82]. The engineering of baculovirus expression systems presented the real breakthrough in application of insect cells for recombinant protein expression[83]. In this context, recombinant proteins have been produced for pharmacological characterization and drug screening purposes, but especially baculovirus has been the second most used production system in structural biology after *E. coli*. Alternatively to baculovirus, stable inducible expression systems in *Drosophila* Schneider-2 cells have also been established[84].

3.4.1   Baculovirus

The most frequently used system for expression in insect cells is based on baculovirus (*Autographa californica*)[83]. A number of modifications have been introduced to facilitate the cloning and virus production procedures. The robust expression obtained from baculovirus vectors have made them attractive for production of various types of human recombinant proteins[85]. Although baculovirus vectors were originally developed for transduction of insect cells, modified vectors with mammalian-specific promoters has allowed efficient expression in different mammalian cells[86]. This approach has, however, mainly served functional studies and drug screening programs as relatively high virus concentrations are required to established efficient transduction rates in mammalian cells, which make large-scale applications unfeasible. In contrast, the scale-up procedure in insect cells is straight forward and presents a sensible alternative to bacterial and yeast expression for production of large quantities of recombinant protein for structural studies. Scale-up has been conducted in several insect cell lines, preferentially in *Spodoptera frugiperda* (Sf9 and Sf21 cells), *Mamestra brassica* and *Trichoplusia ni* (High Five) cells[87].

Baculovirus vectors have been applied for the expression of several membrane proteins and especially GPCRs. For instance, rhodopsin was expressed at yields of 4-6 mg/L of which 80% represented functional receptor[88]. Optimal expression has been obtained under the following conditions[87]: bioreactor cultures with control of stirring speed, oxygen supply, removal of CO_2 and ambient temperature; dividing cells are infected at a low MOI; cells are harvested before they start to disintegrate. In an attempt to improve folding and transport of membranes proteins to the plasma membrane the signal sequence from the influenza hemagglutinin gene was placed in front of the human β2 adrenergic receptor, which resulted in approximately two-fold increase in receptor expression[89]. In another study, 16 human GPCRs were expressed in three baculovirus-infected cell lines and monitored by radioligand binding assays[90]. The expression levels varied considerably from 1 to 250 pmol/mg protein. Recently, the C-terminally His-tagged human histamine H1 receptor was expressed in Sf9 cells at levels of 30–40 pmol per 10^6 cells[91].

Non-GPCR membrane proteins such as ion channels and transporter proteins have also been successfully expressed in baculovirus-insect cell systems. For instance, the voltage-gated AKT1 K^+ channel[92] and the NaPi-2 and NaSi-1 cotransporters[93] demonstrated functional activity when expressed in insect cells. Furthermore, a membrane spanning domain of the cystic fibrosis transporter (CFTR)[94] and the Na^+-K^+-ATPase[95] revealed functional activity.

3.4.2   Other insect cell systems

In addition to baculovirus other insect cell-based expression systems have been developed. Typically, *Drosophila* Schneider cells have been applied for heterologous expression of cytosolic, membrane and secreted proteins[84]. The generation of stable Schneider SL-3 cell lines is relatively easy and cheap. To verify the capacity of post-translational modifications and protein transport in insect cells the VIP36 (vesicular integral membrane protein) was expressed in SL-3 cells[84]. In parallel, a truncated form of VIP36 served as the model for a secreted protein and annexin XIIIb was studied to evaluate myristoylation in SL-3 cells. The secreted truncated VIP36 was N-glycosylated and the N-glycan of the Golgi-localized full-length VIP36 was endo-H resistant. Moreover, annexin XIIIb was myristyolated suggesting that SL-3 cells can be considered as suitable hosts for mammalian proteins.

Drosophila cells were used for the large scale production of human interleukin 5 (IL5) and soluble forms of its receptor alpha subunit[96]. The deglycosylated form of IL5 was active and a 2.6 Å resolution crystal structure determined from the purified recombinant protein. Also GPCRs have been subjected to expression studies in Drosophila cells[97]. In this context, the human mu opioid receptor (hMOR) was expressed in Schneider 2 cells with an N-terminal EGFP-tag. The recombinant EGFP-hMOR showed a similar pharmacological profile as detected in mammalian cells and functionality was demonstrated by cAMP stimulation and [^{35}S] GTPγS binding. Comparison of binding data and quantitative EGFP fluorescence intensity analysis indicated that a relatively large number of the receptors did not present high-affinity binding, which might be at least partly due to retention of the receptors in intracellular structures.

3.5 Expression in Mammalian Cells

The most native environment for expression of mammalian proteins is obviously mammalian host cells. Two strategies have been applied by performing either transient transfection or establishing stable cell lines[98]. Both approaches have their advantages and disadvantages. Transient transfection provides faster expression and generally higher yields, whereas although time-consuming and relatively labor-intensive stable expression presents the means of obtaining clones for long-term use. One of the bottlenecks for using mammalian expression has been the expensive cell culture components and the more demanding scale-up procedure compared to bacterial and yeast systems. The choice of promoter and cell line plays also a significant role in obtaining high levels of expression and it was demonstrated that the full-length CMV promoter was superior[99].

3.5.1 Transient expression

A large number of mammalian expression vectors have been engineered for transient expression in mammalian cell lines such as BHK-21, CHO-K1, COS-7 and HEK293[100]. Methods have been developed for large scale transient expression[101]. Recently, CHO-K1-S suspension cells cultured in serum-free medium generated mg/L quantities of bioactive antibody[102].

A number of GPCRs have been transiently expressed in mammalian cells. For instance, the cholecystokinin (CCK) A receptor was expressed transiently in both COS cells[103] and HEK293 cells[104]. Other membrane proteins such as the Na^+- and Cl^--coupled GABA transporter GAT-1 was expressed transiently in Ltk-cells[105], the muI Na^+-channel in HEK293 cells[106] and the GLUT1 and GLUT4 transporters in COS-7 cells[107].

3.5.2 Stable cell lines

Although popular, instability of established mammalian cell lines has been of major concern and has led to the engineering and use of inducible expression vectors[108]. Generally, the expression levels obtained from stable cell lines have been lower than in transient expression[99]. The choice of promoter and host cell line are important factors for optimal expression. In a study where four secreted proteins were stably expressed from different promoters in various cell lines, the highest expression levels were obtained from the myeloproliferative sarcoma virus (MPSV) LTR promoter in CHO-K1 cells[99].

Expression levels for membrane proteins have generally been relatively low in stable cell lines. Typically, the kappa opioid receptor was expressed at 266 fmol/mg in CHO cells[109]. More recently, improved levels in the range of 1 to 20 pmol have been obtained for certain GPCRs, although resulting in only moderate yields of 0.1 mg/L in large scale production. Exceptionally high levels of rhodopsin were produced in an inducible mutant HEK293 suspension cell line1[110]. In this case, up to 6 mg receptor was obtained per liter culture. In addition to GPCRs the serotonin transporter (SERT) protein has been expressed in stable cell lines[111]. For instance, the cold-inducible pCytTS system based on the Sindbis virus replicase[112] generated 250,000 SERT receptors per cell, whereas using the tetracycline-inducible T-Rex system produced 400,000 copies per cell[111].

3.5.3   Viral vectors

Viral mammalian vectors have been frequently used for recombinant expression. Generally viruses possess strong promoters from which high levels of expression can be obtained. Although the host range varies, many viruses such as adenoviruses, alphaviruses, lentiviruses and vaccinia viruses can infect many different types of mammalian cells. As previously mentioned, also baculovirus vectors have been successfully used for transduction of mammalian cells[86]. The drawback of viral vectors, especially those naturally infecting mammalian cells is the obvious safety concerns related to their use. However, these issues have been thoroughly addressed by the engineering of deletion mutant and replication-deficient vectors.

Adenovirus vectors are most commonly used for gene expression *in vitro* and *in vivo*[113]. Various replication-deficient and –competent adenovirus vectors have been engineered to express a large number of recombinant proteins. For example, when the nonstructural glycoprotein NS1 from tickborne encephalitis virus (TBEV) was expressed from a CMV major early-immediate promoter in a replication-deficient adenovirus vector up to 25% of the total protein was represented by NS1[114]. Several GPCRs have been expressed from adenovirus vectors. For instance, the mu (MOR) and kappa opioid receptors (KOR) were expressed in CHO cells from replication-deficient adenovirus vectors as fusion proteins with GFP or as such[115]. The pharmacological properties of the recombinant receptors were identical to observations in native tissue. In comparison to expression in stable cell lines, 3 fold higher expression levels (B_{max} values of 3 pmol/mg) were obtained. Likewise, the β2 adrenergic receptor was expressed in rabbit myocytes at a level of 4 pmol/mg receptor[116].

Vaccinia virus expression systems have been engineered as hybrid vectors with bacteriophage RNA polymerases, especially applying T7 phages[117] and as replication-deficient vaccinia virus vectors[118]. Hundreds of foreign genes have been expressed from vaccinia virus vectors[119]. Several GPCRs have been expressed in mammalian cells as recombinant proteins using vaccinia virus vectors. In this context, it was demonstrated that the neuropeptide Y (NPY) receptor was localized in the plasma membrane and saturation binding experiments suggested that 5–10 million binding sites existed per cell[120]. Also the human dopamine D2 and D4 receptors were expressed from vaccinia virus vectors in rat-1 cells in a functional form[121]. The GABA transporter was expressed in HeLa cells infected with vaccinia virus demonstrating similar expression levels to transient expression[105].

Replication-deficient **lentivirus** vectors generally based on the human immuno-deficiency virus 1 (HIV-1) have been widely applied to gene expression in cell lines and *in vivo* especially in cases were long-term expression is advantageous[122]. The expression from lentivirus vectors is generally driven by an internal cassette as lentiviral promoters are inefficient in human cells and biosafety concerns require control mechanisms[123]. Efficient reporter gene expression (GFP and lacZ) was obtained both in cell lines and *in vivo* from the constitutive CMV-1E promoter[122,124]. Lentivirus vectors have also been used for the expression of membrane proteins. The human retinal pigment epithelium (RPE) retinal GPCR was expressed in COS-7 and ARPE-19 (retinal pigment epithelium) cells from replication-deficient vectors[125]. The expression was verified by immunodetection and [^{3}H] all-*trans*-retinal binding showing 100 fold higher expression levels in ARPE-19 cells than in COS-7 cells. The long-term expression was demonstrated by stable expression up to 6 months. Tranzyme Pharma has recently developed the lentivirus-based TranzExpression Technology (TexT™), which has been successfully evaluated for a number of GPCRs (*www.tranzyme.com*).

Semliki Forest virus (SFV), a single strand RNA virus with an envelope structure[126] has been commonly used for *in vitro* and *in vivo* heterologous gene expression[127]. SFV is particularly well suitable for structural genomics approaches due to its fast high-titer virus production and broad host range. A variety of topologically different proteins have been expressed from SFV vectors including cytoplasmic, membrane and secreted proteins[128]. SFV vectors have frequently been used for overexpression of GPCRs and ion channels and large-scale production has yielded up to 10 mg/L receptor for structural studies[129]. As described below 101 GPCRs were overexpressed with a high success rate from SFV vectors in the MePNet consortium[130].

4.   PROTEIN PURIFICATION

A basic requirement for any structural characterization of recombinantly expressed proteins is purification to high homogeneity. For soluble proteins this procedure is straight forward whereas for membrane proteins more complicated steps including solubilization and separation of protein and lipid components are necessary[131]. To facilitate the solubilization process detergents are applied. There are a number of different detergents and it seems that each target protein has to be screened for appropriate detergents[132]. The use of detergents is reviewed elsewhere[133]. Various biochemical purification methods have been developed, which are briefly summarized below.

4.1   Affinity Tag Purification

The most commonly used approach for protein purification is based on immobilized metal affinity chromatography (IMAC). To facilitate the purification of recombinant proteins from the mixture of other cellular proteins different purification tags have been engineered into the constructs. Histidine (His) tags, which bind to chromatographic media charged with Ni^{2+} have been extensively applied[134] and has allowed a one-step purification procedure. Successful positioning of the His-tag at the N- or C-terminus or even within the recombinant protein sequence has been demonstrated by numerous examples.

In addition to His-tags an eight amino acid streptavidin binding sequence (Strep) has been applied and has proven excellent for large scale purification[135]. Also biotin-, FLAG- and hemagglutinin tags are possible alternatives.

4.2   Other Means

In addition to using tags for recombinant protein purification classical biochemical methods including ammonium sulfate precipitation and sucrose gradients are possible. However, these methods require large quantities of available material, which is not the case for membrane proteins. Ion exchange chromatography can also be applied for protein purification. Gel filtration or size exclusion chromatography, which separates the proteins based on their molecular weights, has been commonly used for purification purposes. Additionally, hydrophobic interaction and reverse-flow chromatography can be applied.

5.   STRUCTURE DETERMINATION

Different approaches are possible for collecting structural information on proteins. The highest structure resolution can generally be obtained by x-ray crystallography, but the drawback of this approach is that high quality crystals are required. Although this approach is more or less a routine procedure for soluble proteins today it has only been successful for a limited number of membrane proteins. Other methods such as nuclear magnetic resonance (NMR) have therefore proven useful although limitations have been seen in relation to structure resolution and only recently improved methodology has opened NMR for larger proteins. Finally, structure determination can be exercised by atomic force microscopy and electron microscopy techniques, which can be applied on crude samples but on the other hand results in lower resolution.

5.1   X-Ray Crystallography

One of the cornerstones in structure determination is obviously X-ray crystallography, which has allowed obtaining structure resolution below 2.0 Å. The process includes crystallization of highly purified protein samples, measurement of crystal diffraction, solving phase determination problems, phase and electron density calculations and model building and refinement. The crystallization procedure has during the last few years experienced major development with the introduction of automation and miniaturization[136]. The crystallization process has been strongly facilitated by the reduced volumes (nanoliter). Also the development of high density crystallization microplates in 96 or higher format has contributed to the high throughput nature[137]. It has therefore become possible to optimize solution variables such as pH, ionic strength, temperature, and concentrations of salts and detergents. This approach has allowed the establishment of up to 100,000 crystallization trials per day. Important issues are the harvesting and storage of crystals to be analyzed at synchrotron radiation facilities, data collection at beam lines and characterization of obtained crystals. It is anticipated that the miniaturization of the crystal screening process will result in production of smaller crystals and will require the use of micro-diffractometer technologies.

5.2 Nuclear Magnetic Resonance

Nuclear magnetic resonance (NMR) technologies have generally been seen as a complementary technology to X-ray crystallography[138]. Although the resolution has been poorer and the molecular weight range has set limitations, NMR has routinely been used for the identification and evaluation of chemical leads[139]. Furthermore, NMR requires extensive isotope labeling of the protein, which is expensive and time consuming. Recent development in probe technology, software and NMR methodology itself has made it possible to obtain high resolution structures and increase the size of the studied proteins. For instance, generation of iterative protein-ligand complexes has become feasible with NMR[140]. Novel solid state and solution NMR methods have further provided means for utilization of NMR in structural biology on membrane proteins[141].

5.3 Electron Microscopy and AFM

Although the use of electron microscopy (EM) and atomic force microscopy (AFM) have not seen the same rapid progress as X-ray crystallography recently, these methods can be applied to extract atomic resolution information on proteins[142]. Cryoelectron microscopy of reconstituted membrane proteins in 2D crystals has been conducted for bacteriorhodopsin[143] and aquaporin AQP1[144]. Although the resolution was at 3.5 Å it made it possible to define the atomic structure, subsequently confirmed by X-ray crystallography. Additionally, AFM can be applied on native and reconstituted membranes in aqueous solutions, which has allowed the monitoring of polypeptide loops. For instance, AFM has been used to study disc membranes of vertebrate photoreceptor rod outer segments, where rhodopsin harbors 50% of the surface space. It could be demonstrated that rhodopsin was present in dimers and higher oligomeric forms[145].

6. STRUCTURAL GENOMICS NETWORKS

A large number of both national and international networks have been established to facilitate the interaction between experts in different areas such as protein expression, purification and crystallography, required for structural biology. Working in large consortia allows parallel studies on many targets, which aids significantly in understanding issues related to protein expression and the feasibility of structural approaches. The demands on network coordination are, however, high as the information flow between the different scientists and institutes is essential for achievement of success. Table 1 presents a comprehensive list of existing networks.

6.1   Networks on Soluble Proteins

Many of the established networks have focused their activities entirely on the structural biology of soluble proteins. This strategy is fully understandable as the success rate of structure resolution is very high and a large number diffracting high quality crystals can be obtained and many structures can be solved within a short time period. Soluble proteins are also well expressed in *E. coli* and the purification procedure is relatively straight forward. Quite a few networks have also focused on thermostable proteins, which are often stable at conditions were other proteins are easily degraded facilitating the purification procedure significantly.

6.2   Networks on Membrane Proteins

The importance of membrane proteins as drug targets is reflected by the number of networks that have included membrane proteins in their target list. In many instances whole genomes are studied, which is the case for structural genomics programs on for example *Mycobacterium tuberculosis*[146] and *Thermus thermophilus*[147]. Alternatively, the targets may represent certain types of proteins or protein families. Among networks studying membrane proteins, the most popular targets are GPCRs and ion channels due to their importance in drug discovery. Networks such as the RIKEN Structural Genomics Initiative, The Joint Center for Structural Genomics, Swegene, and the Swiss National Center for Competence in Research all have GPCRs in their programs.

The EU-funded E-MeP (European Membrane Proteins) uniquely focuses on membrane proteins. Among the 100 prokaryotic targets are ABC transporters and other bacterial membrane proteins. E-MeP also studies 200 eukaryotic proteins, of which 100 are GPCRs and 100 ion channels, transporters and other integral membrane proteins. An initial expression evaluation of eukaryotic membrane proteins is carried out in *E. coli*, *L. lactis*, *S. cerevisiae*, *P. pastoris*, baculovirus and SFV.

6.2.1   MePNet – Structural genomics on GPCRs

As an example of a program uniquely dedicated to GPCRs, a brief summary of the MePNet program is presented[148]. MePNet was established in 2001 through private funding from more than 30 pharmaceutical and biotech companies interested in structural biology on GPCRs. The aim of the program was to subject 100 GPCRs to structural biology by developing technologies for expression, purification and crystallization in a high throughput format. The target GPCRs were selected based on ligand availability, representation of GPCR families and subtypes and relevance to human disease. Initially, expression evaluation was carried out in three expression systems based on *E. coli*, *P. pastoris* and SFV vectors. The expression in bacteria was targeted to inclusion bodies, whereas the GPCRs in yeast and mammalian cells were aimed at membranes. Expression in *E. coli* resulted in a success rate of 46% measured by immunodetection, whereas in yeast and mammalian cells

94% and 96%, respectively, of the GPCRs were successfully expressed. Overall, more than 60 different GPCRs were expressed at structural biology compatible levels, i.e. at least 1 mg/L at least in one of the three systems. Further expression optimization improved the binding activity for *P. pastoris* and SFV, leading to the highest binding values of 180 and 287 pmol/mg[130], respectively. The yields for well-expressed GPCRs in *E. coli* were further improved in fermentor cultures resulting in up to 350 mg/L. A limited number of GPCRs have been purified and subjected to refolding attempts. Similarly, selected GPCRs from yeast and mammalian cells have been solubilized and purified. Crystallization attempts on the first purified GPCRs are in progress. MePNet launched in 2005 the second phase of its program (MePNet2), which will have a strong focus on crystallography. The MePNet teams are confident that several high resolution structures on GPCRs will be available in the near future.

7.   CONCLUSIONS

In summary, structural genomics approaches have presented fruitful ways to provide quickly novel structural information. Studies on whole genomes, topologically defined types of proteins and protein families in parallel have provided invaluable information. Technology development has played a key role in this process leading to improved expression vectors and systems, as well as automation and miniaturization of purification and crystallization methods. However, systematic approaches have also required large resources and broad expertise and in this context it has been appropriate to establish structural genomics networks. As technology improvement has been achieved it has also become feasible to study membrane proteins. Not only do they represent some 30% of the genomes for various organisms, but they are also targets for 70% of current drugs and potentially interesting novel targets for new medicines.

8.   FUTURE ASPECTS

Structure determination has been highly successful for a large number of soluble proteins in studies on individual proteins and in parallel studies on a large number of targets within structural genomics networks. From a historical aspect there is a strong similarity today for membrane proteins what was experienced for soluble proteins in the 1970s. At the beginning of that decade the first high resolution structures had been obtained and even in 1979 less than 100 structures were available. The dramatic technology improvement seen during the last two decades of the 20th century resulted in an exponential growth in the number of solved structures, reaching a number of over 30,000 today. If we manage to achieve a similar development of technologies for membrane proteins we might finally be able to soon conquer the last frontier in structural biology.

9. REFERENCES

1. Lombardino, J.G. & Lowe III, J.A. The role of the medicinal chemistry in drug discovery – then and now. *Nat. Rev. Drug Discov.* **3**, 853–62 (2004).
2. Blundell, T.L. Structure-based drug design. *Nature* **384S**, 23–6 (1996).
3. Campbell, S.F. Science, art and drug discovery: A personal perspective. *Clin. Sci.* **99**, 255–60 (2000).
4. Blundell, T.L. & Patel, S. High-throughput X-ray crystallography for drug discovery. *Curr. Opin. Pharmacol.* **4**, 490–6 (2004).
5. Blundell, T.L., Jhoti, H. & Abell, C. High throughput crystallography for lead discovery in drug design. *Nat. Rev. Drug Discov.* **1**, 45–54 (2002).
6. Nienaber, V.L., Richardson, P.L., Klighofer, V., Bouska, J.J., Giranda, V.L. & Greer, J. Discovering novel ligands for macromolecules using X-ray crystallography screening. *Nat. Biotechnol.* **18**, 1105–8(2000).
7. Blundell, T.L., Abell, C., Cleasby, H., Hartshorn, M.J., Tickle, I.J., Parasini, E. & Jhoti, H. High throughput X-ray crystallography for drug discovery, in *Drug Design, Cutting Edge Approaches*, D. Flower, ed., Royal Society Chemistry, London, pp 53–9 (2002).
8. Scharff, A. & Jhoti, H. High-throughput crystallography to enhance drug discovery. *Curr. Opin. Chem. Biol.* **7**: 340–5 (2003).
9. Huth, J.R., Yu, L., Collins, I., Mack, J., Mendoza, R., Isaac, B., Braddock, D.T., Muchmore, S.W., Comess, K.M., Fesik, S.W., Clore, G.M., Levens, D. & Hajduk, P.J. NMR-driven discovery of benzoylanthranilic acid inhibitors of far upstream element binding protein binding to the human oncogene c-myc promoter. *J. Med. Chem.* **47**, 4851–7 (2004).
10. Wallis, R.M., Corbin, J.D., Francis, S.H. & Ellis, P. Tissue distribution of phosphodiesterase families and the effects of sildenafil on tissue cyclic nucleotides, platelet function, and the contractile responses of trabeculae carneae and aortic rings in vitro. *Am. J. Cardiol.* 83, 3C–12C (1999).
11. Goldstein, I., Lue, T.F., Padma-Nathan, H., Rosen, R.C., Steers, W.D. & Wicker, P.A. Oral sildenafil in the treatment of erectile dysfunction. Sildenafil Study Group. *N. Engl. J. Med.* **338**, 1397–404 (1998).
12. Hardy, L.W. & Malikayil, A. The impact of structure-guided drug design on clinical agents. *Curr. Drug. Discov.* **3**, 15–20 (2003).
13. Stoll, V., Qin, W., Stewart, K.D., Jakob, C., Park, C., Walter, K., Simmer, R.L., Helfrich, R., Bussiere, D., Kao, J., Kempf, D., Sham, H.L. & Norbeck, D.W. X-ray crystallographic structure of ABT-378 (lopinavir) bound to HIV-1 protease. *Bioorg. Med. Chem.* **10**, 2803–6 (2002).
14. Varghese, J.N. Development of neuroaminidase inhibitors as anti-influenza virus drugs. *Drug Dev. Res.* **46**, 176–96 (1999).
15. Schindler, T., Bornmann, W., Pellicena, P., Miller, W.T., Clarkson, B. & Kuriyan, J. 2000., Structural mechanism for STI-571 inhibition of abelson tyrosine kinase. *Science* **289**, 1938–42 (2000).
16. Kumar, C.C. & Madison, V. AKT crystal structure and AKT-specific inhibitors. *Oncogene* **24**, 7493–501 (2005).
17. Unwin, N. Structure and action of the nicotinicacetylcholine receptor explored by electron microscopy. FEBS Lett. **555**, 91–5 (2003).
18. Palczewski, K., Kumasaka, T., Hori, T., Behnke, C.A., Motoshima, F., Fox, B.A., Le Trong, I., Teller, D.C., Okada, T., Stenkamp, R.E., Yamamoto, M., & Miyano, M. 2000., Crystal structure of rhodopsin: A G protein-coupled receptor. *Science* **289**: 739–45 (2000).
19. Wimalawansa, S.J. Purification and biochemical characteriza-tion of the neuropeptide Y2 receptor. *J. Biol. Chem.* **270**, 18523–30 (1995).
20. Katzen, F., Chang, G. & Kudlicki, W. The past, present and future of cell-free protein synthesis. *Trends Biotechnol.* **23**, 150–6 (2005).
21. Yokoyama, S. Protein expression systems for structural genomics and proteomics. *Curr. Opin. Chem. Biol.* **7**, 39–43 (2003).

22. Klammt, C., Löhr, F., Schäfer, B., Haase, W., Dötsch, V., Rüterjahns, H., Glaubitz, C. & Bernhard, F. High level cell-free expression and specific labeling of integral membrane proteins. *Eur. J. Biochem.* **271**, 568–80 (2004).

23. Ishihara, G., Goto, M., Saeki, M., Ito, K., Hori, T., Kigawa, T., Shirouzu, M. & Yokoyama, S. Expression of G protein coupled receptors in a cell-free translational system using detergents and thioredoxin-fusion vectors. *Prot. Expr. Purif.* **41**, 27–37 (2005).

24. Guo, Z.D., Suga, H., Okamura, M., Takeda, S. & Haga, T. Receptor-G fusion proteins as a tool for ligand screening. *Life Sci.* **68**, 2319–27 (2001).

25. Furukawa, H. & Haga, T. Expression of functional M2 muscarinic acetylcholine receptor in Escherichia coli. *J. Biochem.* **127**, 151–61 (2000).

26. Tucker, J. & Grisshammer, R. Purification of a rat neurotensin receptor expressed in Escherichia coli. *Biochem. J.* **317**, 891–9 (1996).

27. Miroux, B. & Walker, J.E. Over-production of proteins in Escherichia coli: Mutant hosts that allow synthesis of some membrane proteins and globular proteins at high levels. *J. Mol. Biol.* **260**, 289–98 (1996).

28. La Vallie, E.R. & McCoy, J.M. Gene fusion expression systems in Escherichia coli. *Curr. Opin. Biotechnol.* **6**, 501–6 (1995).

29. Walhout, A.J., Temple, G.F., Brasch, M.A., Hartley, J.L., Lorson, M.A., van den Heuvel, S. & Vidal, M. GATEWAY recombinational cloning: application to the cloning of large numbers of open reading frames or ORFeomes. *Methods Enzymol.* **328**, 575–92 (2000).

30. Hunt, I. From gene to protein: A review of new and enabling technologies for multi-parallel protein expression. *Prot. Expr. Purif.* **40**, 1–22 (2005).

31. Mergulhao, F.J., Summers, D.K. & Monteiro, G.A. Recombinant protein secretion in Escherichia coli. *Biotechnol. Adv.* **23**, 177–202 (2005).

32. Saidijam, M., Bettaney, K.E., Szakonyi, G., Psakis, G., Shibayama, K., Suzuki, S., Clough, J.L., Blessie, V., Abu-Bakr, A., Baumberg, S., Meuller, J., Hoyle, C.K., Palmer, S.L., Butaye, P., Walravens, K., Patching, S.G., O'reilly, J., Rutherford, N.G., Bill, R.M., Roper, D.I., Phillips-Jones, M.K. & Henderson, P.J. Active membrane transport and receptor proteins from bacteria. *Biochem. Soc. Trans.* **33**, 867–72 (2005).

33. Weiss, H.M. & Grisshammer, R. Purification and characterization of the human adenosine A(2a) receptor functionally expressed in Escherichia coli. *Eur. J. Biochem.* **269**, 82–92 (2002).

34. Kiefer, H. In vitro folding of alpha-helical membrane proteins. *Biochim. Biophys. Acta* **1610**, 57–62 (2003).

35. Lopez de Maturana, R., Willshaw, A., Kuntzsch, A., Rudolph, R. & Donnelly, D. The isolated N-terminal domain of the glucagon-like peptide-1 (GLP-1) receptor binds exendin peptides with much higher affinity than GLP-1. *J. Biol. Chem.* **278**, 10195–200 (2003).

36. Baneres, J.L., Martin, A., Hullot, P., Girard, J.P., Rossi, J.C. & Parello, J. Structure-based analysis of GPCR function: conformational adaptation of both agonist and receptor upon leukotriene B4 binding to recombinant BLT1. *J. Mol. Biol.* **329**, 801–14 (2003).

37. Baneres, J.L., Mesnier, D., Martin, A., Joubert, L., Dumuis, A. & Bockaert, J. Molecular characterization of a purified 5-HT4 receptor: a structural basis for drug efficacy. *J. Biol. Chem.* **280**, 20253–60 (2005).

38. Kunji, E.R., Slotboom, D.J. & Poolman, B. Lactococcus lactis as host for overproduction of functional membrane proteins. *Biochim. Biophys. Acta* **1610**, 97–108 (2003).

39. de Ruyter, P.G., Kuipers, O.P. & de Vos, W.M. Controlled gene expression systems for Lactococcus lactis with the food-grade inducer nisin. *Appl. Environ. Microbiol.* **62**, 3662–7 (1996).

40. Monne, M., Chan, K.W., Slotboom, D.J. & Kunji, E.R. Functional expression of eukaryotic membrane proteins in Lactococcus lactis. *Protein Sci.* **14**, 3048–56 (2005).

41. Palva, I., Lehtovaara, P., Kaariainen, L., Sibakov, M., Cantell, K., Schein, C.H., Kashiwagi, K. & Weissmann, C. Secretion of interferon by Bacillus subtilis. *Gene* **22**, 229–35 (1983).

42. Lundstrom, K., Palva, I., Kaariainen, L., Garoff, H., Sarvas, M. & Pettersson, R.F. Secretion of Semliki Forest virus membrane glycoprotein E1 from Bacillus subtilis. *Virus Res.* **2**, 69–83 (1985).

43. Palva, I. Molecular cloning of alpha-amylase gene from Bacillus amyloliquefaciens and its expression in B. subtilis. *Gene* **19**, 81–7 (1982).

44. Phuong Phan, T.T., Nguyen, H.D., and Schumann, W. Novel plasmid-based expression vectors for intra- and extracellular production of recombinant proteins in Bacillus subtilis. *Prot. Expr. Purif.*, *in press*.

45. Luecke, H., Schobert, B., Cartailler, J.P., Richter, H.T., Rosengarth, A., Needleman, R. & Lanyi, J.K. Coupling photoisomerization of retinal to directional transport in bacteriorhodopsin. *J. Mol. Biol.* **300**, 1237–55 (2000).

46. Turner, G.J., Reusch, R., Winter-Vann, A.M., Martinez, L. & Betlach, M.C. Heterologous gene expression in a membrane-protein-specific system. *Prot. Expr. Purif.* **17**, 312–23 (1999).

47. Bartus, C.L., Jaakola, V.P., Reusch, R., Valentine, H.H., Heikinheimo, P., Levay, A., Potter, L.T., Heimo, H., Goldman, A. & Turner, G.J. Downstream coding region determinants of bacterio-opsin, muscarinic acetylcholine receptor and adrenergic receptor expression in Halobacterium salinarum. *Biochim. Biophys. Acta.* **1610**, 109–23 (2003).

48. Melmer, G. Biopharmaceuticals and the industrial environment, discovery, in *Production of Recombinant Proteins – Novel Microbial and Eukaryotic Expression Systems,* G. Gellissen, ed., Wiley-VCH, Weinheim, pp 361–383 (2005).

49. Harford, N., Cabezon, T., Colau, B., Delisse, A.M., Rutgers, T. & De Wilde, M. Construction and characterization of a Saccharomyces cerevisiae strain (RIT4376) expressing hepatitis B surface antigen. *Postgrad. Med. J.* **63 Suppl 2**, 65–70 (1987).

50. Giga-Hama, Y. & Kumagai, H. Expression system for foreign genes using the fission yeast Schizosaccharomyces pombe. *Biotechnol. Appl. Biochem.* **30**, 235–44 (1999).

51. Macauley-Patrick, S., Fazenda, M.L., McNeil, B. & Harvey, L.M. Heterologous protein production using the Pichia pastoris expression system. *Yeast* **22**, 249–70 (2005).

52. Gellissen, G., Kunze, G., Gaillardin, C., Cregg, J.M., Berardi, E., Veenhuis, M. & van der Klei, I. New yeast expression platforms based on methylotrophic Hansenula polymorpha and Pichia pastoris and on dimorphic Arxula adeninivorans and Yarrowia lipolytica – A comparison. *FEMS Yeast Res.* **5**, 1079–96 (2005).

53. Goffeau, A., Barrell, B.G., Bussey, H., Davis, R.W., Dujon, B., Feldmann, H., Galibert, F., Hoheisel, J.D., Jacq, C., Johnston, M., Louis, E.J., Mewes, H.W., Murakami, Y., Philippsen, P., Tettelin, H. & Oliver, S.G. Life with 6000 genes. *Science* **274**, 563–7 (1996).

54. Gellissen, G. & Hollenberg, C.P. Application of yeasts in gene expression studies: a comparison of Saccharomyces cerevisiae, Hansenula polymorpha and Kluyveromyces lactis – A review. *Gene* **190**, 87–97 (1997).

55. Valenzuela, P., Medina, A., Rutter, W.J., Ammerer, G. & Hall, B.D. Synthesis and assembly of hepatitis B virus surface antigen particles in yeast. *Nature* **298**, 347–50 (1982).

56. Cabezon, T., De Wilde, M., Herion, P., Loriau, R. & Bollen, A. 1984., Expression of human alpha 1-antitrypsin cDNA in the yeast Saccharomyces cerevisiae. *Proc. Natl. Acad. Sci. U S A* **81**: 6594–8 (1984).

57. Hinnen, A., Meyhack, B. & Tsapis, R. High expression and secretion of foreign proteins in yeast, in *Gene Expression in Yeast: Foundation for Biotechnical and Industrial Fermentation Research*, M. Korhola, E. Väisänen, eds., Kauppakirjapaino, Helsinki, pp 157–66 (1983).

58. Sakai, A., Shimizu, Y. & Hishinuma, F. Isolation and characterization of mutants which show an oversecretion phenotype in Saccharomyces cerevisiae. *Genetics* **119**, 499–506 (1988).

59. Wang, Z. Controlled expression of recombinant genes and preparation of cell-free extracts in yeast. *Methods Mol. Biol.* **313**, 317–32 (2005).

60. David, N.E., Gee, M., Andersen, B., Naider, F., Thorner, J. & Stevens, R.C. Expression and purification of the Saccharomyces cerevisiae alpha-factor receptor (Ste2p), a 7-transmembrane-segment G protein-coupled receptor. *J. Biol. Chem.* **272**, 15553–61 (1997).

61. Andersen, B. & Stevens, R.C. The human D1A dopamine receptor: heterologous expression in Saccharomyces cerevisiae and purification of the functional receptor. *Prot. Expr. Purif.* **13**, 111–9 (1998).

62. Sizmann, D., Kuusinen, H., Keranen, S., Lomasney, J., Caron, M.G., Lefkowitz, R.J. & Keinanen, K. Production of adrenergic receptors in yeast. *Receptors Channels* **4**, 197–203 (1996).

63. Jidenko, M., Nielsen, R.C., Sorensen, T.L., Moller, J.V., le Maire, M., Nissen, P. & Jaxel, C. Crystallization of a mammalian membrane protein overexpressed in Saccharomyces cerevisiae. *Proc. Natl. Acad. Sci. U S A* **102**, 11687–91 (2005).

64. Grallert, B., Nurse, P. & Patterson, T.E. A study of integrative transformation in Schizosaccharomyces pombe. *Mol. Gen. Genet.* **238**, 26–32 (1993).

65. Gemmill, T.R., and Trimble, R.B. Overview of N- and O-linked oligosaccharide structures found in various yeast species. *Biochim. Biophys. Acta* **1426**, 227–37 (1999).

66. Broker, M., and Bauml, O. New expression vectors for the fission yeast Schizosaccharomyces pombe. *FEBS Lett.* **248**, 105–10 (1989).

67. Sander, P., Grunewald, S., Maul, G., Reilander, H. & Michel, H. Constitutive expression of the human D2S-dopamine receptor in the unicellular yeast Saccharomyces cerevisiae. *Biochim. Biophys. Acta* **1193**, 255–62 (1994).

68. Presland, J. & Strange, P.G. Pharmacological characterization of the D2 dopamine receptor expressed in the yeast Schizosaccharomyces pombe. *Biochem. Pharmacol.* **56**, 577–82 (1998).

69. Arkinstall, S., Edgerton, M., Payton, M. & Maundrell, K. Co-expression of the neurokinin NK2 receptor and protein components in the fission yeast *Schizosaccharomyces pombe. FEMS Lett.* **375**, 183–7 (1995).

70. Cereghino, J.L. & Cregg, J.M. Heterologous protein expression in the methylotrophic yeast Pichia pastoris. *FEMS Microbiol. Rev.* **24**, 45–66 (2000).

71. Thomas, L. & Crawford, D.L. Cloning of clustered Streptomyces viridosporus T7A lignocellulose catabolism genes encoding peroxidase and endoglucanase and their extracellular expression in Pichia pastoris. *Can. J. Microbiol.* **44**, 364–72 (1998).

72. Cote, H.C., Pratt, K.P., Davie, E.W. & Chung, D.W. The polymerization pocket "a" within the carboxyl-terminal region of the gamma chain of human fibrinogen is adjacent to but independent from the calcium-binding site. *J. Biol. Chem.* **272**, 23792–8 (1997).

73. Weiss, H.M., Haase, W., Michel, H. & Reilander, H. Expression of functional mouse 5-HT5A serotonin receptor in the methylotrophic yeast Pichia pastoris: pharmacological characterization and localization. *FEBS Lett.* **377**, 451–6 (1995).

74. Grunewald, S., Haase, W., Molsberger, E., Michel, H. & Reilander, H. Production of the human D2S receptor in the methylotrophic yeast P. pastoris. *Receptors Channels* **10**, 37–50 (2004).

75. Reinhart, C. & Kettler, C. Expression of membrane proteins in yeast, in *Structural Genomics on Membrane Proteins*, K. Lundstrom, ed., CRC Press, Boca Raton, pp 115–52 (2006).

76. Sen, S., Jaakola, V.P., Heimo, H., Engstrom, M., Larjomaa, P., Scheinin, M., Lundstrom, K. & Goldman, A. Functional expression and direct visualization of the human alpha 2B -adrenergic receptor and alpha 2B -AR-green fluorescent fusion protein in mammalian cell using Semliki Forest virus vectors. *Prot. Expr. Purif.* **32**, 265–75 (2003).

77. King, K., Dohlman, H.G., Thorner, J., Caron, M.G. & Lefkowitz, R.J. Control of yeast mating signal transduction by a mammalian beta 2-adrenergic receptor and Gs alpha subunit. *Science* **250**, 121–3 (1990).

78. Karlsson, M., Fotiadis, D., Sjovall, S., Johansson, I., Hedfalk, K., Engel, A. & Kjellbom, P. Reconstitution of water channel function of an aquaporin overexpressed and purified from Pichia pastoris. *FEBS Lett.* **537**, 68–72 (2003).

79. Tornroth-Horsefield, S., Wang, Y., Hedfalk, K., Johanson, U., Karlsson, M., Tajkhorshid, E., Neutze, R. & Kjellbom, P. Structural mechanism of plant aquaporin gating. *Nature, in press*.

80. Parcej, D.N. & Eckhardt-Strelau, L. Structural characterisat-ion of neuronal voltage-sensitive K+ channels heterologously expressed in Pichia pastoris. *J. Mol. Biol.* **333**, 103–16 (2003).

81. Long, S.B., Campbell, E.B. & MacKinnon, R. Crystal structure of a mammalian voltage-dependent Shaker family K$^+$ channel. *Science* **309**, 897–903 (2005).

82. Jarvis, D.L. & Finn, E.E. Biochemical analysis of the N-glycosylation pathway in baculovirus-infected lepidop-teran insect cells. *Virology* **212**, 500–11 (1995).

83. Luque, T. & O'Reilly, D.R. Generation of baculovirusexpression vectors. *Mol Biotechnol.* **13**, 153–63 (1999).

84. Benting, J., Lecat, S., Zacchetti, D. & Simons, K. Protein expression in Drosophila Schneider cells. *Anal. Biochem.* **278**, 59–68 (2000).

85. Luckow, V.A. Baculovirus systems for the expression of human gene products. *Curr. Opin. Biotechnol.* **4**, 564–72 (1993).

86. Kost, T. & Condreay, J.P. Recombinant baculoviruses as expression vectors for insect and mammalian cells. *Curr Opin Biotechnol.* **10**, 428–33 (1999).

87. Bosman, G.J. & De Grip, W.J. Expression of functional membrane proteins in the baculovirus-insect cell system: Challenges and developments, in *Structural Genomics on Membrane Proteins*, K. Lundstrom, ed., CRC Press, Boca Raton, pp 153–67 (2006).

88. Klaassen, C.H.W. & De Grip, W.J. Baculovirus expression system for expression and characterization of functional recombinant visual pigments. *Meth. Enzymol.* **315**, 12–29 (2000).

89. Guan, X.M., Kobilka, T.S. & Kobilka, B.K. Enhancement of membrane insertion and function in a type IIIb membrane protein following introduction of a cleavable signal peptide. *J. Biol. Chem.* **267**, 21995–8 (1992).

90. Akermoun, M., Koglin, M., Zvalova-Iooss, D., Folschweiller, N., Dowell, S.J. & Gearing, K.L. Characterization of 16 human G protein-coupled receptors expressed in baculovirus-infected insect cells. *Prot. Expr. Purif.* **44**, 65–74 (2005).

91. Ratnala, V.R., Swarts, H.G., Van Oostrum, J., Leurs, R., De Groot, H.J., Bakker, R.A. & De Grip, W.J. Large-scale overproduction, functional purification and ligand affinities of the His-tagged human histamine H1 receptor. *Eur. J. Biochem.* **271**, 2636–46 (2004).

92. Gaymard, F., Cerutti, M., Horeau, C., Lemaillet, G., Urbach, S., Ravallec, M., Devauchelle, G., Sentenac, H. & Thibaud, J.B. The baculovirus/insect cell system as an alternative to Xenopus oocytes. First characterization of the AKT1 K+ channel from Arabidopsis thaliana. *J. Biol. Chem.* **271**, 22863–70 (1996).

93. Fucentese, M., Winterhalter, K.H., Murer, H. & Biber, J. Functional expression and purification of histidine-tagged rat renal Na/Phosphate (NaPi-2) and Na/Sulfate (NaSi-1) cotransporters. *J. Membr. Biol.* **160**, 111–7 (1997).

94. Ramjeesingh, M., Ugwu, F., Li, C., Dhani, S., Huan, L.J., Wang, Y., and Bear, C.E. Stable dimeric assembly of the second membrane-spanning domain of CFTR (cystic fibrosis transmembrane conductance regulator) reconstitutes a chloride-selective pore. *Biochem. J.* **375**, 633–41 (2003).

95. Gatto, C., McLoud, S.M. & Kaplan, J.H. Heterologous expression of Na(+)-K(+)-ATPase in insect cells: intracellular distribution of pump subunits. *Am. J. Physiol. Cell Physiol.* **281**, C982–92 (2001).

96. Johanson, K., Appelbaum, E., Doyle, M., Hensley, P., Zhao, B., Abdel-Meguid, S.S., Young, P., Cook, R., Carr, S., Matico, R, et al. Binding interactions of human interleukin 5 with its receptor alpha subunit. Large scale production, structural, and functional studies of Drosophila-expressed recombinant proteins. *J. Biol. Chem.* **270**, 9459–71 (1995).

97. Perret, B.G., Wagner, R., Lecat, S., Brillet, K., Rabut, G., Bucher, B. & Pattus, F. Expression of EGFP-amino-tagged human mu opioid receptor in Drosophila Schneider 2 cells: a potential expression system for large-scale production of G-protein coupled receptors. *Protein Expr. Purif.* **31**, 123–32 (2003).

98. Artelt, P., Morelle, C., Ausmeier, M., Fitzek, M. & Hauser, H. Vectors for efficient expression in mammalian fibroblastoid, myeloid and lymphoid cells via transfection or infection. *Gene* **68**, 213–19 (1988).

99. Xia, W., Bringmann, P., McClary, J., Jones, P.P., Manzana, W., Zhu, Y., Wang, S., Liu, Y., Harvey, S., Madlansacay, M.R., McLean, K., Rosser, M.P., Macrobbie, J., Olsen, C.L. & Cobb, R.R. High levels of protein expression using different mammalian CMV promoters in several cell lines. *Prot. Expr. Purif.* **45**, 115–24 (2006).

100. Makrides, S.C. Vectors for gene expression in mammalian cells, in *Gene Transfer and Expression in Mammalian Cells*, S.C. Makrides, Ed., Elsevier Science B.V., Amsterdam, pp. 9–26 (2003).

101. Meissner, P., Pick, H., Kulangara, A., Chatellard, P., Friedrich, K. & Wurm, F.M. Transient gene expression: recombinant protein production with suspension-adapted HEK293-EBNA cells. *Biotechnol. Bioeng.* **75**, 197–203 (2001).

102. Rosser, M.P., Xia, W., Hartsell, S., McCaman, M., Zhu, Y., Wang, S., Harvey, S., Bringmann, P. & Cobb, R.R. Transient transfection of CHO-K1-S using serum-free medium in suspension: a rapid mammalian protein expression system. *Prot. Expr. Purif.* **40**, 237–43 (2005).

103. Ulrich, C.D., Ferber, I., Holicky, E., Hadac, E., Buell, G. & Miller, L.J. Molecular cloning and functional expression of the human gallbladder cholecystokinin A receptor. *Biochem. Biophys. Res. Commun.* **193**, 204–11 (1993).

104. Reuben, M., Rising, L., Prinz, C., Hersey, S. & Sachs, G. Cloning and expression of the rabbit gastric CCK-A receptor. *Biochim. Biophys. Acta* **1219**, 321–7 (1994).

105. Keynan, S., Suh, Y.J., Kanner, B.I. & Rudnick, G. Expression of a cloned gamma-aminobutyric acid transporter in mammalian cells. *Biochemistry* **31**, 1974–9 (1992).

106. Ukomadu, C., Zhou, J., Sigworth, F.J. & Agnew, W.S. muI Na+ channels expressed transiently in human embryonic kidney cells: biochemical and biophysical properties. *Neuron* **8**, 663–76 (1992).

107. Schurmann, A., Monden, I., Joost, H.G. & Keller, K. Subcellular distribution and activity of glucose transporter isoforms GLUT1 and GLUT4 transiently expressed in COS-7 cells. *Biochim. Biophys. Acta* 1131, 245–52 (1992).

108. Walter, C.A., Humphrey, R.M., Adair, G.M. & Nairn, R.S. Characterization of Chinese hamster ovary cells stably transformed by a plasmid with an inducible APRT gene. *Plasmid* **25**, 208–16 (1991).

109. Prather, P.L., McGinn, T.M., Claude, P.A., Liu-Chen, L.Y., Loh, H.H. & Law, P.Y. Properties of a kappa-opioid receptor expressed in CHO cells: interaction with multiple G-proteins is not specific for any individual G alpha subunit and is similar to that of other opioid receptors. *Brain Res. Mol. Brain Res.* **29**: 336–46 (1995).

110. Reeves, P.J., Kim, J.M. & Khorana, H.G. Structure and function in rhodopsin: a tetracycline-inducible system in stable mammalian cell lines for high-level expression of opsin mutants. *Proc. Natl. Acad. Sci. USA* **99**, 13413–8 (2002).

111. Tate, C.G., Haase, J., Baker, C., Boorsma, M., Magnani, F., Vallis, Y. & Williams, DC. Comparison of seven different heterologous protein expression systems for the production of the serotonin transporter. *Biochim. Biophys. Acta* **1610**, 141–53 (2003).

112. Boorsma, M., Nieba, L., Koller, D., Bachmann, M.F., Bailey, J.E., Renner, W.A. A temperature-regulated replicon-based DNA expression system. *Nat. Biotechnol.* **18**, 429–32 (2000).

113. Rosenfeld, M.A., Yoshimura, K., Trapnell, B.C., Yoneyama, K., Rosenthal, E.R., Dalemans, W., Fukayama, M., Bargon, J., Stier, L.E., Stratford-Perricaudet, L., et al. In vivo transfer of the human cystic fibrosis transmembrane conductance regulator gene to the airway epithelium. *Cell* **68**, 143–55 (1992).

114. Jacobs, S.C., Stephenson, J.R. & Wilkinson, G.W. High-level expression of the tick-borne encephalitis virus NS1 protein by using an adenovirus-based vector: protection elicited in a murine model. *J. Virol.* **66**, 2086–95 (1992).

115. Zhen, Z., Bradel-Tretheway, B.G., Drewhurst, S., and Bidlack, J.M. Transient overexpression of kappa and mu opioid receptors using recombinant adenovirus vectors. *J. Neurosci. Methods* **136**, 133–9 (2004).

116. Drazner, M.H., Peppel, K.C., Dyer, S., Grant, A.O., Koch, W.J. & Lefkowitz, R.J. Potentiation of beta-adrenergic signaling by adenoviral-mediated gene transfer in adult rabbit ventricular myocytes. *J. Clin. Invest.* **99**, 288–96 (1997).

117. Fuerst, T.R., Niles, E.G., Studier, F.W. & Moss, B. Eukaryotic transient-expression system based on recombinant vaccinia virus that synthesizes bacteriophage T7 RNA polymerase. *Proc. Natl. Acad. Sci. USA* **83**, 8122–6 (1986).

118. Tartaglia, J., Cox, W.I., Taylor, J., Perkus, M., Riviere, M., Meignier, B. & Paoletti, E. Highly attenuated poxvirus vectors. *AIDS Res. Hum. Retroviruses* **8**, 1445–7 (1997).

119. Carroll, M.W., Wilkinson, G.W.G. & Lundstrom, K. Mammalian expression systems and vaccination, in *Genetically Engineered Viruses*, C.J.A. Ring, and E.D. Blair, eds., BIOS Scientific Publishers Ltd, Oxford, pp 107–57 (2001).

120. Walker, P, Munoz M, Combe MC, Grouzmann E, Herzog H, Selbie L, Shine J, Brunner HR, Waeber B, Wittek R. High level expression of human neuropeptide Y receptors in mammalian cells infected with a recombinant vaccinia virus. *Mol. Cell Endocrinol.* **91**, 107–12 (1993).

121. Bouvier, M., Chidiac, P., Hebert, T.E., Loisel, T.P., Moffett, S. & Mouillac, B. Dynamic palmitoylation of G-protein-coupled receptors in eukaryotic cells. *Methods Enzymol.* **250**, 300–14 (1995).

122. Naldini, L., Blomer, U., Gallay, P., Ory, D., Mulligan, R., Gage, F.H., Verma, I.M. & Trono, D. In vivo gene delivery and stable transduction of nondividing cells by a lentiviral vector. *Science* **272**, 263–67 (1996).

123. Gasmi, M. & Wong-Staal, F. Virus-based vectors for gene expression in mammalian cells: Lentiviruses, in *Gene Transfer in Mammalian Cells*, S.C. Makrides, ed., Elsevier Science B.V., Amsterdam, pp 251–64 (2003).

124. Kafri, T., Blomer, U., Peterson, D.A., Gage, F.H. & Verma, I.M. Sustained expression of genes delivered directly into liver and muscle by lentiviral vectors. *Nat. Genet.* **17**, 314–7 (1997).

125. Yang, M., Wang, X.G., Stout, J.T., Chen, P., Hjelmeland, L.M., Appukuttan, B. & Fong, H.K. Expression of a recombinant human RGR opsin in Lentivirus-transduced cultured cells. *Mol. Vis.* **6**, 237–42 (2000).

126. Strauss, J.H. & Strauss, E.G. The alphaviruses: gene expression, replication, and evolution. *Microbiol. Rev.* **58**, 491–562 (1994).

127. Lundstrom, K., Schweitzer, C., Rotmann, D., Hermann, D., Schneider, E.M. & Ehrengruber, M.U. Semliki Forest virus vectors: efficient vehicles for in vitro and in vivo gene delivery. *FEBS Lett.* **504**, 99–103 (2001).

128. Blasey, H.D., Brethon, B., Hovius, R., Tairi, A.P., Lundstrom, K., Rey, L. & Bernard, A.R. Large-scale transient 5-HT3 receptor production with the Semliki Forest virus system. *Cytotechnol.* **32**, 199–208 (2000).

129. Lundstrom, K. Semliki Forest virus vectors for rapid and high-level expression of integral membrane proteins. *Biochim. Biophys. Acta* **1610**, 90–6 (2003).

130. Hassaine, G., Wagner, R., Kempf, J., Cherouati, N., Hassaine, N., Prual, C., André, N., Reinhart, C., Pattus, F. & Lundstrom, K. Semliki Forest Virus vectors for overexpression of 101 G Protein-coupled receptors in mammalian host cells. *Prot. Purif. Expr.* **45**, 343–51 (2006).

131. Lee, A.G. How lipids interact with an intrinsic membrane protein: the case of the calcium pump. *Biochim. Biophys. Acta* **1376**, 381–90 (1998).

132. Keyes, M.H., Gray, D.N., Kreh, K.E. & Sanders, C.R. Solubilizing detergents for membrane proteins, in Methods and Results in Crystallisation of Membrane Proteins, S. Iwata, ed., IUL, San Diego, pp. 17–38 (2003).

133. Byrne, B. & Jormakka, M. Solubilization and purification of membrane proteins, in *Structural Genomics on Membrane Proteins*, K. Lundstrom, ed., CRC Press, Boca Raton, pp 179–98 (2006).

134. Rumbley, J.N., Furlong Nickels, E. & Gennis, R.B. One-step purification of histidine-tagged cytochrome bo3 from Escherichia coli and demonstration that associated quinone is not required for the structural integrity of the oxidase. *Biochim. Biophys. Acta* **1340**, 131–42 (1997).

135. Schmidt, T.G. & Skerra, A. One-step affinity purification of bacterially produced proteins by means of the "Strep tag" and immobilized recombinant core streptavidin. *J. Chromatograph. A.* **676**, 337–45 (1994).

136. Abola, E., Kuhn, P., Earnest, T. & Stevens, R.C. Automation of X-ray crystallography. *Nat. Struct. Biol.* **7**, 973–7 (2000).

137. Tickle, I., Shariff, A., Vinkovic, M., Yon, J. & Jhoti, H. High-trhoughput protein crystallography and drug discovery. *Chem. Soc. Rev.* **33**, 558–65 (2004).

138. Shuker, S.B., Hajduk, P.J., Meadows, R.P. & Fesik, S.W. 1996., Discovering high-affinity ligands for proteins: SAR by NMR. *Science* **274**, 1531–34 (1996).

139. Moore, J.M. NMR techniques for characterization of ligand binding: utility for lead generation and optimization in drug discovery. *Biopolymers* 51, 221–43 (1999).

140. Powers, R. Applications of NMR to structure-based drug design in structural genomics. *J. Struct. Funct. Genomics* 2, 113–23 (2002').

141. Xie, X.-Q. Membrane protein NMR, in *Structural Genomics on Membrane Proteins*, K. Lundstrom, ed., CRC Press, Boca Raton, pp 211–59 (2006).

142. Engel, A. Electron microscopy and atomic force microscopy of reconstituted membrane proteins, in *Structural Genomics on Membrane Proteins*, K. Lundstrom, ed., CRC Press, Boca Raton, pp 300–20 (2006).

143. Henderson, R., Baldwin, J.M., Ceska, T.A., Zemlin, F., Beckmann, E. & Downing, K.H. Model for the structure of bacteriorhodopsin based on high-resolution electron cryo-microscopy. *J. Mol. Biol.* **213**, 899–929 (1990).
144. Murata, K., Mitsuoka, K., Hirai, T., Walz, T., Agre, P., Heymann, J.B., Engel, A. & Fujiyoshi, Y. Structural determinants of water permeation through aquaporin-1. *Nature* **407**, 599–605 (2000).
145. Fotiadis, D., Liang, Y., Filipek, S., Saperstein, D.A., Engel, A.& Palczewski, K. Atomic-force microscopy: Rhodopsin dimers in native disc membranes. Nature 421, 127–8 (2003).
146. Goulding, C.W., Perry, L.J., Anderson, D., Sawaya, M.R., Cascio, D., Apostol, M.I., Chan, S., Parseghian, A., Wang, S.S., Wu, Y., Cassano, V., Gill, H.S. & Eisenberg, D. Structural genomics of Mycobacterium tuberculosis: a preliminary report of progress at UCLA. Biophys. Chem. 105, 361–70 (2003).
147. Yokoyama, S., Hirota, H., Kigawa, T., Yabuki, T., Shirouzu, M., Terada, T., Ito, Y., Matsuo, Y., Kuroda, Y., Nishimura, Y., Kyogoku, Y., Miki, K., Masui, R. & Kuramitsu, S. Structural genomics projects in Japan. Nat. Struct. Biol. 7, Suppl. 943–5 (2000).
148. Lundstrom, K. Structural genomics on membrane proteins: The MePNet approach. Curr. Opin. Drug Discov. Dev. 7, 342–6 (2004).

CHAPTER 2

RNA INTERFERENCE IN HAEMATOPOIETIC AND LEUKAEMIC CELLS

MARIA THOMAS, NATALIA MARTÍNEZ SORIA AND OLAF HEIDENREICH

Dept. Molecular Biology, Interfaculty Institute for Cell Biology, Eberhard Karls University Tuebingen, Germany

Abstract: The haematopoietic system is currently the best characterized mammalian differentiation system. On the one hand, pathologic disturbance of the differentiation programmes results in the development of haemopoietic malignancies such as leukaemias and lymphomas. On the other hand, systematic interference with haemopoietic differentiation is a prerequisite for *ex vivo* expansion of stem and progenitor cells during, for instance, stem cell transplantation. RNA interference (RNAi) provides exciting options for the molecular dissection of processes relevant for stem cell maintenance and leukaemogenesis. However, in comparison to most adhesive cell types, haematopoietic cells require special techniques for the successful application of RNAi, particularly for the delivery of RNAi-triggering molecules. Nevertheless, RNAi has not only been proven to enable functional analysis of haemopoietic processes, but may also contribute to the therapy of haemopoietic malignancies

1. INTRODUCTION

The haematopoietic system is probably the best understood cellular differentiation system. A small number of haemopoietic stem cells (HSCs) generate large quantities of mature blood cells such as erythrocytes, granulocytes, monocytes and lymphocytes. It is involved in many different processes including oxygen transport, immune defence and tissue homeostasis. During haematopoiesis, the generation of blood cells, differentiation and proliferation do not exclude each other, but are tightly linked, thus enabling this tremendous amplification from a few HSCs to the billions of mature blood cells. This amplification requires several layers of increasingly differentiated progenitor cells, which still have proliferation potential. However, only the stem cells contain self-renewal potential: only stem cell division

29

M. Al-Rubeai and M. Fussenegger (eds.), Systems Biology, 29–48.
© 2007 *Springer.*

generates at least one daughter cell, which is identical with the parental stem cell. In contrast, progenitor cells have, at best, a very limited self-renewal capacity. As a consequence, progenitor cells do expand in cell number both *ex vivo* and *in vivo*, but this expansion is transient due to a depletion of the proliferative active cell pool over time. Therefore, only HSCs can reconstitute long-term haematopoiesis, which is an essential requirement for bone marrow transplantation[1].

Disturbing the balance between proliferation and differentiation may have severe consequences for the haematopoietic system culminating in different kinds of anaemia, leukaemia and lymphoma. In particular, malignancies may either originate in the stem cell or the progenitor cell compartment. However, in contrast, to an HSC, a progenitor cell must acquire self-renewal potential for leukaemic transformation. Thus, a leukaemic stem cell may either arise from an HSC, or a more committed progenitor cell[2]. For instance, leukaemias are frequently associated with recurrent chromosomal rearrangements resulting in the generation of leukaemic fusion genes such as *BCR/ABL* or *AML1/MTG8*[3]. Since these fusion genes are exclusively expressed in preleukaemic and leukaemic cells, they would be ideal targets for antileukaemic treatment strategies, if they were still essential for leukaemic maintenance, and if they were amenable to a sufficiently specific and efficient inhibition by low molecular weight compounds. However, a central function for the completely transformed leukaemic (stem) cell has been only demonstrated for a few fusion proteins. Moreover, even the BCR-ABL inhibitor Imatinib mesylate (Gleevec, STI-571), which is now routinely used in the therapy of chronic myeloid leukaemia (CML), does not only inhibit this fusion protein, but also ABL kinase and other protein kinases such as KIT and PDGF receptor A and B[4]. Thus, it is necessary to develop not only tools for the specific examination of haemopoietic (fusion) gene functions, but also to translate these findings into new, more leukaemia-specific therapeutic strategies.

A more and more often used approach for the study of gene functions in haematopoiesis is RNA interference (RNAi), which allows for a sequence-dependent knock-down of a given target RNA and protein[5,6]. Currently, driven by the lack of suitable low molecular weight inhibitors, RNAi is also actively explored for treating haematopoietic diseases including leukaemias[7]. In this chapter, we will discuss the application of RNA interference-based strategies for the modulation of gene expression in haematopoietic cells.

2.   RNA INTERFERENCE

The term RNA interference (RNAi) was introduced for the double-stranded RNA (dsRNA)-dependent gene suppression in the nematode *Caenorhabditis elegans*[8]. dsRNA leads to mRNA downregulation in a large amount of organisms, and depending on the organism, it has received a different names. In plants, this process is named cosuppression or post-transcriptional gene silencing (PTGS), in fungi quelling, and in animals RNA interference. RNAi is mediated by small RNAs of

21 to 23 nucleotides in length, which are the result of dsRNA processing. Dependent on their origin, these tiny RNAs can be divided in two classes, the small interfering RNAs (siRNAs) and the microRNAs (miRNAs). siRNAs are of exogenous origin; they can be derived from double-stranded viral RNAs, or can be artificially introduced into the cell. In contrast, the precursors of miRNAs are encoded by the genome. Both classes of these small regulatory RNAs can induce posttranscriptional silencing by cleaving complementary mRNA, or by inhibiting mRNA translation[9]. The choice between cleavage and translational inhibition depends on the grade of homology with the target mRNA. The higher the homology, the more likely becomes mRNA cleavage. Furthermore, miRNAs and siRNAs may also affect epigenetic modifications such as histone and DNA methylation and histone deacetylation[10].

2.1   Small Interfering RNAs

In the cytoplasm, dsRNAs are cleaved into small RNAs by dsRNA-specific ribonuclease Dicer, which belongs to the RNase III family. *Dicer* genes are evolutionary very conserved and are present in all eukaryotes with the exception of budding yeast. The number of genes encoding Dicer-like proteins varies from four in *Arabidopsis* to one in vertebrates. Dicer digests dsRNA into pieces of 21–23 nucleotides in length, with 2 to 3 nucleotide long 3′-overhangs[11]. The length of the siRNAs is defined by the distance between the PAZ and RNAse III domains of Dicer[12]. Dicer is a 220 kD large protein containing a DExH/DEAH RNA helicase/ATPase domain, a PAZ (Piwi/Argonaute/Zwille) domain, two RNase III-like domains, a dsRNA-binding domain (RBD) and a DUF283 domain (Domain of unknown function 283).

One strand of the siRNA becomes part of the RNA-induced silencing complex (RISC), whereas the other one is degraded. The selective loading of this strand onto RISC proceeds under participation of Dicer and another RNA-binding protein, TRBP (human immunodeficiency virus transactivating response RNA-binding protein)[13,14]. Once part of RISC, the remaining siRNA strand serves as a guide strand to target RISC to complementary RNA sites. Since its discovery[15], several RISC components have been identified, among them the nuclease Ago2, which mediates the cleavage of the complementary target RNA strands[16]. For that reason, Ago2 is also named "Slicer". The target RNA cleavage occurs between residues base paired to nucleotides 10 and 11 of the siRNA, this cleavage itself does not require ATP. The guide siRNA remains associated with the complex, allowing it to carry out multiple rounds of RNA cleavage[17].

2.2   Micro RNAs

Unlike their exogenously derived siRNA cousins, miRNAs are encoded in the genome of plants and animals[18]. Frequently, the sequences encoding miRNAs are located in fragile sites in human chromomes, or in cancer-associated genomic regions[19]. miRNAs are transcribed by RNA polymerase II; they can be expressed independently, in clusters or as part of introns[20]. Like all other Pol II transcripts, primary miRNA transcripts (pri-miRNAs) contain 5′-cap structures as well as 3′-poly (A) tails[21]. The pri-miRNA is processed by the RNase III type nuclease Drosha yielding pre-miRNAs, intermediates of some 70 nucleotides in length, which form 25–30 bp long stem-loop structures[22]. The pre-miRNAs are exported to the cytoplasm as a complex with Exportin-5 and Ran-GTP[23], followed by Dicer-mediated processing to the mature miRNA and association with RISC[24–26].

miRNAs regulate gene expression by binding to the 3′-UTR of the target mRNA, either triggering mRNA degradation, or sequestering the mRNAs into P bodies, cytoplasmically located storage structures for untranslated mRNAs[25,27,28]. Interestingly, translational inhibition depends on the 5′-cap of the mRNA; cap-independent mechanisms are not subject to miRNA-mediated silencing[29].

The fact that miRNAs have not yet been found in single cell eukaryotes such as yeast, and the increasingly large numbers of miRNAs identified in plants and animals suggests a central role of miRNAs in developmental pattern formation including differentiation, proliferation and apoptosis. Indeed, miRNAs regulate the differentiation of different tissues including hematopoietic differentiation[30], and may be involved in tumourigenesis[31,32].

3.   siRNA EFFICIENCY

The first challenge for a successful RNAi experiment is the design of the efficient siRNA. siRNAs may be synthesized by chemical synthesis, by in vitro transcription, or, alternatively, ectopically expressed as shRNAs. The choice of the target site crucially affects siRNA efficacy not only by its secondary structure-dependent accessibility, but it also determines intrinsic properties of the homologous siRNA. siRNAs shorter than 21 bp and longer than 25 bp are inefficient in initiating RNAi[33]. At least one siRNA terminus should carry a 3′-overhang of two nucleotides in length. In contrast to earlier reports, the sequences of these overhangs are not important for siRNA efficacy. Biochemical and functional studies have identified some intrinsic siRNA parameters, which are critical for the overall efficiency of RNAi. Analysis of effective synthetic siRNAs have shown that whichever strand is most easily unwound in 5′-3′ direction will be preferentially assembled within RISC[34]. Although the studies focused on synthetic siRNAs, these features should equally apply to siRNAs derived from shRNAs. Several web-based recourses exist to aid siRNA design. Helpful web-based tools are provided, for instance, by the Whitehead Institute (*http://jura.wi.mit.edu/bioc/siRNAext/*).

To identify siRNA-specific features likely to contribute to efficient target RNA cleavage, Reynolds et al performed a systematic analysis of 180 siRNAs targeting every other position of two mRNAs[35]. Eight characteristics associated with siRNA functionality were identified: low G/C content, a bias towards low internal stability at the 3′-terminus of the sense strand, lack of inverted repeats and sense strand base preferences at positions 3,10,13,19. They also showed that application of an algorithm incorporating all eight criteria substantially improved the selection of potent siRNAs.

In addition to the siRNA nucleotide composition, silencing efficacy is also affected by the secondary structure of the target mRNA[36]. SiRNA activity may be severely compromised when its target sequence is part of a highly structured RNA. Luo and Chang found the number of hydrogen bonds formed by the target region to be a useful parameter in siRNA design[37]. The presence of a large number of unpaired nucleotides may indicate highly accessible target sites. In contrast, a high number of base-paired bases marks a target site, which is less accessible for the siRNA.

Patzel et al examined the potential role of secondary structures of the antisense strand of siRNA in RNAi[38]. The authors described a strong inverse correlation between the degree of guide RNA secondary structure formation and gene silencing by siRNAs. Unstructured guide strands mediated the strongest silencing, whereas structures with base-paired ends were inactive. Thus, the availability of terminal nucleotides within guide structures determines the strength of silencing. A to G and C to U base exchanges, which involve wobble base-pairing with the target and, thus, preserved target complementarity, turned inactive into active guide structures, thereby expanding the space of functional siRNAs.

Another possibility to identify efficient and specific siRNA is an siRNA walk along the mRNA target sequence. We used this approach for targeting the leukaemic fusion gene MLL-AF4[39]. Each siRNA in this scan targeted every other position of the breakpoint fusion site. A shift by a single nucleotide can substantially interfere with siRNA efficiency. With this approach, we identified an optimal siRNA for MLL-AF4 suppression without affecting the unfused alleles MLL or AF4.

In naturally occurring RNAi, pools of siRNAs targeting different regions of target RNA transcripts are generated intracellularly through Dicer-mediated processing of larger dsRNAs. Pools may be advantageous, as there is an increased probability of at least one highly effective siRNA being present within the population. Such siRNA pools may be generated *in vitro* by digesting double-stranded RNA with Dicer or with RNase III[40,41].

4.   siRNA SPECIFICITY

The functional analysis of a gene product of interest as well as a possible therapeutic intervention with siRNAs depends crucially on the sequence specificity and, thus, target specificity of the used siRNA. Initial transcriptome studies suggested a remarkable sequence-specificity of the examined siRNAs[42,43]. Furthermore, single point mutations were shown to severely compromise siRNA efficacy thus further supporting the idea of a high target specificity of siRNAs[44]. However, soon other groups reported contradictory findings suggesting that siRNAs may cause rather substantial off-target effects with down-modulating mRNAs containing short, 11 nucleotide long cores of sequence similarity to the applied siRNAs[45]. A recent study suggests that siRNAs may trigger the degradation of mRNAs bearing only a homology stretch of 7 nucleotides in length[46]. Furthermore, siRNAs may not induce RISC-mediated cleavage of an imperfectly homologous mRNA, but may still interfere with its translation[47]. In line with these findings, siRNAs may cause changes in expression levels of unintentionally targeted proteins[48].

Recently, several systematic studies were performed to examine essential parameters of siRNA specificity. First of all, computational analysis suggests that the generally chosen siRNA length of 21 nucleotides represents a optimal compromise between gene specificity and off-target reactivity[49]. Nevertheless, as indicated by the aforementioned studies, siRNA can be rather tolerant to single mismatches. In particular, mismatches located on the termini of the target-RNA-siRNA duplex hardly affect siRNA efficacy. In contrast, some wobble mutations may even enhance siRNA activity[50]. Centrally located mismatches, however, are more likely to impair siRNA activity[51].

A further limitation for the target specificity of siRNAs is a possible induction of interferon response-associated gene expression. In general, double-stranded RNA of more than 50 base pairs in length may potently trigger interferon responses by binding to, for instance, Toll-like receptors (TLRs) or protein kinase R (PKR). Initially, due to their short length, siRNAs were considered not to induce such interferon responses[52]. However, recent studies suggest that some siRNAs and shRNAs may activate interferon response genes[53-55]. This activation is particularly strong in plasmacytoid dendritic cells, a class of antigen presenting cells involved in interferon production. In these cells, the binding of siRNAs to TLR7 activates expression of interferon-α[56]. Interestingly, this immune-stimulating activity of the siRNAs is sequence-dependent: GU-rich sequence elements have been identified as a prerequisite for interferon induction[56,57]. Moreover, this stimulation is dependent on the presence of transfection agents. Naked siRNAs do not trigger this immune response *in vivo*[58].

What are meaningful controls for such sequence-related and unrelated side effects of siRNAs? The most commonly used controls is the transfection with a sequence-unrelated siRNA such as a scrambled siRNA or a luciferase siRNA. This setting controls for general, siRNA transfection-related effects, but does not allow the detection of sequence-specific effects such as inhibition of specific genes or induction of interferon responses. A better control for sequence-related effects are siRNAs with one or two nucleotide changes. However, due to the sequence promiscuity of siRNAs, such mismatch control siRNAs may still be able to inhibit the expression of the intended target gene. Furthermore, the introduced mutations should not be placed into sequence elements, which are suspected to trigger immune responses. Therefore, it is advisable to directly control for interferon responses by monitoring a possible induction of interferon response-associated genes such as *STAT1* or 2′, 5′-oligoadenylate synthetase 1 (*OAS1*). Furthermore, a complementing alternative to control the target specificity of a given siRNA are rescue experiments by ectopically expressing an siRNA-insensitive version of the target gene. This may be achieved by, for instance, using homologous genes from other species, which do not contain the siRNA target sequence[59]. However, if the phenotype is crucially dependent on the expression level of a target protein and does not tolerate higher than "normal" expression, such rescue experiments are difficult to perform. In this case, it may be possible to examine siRNA specificity by using two cell types, which express different isoforms or the target mRNA. With such an approach, we proved the specificity of siRNAs targeting the leukaemic fusion gene *MLL-AF4*, which is generated by the chromosomal translocation t(4;11)[39]. In contrast to fusion genes such as *AML1/MTG8*, the fusion site of *MLL-AF4* mRNA varies among patients. We designed siRNAs targeting two different variants expressed in the corresponding t(4;11)-positive leukaemic cell lines. We could show that the phenotype generated by the MLL-AF4 siRNAs was strictly connected with the specific depletion of that MLL-AF4 variant, which was homologous to the corresponding siRNA. This approach demonstrates the exclusive specificity of these MLL-AF4 siRNAs, and, thus, the direct dependence of the antileukaemic effects of these siRNAs on the inhibition of MLL-AF4 expression. Similarly, if depletion of a given target protein yields similar consequences in related cell types obtained from different organisms, such a specificity analysis should be feasible for target genes not containing any genetic alteration.

5.   DELIVERY OF siRNAs AND shRNAs

5.1   siRNA Delivery

Due to their size and negative charge, siRNAs cannot efficiently cross membranes. Thus, for efficient cellular uptakes, vectors such as cationic lipids (for siRNAs) or recombinant viruses (for shRNA expression) are required. The first successful delivery of siRNAs to mammalian cells was performed by cationic lipid-mediated lipofection[52]. Most haematopoietic cell types are suspension cells, which are difficult to transfect by lipofection. Nevertheless, several reports describe siRNA delivery to haematopoietic cell lines using cationic lipids[60,61]. However, in our hands, lipofection yielded only poor siRNA transfection efficiencies in the leukaemic cell lines tested by us.

Electroporation is commonly used for the transient transfection of suspension cells. The application of a short high voltage pulse depolarizes cell membranes, which leads to the formation of holes in the membrane thereby facilitating the uptake of, for instance, siRNAs. Currently, two different types of electroporation are applied for siRNA delivery (see Figure 1). The more commonly used approach discharges the condensator completely via the cell suspension. In an alternative system, a short rectangle pulse is applied for a defined time period to the sample. The major advantage of this setting compared to the previous one is the independence of the time and voltage parameters from the cell culture volume in the electroporation cuvette: Volumes ranging from $100\,\mu l$ up to $800\,\mu l$ can be efficiently electroporated using the very same parameters. With this approach, up to 10^7 cells including primary haematopoietic progenitor cells or dendritic cells can be transfected with siRNAs with efficiencies well above 90% and very limited cell death of less than 10%[62]. This technique has been successfully used for siRNA-mediated down-modulation

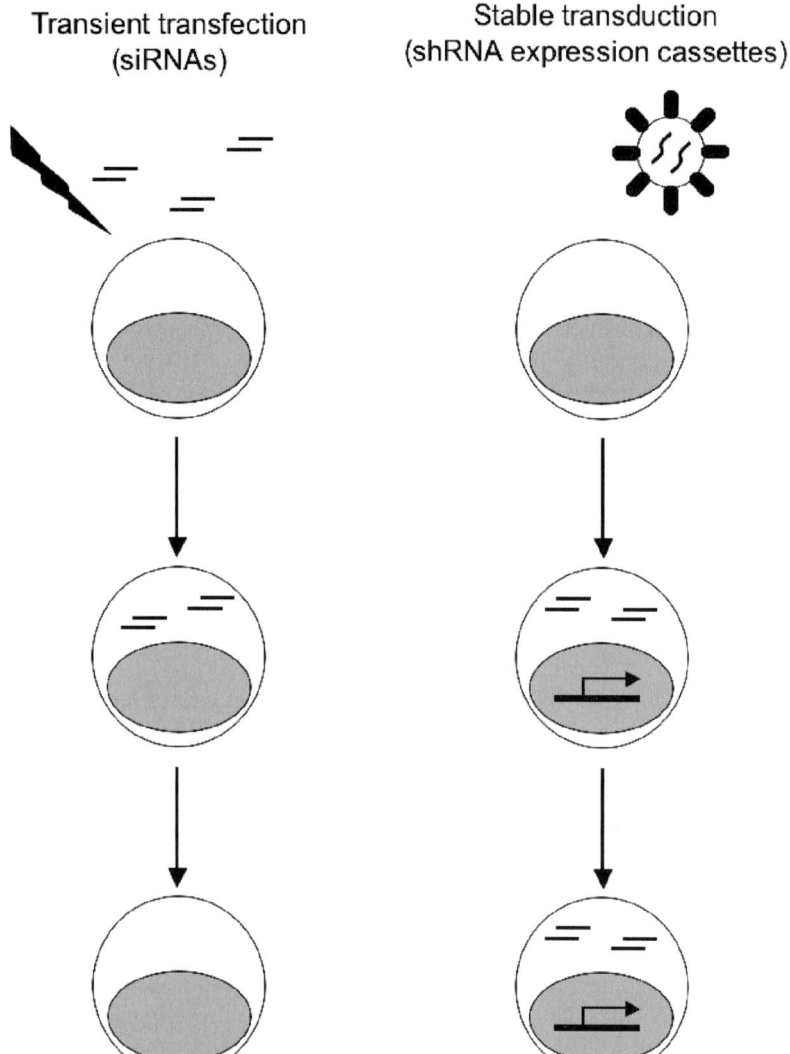

Figure 1. Delivery of siRNAs and shRNAs. Transfection of haemopoietic cells by, for instance, electroporation, delivers preformed siRNAs into the cytoplasm, where one strand can become part of RISC. Dependent on the cell proliferation rate and siRNA turnover, RNAi will decline over time. In contrast, stable transfection by retroviral transduction may result in a stable intracellular expression of shRNAs, which will be processed in the cytoplasm to siRNAs

of oncogenic fusion proteins such as AML1/MTG8[63], BCR-ABL[64] or MLL-AF4[39] in both leukaemic cell lines and in primary leukaemic blasts. Alternatively, several groups applied an electroporation system originally developed for the efficient transfer of nucleic acids to the cell nucleus. Even though siRNAs do not require nucleofection due to their cytoplasmic location, this electroporation system allows for an efficient siRNA delivery to haematopoietic cells under mild conditions[65].

The term of gene suppression by transiently transfected siRNAs is mainly limited by dilution due to cell proliferation[66]. Gene silencing is seen just for few days in fast replicating cells; this time may extend to weeks in slowly proliferating or non-dividing cells. A long half-life of the target protein in combination with a high proliferation rate may attenuate siRNA efficacy. In order to reduce levels of long living proteins, repetitive siRNA transfections may be required, which may be associated, particularly in the case of lipofection, with transfection-associated toxicity. This, however, is not the case with electroporation. In our hands, due to the rather mild conditions, recurrent electroporations are well tolerated by most haematopoietic cell lines.

5.2   shRNA Delivery

An alternative to the transient application of preformed siRNAs is the stable intra-cellular generation of shRNAs. In this case, target cells are stably transfected or transduced by plasmid or viral vectors containing an shRNA expression cassette. Mostly, polymerase III-dependent promoters such as the U6 or the H1 promoter are used, which are constitutively expressed in all cell types[67]. However, just the stable expression of shRNAs limits their applicability for suppressing genes essential for cell survival and proliferation, as this will result in a strong counter-selection against higher shRNA levels during the selection of stably transfected cells. For instance, cells dependent on the leukaemic fusion protein BCR-ABL have been shown not tolerating long-term expression of an anti-BCR-ABL shRNA[68]. For that reason, several inducible shRNA expression systems have been developed. Tetracyclin- or ecdysone-responsive shRNA expression systems permit an inducible regulation of RNAi thereby facilitating the targeting of genes crucial for cell proliferation and survival[69–71].

An interesting alternative to Pol III-mediated shRNA expression is the use of Pol II-dependent expression systems. In this case, siRNA sequences are introduced into the backbone of the precursor sequence of miRNA 30a[72]. This strategy yielded a comparable degree of knock-down compared to Pol III-dependent shRNA expression. However, it permits the use of conventional Pol II promoters such as CMV or even inducible promoters to drive siRNA expression. Such a system will be particularly attractive, if a Tet-responsive cell line using a Tet-VP16 activator fusion protein is available. Such a cell line can be used for inducible, Pol-II-mediated siRNA expression, but not for tet-inducible Pol III-mediated shRNA expression, since the latter system requires a nonfused Tet repressor protein.

A prerequisite for long-term shRNA or siRNA expression is the stable trans-fection of the target cell, preferentially by incorporation of the expression cassette into the cell genome. Unfortunately, haematopoietic cell lines are hard to transfect by conventional plasmid transfection methods. For that reason, oncoretroviral and lentiviral transduction have become the method of choice for the delivery of shRNA expression units. In contrast to oncoretroviruses, lentiviral particles are able to efficiently transduce slowly proliferating or even non-dividing cells. For that reason, pseudotyped lentiviruses are increasingly used for the transduction of haematopoietic stem cells. ShRNA expression cassettes may be either introduced between the viral long terminal repeats (LTR) yielding a so called single copy vector. In this case, viral integration introduces a single copy of the shRNA gene into the genome. This approach has been successfully used for both Pol III- and Pol II-dependent shRNA expression[72–74]. Alternatively, the rather small Pol III-dependent cassettes may be introduced into the 3′-LTR of oncoretro- and lentiviral vectors yielding a double copy vector. During reverse transcription, the 3′-LTR including the shRNA cassette are duplicated. Thus, integration of this vector introduces two copies of the shRNA into the host cell genome resulting in increased levels of shRNA expression and improved gene suppression[68,75–77].

SiRNA delivery by SV40 pseudovirions may become an interesting alternative to retroviral transduction. Nuclear extracts containing ectopically expressed SV40 VP1 capsid protein was used for in vitro packaging of siRNAs or shRNA-encoding plasmids followed by transfection of mammalian cells. This approach yielded an efficient suppression of cotransfected *GFP* in a lymphoblastoid cell line, and reduction of GFP levels in HeLa cells stably expressing *GFP*[78].

6.   TARGETING TRANSCRIPTIONAL MODULATORS WITH siRNAs

6.1   AML1/MTG8

The chromosomal translocation t(8;21) occurs in about 10%–15% of all de novo acute myeloid leukemia (AML) patients[79]. It fuses the DNA binding domain of the transcription factor AML1 located on chromosome 21 (also called RUNX1) to the almost complete open reading frame of MTG8 (ETO, CBFA2T1, RUNXT1) on chromosome 8. Whereas AML1 is an essential transcription factor for definitive haematopoiesis[80], MTG8 is part of histone deacetylase containing complexes[81]. The resulting fusion protein AML1/MTG8 is a transcriptional repressor of haematopoietic gene expression and was shown to interfere with signal transduction pathways controlling differentiation and proliferation[79].

We examined the efficacy and specificity of siRNAs homologous to the fusion site of the *AML1/MTG8* transcript[63,82]. Such siRNAs efficiently and specifically suppressed *AML1/MTG8* without interfering with nonfused *AML1* expression. The depletion of AML/MTG8 lead to the increased susceptibility of both t(8;21)-positive cell lines Kasumi-1 and SKNO-1 toward cytokine-driven induction of myeloid differentiation, impaired clonogenicity and a senescence-associated G1 cell cycle arrest. These results support a central function of AML1/MTG8 not only in the expansion of haematopoietic progenitor cells but also in the maintenance of the leukaemic phenotype. The observed synergism between differentiation-inducing agents such as TGFβ and vitamin D$_3$ and AML1/MTG8 siRNAs in combination with the antiproliferative effects of AML1/MTG8 depletion suggest that AML1/MTG8 siRNAs are not only useful tools for the functional analysis of this fusion gene, but that they may also have therapeutic potential.

6.2   MLL-AF4

The human homologue of *Trithorax, MLL* encodes a protein with histone methylase activity and DNA binding capacity. MLL is part of a large protein complex involved in the regulation of DNA and histone modifications[83,84]. Chromosomal rearrangements affecting the mixed lineage leukaemia (MLL) gene on chromosome 11 are frequently involved in leukaemogenesis including the development of therapy-related leukaemias. In particular, translocations resulting in MLL fusion genes cause, dependent on the partner gene, acute leukaemias of both the myeloid and the lymphoblastic type[85]. The most common MLL translocation is the t(4;11), which produces MLL-AF4 fusion protein. This translocation accounts for about 40% of all MLL translocations, including the majority of infant leukaemias[86]. The molecular mechanisms of MLL-associated leukaemogenesis are currently intensively studied. Despite the diversity of the fusion proteins, characteristic features of MLL-associated leukaemias are inappropriate expression of a subset of homeotic genes, such as *MEIS1* and *HOXA9*, and the overexpression of FMS-like tyrosine kinase 3 (FLT3)[87]. Patients with MLL-AF4 leukaemia almost always develop acute lymphoblastic leukaemia (ALL) and have a particularly poor prognosis with current therapy protocols. Therefore, it is very important to develop novel treatment strategies for this type of leukaemia.

Recently, we showed that transient inhibition of MLL-AF4 expression with small interfering RNAs impaired the proliferation and clonogenicity of t(4;11)-positive human leukaemic cell lines[39]. Decrease of MLL-AF4 levels also induced apoptosis associated with caspase-3 activation and diminished BCL-X_L expression. The depletion of MLL-AF4 fusion was also accompanied by a decreased expression of the homeotic genes *HOXA7, HOXA9* and *MEIS1* as well as of the haematopoietic stem cell marker CD133 indicating haematopoietic differentiation. Finally, siRNA-mediated MLL-AF4 suppression seriously compromised the leukaemic engraftment in xenotransplanted SCID mice. Transfection of leukaemic cells with MLL-AF4 siRNAs reduced leukemia-associated morbidity and mortality suggesting that MLL-AF4 depletion negatively affects leukemia-initiating cells. Thus, MLL-AF4 is important for maintenance of this highly aggressive leukemia. Since the translocation t(4;11) is predictive for a poor clinical outcome, the siRNA approach might become an option for developing new strategies for the treatment of this so far chemoresistant infant leukemia.

7.   TARGETING KINASES WITH siRNAs

7.1   BCR-ABL

Chronic myeloid leukemia (CML) progresses from a chronic phase associated with myeloid expansion to life-threatening blast crisis. In the vast majority of cases, the leukaemic cells carry an abnormal "Philadelphia" chromosome. Moreover, some 4% of all ALL cases are also positive for this chromosome. The underlying translocation involves the reciprocal transfer of the 3' Abelson proto-oncogene (ABL) sequence of chromosome 9 to variable locations in the 3'-breakpoint cluster region (BCR) of chromosome 22. Dependent on the fusion site, the resulting fusion gene BCR-ABL is transcribed into two different mRNAs encoding p210 or p190 BCR-ABL proteins. Both isoforms exhibit constitutively active tyrosine kinase activity. Treatment with imatinib mesylate, a small molecule inhibitor of BCR-ABL and other kinases such as PDGFR and c-KIT, has resulted in complete cytogenetic remission in a majority of chronic phase CML patients, and in fewer patients with accelerated and blast phase disease. However, as patients may relapse mainly due to the presence of imatinib-resistant mutants, further CML therapies are required[88].

A chimaeric BCR-ABL gene with either b2a2 or b3a2 junction is present only in CML cells and represents a fundamental target to set up molecular therapies. Different approaches were undertaken to deactivate the fusion protein, such a single-stranded antisense oligonucleotides, which, however, provided conflicting results concerning specificity and functionality. In contrast, many different groups reported the successful suppression of BCR-ABL by RNAi[64,76,89–93]. According to the number of publications dealing with the application of BCR-ABL siRNAs, this fusion protein counts as a top favorite among the rest of known leukaemic fusion.

Summarizing the data obtained by different groups, siRNAs targeting the fusion breakpoint site of BCR-ABL decreased the expression of the fusion gene without affecting BCR and ABL1 expression. The depletion of BCR-ABL was correlated with a decreased leukaemic proliferation and an increased extent of apoptosis. Scherr and colleagues addressed the question, whether RNAi may be exploited for therapeutic application besides its well established value as an analytical tool for functional genomics. The lentiviral gene transfer strategy was evaluated as an alternative to the exogenous delivery of chemically synthesized siRNAs. The stable expression of shRNAs resulted in sustained reduction of BCR-ABL and, conse- quently, inhibited proliferation, decreased cell survival and clonogenic capacity of CD34+ primary CML cells[68]. The group of Whitey targeted the b3a2 variant in primary chronic phase of CML[94]. The siRNA treatment led to a decreased expansion of granulocyte-macrophage progenitors expressing the targeted b3a2 variant without affecting the b2a2 variant expressed in other cells. Moreover, this work provided additional evidence for the role of BCR-ABL in driving aberrant amplification of myeloid progenitors.

All the experiments were performed using siRNAs, which target the fusion breakpoint site of BCR-ABL. Ohba and colleagues chose a downstream sequence of the ABL1 part instead of the fusion site as a target for the siRNA[93]. In this case, the siRNA interfered with both BCR-ABL and ABL1 expression. This siRNA markedly decreased target mRNA levels leading to the inhibition of protein tyrosine kinase activity and suppression of cell proliferation. A microarray analysis showed a cross-talk between siRNA-mediated suppression of BCR-ABL and expression of several apoptosis and cell proliferation factors.

Wohlbold and colleagues were the first to examine possible synergisms between BCR-ABL siRNAs and established treatment approaches[90]. They used an siRNA construct homologous to the b3a2 breakpoint to silence *BCR-ABL* gene expression in a murine and a human *BCR-ABL*-dependent cell line. They showed that interference with *BCR-ABL* expression enhanced the cellular sensitivity towards gamma irradiation and imatinib mesylate, an established tyrosine kinase inhibitor approved for CML therapy. Moreover, siRNA-mediated BCR-ABL depletion restored imatinib sensitivity in cells carrying *BCR-ABL* mutants thereby sensitizing the cells for this inhibitor. Depleting BCR-ABL protein using siRNA antagonized the two major mechanisms of imatinib resistance, namely BCR-ABL overexpression and the occurrence of point mutations. Thus, a possible application of antileukaemic siRNAs as a supplementary therapy could be of clinical significance. In a follow-up study Wohlbold and colleagues demonstrated that all common *BCR-ABL* transcript variants could be successfully targeted with siRNAs being homologous to the corresponding fusion sites[95].

7.2   TEL-PDGFβR

The TEL-PDGFβR fusion was identified as a consequence of t(5;12)(q33;p13) chromosomal translocation, a recurring cytogenetic abnormality associated with chronic myelomonocytic leukaemia (CMML) that is characterized by abnormal myelopoiesis with eosinophilia, myelofibrosis and frequent progression to acute myeloid leukemia. The resulting fusion has a tyrosine kinase functions and in line with BCR-ABL is a well validated therapeutic target in human leukemia. Small molecule inhibitors such as imatinib mesylate are an effective therapy against TEL-PDGFβR-associated leukemia. However, imatinib is not curative as a single agent, and cases of clinical resistance occur quite often. Chen and colleagues developed a retroviral system for stable expression of siRNA directed to the unique junction sequence of this fusion tyrosine kinase TEL-PDGFβR in transformed haematopoietic cells[76]. The siRNAs were transcribed using an H1-promoter-based short hairpin RNA expression system in self-inactivating retroviral vector. The suppression of TEL-PDGFβR affected signal transduction through PI3 kinase and MTOR (mammalian target of rapamycin) and consequently lead to a reduced proliferation of TEL-PDGFβR transformed cells. Additionally, TEL-PDGFβR siRNAs, analogously to BCR-ABL siRNAs, sensitized cells expressing either wild-type or mutated, chemoresistant version to the small molecular inhibitors. To evaluate the therapeutic efficacy of siRNAs *in vivo*, TEL-PDGFβR transformed cells with or without active siRNA were injected into the tail vein of nude mice. Whereas cells stably expressing TEL-PDGFβR alone caused tumor development and death with a median survival of 24 days, expression of TEL-PDGFβR siRNA resulted in a significant prolongation in survival with a median latency of 41 days.

7.3   LYN

Leukaemic cells expressing *BCR-ABL* are resistant to apoptotic stimuli and are relatively insensitive to chemotherapy. SRC kinases such as *LYN* form signalling complexes with BCR-ABL, and are activated by this oncoprotein[96]. The increased kinase activity of LYN in blast crisis cells versus normal ones suggests LYN as a putative therapeutic target. Ptasznik at al studied the consequences of siRNA-mediated LYN depletion on the survival of imatinib mesylate-resistant chronic myelogenous leukaemia (CML) and acute lymphoblastic leukaemia (ALL) blast crisis cells[65]. Notably, LYN siRNA led to a five- to tenfold reduction of LYN protein in both normal and *BCR-ABL* expressing cells. However, only BCR-ABL-positive CML and ALL blasts underwent apoptosis within 48h after the siRNA treatment, whereas normal cells remained viable. The authors noted that ALL blasts were more affected by LYN depletion than CML blasts; however, both types of blasts were dependent on LYN kinase for growth and survival. These data further suggest Lyn and possibly other Src kinases as promising targets in so far chemoresistant BCR-ABL-positive, in particular lymphoblastic, leukaemias.

7.4   FLT3

FMS-like tyrosine kinase 3 (flt3) is a receptor tyrosine kinase that is constitutively activated in $\sim 30\%$ of acute myelogenous leukaemia (AML) patients. Constitutively active FLT3 is associated with adverse prognosis[97]. Therefore, the development of inhibitors specifically targeting FLT 3 has been of substantial interest. Walters et al used an siRNA approach to down-regulate FLT 3 expression in two human leukaemia cell lines, which expressed a FLT3 mutant containing an activating internal tandem duplication (ITD) in the juxtamembrane domain[98]. Treatment with FLT3 siRNA resulted in diminished phosphorylation of several downstream molecules, inhibited growth inhibition and induced apoptosis. The combination of FLT siRNAs together with the specific FLT3 inhibitor MLN518 led to synergistic effects on cell proliferation and apoptosis induction.

8.   RNAi *IN VIVO*

Results from both cell culture and animal models suggest that RNAi may have therapeutic potential, particularly for the treatment of diseases such as viral infections and cancer. Both, exogenously added siRNAs and ectopically expressed shRNAs allow the direct targeting of disease-associated transcripts. The examples discussed in the previous sections strongly suggest that siRNA-mediated oncogene suppression may become a promising option for antileukaemic therapy. However, there are major obstacles to overcome such as limited specificity, induction of interferon responses, emergence of escape mutations or inefficient systemic delivery of siRNA *in vivo*[5,7]. Only if these challenges are successfully solved, RNAi based technologies will fulfil their promises in cancer therapy.

While specificity, interferon response induction or siRNA sequence homology could be addressed in the cell culture, the most challenging problem for the therapeutic use of siRNAs or shRNAs is their efficient delivery to tumour and leukaemic tissues. Transfection of such RNAi modulators *ex vivo* has been shown to be a suitable short-cut for their *in vivo* application. For instance, we and others injected leukaemic cells electroporated with MLL-AF4 siRNAs, or cells stably expressing TEL-PDGFβR siRNAs into immunodeficient mouse strains[39,76]. Xenotransplantation of fusion protein-depleted cells prolonged survival of transplanted mice and, in the case of MLL-AF4 suppression, reduced mortality. These findings suggest that pretreatment with antileukaemic siRNAs may compromise engraftment and expansion of leukaemic cells *in vivo*. On this basis, one may envisage the application of such siRNA treatments in purging protocols during the course of autologous stem cell transplantations to reduce the leukaemic contamination of a transplant.

To date, only very few studies address the problem of systemic siRNA delivery to haematopoietically relevant organs. siRNAs are unable to cross cellular membranes, are rather unstable in blood serum and are rapidly secreted via the renal system[99]. In spite of these limitations, techniques were developed for the systemic delivery of naked siRNAs in mouse models. However, approaches such as high-pressure, high-volume intravenous injection of synthetic siRNAs, the so-called hydrodynamic delivery, are of limited clinical use because of the severe side effects[7,100].

To protect siRNAs against serum nucleases and to enhance their cellular uptake, Urban-Klein and colleagues used complexes of linear low molecular weight polyethylenimine (PEI) with siRNA molecules[101]. The PEI complexation almost completely protected the siRNA against degradation in fetal calf serum. In a subcutaneous tumour mouse model, intraperitoneal (i.p.) injection of PEI/ERBB2 siRNA complexes led to siRNA delivery into tumour tissue and delayed tumour growth. The group of Schiffelers used a similar approach for systemic siRNA delivery[102]. In this case, VEGFR2 siRNAs were complexed with a conjugate consisting of a PEI and a polyethylene glycol (PEG) moiety. PEG shields the positive charges of PEI resulting in a prolonged circulation time and a reduced unspecific cell adhesion of these nanoplexes. Cellular uptake was achieved by linking an RGD peptide ligand to distal end of PEG. Intravenous injection of these nanoplexes yielded efficient siRNA delivery to tumours and interfered with tumour angiogenesis and growth.

An alternative to the use of ligand-containing nanoplexes is the direct conjugation of ligands such as a cholesterol moiety to an siRNA[103]. Chemically modified siRNAs containing 2′-modified nucleotides on selected positions in combination with a cholesterol at the 3′end of the sense strand inhibit apolipoprotein B (ApoB) expression after intravenous injection in mice. Administration of these chemically modified siRNAs resulted in silencing of the apoB messenger RNA in liver and jejunum, decreased plasma levels of ApoB protein and reduced total cholesterol. Furthermore, siRNA-mediated mRNA degradation including the specific cleavage of the apoB mRNA was demonstrated in this *in vivo* study. However, no studies using such modified siRNAs have been reported so far for haemopoietic tissues.

Antibody-based therapies have become an important option for the treatment of leukaemias. In particular, a construct consisting of a calicheamicin moiety linked to humanized antibody targeting the myeloid antigen CD33 (gemtuzumab ozogamicin, mylotarg) has been approved for treatment of certain types of AML[104]. It is therefore tempting to think of antibody-mediated *in vivo* delivery of siRNAs. For instance, linking a positively charged protamine moiety to the heavy chain FAB fragment of an HIV envelope antibody allowed for the specific and efficient siRNA delivery to HIV envelope-expressing cells both in cell culture and in murine tumour models[105]. Antibody-directed cell type specific siRNA delivery may become very popular due to its specificity for cells expressing the corresponding cell-surface receptor[106]. Thus, in principal, the use of FABs targeting haematopoietic surface molecules such as CD33 is a very promising approach for the systemic delivery of antileukaemic siRNAs to leukaemic tissues.

9.   OUTLOOK

Both siRNAs or shRNAs have quickly become standard tools for the functional analysis of genes involved in haematopoiesis and leukaemogenesis. The next challenge for this technology is its transfer into therapy. The big advantage of siRNAs compared to other molecular approaches such as therapeutic antibodies is their relatively easy and straightforward design, their quick generation and evaluation in cell culture. However, inefficient systemic delivery of siRNAs or shRNAs is still a major obstacle for their use in combatting disease including haematopoietic malignancies. Nevertheless, the recent progress made in this field may get the therapeutic application of siRNAs in reach. In combination with efficient delivery agents such as nanoplexes or antibody-based conjugates, siRNAs will provide a specific, but still flexible and, thus, promising therapeutic tool for the treatment of haematological malignancies.

10.   REFERENCE

1. Orkin, S. H. Diversification of haematopoietic stem cells to specific lineages. *Nat Rev Genet* **1**, 57–64 (2000).
2. Passegue, E., Jamieson, C. H., Ailles, L. E. & Weissman, I. L. Normal and leukemic hematopoiesis: are leukemias a stem cell disorder or a reacquisition of stem cell characteristics? *Proc Natl Acad Sci U S A* **100**, 11842–9 (2003).
3. Look, A. T. Oncogenic transcription factors in the human acute leukemias. *Science* **278**, 1059–64 (1997).
4. Pardanani, A. & Tefferi, A. Imatinib targets other than bcr/abl and their clinical relevance in myeloid disorders. *Blood* **104**, 1931–9 (2004).
5. Sledz, C. A. & Williams, B. R. RNA interference in biology and disease. *Blood* **106**, 787–94 (2005).
6. Scherr, M., Morgan, M. A. & Eder, M. Gene silencing mediated by small interfering RNAs in mammalian cells. *Curr Med Chem* **10**, 245–56 (2003).
7. Borkhardt, A. & Heidenreich, O. RNA interference as a potential tool in the treatment of leukaemia. *Expert Opin Biol Ther* **4**, 1921–9 (2004).

8. Fire, A. et al. Potent and specific genetic interference by double-stranded RNA in Caenorhabditis elegans. *Nature* **391**, 806–11 (1998).

9. Martinez, J. & Tuschl, T. RISC is a 5′ phosphomonoester-producing RNA endonuclease. *Genes Dev* **18**, 975–80 (2004).

10. Jaronczyk, K., Carmichael, J. B. & Hobman, T. C. Exploring the functions of RNA interference pathway proteins: some functions are more RISCy than others? *Biochem J* **387**, 561–71 (2005).

11. Bernstein, E., Caudy, A. A., Hammond, S. M. & Hannon, G. J. Role for a bidentate ribonuclease in the initiation step of RNA interference. *Nature* **409**, 363–6 (2001).

12. Macrae, I. J. et al. Structural basis for double-stranded RNA processing by Dicer. *Science* **311**, 195–8 (2006).

13. Chendrimada, T. P. et al. TRBP recruits the Dicer complex to Ago2 for microRNA processing and gene silencing. *Nature* **436**, 740–4 (2005).

14. Haase, A. D. et al. TRBP, a regulator of cellular PKR and HIV-1 virus expression, interacts with Dicer and functions in RNA silencing. *EMBO Rep* **6**, 961–7 (2005).

15. Hammond, S. M., Bernstein, E., Beach, D. & Hannon, G. J. An RNA-directed nuclease mediates post-transcriptional gene silencing in Drosophila cells. *Nature* **404**, 293–6 (2000).

16. Meister, G. et al. Human Argonaute2 mediates RNA cleavage targeted by miRNAs and siRNAs. *Mol Cell* **15**, 185–97 (2004).

17. Filipowicz, W. RNAi: the nuts and bolts of the RISC machine. *Cell* **122**, 17–20 (2005).

18. Pasquinelli, A. E. et al. Conservation of the sequence and temporal expression of let-7 heterochronic regulatory RNA. *Nature* **408**, 86–9 (2000).

19. Calin, G. A. et al. Human microRNA genes are frequently located at fragile sites and genomic regions involved in cancers. *Proc Natl Acad Sci U S A* **101**, 2999–3004 (2004).

20. Lee, Y. et al. MicroRNA genes are transcribed by RNA polymerase II. *Embo J* **23**, 4051–60 (2004).

21. Cai, X., Hagedorn, C. H. & Cullen, B. R. Human microRNAs are processed from capped, polyadenylated transcripts that can also function as mRNAs. *Rna* **10**, 1957–66 (2004).

22. Lagos-Quintana, M., Rauhut, R., Lendeckel, W. & Tuschl, T. Identification of novel genes coding for small expressed RNAs. *Science* **294**, 853–8 (2001).

23. Yi, R., Qin, Y., Macara, I. G. & Cullen, B. R. Exportin-5 mediates the nuclear export of pre-microRNAs and short hairpin RNAs. *Genes Dev* **17**, 3011–6 (2003).

24. Hutvagner, G. et al. A cellular function for the RNA-interference enzyme Dicer in the maturation of the let-7 small temporal RNA. *Science* **293**, 834–8 (2001).

25. Hutvagner, G. & Zamore, P. D. A microRNA in a multiple-turnover RNAi enzyme complex. *Science* **297**, 2056–60 (2002).

26. Gregory, R. I., Chendrimada, T. P., Cooch, N. & Shiekhattar, R. Human RISC couples microRNA biogenesis and posttranscriptional gene silencing. *Cell* **123**, 631–40 (2005).

27. Liu, J., Valencia-Sanchez, M. A., Hannon, G. J. & Parker, R. MicroRNA-dependent localization of targeted mRNAs to mammalian P-bodies. *Nat Cell Biol* **7**, 719–23 (2005).

28. Yekta, S., Shih, I. H. & Bartel, D. P. MicroRNA-directed cleavage of HOXB8 mRNA. *Science* **304**, 594–6 (2004).

29. Humphreys, D. T., Westman, B. J., Martin, D. I. & Preiss, T. MicroRNAs control translation initiation by inhibiting eukaryotic initiation factor 4E/cap and poly(A) tail function. *Proc Natl Acad Sci U S A* **102**, 16961–6 (2005).

30. Chen, C. Z., Li, L., Lodish, H. F. & Bartel, D. P. MicroRNAs modulate hematopoietic lineage differentiation. *Science* **303**, 83–6 (2004).

31. Calin, G. A. et al. A MicroRNA signature associated with prognosis and progression in chronic lymphocytic leukemia. *N Engl J Med* **353**, 1793–801 (2005).

32. Eder, M. & Scherr, M. MicroRNA and lung cancer. *N Engl J Med* **352**, 2446–8 (2005).

33. Elbashir, S. M., Lendeckel, W. & Tuschl, T. RNA interference is mediated by 21- and 22-nucleotide RNAs. *Genes Dev* **15**, 188–200 (2001).

34. Khvorova, A., Reynolds, A. & Jayasena, S. D. Functional siRNAs and miRNAs exhibit strand bias. *Cell* **115**, 209–16 (2003).

35. Reynolds, A. et al. Rational siRNA design for RNA interference. *Nat Biotechnol* **22**, 326–30 (2004).

36. Kretschmer-Kazemi Far, R. & Sczakiel, G. The activity of siRNA in mammalian cells is related to structural target accessibility: a comparison with antisense oligonucleotides. *Nucleic Acids Res* **31**, 4417–24 (2003).

37. Luo, K. Q. & Chang, D. C. The gene-silencing efficiency of siRNA is strongly dependent on the local structure of mRNA at the targeted region. *Biochem Biophys Res Commun* **318**, 303–10 (2004).

38. Patzel, V. et al. Design of siRNAs producing unstructured guide-RNAs results in improved RNA interference efficiency. *Nat Biotechnol* **23**, 1440–4 (2005).

39. Thomas, M. et al. Targeting MLL-AF4 with short interfering RNAs inhibits clonogenicity and engraftment of t(4;11)-positive human leukemic cells. *Blood* **106**, 3559–3566 (2005).

40. Yang, D. et al. Short RNA duplexes produced by hydrolysis with Escherichia coli RNase III mediate effective RNA interference in mammalian cells. *Proc Natl Acad Sci U S A* **99**, 9942–7 (2002).

41. Kawasaki, H., Suyama, E., Iyo, M. & Taira, K. siRNAs generated by recombinant human Dicer induce specific and significant but target site-independent gene silencing in human cells. *Nucleic Acids Res* **31**, 981–7 (2003).

42. Semizarov, D. et al. Specificity of short interfering RNA determined through gene expression signatures. *Proc Natl Acad Sci U S A* **100**, 6347–52 (2003).

43. Chi, J. T. et al. Genomewide view of gene silencing by small interfering RNAs. *Proc Natl Acad Sci U S A* **100**, 6343–6 (2003).

44. Martinez, L. A. et al. Synthetic small inhibiting RNAs: efficient tools to inactivate oncogenic mutations and restore p53 pathways. *Proc Natl Acad Sci U S A* **99**, 14849–54 (2002).

45. Jackson, A. L. et al. Expression profiling reveals off-target gene regulation by RNAi. *Nat Biotechnol* **21**, 635–7 (2003).

46. Lin, X. et al. siRNA-mediated off-target gene silencing triggered by a 7 nt complementation. *Nucleic Acids Res* **33**, 4527–35 (2005).

47. Saxena, S., Jonsson, Z. O. & Dutta, A. Small RNAs with imperfect match to endogenous mRNA repress translation. Implications for off-target activity of small inhibitory RNA in mammalian cells. *J Biol Chem* **278**, 44312–9 (2003).

48. Scacheri, P. C. et al. Short interfering RNAs can induce unexpected and divergent changes in the levels of untargeted proteins in mammalian cells. *Proc Natl Acad Sci U S A* **101**, 1892–7 (2004).

49. Qiu, S., Adema, C. M. & Lane, T. A computational study of off-target effects of RNA interference. *Nucleic Acids Res* **33**, 1834–47 (2005).

50. Holen, T. et al. Tolerated wobble mutations in siRNAs decrease specificity, but can enhance activity in vivo. *Nucleic Acids Res* **33**, 4704–10 (2005).

51. Du, Q., Thonberg, H., Wang, J., Wahlestedt, C. & Liang, Z. A systematic analysis of the silencing effects of an active siRNA at all single-nucleotide mismatched target sites. *Nucleic Acids Res* **33**, 1671–7 (2005).

52. Elbashir, S. M. et al. Duplexes of 21-nucleotide RNAs mediate RNA interference in cultured mammalian cells. *Nature* **411**, 494–8 (2001).

53. Bridge, A. J., Pebernard, S., Ducraux, A., Nicoulaz, A. L. & Iggo, R. Induction of an interferon response by RNAi vectors in mammalian cells. *Nat Genet* **34**, 263–264 (2003).

54. Sledz, C. A., Holko, M., de Veer, M. J., Silverman, R. H. & Williams, B. R. Activation of the interferon system by short-interfering RNAs. *Nat Cell Biol* **5**, 834–9 (2003).

55. Kim, D. H. et al. Interferon induction by siRNAs and ssRNAs synthesized by phage polymerase. *Nat Biotechnol* **22**, 321–5 (2004).

56. Hormes, R. et al. The subcellular localization and length of hammerhead ribozymes determine efficacy in human cells. *Nucleic Acids Res.* **25**, 769–775 (1997).

57. Judge, A. D. et al. Sequence-dependent stimulation of the mammalian innate immune response by synthetic siRNA. *Nat Biotechnol* **23**, 457–62 (2005).

58. Heidel, J. D., Hu, S., Liu, X. F., Triche, T. J. & Davis, M. E. Lack of interferon response in animals to naked siRNAs. *Nat Biotechnol* **22**, 1579–82 (2004).

59. Kittler, R. et al. RNA interference rescue by bacterial artificial chromosome transgenesis in mammalian tissue culture cells. *Proc Natl Acad Sci U S A* **102**, 2396–401 (2005).

60. Wianny, F. & Zernicka-Goetz, M. Specific interference with gene function by double-stranded RNA in early mouse development. *Nat Cell Biol* **2**, 70–5 (2000).
61. Kasashima, K., Sakota, E. & Kozu, T. Discrimination of target by siRNA: designing of AML1-MTG8 fusion mRNA-specific siRNA sequences. *Biochimie* **86**, 713–21 (2004).
62. John, M., Geick, A., Hadwiger, P., Vornlocher, H. P. & Heidenreich, O. in *Current protocols in molecular biology* (eds. Ausubel, F. M. et al.) 26.2.1–26.2.14 (John Wiley & Sons, New York, 2003).
63. Heidenreich, O. et al. AML1/MTG8 oncogene suppression by small interfering RNAs supports myeloid differentiation of t(8;21)-positive leukemic cells. *Blood* **101**, 3157–63 (2003).
64. Scherr, M. et al. Specific inhibition of bcr-abl gene expression by small interfering RNA. *Blood* **101**, 1566–1569 (2003).
65. Ptasznik, A., Nakata, Y., Kalota, A., Emerson, S. G. & Gewirtz, A. M. Short interfering RNA (siRNA) targeting the Lyn kinase induces apoptosis in primary, and drug-resistant, BCR-ABL1(+) leukemia cells. *Nat Med* **10**, 1187–9 (2004).
66. Bartlett, D. W. & Davis, M. E. Insights into the kinetics of siRNA-mediated gene silencing from live-cell and live-animal bioluminescent imaging. *Nucleic Acids Res* **34**, 322–33 (2006).
67. Tuschl, T. Expanding small RNA interference. *Nat Biotechnol* **20**, 446–8 (2002).
68. Scherr, M., Battmer, K., Schultheis, B., Ganser, A. & Eder, M. Stable RNA interference (RNAi) as an option for anti-bcr-abl therapy. *Gene Ther* **12**, 12–21 (2005).
69. Lin, X. et al. Development of a tightly regulated U6 promoter for shRNA expression. *FEBS Lett* **577**, 376–80 (2004).
70. Gupta, S., Schoer, R. A., Egan, J. E., Hannon, G. J. & Mittal, V. Inducible, reversible, and stable RNA interference in mammalian cells. *Proc Natl Acad Sci U S A* **101**, 1927–32 (2004).
71. Czauderna, F. et al. Inducible shRNA expression for application in a prostate cancer mouse model. *Nucleic Acids Res* **31**, e127 (2003).
72. Zhang, D. E. et al. CCAAT enhancer-binding protein (C/EBP) and AML1 (CBF alpha2) synergistically activate the macrophage colony-stimulating factor receptor promoter. *Mol Cell Biol* **16**, 1231–40 (1996).
73. Li, K., Lin, S. Y., Brunicardi, F. C. & Seu, P. Use of RNA interference to target cyclin E-overexpressing hepatocellular carcinoma. *Cancer Res* **63**, 3593–7 (2003).
74. Stegmeier, F., Hu, G., Rickles, R. J., Hannon, G. J. & Elledge, S. J. A lentiviral microRNA-based system for single-copy polymerase II-regulated RNA interference in mammalian cells. *Proc Natl Acad Sci U S A* **102**, 13212–7 (2005).
75. Scherr, M., Battmer, K., Ganser, A. & Eder, M. Modulation of gene expression by lentiviral-mediated delivery of small interfering RNA. *Cell Cycle* **2**, 251–7 (2003).
76. Chen, J. et al. Stable expression of small interfering RNA sensitizes TEL-PDGFbetaR to inhibition with imatinib or rapamycin. *J Clin Invest* **113**, 1784–91 (2004).
77. Zheng, X. et al. Gamma-catenin contributes to leukemogenesis induced by AML-associated translocation products by increasing the self-renewal of very primitive progenitor cells. *Blood* **103**, 3535–43 (2004).
78. Kimchi-Sarfaty, C. et al. Efficient delivery of RNA interference effectors via in vitro-packaged SV40 pseudovirions. *Hum Gene Ther* **16**, 1110–5 (2005).
79. Downing, J. R. The AML1-ETO chimaeric transcription factor in acute myeloid leukaemia: biology and clinical significance. *Br J Haematol* **106**, 296–308 (1999).
80. Okuda, T., van Deursen, J., Hiebert, S. W., Grosveld, G. & Downing, J. R. AML1, the target of multiple chromosomal translocations in human leukemia, is essential for normal fetal liver hematopoiesis. *Cell* **84**, 321–30 (1996).
81. Gelmetti, V. et al. Aberrant recruitment of the nuclear receptor corepressor-histone deacetylase complex by the acute myeloid leukemia fusion partner ETO. *Mol Cell Biol* **18**, 7185–91 (1998).
82. Martinez, N. et al. The oncogenic fusion protein RUNX1-CBFA2T1 supports proliferation and inhibits senescence in t(8;21)-positive leukaemic cells. *BMC Cancer* **4**, 44 (2004).
83. Nakamura, T. et al. ALL-1 is a histone methyltransferase that assembles a supercomplex of proteins involved in transcriptional regulation. *Mol Cell* **10**, 1119–28 (2002).

84. Yokoyama, A. et al. Leukemia proto-oncoprotein MLL forms a SET1-like histone methyltrans-ferase complex with menin to regulate Hox gene expression. *Mol Cell Biol* **24**, 5639–49 (2004).

85. Pui, C. H., Schrappe, M., Ribeiro, R. C. & Niemeyer, C. M. Childhood and adolescent lymphoid and myeloid leukemia. *Hematology (Am Soc Hematol Educ Program)*, 118–45 (2004).

86. Rowley, J. D. The role of chromosome translocations in leukemogenesis. *Semin Hematol* **36**, 59–72 (1999).

87. Armstrong, S. A. et al. Inhibition of FLT3 in MLL. Validation of a therapeutic target identified by gene expression based classification. *Cancer Cell* **3**, 173–83 (2003).

88. Barthe, C., Cony-Makhoul, P., Melo, J. V. & Mahon, J. R. Roots of clinical resistance to STI-571 cancer therapy. *Science* **293**, 2163 (2001).

89. Wilda, M., Fuchs, U., Wossmann, W. & Borkhardt, A. Killing of leukemic cells with a BCR/ABL fusion gene by RNA interference (RNAi). *Oncogene* **21**, 5716–24 (2002).

90. Wohlbold, L. et al. Inhibition of bcr-abl gene expression by small interfering RNA sensitizes for imatinib mesylate (STI571). *Blood* **102**, 2236–9 (2003).

91. Zhelev, Z. et al. Suppression of bcr-abl synthesis by siRNAs or tyrosine kinase activity by Glivec alters different oncogenes, apoptotic/antiapoptotic genes and cell proliferation factors (microarray study). *FEBS Lett* **570**, 195–204 (2004).

92. Rapozzi, V. & Xodo, L. E. Efficient silencing of bcr/abl oncogene by single- and double-stranded siRNAs targeted against b2a2 transcripts. *Biochemistry* **43**, 16134–41 (2004).

93. Ohba, H. et al. Inhibition of bcr-abl and/or c-abl gene expression by small interfering, double-stranded RNAs: cross-talk with cell proliferation factors and other oncogenes. *Cancer* **101**, 1390–403 (2004).

94. Withey, J. M. et al. Targeting primary human leukaemia cells with RNA interference: Bcr-Abl targeting inhibits myeloid progenitor self-renewal in chronic myeloid leukaemia cells. *Br J Haematol* **129**, 377–80 (2005).

95. Wohlbold, L. et al. All common p210 and p190 Bcr-abl variants can be targeted by RNA inter-ference. *Leukemia* (2004).

96. Danhauser-Riedl, S., Warmuth, M., Druker, B. J., Emmerich, B. & Hallek, M. Activation of Src kinases p53/56lyn and p59hck by p210bcr/abl in myeloid cells. *Cancer Res* **56**, 3589–96 (1996).

97. Gilliland, D. G. & Griffin, J. D. The roles of FLT3 in hematopoiesis and leukemia. *Blood* **100**, 1532–1542 (2002).

98. Walters, D. K., Stoffregen, E. P., Heinrich, M. C., Deininger, M. W. & Druker, B. J. RNAi-induced down-regulation of FLT3 expression in AML cell lines increases sensitivity to MLN518. *Blood* **105**, 2952–4 (2005).

99. Braasch, D. A. et al. Biodistribution of phosphodiester and phosphorothioate siRNA. *Bioorg Med Chem Lett* **14**, 1139–43 (2004).

100. Song, E. et al. RNA interference targeting Fas protects mice from fulminant hepatitis. *Nat Med* **9**, 347–51 (2003).

101. Urban-Klein, B., Werth, S., Abuharbeid, S., Czubayko, F. & Aigner, A. RNAi-mediated gene-targeting through systemic application of polyethylenimine (PEI)-complexed siRNA in vivo. *Gene Ther* (2004).

102. Schiffelers, R. M. et al. Cancer siRNA therapy by tumor selective delivery with ligand-targeted sterically stabilized nanoparticle. *Nucleic Acids Res* **32**, e149 (2004).

103. Soutschek, J. et al. Therapeutic silencing of an endogenous gene by systemic administration of modified siRNAs. *Nature* **432**, 173–8 (2004).

104. Linenberger, M. L. CD33-directed therapy with gemtuzumab ozogamicin in acute myeloid leukemia: progress in understanding cytotoxicity and potential mechanisms of drug resistance. *Leukemia* **19**, 176–82 (2005).

105. Song, E. et al. Antibody mediated in vivo delivery of small interfering RNAs via cell-surface receptors. *Nat Biotechnol* **23**, 709–17 (2005).

106. Vornlocher, H. P. Antibody-directed cell-type-specific delivery of siRNA. *Trends Mol Med* **12**, 1–3 (2006).

CHAPTER 3

GENOMICS AND PROTEOMICS OF CHINESE HAMSTER OVARY (CHO) CELLS

Understanding CHO Cell Culture at the Molecular Level

PETER MORIN NISSOM AND ROBIN PHILP

*Bioprocessing Technology Institute, A*Star, #06-01 Centros, 20 Biopolis Way, Singapore 138668*

Abstract: This chapter provides an overview on the use of high throughput screening technologies like microarray and proteomics, to studying bioprocess issues in cell cultures. In bioprocessing, these tools may be used to study the genetic circuitry underlying cellular metabolism, growth, death and protein glycosylation during high cell density production processes and protein-free media adaptation. These studies can provide valuable insights into the crucial pathways involved in energy metabolism, product formation and cellular death as well as identify potential gene targets for engineering cell lines with desirable characteristics for biologics production

Keywords: microarray, proteomics, proteome, CHO, Chinese hamster ovary

1.   INTRODUCTION

Cell engineering of industrial cell lines, leading to enhanced traits such as improved viability, increased production of recombinant proteins, robustness and a host of other desirable characteristics, requires an understanding at the molecular level of how cells operate during cell culture. This requires a knowledge of which genes are expressed and subsequently translated into proteins. Gene expression is usually studied at the mRNA level using high throughput technologies like microarray while the protein levels are studied using proteomics. Gene expression studies measures message (mRNA) abundance but not actual protein levels. There may not be a good correlation between mRNA abundance and protein levels, as was shown in the study by Gygi et al[1] on protein and mRNA abundance in yeast and Anderson and Seilhamer[2], with regards to the human liver. While there is a general lack of correlation at any given time point within the total proteome and transcriptome,

49

M. Al-Rubeai and M. Fussenegger (eds.), Systems Biology, 49–68.
© 2007 *Springer.*

however, there may be correlation in functional subsets of genes/proteins. Subsets which do not correlate may represent pathways regulated by post-translational modification[3].

In recent years, with advances in genomics and proteomics, genomic-scale analyses have become a more practical approach as opposed to traditional biochemical assays which focused on analyzing a single reaction at a time. Microarrays permit rapid, large scale analyses of samples. The ability to analyze many samples simultaneously is perhaps the most attractive feature of this technology. A very well established use of DNA microarrays is to create transcription profiles which is a measure of gene expression. Microarrays are patterns of cDNA sequences deposited on a substrate, usually glass. These arrays of DNA are then probed with fluorescently labeled mRNAs extracted from cells. The mRNAs will hybridize to their immobilized complementary DNA on the chip (Figure 1). The fluorescence intensities give a relative measure of gene expression.

The proteome is defined as the complement of proteins that is present in a cell at a given moment in time and under a particular set of environmental conditions. Whilst the genome is a relatively static entity the proteome, in contrast, is a highly dynamic system and constantly undergoing change. Proteomics, or the study of the proteome, is analogous to genomics but is much more complicated due to the greater number of proteins present than are the genes that code for them. This is due to the presence of protein isoforms as well as posttranslational modifications of proteins such as phosphorylation, glycosylation, acetylation and a myriad of other known chemical modifications – each giving rise to a different and distinguishable entity.

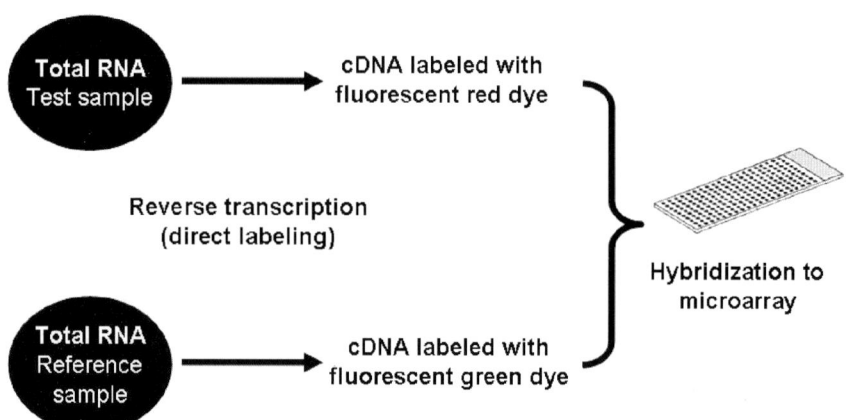

Figure 1. **Microarray analysis workflow for a typical pairwise comparison of two samples.** The total RNA or mRNA of a reference sample is reverse transcribed in the presence of a green fluorescent dye (Cyanine 3) into cDNA. Similarly, the total RNA or mRNA of a test sample reversed transcribed in the presence of a red fluorescent dye (Cyanine 5). The two samples of red and green labeled cDNAs (targets) are then pooled and hybridized onto cDNA probes arrayed onto microscope slides. The location and extent of hybridization of the probes to their targets is a quantitative measure of the identity and gene expression level of the gene in the samples

Further, proteins are present in the cell at concentrations ranging over several orders of magnitude resulting in a major challenge to the analytical processes employed. However, using proteomics we can look at the products of these genes and focus on the proteins that are actually present in the cells at the time of sampling as well as identifying some of the posttranslational modifications that they undergo, thus allowing us to investigate the response of the cells to different stimuli.

2.   GENE EXPRESSION PROFILING IN CHO CELLS

The Chinese hamster ovary (CHO) cells are currently the host cell line of choice for recombinant protein production in industry. These cells produce glycoproteins with similar oligosaccharide structures to those produced in humans[4-6]. One of the key concerns in improving heterologous protein expression in CHO cells, as well as in other industrial cell lines, has been a desire to understand the relationship between gene expression and observed phenotype. This has been made possible with the publication of CHO EST sequences by Wlaschin et al,[7] marking the first development of CHO cDNA microarrays (Figure 2) as tools to monitor gene expression in CHO cells. Use of these CHO cDNA microarrays in transcriptome profiling experiments, have resulted in some interesting outcomes, with respect to enhancing viability in CHO cells.

Wong et al[8] reported, the successful identification of key anti-apoptotic genes and pathways in fed-batch and batch CHO cell cultures. During periods of high

Figure 2. **A section of the CHO 15K DNA microarray**. The array was generated by spotting purified PCR products of CHO cDNA clones. Test RNA was labelled with Cy5 (red) and the reference channel corresponds to Cy3 (green)-labelled control DNA

viability, most pro-apoptotic genes were downregulated but upon loss in viability, several early pro-apoptotic signaling genes were upregulated. At later stages of viability loss, late pro-apoptotic effector genes such as *caspases* and *DNases* were detected as being upregulated. This sequential-like upregulation of apoptotic genes showed that microarray could be used as a tool to study apoptosis. It was found that in both batch and fed-batch cultures, receptor- and mitochondrial-mediated apoptosis pathways play important roles in apoptosis induction. By deciphering these pathways, genes such as *Fadd*, *Pdcd6*, *Bad* and *Bim* that seem to play key roles in both batch and fed-batch CHO cell cultures were identified. Insights from such gene expression profilings provide us with a greater understanding of the regulatory circuitry of apoptosis relevant to cell culture processes and subsequent formulation of strategies towards cell death prevention.

The development of CHO cDNA microarrays is critical if one is to use this technology to understand gene expression in CHO cells. Cross species microarrays (rat and mouse cDNA microarrays) to study CHO cells have yielded unsatisfactory information because of the difference in sequence homologies[7,9].

Microarray technology has matured over the years, since it was first introduced. Researchers have a choice of using Affymetrix suite (www.affymetrix.com), which comes as a complete package of printed chips, hybridization stations and analysis or the traditional glass cDNA microarray chips, which is also available as oligo printed slides. In this chapter, we describe the technology as applied to cDNA microarray, also commonly known as "two-color microarrays". These are widely available and fairly affordable.

2.1   Experimental Design and Sampling

Answering an important question in bioprocessing using high throughput technologies like microarrays and proteomics, requires the right experimental design. A given question may be approached using many possible designs depending on resources, budget and the purpose of the experiment. A comprehensive discussion of microarray experimental designs is available in publications by Yang and Speed[10], Churchill[11] and Kerr and Churchill[12]. Here we describe the design that is suitable for addressing common issues associated with mammalian cell cultures in a bioprocess environment.

In this example, we describe the 'reference design' (Figure 3). In this design, each experimental sample (red) is hybridized to a common reference sample (green). This design is appropriately performed in a time course analysis, where the reference is the innoculum sample and is referred to as time zero (Figure 4). Such a design requires sufficient quantity of reference sample. The advantages of such a design is that it is easily applied to other experiments which uses the same common reference and reduces technical errors and variation[11,12]. A reliable alternative is a common reference obtained by pooling all samples.

A typical microarray experiment is usually made up of three technical replicates, in which one of these replicates is a dye-swap. The experiment is repeated with

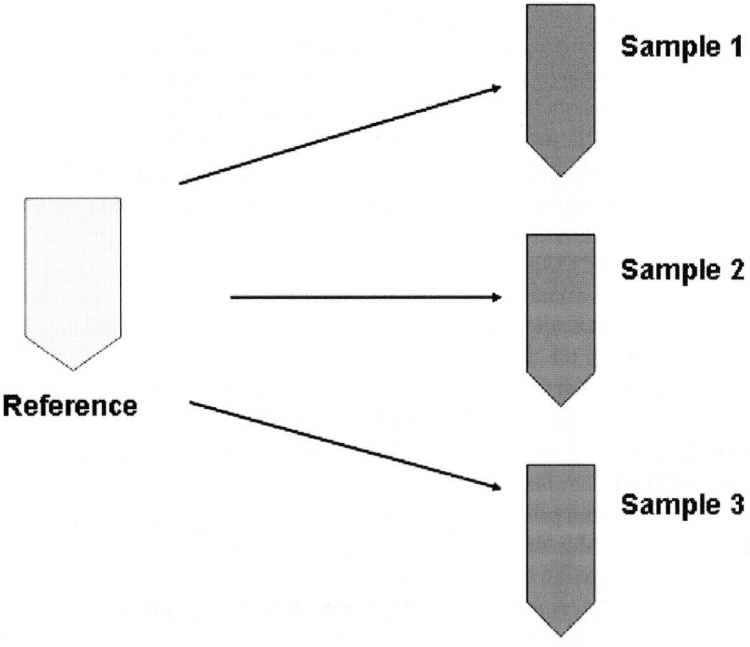

Figure 3. **The "Reference design".** In this design, a single sample is used as the reference sample. This needs to be available in sufficient quantities to permit many replicate hybridization

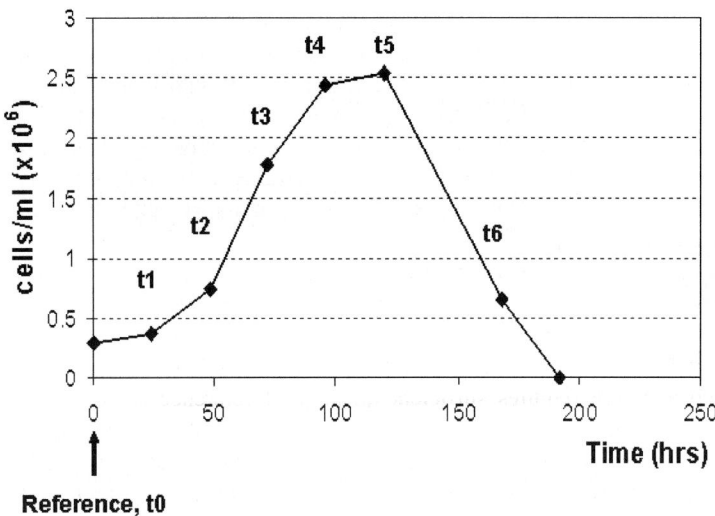

Figure 4. **A time analysis profile using the reference design.** Sample T6 is usually not studied as total RNA yield is generally poor due to poor cell viability

fresh culture or samples to generate biological replicates (Figure 5). This serves to generate sufficient data points for statistical analysis. The number of replicates or microarrays to perform depends on the goals of the study, the resources and the proficiency of the scientist or technician involved. As a rule of thumb, a design with 3 microarrays per experiment is adequate if one aims to find large differences between two samples. A bigger size of 5 microarrays (or more) per condition is necessary to find smaller differences. To do meaningful clustering requires at least 20 replicates or more[13]. In a time profile analysis, using the "reference design", the

(a)

(b)

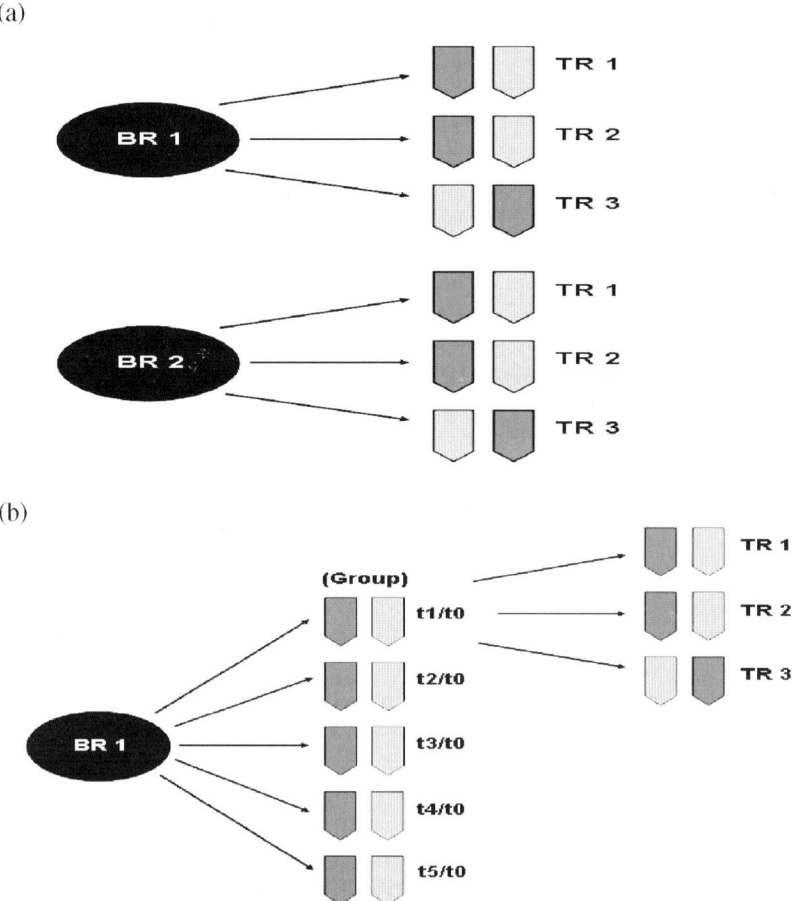

Figure 5. **Replicates in a typical microarray experiment.** (a) A biological replicate (BR) can be considered as an experimental group in which a comparison is being made between a test sample and a reference sample. Each comparison is considered as a technical replicate (TR). (b) In a time profile analysis as shown in figure 4, thus, the time profile itself is a biological replicate, inside which there are many small groups, each made up of three or more technical replicates

number of replicates can become quite large (Figure 5b), hence one has to decide on the number of time points to study.

2.2   Methodology

Microarray expression profiling is a process based on hybridization of nucleic acid molecules to their complementary molecules. These fluorescently labeled nucleic acid molecules act as probes to identify target sequences that base-pair with one another.

Lets consider a typical cell culture experiments consisting of two samples: Sample 1, a culture harvested at exponential growth phase, and Sample 2, a culture at stationary or death phase. Both these two samples contain an identical set of genes, for example, *β-actin, BCL2* or *Eef1α*. We are interested in determining the expression profile of these four genes in the two culture samples. The first task is to isolate mRNA from each individual sample and to generate cDNAs incorporating fluorescent tags. Different tags (red and green) are used so that the samples can be differentiated in subsequent steps.

The samples are then pooled and incubated with the probe sequences immobilized on the microarray. After this hybridization step is complete, the microarray is read in a scanner. The fluorescent tags are excited by the laser, and the microscope and camera work together to create a digital image of the array. Each spot on an array is associated with a particular gene. The intensity of the individual spots provide an estimate of the expression level of the gene(s) in the test and control DNA.

2.2.1   *Isolation of total RNA from cultured CHO cells*

There are many methods for isolating good quality RNA from cells. The important points to remember are: 1. Release RNA from cells, 2. Protect the RNA from RNAses, 3. Separate RNA from the mixture of proteins, lipids, cellular debris, carbohydrates and genomic DNA. The most ideal method to use for cell culture experiments is the acid phenol and guanidium thiocyanate method, first described by Chomczynski and Sacchi[14]. This mixture is sold commercially under various brands like Trizol (Invitrogen) and RNAWiz (Ambion). Large samples, such as those collected from bioreactors or shake flasks can be safely and rapidly processed to yield high quality intact total RNA suitable for microarray analysis. Precautions must be taken to prevent RNase contamination and all materials used must be RNase free grade.

Isolating good quality total RNA using the Trizol reagent method involves some critical steps which should be performed with care. An initial step to lyse the cells in the presence of trizol is performed, whereby cells are repeatedly sheared with a syringe (Figure 6a). Thereafter, chloroform is added and the mixture is separated by centrifugation into three distinct layers, the aqueous phase, containing the RNA, a middle layer of genomic DNA and an organic layer of mostly proteins, phenol and solvents (Figure 6b). The aqueous layer containing the total RNA must then be carefully transferred to new tube. Care must be taken to avoid disturbing the middle

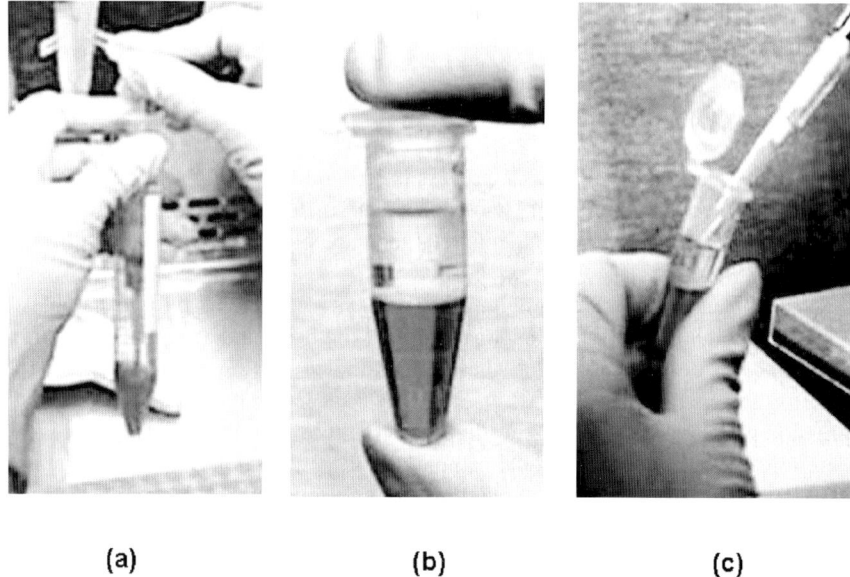

(a)   **(b)**   **(c)**

Figure 6. **Isolation of total RNA from cultured mammalian cells using Trizol method.** (a) Performing cell lysis of cultured mammalian cells in the presence of using a syringe. (b) Separation of trizol/chloroform/lysed mammalian cells mixture after centrifugation. Three separate layers can be seen. The top aqueous layer contains total RNA, the middle interface is mainly genomic DNA while bottom organic layer is the trizol components and proteins. (c) The top layer should be carefully removed into a fresh RNase free tube. Care must be taken to avoid disturbing the middle layer

layer and carry over of the organic layer (Figure 6c). The total RNA is recovered by isopropanol precipitation, washed and dissolved in nuclease free water. Good quality RNA is obtained by washing the pellet at least thrice with 75% ethanol. The RNA pellet can be stored in 75% ethanol at $-20\,^\circ$C.

2.2.2   Labeling of cDNA with fluorescent dyes and hybridization

Two colour microarrays are generated by hybridization of cDNA labeled with two dyes, namely Cyanine 3, green and Cyanine 5, red, to probes printed on glass slides. Labeling of the targets (genes) proceed via a reverse transcription reaction whereby the mRNA is converted to cDNA in a reaction catalysed by reverse transcriptase and in the presence of deoxynucleotides tagged to the flourescent probes. These deoxynucleotides are usually uridines (dUTP) or cytidines (dCTP). The two labeled samples are pooled and incubated with a microarray. The labeled molecules bind to the sites on the array corresponding to the genes expressed in each sample.

2.2.3   Image and data analysis

The labeled probes that are bound to the each element (cDNA) on the microarray is visualized by fluorescence detection using laser scanning devices, such as the Axon

4000B (Molecular Devices, Sunnyvale, CA) or the Scanarray Gx (Perkin Elmer, Wellesley, MA). The fluorescent emission is converted into a16-bit TIFF image file (e.g Figure 2) which can then be quantitated and analysed using image analysis softwares which comes packaged with the scanning instruments, for example, Genepix Pro (Molecular Devices, Sunnyvale, CA). These software packages contain algorithms that normalizes the data extracted from these images. The output from these image analysis softwares are usually intensity values of the green (reference sample) and red (test) channels. The ratios of these values represent the relative gene expression of the genes in the test sample over the reference sample. These values can be further analysed using advanced softwares that permit clustering, pathway analysis and higher order relationships like gene regulatory networks. There are many excellent publications on the topic such as Quackenbush[15], who reviewed the different strategies of data analysis and computational software for normalization, statistical analysis as well as clustering algorithms. Leung and Cavalieri[16] provided an overview of the latest trend on various steps in the microarray data analysis process, which includes experimental design, data standardization, image acquisition and analysis, normalization, statistical significance inference, exploratory data analysis, class prediction and pathway analysis.

3.   PROTEOMICS OF CHO CELLS

With the advent of genomics and the ability to sequence the complete genome of an organism there is now a massive amount of data available. To date the sequencing and assembly of hundreds of genomes has been completed and the emphasis has now shifted to study the proteins that are coded by these genes. The term proteome refers to the total complement of proteins that are expressed by a particular organism under a particular set of conditions. The genome is a fairly static entity which has the potential to code for many proteins depending on its size. However, the proteome, in contrast, is a highly dynamic system which can alter the number of proteins expressed depending on the conditions that the cell is under. For example, the human genome has been shown to consist of 35–40,000 genes but these can be translated into perhaps ten times that number of proteins due to iso-forms and alternate splicing as well as the inclusion of post translational modifications. It is for this reason that proteomics or functional proteomics has much to offer in complementation to the use of other techniques such as microarray.

The use of the term 'functional proteome' better describes the proteins that are expressed and reflects the conditions that the cell is under and the influence on transcription. Proteomics or the study of the proteome, has been used significantly in bio-medical research where comparisons have been made between normal tissue or cells to those that have undergone some change such as in cancer or other diseases. Proteomics has been applied to many types of samples such as liver, heart, brain, kidney and body fluids such as serum and cerebro spinal fluid. However, there is now an increasing use of proteomics applied to the area of bioprocessing and its use in better understanding processes that will enable an enhancement of production, lower costs and faster delivery time of protein based therapeutics.

Genomics has begun to revolutionize the approach to research and development in all of the life sciences and particularly in the area of drug discovery and pharmaceutical based research. Technologies such as systems biology and structural biology are now expanding our understanding of biological structure and function, while genome-wide transcriptomics and proteomics are providing us with the tools to study the dynamic processes that underlie the biochemical and physiological processes of the cell. Together these techniques provide a systematic approach to study in-depth the pathways that are responsible for the control of cellular processes.

During the culture of mammalian cells nutrients that are consumed during growth, are converted to metabolites amongst which lactate and ammonia are two of the highest. These metabolites accumulate in the cell and cause an adverse effect on cell growth resulting in low cell density. Studies have shown that by controlled nutrient feeding, which maintains lowered concentrations of glucose and glutamine, the cells are able to maintain a higher growth rate. Using a proteomics approach Seow et al[17] showed that a number of proteins were shown to be differentially expressed in the altered metabolic state. This information can then be used as potential biomarkers for the design and monitoring of different growth conditions.

One of the biggest challenges to the industry is to be able to control and increase the expression of a given protein being produced by the cells. It was shown by Palermo et al[18] and Oh et al[19] that treatment of CHO and other cell types with sodium butyrate increases the levels of recombinant protein though in a rather unpredictable manner. Further work by Van Dyk et al[20] used proteomics to study the elevated production of recombinant growth hormone in CHO in response to the media additives sodium butyrate and zinc sulphate. Cultivation of cells at a lowered temperature has been shown to cause a growth arrest of cells, predominantly in the G1 phase of the cell cycle, with a concomitant increase of specific productivity of up to 1.7 fold. Kaufmann et al[21] used a proteomics approach based on 2-DE to identify a number of proteins that are involved in the cells response to lowered temperature when expressing secreted alkaline phosphatase (SEAP) in the CHO cell line XM111-10.

There have been many approaches made to utilize the power of proteomics in identifying proteins that are changing in response to environmental conditions of the cell. A growing number of reference maps based on the use of 2-DE and for an increasing number of organisms such as the work of M. Hecker[22] which describes detailed work on building a proteomic map of *Bacillus subtilis* are now starting to be made available. A more complete reference to this and other such projects are available at *http://tw.expasy.org/ch2d/2d-index.html*. The establishment of a two-dimensional map of CHO cells by Champion et al[23] has begun to provide a reference for further studies of this industrially relevant cell line. Continued work in this area by Baik et al[9]; Lee et al[24]; Van Dyk et al[20] and Chen et al[25] illustrate the importance of this technique and its application to bioprocessing. Provided access to the completed work is available publicly it will endow researchers in this field with valuable information on the proteins that are expressed under a given set of circumstances and their relative abundance. However, to fully utilize such a reference map it is essential to be able to reproduce the exact conditions that were used in the original separation. Unlike microarray that makes use of minimum information about microarray experiment (MIAME)[26] for the publication of microarray-based expression data, which ensures that the data can be easily interpreted and that results derived from its analysis can be independently verified, proteomics experiments are not yet under any such guideline. In an article published by Taylor et al[27] a proposal has been put forward to establish a framework that would significantly help in the handling, exchange and dissemination of proteomic data. There have been attempts to provide protocols for comparing gel image data across the internet[28] which should allow such comparisons to be made but still requires a concerted effort by researchers in this field to comply with certain standards. A recent public repository for proteomics data has also been established to cater to the proteomics community[29].

Generally the preparation of a static 2-DE map or protein expression map (PEM) of an organism's proteome, whilst a valuable catalogue of expressed proteins, does not provide more than a reference for further experimental studies. In order to measure the dynamic changes seen in the cell under different environmental or physiological conditions it is necessary to compare PEMs of samples obtained in different conditions or over time courses using sophisticated image analysis tools. Data resulting from these studies can provide quantitative changes of specific protein expression and therefore, when coupled with identification using mass spectrometry, protein sequencing or western analysis, a more detailed profile of physiological changes. Integrating the output with that of microarray for example now allows the construction of a highly detailed picture of the cells metabolic and physiological status.

A molecular understanding of how cells operate during cell culture requires a knowledge of which genes are expressed as well as those that are translated into proteins. Approaches using proteomics involves a number of steps including the initial cell disruption to release proteins, a separation technique that will provide isolation of single proteins from a complex mixture, a method for measuring relative quantification and the final identification of the protein(s) being studied. This section reviews current techniques available for the study of the CHO cell proteome and describes the techniques most suitable for rapid identification of proteins isolated from the CHO cell lysate as well as recommended techniques for quantitative and comparative studies of CHO proteome.

3.1   Gel Based Methods: Two-Dimensional Gel Electrophoresis (2-DE)

The technique frequently used in proteomics is two-dimensional gel electrophoresis (2-DE) first described by O'Farrell[30] and which has been developed extensively over the proceeding thirty years. This technique uses a combination of isoelectric focusing (IEF) and sodium dodecyl sulphate-polyacrylamide gel electrophoresis

(SDS-PAGE) to separate proteins firstly by their iso electric point (pI) and then by their molecular weight. Images of the stained, separated proteins can then be subjected to analysis using computer software to identify those proteins that are shown to alter their expression. This is done by measuring the relative intensity of each protein 'spot' after normalization and comparing to a control. Those proteins shown to exhibit altered expression can then be identified by excising the protein from the gel and analyzing by mass spectrometry. The use of 2-DE has also been described for the creation of proteome maps which can serve as a reference for the study of the proteome of a particular cell type. A good, general reference on this subject is Görg et al[31]. 2-DE gel electrophoresis is a major component of proteomics and is ideal for large scale identification of every protein in a sample. This is most suited to an investigation involving the global protein expression of an organism whose genomes have been fully sequenced. 2-DE technology is very well established and there are numerous articles and books describing detailed procedures for performing a 2-DE experiment[32–34].

Significant improvements in the ability to accurately match proteins across different samples and perform relative quantification have been made with the introduction of the fluorescence difference gel electrophoresis system (DIGE, GE Healthcare). This system allows up to three samples to be separately labelled at the protein level, mixed together and co-run on a single 2-DE separation. This method is based on the use of three separate cyanine based fluorescent dyes (Cy2, Cy3, Cy5). The ability to co-run three samples on a single gel results in the conditions of separation in both the first and second dimensions being identical for all samples and greatly minimizes variation. The individual samples can be visualized after scanning by using the appropriate wavelength according to the dye used.

3.2   Quantitative Proteomic Profiling Using Mass Spectrometry

More recently there have been a number of newly developed techniques for performing quantifiable comparisons of protein expression based on the use of stable isotope labeling. The stable isotope such as ^{2}H, ^{13}C, or ^{18}O can be introduced into the protein either metabolically during cell growth in the form of an amino acid[35] or as a labeled 'tag'[36,37]. One of the main advantages of using these labeling techniques is that subsequent separation can be performed using alternative techniques particularly chromatography which complements 2-DE and addresses some of the shortfalls associated with that type of separation.

The more recent development of using stable isotopes to label proteins and thus provide a means of relative quantification at the mass spectrometry stage has provided a good complement to gel electrophoresis approaches. The different methods of separation post-labelling, such as liquid chromatography also provides some distinct advantages such as the ability to capture much smaller proteins and even peptides that would otherwise have been lost using electrophoresis. Other advantages include the ability to deal with proteins that are difficult to solubilize such as membrane and high molecular weight proteins.

Two methods are now quite widely used for the differential analysis of protein expression in cells and are described below. There are a number of variations to these approaches which have been reviewed[38,39]. The first, Stable Isotope Labelling by Amino Acids in Culture (SILAC), involves metabolic labelling using an amino acid that incorporates a stable isotope (^{13}C-Arginine or ^{13}C-Lysine) into the proteins and can be fed to one of the test cultures as an essential amino acid[35]. One of the main advantages of this form of labelling is that the control (non-labelled) cells can be mixed with the labelled cells prior to performing any complex sub-cellular fractionation and thus minimizing any biased loss of protein from one pool compared to the other. This advantage more than offsets the fact that it only allows for a binary comparison.

In contrast to SILAC there has been a more recent development for labelling proteins using stable isotopes in a method called Isobaric Tagging for Relative and Absolute Quantitation (iTRAQ)[36]. This method utilizes up to four different mass 'tags' allowing multiplex analysis which could include biological or technical replicates in a single analysis. The tags are amine reactive and label the N-termini of peptides as well as side chain amino groups of lysine residues. The labelling is performed at the peptide level rather than at the protein level. This has the advantage of providing greater coverage of protein sequence as all peptides are labelled. However, due to the labelling being performed only after purification of total protein and proteolytic digestion it is not a method that lends itself readily to studies involving complex sub-cellular preparations prior to the addition of the label.

The iTRAQ reagent has a reporter group, a balance group and a peptide reactive group. The amine reactive group links an iTRAQ reagent isobaric tag to each peptide at the amino-terminus and any lysine residue side chains. The reporter group is a tag with a mass of 114, 115, 116 or 117 Da. The balance group ensures the combined mass of the reporter and balance groups remains constant for all four reagents (Applied Biosystems bulletin). As a result, the labeled peptides are isobaric. To profile protein expression in cells, each separate cellular protein sample is digested with trypsin into peptide pools and reacted with the isobaric mass tags. Each labeled pool is mixed together and fractionated. The fractions are then applied to a second-dimensional separation and the eluting peptides are analysed using mass spectrometry (MS). From the fragmentation it is possible to measure the relative abundance of each of the equivalent peptides based on the mass tag used in labeling. All data files are processed through software to derive protein identification and the relative expression value for each protein. An overview of the iTRAQ methodology as applied to duplex samples is shown in figure 7a. The most significant advantage of this technology is that it allows labeling of up to four different samples within a single experiment. This quadruplex labeling strategy is useful for quantifying proteins from multiplex samples, such as those in a time course study or replicate measurements of the same sample (Figure 7b).

(a)

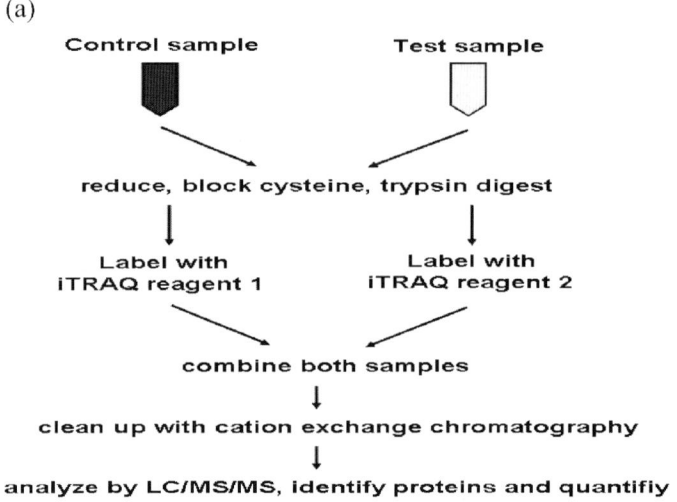

(b)

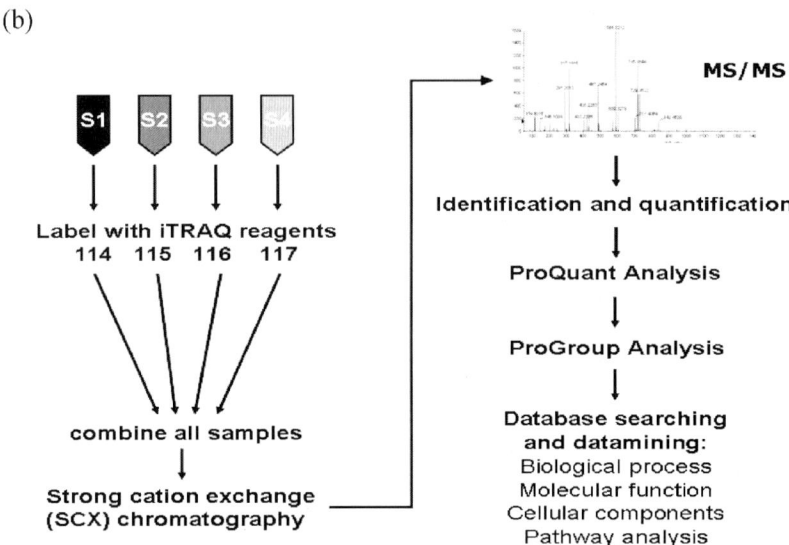

Figure 7. **iTRAQ workflow for quantitative proteomic profiling.** (a) workflow as applied to duplex samples. (b) A multiplex profiling experiment using all four iTRAQ reagents, 114, 115, 116 and 117

3.3   Data Analysis and Interpretation

The use of either SILAC or iTRAQ results in a large amount of raw mass spectrometry data. It is this data that needs to be analysed to firstly identify the protein and secondly to attribute some relative quantitation to the proteins. Software is available to perform this task but it also requires a reasonable amount of manual input and understanding of the processes involved is essential. It is outside the scope of this work to give explicit details of how to perform a full analysis using this software but rather an idea of what it involves. There are excellent guides for packages for iTRAQ data analysis (ProQuant™, Applied Biosytems) and for the use of SILAC (MSQuant, *http://www.pil.sdu.dk/silac_msquant.htm*).

3.4   Mass Spectrometry

Mass spectrometry has become the method of choice for the rapid identification of proteins and posttranslational modifications. Due to the speed and sensitivity that mass spectrometry provides it is now possible to perform separation, identification and relative quantification of complex protein mixtures in studies of protein expression in very short periods of time. It is not the aim of this work to give an exhaustive description of the instruments that are available but to review the main types used in protein based studies with some reference to usage in the author's laboratory. The use of mass spectrometry has been described and reviewed in a number of publications[40-43].

A mass spectrometer consists of an ionisation source, a mass separator or analyzer and a mass detector. The main types of ionisation source used for the analysis of bio-molecules such as peptides and proteins are electrospray ionisation (ESI) and matrix assisted laser desorption ionosation (MALDI). ESI gets its name from the fact that an electrical charge is applied to a liquid flow prior to introduction to the mass spectrometer. The liquid flow is usually in the form of the outlet from a separation by high-performance liquid chromatography (HPLC) and hence often referred to as LC-MS or LC-MS/MS. The electrical charge applied to the liquid flow, usually at a level of between 2 and 5 kilovolts, results in the formation of a fine mist or spray consisting of highly charged droplets which enter the mass spectrometer. A combination of vacuum and a flow of gas, usualy nitrogen, will cause a rapid desolvation of the droplets leaving the charged peptide or protein ions to enter the instrument. ESI will result in the formation of multiply-charged ions which is an important feature since the mass spectrometer measures mass to charge ratio (m/z). This feature provides the ability to measure the mass of very large molecules using an instument with a fairly small mass range (*e.g.*, 50-4000 m/z).

In contrast, MALDI results predominantly in the formation of singly charged species but has the advantage of being a very rapid form of ionisation. Samples, usually mixtures of peptides from a proteolytic digest of a protein, are mixed with a chemical matrix which, upon drying, forms a layer of co-crystals. A laser beam is then directed at the sample which is present as a small dried spot about 1 mm in diameter on a sample stage. As the laser beam strikes the sample the matrix acts to transfer the energy and convert it to heat thus causing the sample to vapourize. The plume of charged ions that is formed can then enter the mass spectrometer.

The most common type of mass analyzers are ion-traps, quadrupole/triple quadrupole, time-of-flight (TOF) or a combination of quadrupole/TOF. The ESI type of ionisation source is more commonly used in conjunction with the ion-trap, quadrupole and quadrupole/TOF analyzers and the MALDI ion source with the

TOF analyzer. Other combinations have been described and the reader is directed to specific journals for reference to these instruments. There are also newer developments of mass analyzers that have been described such as Fourier Transform-Ion Cyclotron Resonance (FT-ICR)[44] and the Orbitrap[45]. These instruments provide very high resolution and mass accuracy and will play an important role in specific aspects of protein analysis, particularly in posttranslational modifications.

The final decision on which instrument to purchase depends on many factors including the expected throughput, complexity of the samples and budget. The description below is for a nanoLC-MS/MS system which provides great flexibility and can be used for a number of analysis types including the identification of proteins based on a trytic digestion and subsequent analysis of peptides at the MS and MS/MS level, identification of phosphorylation sites, analysis of complex peptide mixtures using multidimensional protein identification technology (MudPIT, Washburn MP, 2001), quantification studies using stable isotope labelling (SILAC, ICAT, iTRAQ) and molecular weight determination of whole proteins.

3.5   Search Engines and Databases

In order to identify a protein from the acquired MS/MS profile it is necessary to use a search engine which extracts the required information from the raw data file and searches it against a protein database to find the best match. There are a number of available search engines, both commercial and open source, for example, as shown in Table 1.

An appropriate database is required for the search engine to searh against and is usually structured in the FASTA format. A number of databases are available for download for use with the search engine. Some of these have been specifically developed for mass spectrometry use whilst others are for general use depending on application. Some of the more commonly used protein databases are listed in Table 2.

3.6   Functional Analysis

The significant differentially regulated proteins are subjected to database searching and data mining for important data needed for functional analysis. Essential data such as gene names, species, Gene Ontology (biological processes, molecular functions, and cellular locations), sequence descriptions and KEGG pathways need to be

Table 1. Some commonly used search engines

Name	Type	URL
MASCOT	Commercial	www.matrixscience.com
SEQUEST	Commercial	www.thermo.com
Phenyx	Commercial	www.phenyx-ms.com
Spectrum Mill	Commercial	www.agilent.com
X! Tandem	Open source	www.thegpm.org/TANDEM/

Table 2. Some commonly used databases available for download for use with the search engines

Name	Description	URL
MSDB	Composite, non-identical database built from a number of sources	http://csc-fserve.hh.med.ic.ac.uk/ msdb.html
UniProt	Compiled from Swiss-Prot, TrEMBL and PIR	www.ebi.uniprot.org
IPI	Minimally redundant, featured species	www.ebi.ac.uk/IPI/IPIhelp
NCBInr	Non redundant	http://www.ncbi.nih.gov/Ftp/

thoroughly searched to obtain information on these differentially expressed proteins. Several web-based applications and tools can be utilized for this purpose and they are listed below:

European Bioinformatics Institute (EBI) Database
International Protein Index (IPI)
http://www.ebi.ac.uk/IPI
Mouse Genome Informatics (MGI)
http://www.informatics.jax.org
AmiGO
http://www.godatabase.org/cgi-bin/amigo/go.cgi
FatiGO
http://fatigo.bioinfo.cnio.es

4.   CONCLUSION

It is clear that CHO genomics and proteomics offer a lot of potentials, in terms of understanding biological processes that occur in CHO cells in a bioprocess environment. Baik et al[9], for example, combined transcriptome and proteome analysis to study the molecular behaviour of erythropoietin producing CHO cells under low temperature culture. These workers used commercially available rat and mouse cDNA microarrays. Microarray analysis showed that cellular processes such as metabolism, transport, and signaling pathways were affected by culture at low temperature. 2-DE technique was employed in the proteomic analysis and results showed that the expression levels of PDI, vimentin, NDK B, ERp57, RIKEN cDNA, phosphoglycerate kinase, and heat shock cognate 71 kDa protein, exhibited a twofold increase at 33 °C whereas those of HSP90-beta and EF2, decreased over twofold.

Gene and protein expression profiling are powerful approaches to understanding mammalian cell cultures[46]. We described protocols for use of microarrays as well as proteomics, however, we have not touched on data analysis, which is a major component of high throughput analysis using microarrays and proteomics. Choosing the appropriate data analysis strategy is influenced by the purpose of the experiment, and the user's knowledge of the biology of the system under investigation. Data analysis takes a major portion of a researcher's time and good access to tools such as pathway analysis softwares e.g GeneMapp[47,48] are essential to make meaningful interpretations of the data. Effective use of these two high throughput techniques require integrating bioinformatics and computational biology.

The integration of all three fields is a fast emerging area, known as system biology. System biology aims to analyze biological data and unravel the function and interaction of the biological systems being studied. This is a major challenge as it involves the integration and interpretation of biological data generated by technologies ranging from microarray and proteomics data to mundane and routine DNA/protein sequencing. The information and knowledge gained from incorporating this discipline into research, however, would certainly pave the way to new innovations and discoveries in the field of mammalian cell culture and bioprocessing technologies.

5.   REFERENCES

1. Gygi, S.P. et al. (1999) Correlation between protein and mRNA abundance in yeast. *Mol Cell Biol* 19 (3), 1720–1730
2. Anderson, L. and Seilhamer, J. (1997) A comparison of selected mRNA and protein abundances in human liver. *Electrophoresis* 18 (3–4), 533–537
3. Conrads, K.A. et al. (2005) A combined proteome and microarray investigation of inorganic phosphate-induced pre-osteoblast cells. *Mol Cell Proteomics* 4 (9), 1284–1296
4. Chu, L. and Robinson, D.K. (2001) Industrial choices for protein production by large-scale cell culture. *Curr Opin Biotechnol* 12 (2), 180–187
5. Andersen, D.C. and Krummen, L. (2002) Recombinant protein expression for therapeutic applications. *Curr Opin Biotechnol* 13 (2), 117–123
6. Goochee, C.F. et al. (1992) Frontiers in Bioprocessing II. (Todd, P. et al., eds.), pp. 198–240, American Chemical Society
7. Wlaschin, K.F. et al. (2005) EST sequencing for gene discovery in Chinese hamster ovary cells. *Biotechnol Bioeng* 91 (5), 592–606
8. Danny Chee Furng Wong et al. (2006) Transcriptional profiling of apoptotic pathways in batch and fed-batch CHO cell cultures. *Biotechnology and Bioengineering*
9. Baik, J.Y. et al. (2006) Initial transcriptome and proteome analyses of low culture temperature-induced expression in CHO cells producing erythropoietin. *Biotechnol Bioeng* 93 (2), 361–371
10. Yang, Y.H. and Speed, T. (2002) Design issues for cDNA microarray experiments. *Nat Rev Genet* 3 (8), 579–588
11. Churchill, G.A. (2002) Fundamentals of experimental design for cDNA microarrays. *Nat Genet* 32 Suppl, 490–495
12. Kerr, M.K. and Churchill, G.A. (2001) Statistical design and the analysis of gene expression microarray data. *Genet Res* 77 (2), 123–128
13. Cui, X. and Churchill, G.A. (2003) Statistical tests for differential expression in cDNA microarray experiments. *Genome Biol* 4 (4), 210
14. Chomczynski, P. and Sacchi, N. (1987) Single-step method of RNA isolation by acid guanidinium thiocyanate-phenol-chloroform extraction. *Anal Biochem* 162 (1), 156–159
15. Quackenbush, J. (2001) Computational analysis of microarray data. *Nat Rev Genet* 2 (6), 418–427
16. Leung, Y.F. and Cavalieri, D. (2003) Fundamentals of cDNA microarray data analysis. *Trends Genet* 19 (11), 649–659
17. Seow, T.K. et al. (2001) Proteomic investigation of metabolic shift in mammalian cell culture. *Biotechnol Prog* 17 (6), 1137–1144
18. Palermo, D.P. et al. (1991) Production of analytical quantities of recombinant proteins in Chinese hamster ovary cells using sodium butyrate to elevate gene expression. *J Biotechnol* 19 (1), 35–47

19. Oh, H.K. et al. (2005) Effect of N-Acetylcystein on butyrate-treated Chinese hamster ovary cells to improve the production of recombinant human interferon-beta-1a. *Biotechnol Prog* 21 (4), 1154–1164

20. Van Dyk, D.D. et al. (2003) Identification of cellular changes associated with increased production of human growth hormone in a recombinant Chinese hamster ovary cell line. *Proteomics* 3 (2), 147–156

21. Kaufmann, H. et al. (1999) Influence of low temperature on productivity, proteome and protein phosphorylation of CHO cells. *Biotechnol Bioeng* 63 (5), 573–582

22. Hecker, M. (2003) A proteomic view of cell physiology of Bacillus subtilis–bringing the genome sequence to life. *Adv Biochem Eng Biotechnol* 83, 57–92

23. Champion, K.M. et al. (1999) A two-dimensional protein map of Chinese hamster ovary cells. *Electrophoresis* 20 (4-5), 994–1000

24. Lee, M.S. et al. (2003) Proteome analysis of antibody-expressing CHO cells in response to hyper-osmotic pressure. *Biotechnol Prog* 19 (6), 1734–1741

25. Chen, Z. et al. (2004) Initial analysis of the phosphoproteome of Chinese hamster ovary cells using electrophoresis. *J Biomol Tech* 15 (4), 249–256

26. Brazma, A. et al. (2001) Minimum information about a microarray experiment (MIAME)-toward standards for microarray data. *Nat Genet* 29 (4), 365–371

27. Taylor, C.F. et al. (2003) A systematic approach to modeling, capturing, and disseminating proteomics experimental data. *Nat Biotechnol* 21 (3), 247–254

28. Ravichandran, V. et al. (2004) Ongoing development of two-dimensional polyacrylamide gel electrophoresis data standards. *Electrophoresis* 25 (2), 297–308

29. Jones, P. et al. (2006) PRIDE: a public repository of protein and peptide identifications for the proteomics community. *Nucleic Acids Res* 34 (Database issue), D659–663

30. O'Farrell, P.H. (1975) High resolution two-dimensional electrophoresis of proteins. *J Biol Chem* 250 (10), 4007–4021

31. Gorg, A. et al. (2000) The current state of two-dimensional electrophoresis with immobilized pH gradients. *Electrophoresis* 21 (6), 1037–1053

32. Westermeier, R. and Naven, T. (2002) *Proteomics in practice: a laboratory manual of proteome analysis.*, Wiley-VCH

33. Liebler, D.C. (2002) *Introduction to proteomics: tools for the new biology.*, Humana Press

34. Appella E et al. (2000) Proteome mapping by two-dimensional polyacrylamide gel electrophoresis in combination with mass spectrometric protein sequence analysis. In *Proteomics in functional genomics* (P. Jolles, H.J., ed.), Birkhauser Verlag

35. Ong, S.E. et al. (2002) Stable isotope labeling by amino acids in cell culture, SILAC, as a simple and accurate approach to expression proteomics. *Mol Cell Proteomics* 1 (5), 376–386

36. Ross, P.L. et al. (2004) Multiplexed protein quantitation in Saccharomyces cerevisiae using amine-reactive isobaric tagging reagents. *Mol Cell Proteomics* 3 (12), 1154–1169

37. Gygi, S.P. et al. (1999) Quantitative analysis of complex protein mixtures using isotope-coded affinity tags. *Nat Biotechnol* 17 (10), 994–999

38. Regnier, F.E. et al. (2002) Comparative proteomics based on stable isotope labeling and affinity selection. *J Mass Spectrom* 37 (2), 133–145

39. Tao, W.A. and Aebersold, R. (2003) Advances in quantitative proteomics via stable isotope tagging and mass spectrometry. *Curr Opin Biotechnol* 14 (1), 110–118

40. Mann, M. et al. (2001) Analysis of proteins and proteomes by mass spectrometry. *Annu Rev Biochem* 70, 437–473

41. Lane, C.S. (2005) Mass spectrometry-based proteomics in the life sciences. *Cell Mol Life Sci* 62 (7–8), 848–869

42. Yates, J.R., 3rd. (2004) Mass spectral analysis in proteomics. *Annu Rev Biophys Biomol Struct* 33, 297–316

43. Ferguson, P.L. and Smith, R.D. (2003) Proteome analysis by mass spectrometry. *Annu Rev Biophys Biomol Struct* 32, 399–424

44. Bruce, J.E. et al. (2000) A novel high-performance fourier transform ion cyclotron resonance cell for improved biopolymer characterization. *J Mass Spectrom* 35 (1), 85–94

45. Hu, Q. et al. (2005) The Orbitrap: a new mass spectrometer. *J Mass Spectrom* 40 (4), 430–443
46. Tian, Q. et al. (2004) Integrated genomic and proteomic analyses of gene expression in Mammalian cells. *Mol Cell Proteomics* 3 (10), 960–969
47. Doniger, S.W. et al. (2003) MAPPFinder: using Gene Ontology and GenMAPP to create a global gene-expression profile from microarray data. *Genome Biol* 4 (1), R7
48. Dahlquist, K.D. et al. (2002) GenMAPP, a new tool for viewing and analyzing microarray data on biological pathways. *Nat Genet* 31 (1), 19–20

CHAPTER 4

THE UNFOLDED PROTEIN RESPONSE

NATALIE STRUDWICK AND MARTIN SCHRÖDER

School of Biological and Biomedical Sciences, University of Durham, Durham DH1 3LE,
United Kingdom

Abstract: Protein folding is the rate-limiting step for secretion of recombinant proteins
in nearly all expression systems. Overexpression of recombinant proteins easily
exhausts the cellular protein folding machinery in the endoplasmic reticulum (ER).
In this ER stress situation a signal transduction network, called the unfolded protein
response (UPR), is activated. The UPR coordinates adaptive responses to this stress
situation and, presumably in response to prolonged ER stress, induces apoptosis.
A molecular understanding of the UPR will enable the engineering of expression
systems with the aim to increase cellular productivities and to improve the cost
effectiveness of these expression systems. In this paper we review our current
molecular understanding of the UPR and its implications for the engineering of
recombinant protein expression systems

Keywords: Endoplasmic reticulum, molecular chaperone, recombinant protein production,
signal transduction, unfolded protein response

1.   INTRODUCTION

Proteins have many applications in industrial processes and as therapeutic agents[1].
The method of choice for production of these proteins is their expression and secretion in heterologous host cells to minimize the risk of contaminating the protein
product with infectious agents[2]. Correct, humanized post-translational modifications, especially the glycosylation pattern, for proteins destined for use in the
clinic are essential to obtain desired characteristics, such as activity[3–9], immunogenicity[10–13], and clearance rates *in vivo*[14–17]. Therefore, these proteins are usually
expressed in mammalian cells, e.g. Chinese hamster ovary (CHO) cells[18], murine
lymphoid cells, such as NS0[19] and SP2/0[18] cells, and human embryonic kidney
(HEK-293) cells[20].

M. Al-Rubeai and M. Fussenegger (eds.), Systems Biology, 69–155.
© 2007 *Springer.*

The efficiency of a recombinant protein production process is characterized by its volumetric product yield. Volumetric product yield is determined by two parameters, the cell specific production rate and the integral of viable cell concentration, usually expressed as cell time per unit volume[21]. Thus, two routes can be chosen to increase volumetric product yields. First, viable cell densities have been increased through optimization of media formulations and the design of feeding regimes[1,22–25], engineering of apoptosis resistant production cell lines[26,27], and metabolic engineering to increase the efficiency of cellular metabolism and to limit the accumulation of toxic metabolic waste products in the culture medium[28–34]. Second, engineering of the complex cellular machinery involved in gene expression improved cell specific production rates. Powerful expression systems, i.e. the dihydrofolate reductase[35–39] or glutamine synthetase[40–42] selection and amplification systems have been developed, and improved through the use of even more powerful promoters[43], chromatin boundary elements[44,45], integrating recombinant DNA into transcriptionally active genomic regions[46], optimization of codon usage[47–51], and use of genomic DNA sequences[52]. Both approaches intricately interact. For example, engineering of metabolism has an effect on cell specific production rates[30,32,33]. Furthermore, increased protein synthesis may require increased metabolic activity[53] and may induce apoptosis[54,55].

Examination of optimized expression systems revealed that gene copy number[56], the mRNA encoding the recombinant protein[57–59], or even intracellular recombinant protein[55,57] are not proportional to cell specific production rates. In these instances recombinant protein remained associated with the cell[60], localized to the endoplasmic reticulum (ER)[61], was found in high molecular weight aggregates in the ER[53], and associated with the molecular chaperone heavy chain binding protein (BiP)/glucose regulated protein of 78 kDa (GRP78)/karyogamy 2 protein (Kar2p)[62]. These observations are hallmarks for a protein folding defect in the ER and activation of an ER stress response[63,64], demonstrating that in current expression systems for secretory proteins the rate- and yield limiting step is maturation of the protein in the ER[55]. At least in one case additional limitations in gene expression, i.e. at a transcriptional or translational level, have been ruled out[58]. For these reasons, engineering of the protein folding machinery in the ER and of cellular signal transduction pathways that regulate the activity of this protein folding machinery, called the unfolded protein response (UPR), promises to further increase cell specific production rates and volumetric product yields. Here we review our current molecular understanding of the ER protein folding machinery and signal transduction by the UPR and highlight implications of this knowledge for the engineering of cell lines with increased cell specific production rates.

2.   PROTEIN FOLDING IN THE ENDOPLASMIC RETICULUM

The endoplasmic reticulum (ER) is an extended network of membranous tubules that form the first part of the secretory pathway. Secretory proteins, synthesized on cytosolic ER bound ribosomes[65], begin their journey to the extracellular space by entering the lumen of the ER through the Sec61p translocon[66]. In the ER optimal conditions for post-translational modification and folding of the nascent peptide chain are maintained. The ER is more oxidizing than the cytosol to facilitate disulfide bond formation[67] and also the major Ca^{2+} store in higher eukaryotes[68]. The total protein concentration within the ER lumen reaches 100 g/l ($\approx$2 mM)[69]. Folding into the native conformation and modification of nascent polypeptides is necessary for export of a secretory protein to the Golgi complex. Thus, the ER provides a quality control system that monitors the folding status of transiting polypeptide chains[70]. Misfolded proteins are retained within the ER until they have folded into their native conformation. If this cannot be achieved within a certain time frame, the protein is retrotranslocated to the cytosol, where it is ubiquitinated, and targeted for degradation by the 26S proteasome[71] in a process termed ER associated degradation (ERAD). The build up of unfolded or incorrectly folded proteins beyond the folding capacity of the ER results in a situation known as ER stress, which triggers a series of cellular signaling events collectively known as the UPR. The UPR is conserved in all eukaryotes and primarily serves a homeostatic function (Figure 1). It reduces the protein folding burden on the ER by increasing the processing capacity and the size of the ER and by decreasing the unfolded protein load through attenuation of transcription of genes encoding secretory proteins[72–74] and attenuation of general

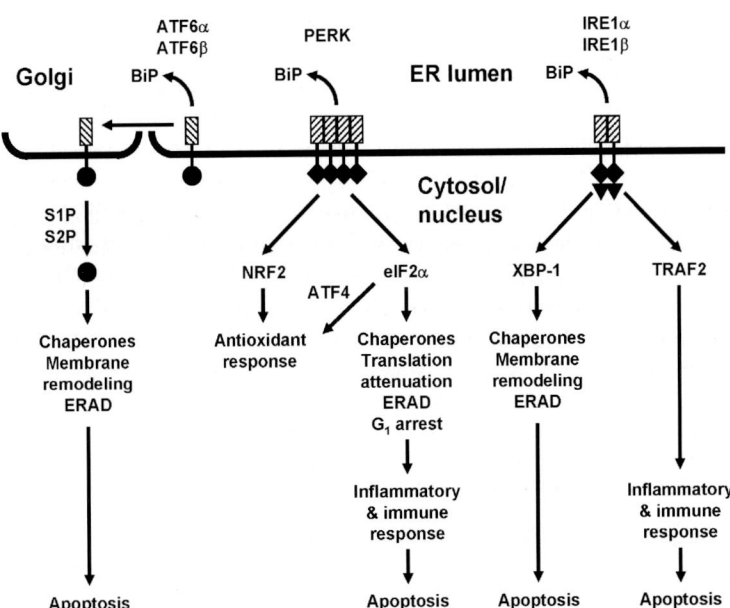

Figure 1. Physiological responses to ER stress coordinated by the UPR. Symbols: hatched rectangles – ER luminal domains of ATF6, IRE1, and PERK, circle – bZIP transcription factor domain of ATF6, square – protein kinase domain, triangle – domain with homology to RNase L. Reprinted from *Current Molecular Medicine*, copyright 2006, with permission from Bentham Science Publishers

translation. The ER is also the location for the biosynthesis of sterols, cholesterol and lipids. Recent work has shown that lipid metabolism and the UPR are intricately linked, even if molecular details of these interactions remain to be uncovered. In higher eukaryotes the UPR gained control over an alarm response by regulating the activity of the transcription factor nuclear factor κB (NF-κB) to induce inflammation and immune genes to combat viral infections[75] associated with a massive boost in glycoprotein synthesis. Presumably in response to severe or prolonged ER stress, the UPR activates the apoptotic program to eliminate unhealthy cells from an organism or a population[75]. Finally, the cell interprets the level of UPR activation as a measure for its own metabolic activity and metabolic state[76–78].

2.1   ER Stress

ER stress is the imbalance between the unfolded protein load and the protein folding capacity of the ER. A number of events results in the accumulation of unfolded proteins. Inhibition of post-translational modifications within the ER results in accumulation of incorrectly folded proteins. Examples are disruption of N-linked glycosylation by glucose deprivation[79], by inhibition of UDP-N-acetyl-glucosamine-1-phosphate transferase by tunicamycin[80–84], or incorporation of D glucose analogs, such as 2-deoxy-D-glucose, into dolicholpyrophosphate bound core oligosaccharides[85–87], and impairment of disulfide bond formation due to disturbance of the redox status within the ER[88]. Similarly, abnormal Ca^{2+} levels induce ER stress because of the Ca^{2+} dependency of many ER-resident chaperones[89–91]. In addition, the folding of some proteins, i.e. apo-α-lactalbumin, is Ca^{2+} dependent[92]. Expression of mutant, folding incompetent proteins causes ER stress[93], as does the overexpression of particularly large or extensively modified proteins such as blood coagulation factor VIII[94,95]. In fact, even overexpression of small and relatively simple proteins such as antithrombin III is sufficient to cause aggregation and retention of the unfolded recombinant protein in the ER[53,55]. Physiological conditions in which increased secretory activity causes ER stress and an UPR are differentiation of B cells into antibody secreting plasma cells[96,97] and the response of plants to microbial infections[98]. Finally, impairment of the proteasome may lead to a backlog of incorrectly folded proteins awaiting degradation and hence an increase in the unfolded protein concentration within the ER[99].

2.2   Molecular Chaperones and Protein Foldases

Proteins are targeted to the ER by a hydrophobic signal sequence. An aqueous channel, the Sec61p translocon complex allows the protein to cross the ER membrane into the luminal space[66]. A very early event in the maturation process of peptides involves the removal of the signal sequence by signal peptidases. The signal sequence is likely to be present during initial protein folding events, as bacterial signal peptidases cleave following translation of 80% of the polypeptide chain[201,202]. The nature of the signal sequence modulates the timing and efficiency of both signal sequence removal and *N*-linked glycosylation[203]. Delayed cleavage of the signal peptide sequence prolongs interaction of proteins with the ER protein folding machinery[204]. Protein folding in the ER is assisted by the presence of two classes of proteins, protein foldases, such as protein disulfide isomerases (PDI) and peptidyl prolyl *cis-trans* isomerases, and molecular chaperones, of the heat shock protein (HSP) 70 and HSP90 families, and the lectin chaperones calnexin (CNX) and calreticulin (CRT) (Table 1).

Table 1. ER resident molecular chaperones, foldases, lectins, and *N*-linked oligosaccharide modifying enzymes. Reprinted, with permission, from the *Annual Review of Biochemistry*, Volume 74 ©2005 by Annual Reviews www.annualreviews.org.

Class and name	Function	Reference
Chaperones, HSP70 class		
BiP/GRP78/Kar2p	Chaperone, translocation, folding sensor	100–105
Lhs1p/Cer1p/Ssi1p/GRP170	Chaperone	106–110
Co-chaperones, DnaJ-like, HSP40 class (BiP ATPase stimulation)		
ERdj1/MTJ1		111
ERdj3/HEDJ/Scj1p		112–115
ERdj4		116
ERdj5		117
Jem1p		118
Sec63p	Translocation	119–121
Chaperones, GrpE-like (nucleotide exchange factor for BiP)		
BAP		122
Sil1p/Sls1p		123–125
Chaperones, HSP90 class		
GRP94/endoplasmin/tumorrejection antigen gp96/ ERp99	Chaperone	126–129
Lectins		
calmegin	Glycoprotein quality control	130
calnexin	Glycoprotein quality control	131, 132
calreticulin	Glycoprotein quality control	133, 134
Mnl1p/Htm1p/EDEM	Glycoprotein degradation	135–137
EDEM-2	Glycoprotein degradation	138, 139
Carbohydrate processing enzymes		
UGGT	Folding sensor	140, 141
α-glucosidase I	Removal of terminal D-glucose residues from glycoproteins	142
α-glucosidase II	Removal of terminal D-glucose residues from glycoproteins, release of glycoproteins from calnexin	143, 144
α-mannosidase I	Removal of terminal D-mannose residues, extraction of glycoproteins from calnexin cycle	145, 146
α-mannosidase II	Removal of terminal D-mannose residues, extraction of glycoproteins from calnexin cycle	147, 148

(*continued*)

Table 1. (Continued)

Class and name	Function	Reference
Foldases, subclass disulfide isomerases		
Ero1p/ERO1-Lα, ERO1-Lβ	Oxidoreductase for PDI	149–152
Erv2p	Oxidoreductase for PDI	153, 154
Pdi1p/PDI/ERp59/GSBP/ER calcistorin-PDI		155–160
PDIp	Pancreas specific	161, 162
PDILT	Testis specific	163
P5/CaBP1/Mpd1p		164–167
ERp72/CaBP2		158, 166, 168, 169
ERp61		170
ERp60/Eug1p		168, 171–173
ERp57/GRP58		174
PDIR		175
ERp46		176
ERp44	Retention of ERO1-Lα in ER	177
ERp19		176
TMX		178–180
Mpd2p		181
Eps2p		182, 183
Thioredoxin homology domain containing chaperones		
ERp29/ERp28/Windbeutel/PDI-Dβ		184–187
Foldases, subclass FAD-dependent oxidases		
Fmo1p	FAD dependent oxidase	188
Foldases, peptidyl prolyl cis-trans *isomerases, family: cyclophilins (inhibited by cyclosporine A)*		
S-Cyclophilin		189
SCYLP		190
Cyclophilin B		191–193
ninaA (*Drosophila*)	Opsin folding, chaperone function	194–197
Foldases, peptidyl prolyl cis-trans *isomerases, family: immunophilins (inhibited by FK506)*		
FKBP13		198, 199
FKBP65		200
Other		
ERp49		170

Abbreviations (if not explained in the text): CaBP2 – Ca^{2+} binding protein, Fmo1p – flavin-containing monooxygenase 1 protein, GSBP – glycosylation site binding protein, Mpd1p – multicopy suppressor of *PDI1* deletion, and TMX – transmembrane thioredoxin-related protein.

A large group of ER-resident chaperones are the HSPs, first identified as proteins up-regulated following thermal shock[205–207] in order to prevent heat-induced protein aggregation[208]. Subsequently, these proteins were shown to be essential during normal conditions for a number of steps required for efficient protein folding[208]. HSPs are able to recognize and bind hydrophobic surfaces exposed on unfolded proteins[209,210], which ultimately become buried within the protein core structure in the native conformation[69]. These interactions serve a number of purposes. Firstly, the ER chaperones assist the entry of the protein into the ER *via* gating of the

Sec61p translocon pore in the ER membrane[211-216]. Secondly, the association of chaperones with unfolded proteins prevents inappropriate, unproductive interactions of the protein with itself and other surrounding proteins and maintains the protein in a folding competent state[208]. Thirdly, chaperones constitute an important arm of the ER quality control mechanism, recognizing the presence of unfolded proteins and targeting them for degradation by the proteasome[182,183,217,218]. Many of these chaperones coordinate their activities, forming higher-order complexes[219].

2.2.1   ER resident HSP70 chaperones

All HSP70 chaperones are characterized by a conserved N-terminal 44 kDa ATPase domain, a conserved 15 kD substrate binding domain and a less conserved C-terminal domain[220]. ER luminal HSP70 chaperone family members are BiP and luminal HSP seventy 1 protein (Lhs1p)/chaperone in the ER 1 protein (Cer1p)/Ssi1p/GRP170/oxygen-regulated protein of 150 kDa (ORP150)[106]. The substrate binding domain of the *Escherichia coli* HSP70 chaperone DnaK consists of a peptide binding pocket composed of two sheets each of four antiparallel β-strands and an α-helical lid. In the ADP bound state the lid is closed and the peptide bound with high affinity. Exchange of ADP for ATP in the nucleotide binding domain results in opening of the lid, a lower affinity of DnaK for their substrates, and release of the substrate[222,223]. BiP preferentially binds to short hydrophobic peptides, i.e. those forming β-strands in the hydrophobic protein core[209,210] with low affinity (1-100 mM)[223], as does Lhs1p[224]. Substrate binding stimulates the ATPase activity[210,225-228], returning the HSP70 chaperone back into the ADP bound, high affinity state. Thus, protein folding requires cycling of HSP70 chaperones through several cycles of ATP hydrolysis (Figure 2A). Indeed, depletion of cellular ATP levels inhibits the folding and secretion of many secretory proteins[229-232]. Cochaperones stimulate the ATPase activity of HSP70 chaperones and catalyze their ADP ATP exchange reaction. The J domain of HSP40 cochaperones interacts with HSP70 chaperones, stimulates their ATPase activity[233,234], increases the stability of the chaperone substrate complex[233], and influences their substrate specificity[233,234]. HSP40 cochaperones of the ER are ER DnaJ 1 (ERdj1)/murine DnaJ-like protein 1 (MTJ1)[111,235], ERdj3[236]/human ER associated DnaJ (HEDJ)[237]/*Saccharomyces cerevisiae* DnaJ 1 protein (Scj1p)[112,113], microvascular endothelial differentiation gene 1 (MDG1)[238]/ERdj4[116], ERdj5[117], DnaJ like protein of the ER membrane (Jem1p)[118], and secretory 63 protein (Sec63p)[119-121,239]. GrpE cochaperones, such as BiP associated protein (BAP)[122] and suppressor of Ire1/Lhs1 synthetic lethality 1 protein (Sil1p)[125,240]/synthetic lethal with the 7 S RNA mutation 1 protein (Sls1p)[123,124,241], are nucleotide exchange factors that catalyze ADP ATP exchange on HSP70 chaperones[223]. Their function is essential as HSP70s have an ~6 fold greater affinity for ADP than for ATP[242]. The function of cytosolic HSP70s, where the ATP concentration (usually 1-2 mM[242]) by far exceeds their K_m for ATP (usually 1–2 μM[242]), is not limited by ATP availability. In the ER, the amount of ATP may be limiting for function of ER resident HSP70s, because ATP is transported *via* an ADP/AMP antiport mechanism into the ER[243]. The two ER luminal HSP70 chaperones BiP and

A

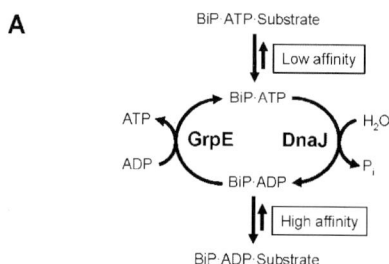

B

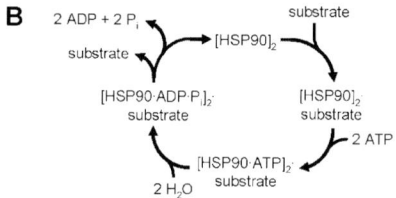

Figure 2. Chaperone ADP ATP cycles. A BiP ADP ATP cycle[220]. B HSP90 ADP ATP cycle[221]. Abbreviation: P_i – inorganic phosphate. Reprinted from *Current Molecular Medicine*, copyright 2006, with permission from Bentham Science Publishers

Lhs1p work on unfolded substrates in a coordinated manner. Lhs1p is a nucleotide exchange factor for BiP, and, in turn, BiP stimulates the ATPase domain of Lhs1p[219], showing that these two chaperones coordinate their work on the same substrate molecule.

HSP70 chaperones are able to form stable dimeric and even oligomeric complexes[226,242,244–246]. ATP bound BiP is monomeric and interacts with substrates[244,246]. The oligomeric form is stabilized by ADP and posttranslational modifications, such as phosphorylation in its peptide binding domain[244,247–250] and ADP ribosylation[223,251–253] and does not interact with substrates. Unfolded proteins sequester monomeric BiP out of its oligomeric form[246,254]. Therefore, oligomerized BiP may constitute a reserve pool from which the chaperone is recruited to the monomeric pool during ER stress. These events may be the first events in sensing unfolded proteins and ultimately trigger activation of the UPR.

BiP also plays an important role in gating the aqueous pore formed by the Sec61p translocon and therefore controls protein influx into the ER[216]. The ADP bound form of BiP seals the translocon from the luminal side[214], only allowing pore opening when the translocated polypeptide chain reaches ∼70 amino acids in length[214,255]. Only ATP bound BiP will allow pore opening and subsequent protein passage[216].

2.2.2   The ER resident HSP90 chaperone GRP94

The third HSP chaperone within the ER is the ER resident member of the HSP90 family, GRP94[126,127,256,257]/adenotin[258,259]/endoplasmin[128]/tumor rejection antigen glycoprotein of 96 kDa (gp96)[260,261]/ER protein of 99 kDa (ERp99)[262]/HSP108[263]/Ca^{2+} binding protein 4[164]/protein kinase of 80 kDa[264]. There is no functional equivalent within yeast[257], yet in mammalian cells, GRP94 is one of the most abundant ER proteins[128]. The N-terminal nucleotide binding domain of GRP94 is homologous to its cytosolic counterpart, HSP90[257], while the middle of the protein displays an imperfect basic leucine zipper region. The C-terminus is essential for both binding of client proteins and dimerization[265]. The substrates for GRP94 are advanced folding intermediates and incompletely assembled proteins[266]. Thus, GRP94 displays more substrate specificity than BiP[257] and acts, like its cytosolic counterpart, after BiP in protein folding[266]. The mechanism by which GRP94 fulfills its role as an ER chaperone is not fully understood. GRP94 binds ATP with very low affinity ($\sim 100\,\mu M$)[267,268] and has an ATPase activity that is barely above background ATP hydrolysis[269]. Site directed mutagenesis of residues predicted to be involved in adenine nucleotide binding and hydrolysis on the basis of the crystal structure of the N-terminal domain of HSP90 abolished ATP binding and hydrolysis by the closely related yeast HSP90[270,271] (Figure 2B). Furthermore, yeast cells carrying these mutations were inviable, showing that an ATP hydrolysis cycle is required for cytosolic HSP90 function *in vivo*[270,271]. However, because of the very low ATPase activity of GRP94 and the current lack of co-chaperones that stimulate its ATPase activity, it has not been established whether ATP hydrolysis cycles regulate the function of GRP94. Ligand binding may simply serve to deactivate GRP94[272,273], and alternative ligands for GRP94 have also been proposed[274].

Cytosolic HSP90 has several cochaperones that regulate its ATPase activity, load substrates onto HSP90, i.e. from HSP70 chaperones, assist in protein folding, i.e. peptidyl prolyl *cis-trans* isomerases, and target substrates to degradation pathways, i.e. the ubiquitin ligase carboxyl terminus of heat shock cognate protein of 70 kDa (HSC70) interacting protein (CHIP)[275–278]. Cochaperones with similar functions have not yet been identified for GRP94. However, GRP94 is found in multiprotein complexes with other ER luminal chaperones and foldases, i.e. BiP, Lhs1p, ERdj3, protein disulfide isomerase (PDI), ERp72, cyclophilin B, and UDP-glucose:glucoprotein glucosyl transferase (UGGT)[279–282]. In analogy to the ER luminal HSP70 chaperones BiP and Lhs1p, any one of these proteins could fulfill the function of a GRP94 cochaperone.

2.2.3   N-linked glycosylation and lectin chaperones

Many secretory and membrane proteins are modified by the addition of the oligosaccharide $Glc_3Man_9GlcNAc_2$ to consensus Asn-X-Ser/Thr residues in a process termed N-linked glycosylation[283,284]. Glycosylation increases the hydrophilic nature of a polypeptide chain, thus increasing its solubility. Glycosylation marks the attachment site as being displayed on the surface of the protein and thus influences protein folding pathways. Oligosaccharides, because of their large hydrated volume, provide a chaperone-like shield, hiding hydrophobic regions from the many other proteins within the ER. The overall stability of the protein is increased by the presence of *N*-glycosylation because of interactions of the oligosaccharide with the peptide backbone[285]. In addition, protein retrotranslocation becomes a more energetically demanding process when proteins are glycosylated, therefore encouraging ER retention until proper folding and maturation has taken place[69]. Finally, the glycosylation status of a protein is monitored by the lectin chaperone machinery as an indicator of protein folding status[286,287] (Figure 3).

Higher eukaryotic cells have three lectin chaperones, p88[131,288]/CNX[289], calmegin[130], and CRT[134,290]/calcium binding protein of 63 kDa (CAB-63)[133]/calregulin[291]/calcium regulated protein of 55 kDa (CRP55)[290,292]/high affinity calcium binding protein (HACBP)[293]. CNX and CRT are ubiquitously expressed. Calmegin is a testis specific protein[130]. Structurally and functionally, the lectin chaperones are very similar except that CNX and calmegin are transmembrane proteins whereas CRT is a soluble protein[294,295]. They consist of a globular lectin binding domain and a long hairpin loop, the P domain, which provides the chaperone function[296]. The lectin chaperones specifically recognize monoglucosylated *N*-glycans[295]. Immediately after the addition of an *N*-glycan to a polypeptide chain, sequential action by α-glucosidases I and II removes two of the three D-glucose residues yielding the monoglucosylated form GlcMan$_9$GlcNAc$_2$[297]. CNX and CRT remain associated with monoglucosylated substrates until the final D-glucose is removed by α-glucosidase II[298]. It remains unclear how α-glucosidase II accesses its

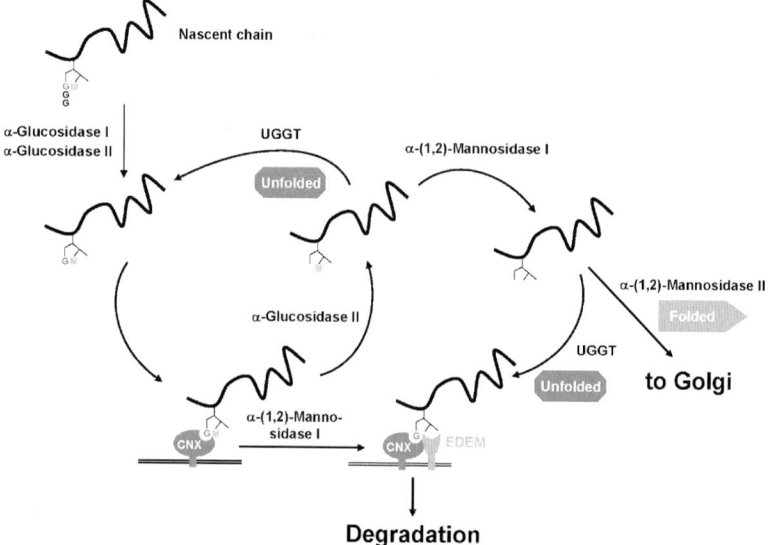

Figure 3. The calnexin/calreticulin cycle. Abbreviations: CNX - calnexin, EDEM - ER degradation-enhancing α-mannosidase-like protein, G – D-glucose, M – D-mannose, and UGGT – uridine diphosphate (UDP)-glucose:glycoprotein glucosyl transferase. Reprinted from Mutation Research, 569, Martin Schröder and Randal J. Kaufman, ER stress and the unfolded protein response, Copyright (2005), with permission from Elsevier

substrate when the substrate is bound to the lectin domain. Nevertheless, the lectins retain the protein within the ER and allow it to fold into its native conformation. Following removal of the remaining D-glucose, UGGT probes the folding status of the protein. UGGT recognizes both the exposed hydrophobic sequence[141] of an unfolded protein and the N-glycan[140]. The denatured state of a protein is likely to display an increased conformational flexibility and hence allows UGGT to access the innermost N-acetylglucoseamine moiety[287,297]. Definitive structural determinants that are recognized by UGGT remain obscure. Completely denatured proteins are poor substrates for UGGT[299], suggesting that UGGT acts after other chaperones such as BiP and only recognizes already relatively well folded proteins. UGGT is also required for productive folding of extensively misfolded proteins and can act before BiP on these substrates[300]. Likewise, the N-glycan reglucosylated by UGGT and unfolded protein determinants recognized by UGGT can be close together or far apart[301–304]. Reglucosylation by UGGT retains the substrate protein in the CNX/CRT cycle for a further attempt at correct folding (Figure 3), while those which have accomplished this task are allowed to exit the ER, often in partnership with cargo receptors, such as the lectin ER Golgi intermediate compartment protein of 53 kDa (ERGIC-53)/LMAN[305], and to continue to the Golgi complex. These cycles of de- and reglucosylation repeat until the protein has reached its native conformation. Some proteins never reach their native conformation and do not pass this quality control mechanism, or, despite being folded into a functional conformation are still recognized as unfolded. A prominent example for the latter case is the ΔF508 mutation in the cystic fibrosis transmembrane conductance regulator (CFTR)[306]. In order that the ER does not become jammed with unfolded proteins, these are targeted for retrograde translocation across the ER membrane for ERAD. Demannosylation of N-glycans by ER resident α1,2-mannosidases slows down the CNX/CRT cycle, because with increasing demannosylation, N-glycans become poorer substrates for α-glucosidase II[307] and UGGT[140]. Demannosylation affects deglucosylation by α-glucosidase II more than reglucosylation by UGGT[140,307], suggesting that the interaction between partially folded proteins and the lectin chaperones is prolonged by initial demannosylation events. However, the affinity of CNX and CRT for N-glycans decreases with increasing demannosylation[308,309], which should counteract the effect of demannosylation on α-glucosidase II and UGGT. In any case, demannosylation to $Man_6GlcNAc_2$ removes the D-mannose residue that is the D-glucose acceptor site, resulting in efficient removal of the protein from the CNX/CRT cycle[310]. This demannosylation event is slow[146]. The mannosidase-like protein homologous to mannosidase I 1 protein (Htm1p)[135]/mannosidase like 1 protein (Mnl1p)[137]/ER-degradation enhancing α-mannosidase-like protein (EDEM)[136,138,311,312] forms a complex with CNX[312] and targets proteins with demannosylated N-glycans for ERAD. Thus, slow demannosylation events define a time window for folding of a nascent, N-glycosylated polypeptide chain through limiting the number of rounds in the CNX/CRT cycle (Figure 3). If the protein folds sufficiently fast, it is allowed to exit the ER, if not, it is retained and targeted for degradation by ERAD.

2.2.4   Protein disulfide isomerases

The ratio between the redox buffers, glutathione (GSH) and glutathione disulfide (GSSG) is critical to maintain the redox potential of the ER, necessary for optimal disulfide bond formation. The highly oxidizing ER environment is maintained by a GSH/GSSG ratio of about 1:1-3:1, which is quite distinct from that of the cytosol where ratios are typically 100:1[67]. Within the ER, the processes of disulfide bond formation, isomerization and reduction are catalyzed by an expanding family of oxidoreductases including PDI, ERp57, ERp72, PDI related protein (PDIR)[175], PDI of the pancreas (PDIp)[161,162], PDI-like protein of the testis (PDILT)[163], and P5[313]. The active site of these enzymes contains a CXXC motif which enables switching between the oxidized disulfide and reduced dithiol forms[314]. The rate of formation of disulfide bonds is limited by the ability of the PDIs to regenerate the disulfide within their active sites[315], a process carried out by the FAD dependent oxidases, ER oxidation 1 protein (Ero1p)[149,150]/ERO1-Lα[151]/ ERO1-Lβ[152] and Erv2p[154]. The final electron acceptor for these enzymes is O_2[316], although peroxide and superoxide contribute to some degree. In yeast, *ERO1* is an essential gene, even under anaerobic conditions, suggesting that an alternative electron acceptor for Ero1p and Erv2p exists[316]. Thus, repeated folding attempts by PDI may generate reactive oxygen species and cause oxidative stress.

The presence of unpaired L-cysteine residues signals the existence of an immature or unfolded protein and provides an additional quality control mechanism within the ER. Exposed sulfhydryl groups are able to interact with PDIs, forming mixed disulfides, resulting in ER retention. Thiol mediated retention has been shown to be vital in avoiding the premature exit of incorrectly assembled proteins including acetylcholinesterase[317] and IgM[318]. In addition to PDI, ERp72 and more recently, ERp44 have been shown to be important for this process[177,319]. The oxidoreductases ER retained plasma membrane ATPase (*pma1*) suppressing 1 protein (Eps1p)[183] and Pdi1p[320] are also able to recognize folding incompetent or slowly folding proteins and to target these for ERAD.

2.2.5   Peptidyl prolyl cis-trans isomerases

A second class of protein foldases found in the ER are peptidyl prolyl *cis-trans* isomerases (PPIs). ER resident PPIs can be divided into two families. Cyclophilins, such as S-cyclophilin[189], secreted cyclophilin-like protein (SCYLP)[190], cyclophilin B[193], and *Drosophila* ninaA[194,321] are inhibited by the drug cyclosporine A. The immunophilins FK506 binding protein of 13 kDa (FKBP13)[322] and FKBP65[200] are inhibited by the immunosuppressant FK506. These enzymes catalyze the otherwise slow *cis-trans* isomerization of peptidyl prolyl bonds. Yeast deleted for all cyclophilins and immunophilins are still viable[323]. Furthermore, mutations in all secretory pathway PPIs of yeast, the cyclophilins cyclosporine sensitive proline rotamase 2 protein (Cpr2p), Cpr4p, Cpr5p, and Cpr8p, and the immunophilin FKBP proline rotamase 2 protein (Fpr2p)/FKBP13, showed no synthetic lethalities with calnexin[323], temperature sensitive *BiP* alleles[323], or the proximal signal transducer of the UPR in yeast, *myo*-inositol requiring 1 (*IRE1*)/ER to nucleus signaling 1

$(ERN1)^{323}$, suggesting that PPIs play a minor role in protein folding *in vivo*. In contrast, the PPI of the parvulin family, essential 1 protein (Ess1p), is an essential protein[323].

3.   SIGNAL TRANSDUCTION BY THE UPR

The protein folding status in the ER is recognized by three mechanisms, interaction of molecular chaperones, such as BiP, with stretches of hydrophobic amino acids on the surface of the protein, interaction of PDIs with exposed, unpaired cysteine residues, and the status of trimming of *N*-glycans in the CNX/CRT cycle. The ER membrane harbors numerous transmembrane spanning proteins with ER luminal stress sensing domains. The role of these proteins is to transduce information regarding the protein folding status in the ER across the ER membrane to the cytosol and the nucleus. Three different types of transmembrane ER proteins are known; activating transcription factor 6 (ATF6)[327,328] is a prototype of transcription factors synthesized as an ER located, inactive type II transmembrane protein, the protein kinase endoribonuclease IRE1[329–332], and double-stranded RNA-activated protein kinase (PKR)-like endoplasmic reticulum kinase (PERK)[333]/pancreatic eukaryotic initiation factor 2α (eIF2α) kinase (PEK)[334,335] (Figure 4).

3.1   Transcription Factors of the ER Membrane

Three classes of transcription factors that are synthesized as inactive ER membrane resident transmembrane proteins have been identified: basic leucine zipper (bZIP) transcription factors such as ATF6α[327,336], ATF6β/cyclic adenosine monophosphate (cAMP) response element binding protein (CREB) related protein (CREB-RP)/G13[337–339], old astrocyte specifically induced substance (OASIS)[340,341], CREB3/Luman[342–344]/long bZIP-1 (LZIP-1)[345]/LZIP-2[345], CREB4[346], and Box B binding factor (BBF2H7)[347], the basic helix loop helix (bHLH) transcription factors sterol response element (SRE) binding protein 1a (SREBP-1a), SREBP-1c, and SREBP-2, and the avian reticuloendotheliosis virus T leukemia (Rel) A transcription factors suppressor of Ty 23 protein (Spt23p) and multicopy suppressor of *gam1* 2 protein (Mga2p) in yeast (Figure 5). These are discussed in the following sections.

3.1.1   ATF6

ATF6 is a type II ER transmembrane protein containing a cytosolic domain that encodes a bZIP transcription factor[327]. Two mammalian homologues exist: ATF6α[327,336] and ATF6β[337–339]. ATF6 is constitutively synthesized in the ER, but upon accumulation of unfolded protein within the ER, translocates to the Golgi complex (Figure 6)[348]. Here, the 90 kDa protein (p90ATF6) is cleaved within its luminal domain by the serine protease site-1 protease (S1P)[327]. This proteolytic event prepares the *N*-terminal luminal domain of the membrane embedded ATF6 for

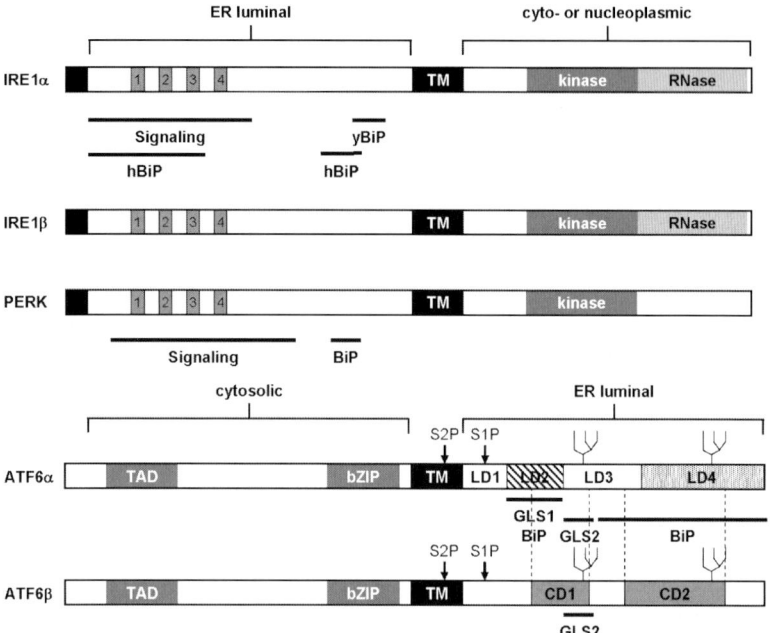

Figure 4. Primary structure of ER stress sensors. Sequences sufficient for signal transduction (called 'signaling'), BiP interaction regions, and GLS1 and GLS2 are indicated by black bars. The unlabeled black box represents the signal peptide, grey boxes labeled 1 – 4 conserved regions in the ER luminal domains of IRE1 and PERK[324], arrows indicate the S1P and S2P cleavage sites, tree like structures conserved *N*-linked oligosaccharides. Abbreviations: CD – conserved domain, hBiP – BiP interaction sites in human IRE1α[325], TAD – transcriptional activation domain, TM – transmembrane domain, and yBiP – BiP interaction site in yeast Ire1p[326]. Drawings are not to scale. Reprinted from *Current Molecular Medicine,* copyright 2006, with permission from Bentham Science Publishers

subsequent cleavage by the metalloprotease site-2 protease (S2P)[327,349,350] in a proteolytic activation process called regulated intramembrane proteolysis (RIP). Interestingly, these proteases are the same enzymes responsible for cleavage of the SREBPs, which are activated when cholesterol is depleted[349]. The resulting 50 kDa protein (p50ATF6), consisting of the cytosolic bZIP region of ATF6, translocates to the nucleus where it activates transcription[348]. Two DNA binding sequences for active p50ATF6 have been identified, the ATF/cAMP response element (CRE)[351] and the ER stress response elements (ERSE) I (CCAAT-N_9-CCACG)[352] and II (ATTGG-N-CCACG)[353]. ATF6 binding to ERSE-I and -II is dependent on binding of the general transcription factor, nuclear factor Y (NF-Y)/CCAAT binding factor (CBF)/CCAAT binding protein (CBP)/heme activator protein 2 protein (Hap2p)/Hap3p/ Hap5p[353,354] to the CCAAT and ATTGG half sides of ERSE-I and –II, respectively. ATF6β was reported to be a repressor of ATF6α through formation of transcriptionally inactive ATF6α · ATF6β heterodimers in the regulation of *BiP*[355]. In contrast, another

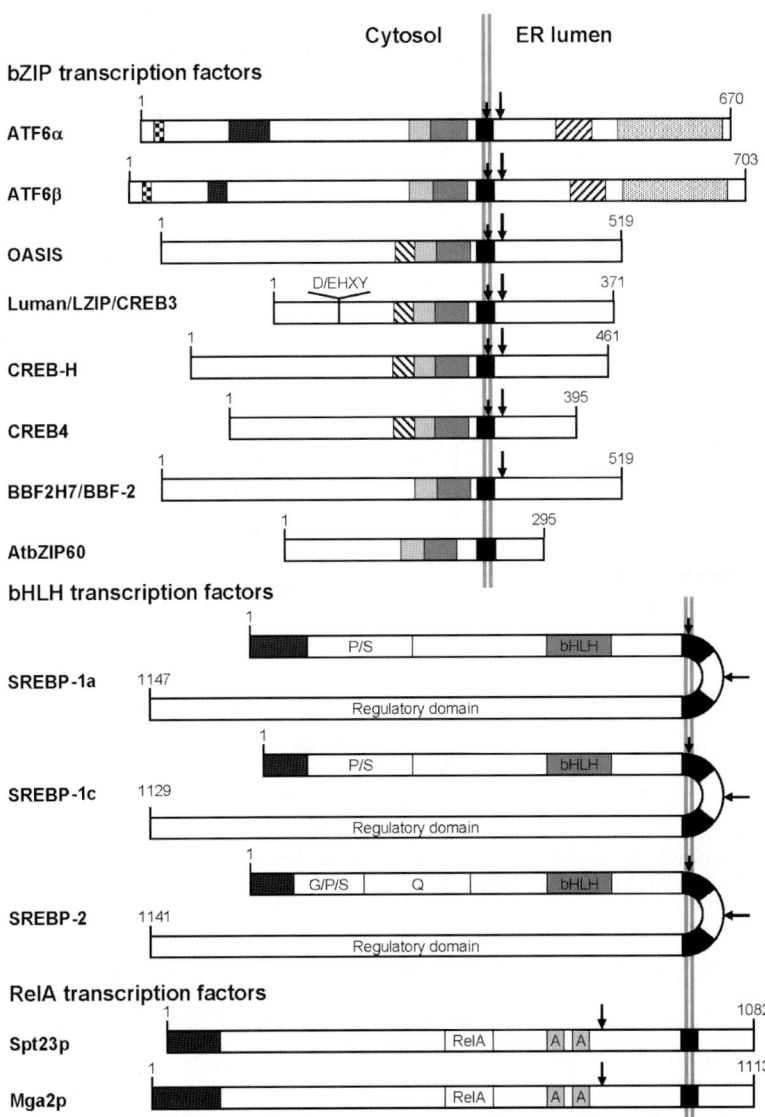

Figure 5. Primary structure of transcription factors synthesized as inactive ER transmembrane precursors. Abbreviations: A – ankyrin repeat, G/P/S – glycine, L-proline, and L-serine rich region, P/S – L-proline and L-serine rich region, Q – L-glutamine rich region, and RelA – RelA homology domain. Symbols: long arrow – S1P cleavage site, short arrow – S2P cleavage site, black boxes – transmembrane domain, dark grey boxes – transcriptional activation domain, intermediate and light grey boxes – bZIP or bHLH domain, grey double line – ER membrane. In Spt23p and Mga2p the arrow indicates the approximate site of proteasomal processing. Homologous regions are shaded, checker boarded or indicated by black diagonal lines on a white background. D/EHXY – host cell factor interaction region

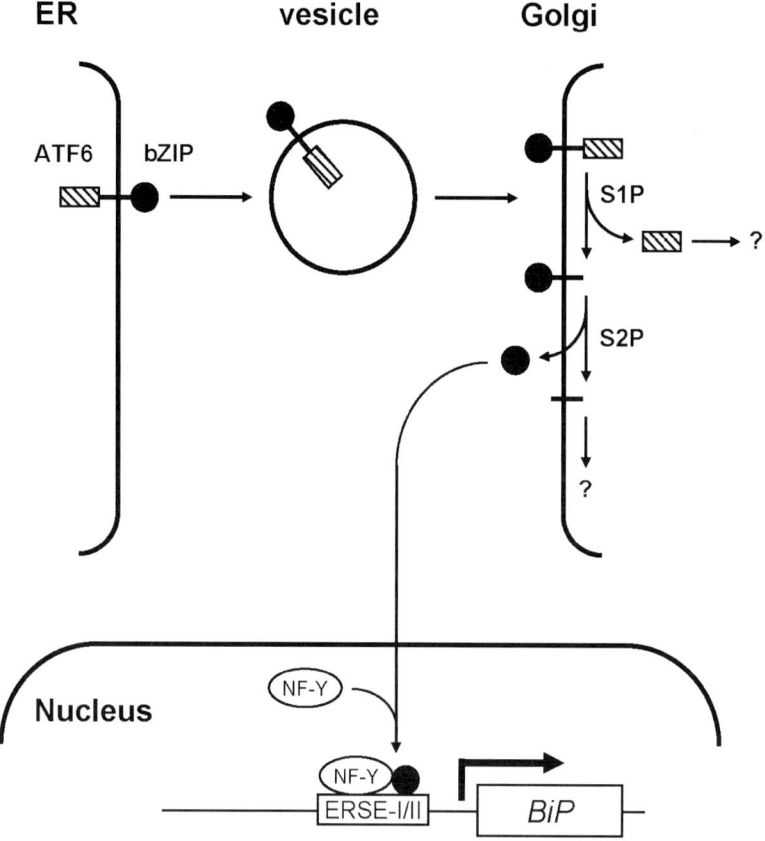

Figure 6. Signal transduction by ATF6. Reprinted from Current Molecular Medicine, copyright 2006, with permission from Bentham Science Publishers

group reported that ATF6β is a transcriptional activator on ERSE-I and the *BiP* promoter[339].

Two Golgi localization sequences (GLS1 and GLS2) were identified in the ER luminal domain of ATF6. BiP binds to GLS1[356], resulting in retention of ATF6 in the ER. Unfolded proteins in the ER sequester BiP from GLS1. When BiP is not bound to GLS1, GLS2 targets ATF6 to the Golgi complex[356]. Thus, BiP may be a negative regulator of the UPR. However, the concept of competitive binding of unfolded proteins to BiP as the cause of BiP dissociation is disputed by recent evidence suggesting that active restarting of the BiP ATPase cycle by ER stress causes dissociation of BiP from ATF6[357]. In addition to BiP binding, ATF6 is retained within the ER *via* interactions with the lectin, CRT[358]. Within the *C*-terminus of ATF6, three *N*-linked glycosylation sites, conserved throughout evolution (Figure 4), act as additional ER homeostasis sensors that allow ATF6 to evade CRT dependent interaction *via* differential glycosylation[358]. Upon ER stress, newly synthesized ATF6 is underglycosylated, abolishing interaction with CRT, hence allowing translocation to the Golgi compartment. Newly synthesized underglycosylated ATF6 would also be retained within the ER via heterodimerization with CRT bound glycosylated ATF6 through their bZIP domains. Hence, fully glycosylated ATF6 is targeted for proteasomal degradation in a S1P/S2P independent manner upon ER stress[359]. Causality has not yet been shown for both mechanisms in activation of ATF6.

3.1.2   OASIS

Similar to ATF6, OASIS is a type II transmembrane ER localized bZIP transcription factor[360,361]. Mutations in putative S1P and S2P cleavage sites in OASIS blocked proteolytic activation of OASIS[340]. However, S1P and S2P both cleave OASIS independent of each other[341]. Cleavage of OASIS by S1P or S2P by genetically or pharmacologically recruiting these proteases to the ER has not been shown. The bZIP domain after translocation to the nucleus activates transcription of BiP through the CRE and ERSE[340]. OASIS displays a marked preference for the CRE[340]. It still remains to be seen if the effects of OASIS are specific to astrocytes as OASIS is expressed in many other tissues[361] and in osteoblasts in osteogenesis imperfecta[362].

3.1.3   SREBPs

The SREBPs are members of a family of membrane localized transcription factors that regulate biosynthesis of cholesterol, triacylglycerols, phospholipids and fatty acids[363]. Two SREBP genes exist, *SREBP1*, which through alternative splicing encodes SREBP-1a and SREBP-1c, and *SREBP2*. SREBP-1a is a potent activator of all SRE containing genes[363]. SREBP-1c preferentially activates genes required for fatty acid and triglyceride metabolism[363] and SREBP-2 genes required for cholesterol synthesis[363]. The full length proteins consist of a cytosolic *N*-terminal bHLH region, two membrane spanning domains, and a regulatory cytosolic *C*-terminal domain (Figure 5). In response to a decrease in sterols, the *N*-terminus is cleaved by S1P and S2P[364]. The liberated *N*-terminal domain translocates to the nucleus where it activates transcription of genes containing SRE in their promoters[363,364]. The soluble bHLH domain is subject to rapid proteolysis to control the level of SREBP mediated transcription[365]. In order for S1P to cleave SREBP, SREBP must translocate from the ER to the Golgi complex[363]. SREBP cleavage-activating protein (SCAP), a large protein with at least eight transmembrane regions, stimulates S1P mediated cleavage of SREBP by escorting SREBPs to the Golgi[366,367] through interaction of its L-tryptophan (W) L-aspartic acid (D) repeat (WD repeat) domain with the *C*-terminus of SREBPs[367]. Through a complex with additional proteins, insulin-induced gene-1 (INSIG-1) and INSIG-2, SCAP and the SREBPs, are retained in the ER when cholesterol levels are high[367–369]. When cholesterol levels drop, a conformational change in SCAP releases the SCAP·SREBP complex from INSIG-1 and INSIG-2, allowing this complex to translocate to the Golgi complex[370].

ATF6 forms a heterodimer with SREBP-2 which binds to the SRE in the promoter of the low density lipoprotein receptor (LDLR)[78]. The ATF6·SREBP2 complex recruits the transcriptional repressor histone deacetylase 1 (HDAC1) to the LDLR promoter[78]. Whether the heterotrimeric ATF6·SREBP-2·HDAC1 complex is necessary for repression of SREBP-2 targets, or whether ATF6 alone is able to recruit HDAC1 to the SRE is not yet clear. Repression of cholesterol synthesis *via* formation of the ATF6·SREBP2 complex may alter membrane fluidity during ER stress. Cholesterol loading of macrophages induces the UPR and apoptosis through activation of the bZIP transcription factor CCAAT enhancer binding protein (C/EBP) homologous protein (CHOP)/growth arrest and DNA damage inducible gene 153 (GADD153)[371], whose activity is regulated by the UPR (cf. below). Although the mechanism by which cholesterol activates the UPR is unclear, ER resident chaperones may be deleteriously affected by dysfunction of ER calcium pumps caused by the presence of cholesterol[371]. Furthermore, the putative cholesterol transfer steroidogenic acute regulatory protein (StAR) related lipid transfer (START) protein, StarD5, involved in lipid transport, metabolism, and signaling, is transcriptionally induced by ER stress[372], suggesting that remodeling of cholesterol metabolism is part of the physiological response to ER stress.

3.1.4   *RelA transcription factors synthesized as ER transmembrane proteins*

The type II transmembrane transcription factors Spt23p and Mga2p are distant relatives of the RelA transcription factors, i.e. NF-κB (Figure 5). Both show considerable homology[373]. *mga2Δ spt23Δ* strains are unsaturated fatty acid auxotrophs[374]. This auxotrophy is rescued by either *MGA2*, *SPT23*, or *OLE1*, the gene encoding the ER membrane protein Δ9 fatty acid desaturase[374], suggesting that both regulate transcription of *OLE1*. In contrast to type II transmembrane bZIP or bHLH transcription factors that are activated by proteolytic cleavage by S1P and S2P, these transcription factors are activated by proteolysis of their *C*-terminus by the proteasome in a process called regulated ubiquitin/proteasome-dependent processing (RUP). This process resembles proteasomal processing of NF-κB1 and NF-κB2. Unsaturated fatty acids inhibit processing of Spt23p to its active form[375]. The ability of an unsaturated fatty acid to inhibit Spt23p processing correlated with its melting temperature, i.e. exposure of cells to the fatty acid with the highest melting temperature did not inhibit processing of Spt23p, suggesting that Spt23p and Mga2p respond to changes in membrane fluidity. Higher eukaryotic homologs for Spt23p and Mga2p have not yet been identified. Furthermore, it is interesting to speculate that ER stress may cause changes in membrane fluidity.

3.2   IRE1

Both IRE1 and PERK are type 1 ER resident transmembrane protein serine/threonine protein kinases (Figure 4). IRE1 also encodes a *C*-terminal domain with homology to RNase L[330,331]. Following the discovery of Ire1p as the proximal signal transducer of the UPR in yeast[331,332], it was shown that *IRE1* is conserved throughout all eukaryotic cells. Mammalian cells encode two homologues, the isoforms *IRE1α*, expressed ubiquitously[376], and *IRE1β*, which is found predominantly within cells of the intestinal epithelia[376,377]. The most *N*-terminal of the three functional domains comprising IRE1 is located within the ER lumen and is thought to be responsible for sensing unfolded proteins. Found within the cytosol are a protein kinase domain and, most *C*-terminally, a RNase domain[331,378,379].

3.2.1   Sensing of unfolded proteins by IRE1

Homooligomerization or homodimerization mediated by its ER luminal domain activates IRE1[378,380]. Interaction between the cytosolic portions of IRE1, albeit to a lesser extent, also contributes to oligomerization[380]. Glycerol gradient sedimentation[103] and gel filtration experiments[325] show that IRE1 can form dimers *in vitro*. This observation is consistent with the finding that two functional RNase domains were necessary for signal transduction by Ire1p *in vivo*[381]. Replacement of the ER luminal domain of yeast Ire1p with the leucine zipper dimerization motif from the bZIP transcription factors MafL and JunL reconstituted functional Ire1p, supporting that the active form of Ire1p is a dimeric species[104]. These chimeras were also regulated by ER stress as was wild type Ire1p, suggesting that a ligand independent mechanism activates Ire1p. Furthermore, when bZIP domains containing point mutations in the leucine zippers known to prevent dimerization of bZIP transcription factors were fused to the transmembrane domain and cytosolic portion of Ire1p, activation of the UPR by these chimeras was not observed[104].

Two models that explain activation of IRE1 are in circulation. In the competition model, IRE1, and PERK, are activated in a manner similar to ATF6 by sequestering the molecular chaperone BiP from their ER luminal domains by increased concentrations of unfolded proteins in the ER lumen[382]. Precedence for such an activation mechanism exists in the bacterial heat shock response[383,384], and the activation of heat shock transcription factor by unfolded proteins in the cytosol of eukaryotic cells[385]. Experimentally, BiP and the luminal domains of IRE1 and PERK form a stable complex in unstressed cells[103,105,324]. Upon ER stress this complex dissociates[103,105]. Consistent with these data is the observation that overexpression of BiP interferes with activation of IRE1 and PERK[103,386]. In IRE1α, the sequences required for signaling overlap with BiP binding sites (Figure 4), suggesting that BiP may mask a crucial dimerization motif[324]. In PERK, BiP binding sites and regions required for signaling are distinct (Figure 4)[387]. Thus, BiP may inhibit oligomerization of PERK through inducing a conformational change in its ER luminal domain, or sterically interfere with interaction of two PERK molecules. BiP preferentially binds to stretches of hydrophobic amino acid residues. Thus, to explain the preference for IRE1 and PERK to homodi- or homooligomerize, and not to fall prey to the, presumably, vast excess of unfolded proteins in the ER during ER stress, a mechanism must exist that provides for higher affinity interactions between the ER luminal domains of IRE1 and PERK with themselves than with the other stress sensors or unfolded proteins.

A positive priming role for release of BiP from the luminal domains of ATF6 and IRE1 was recently suggested. *In vitro*, addition of ATP to ATF6·BiP complexes induced release of BiP from ATF6. In contrast to an unfolded protein substrate for

BiP, ER stress specifically dissociated BiP from ATF6, and not from the unfolded substrate. These data suggest that restarting of the BiP ADP ATP cycle is involved in release of BiP from ATF6[357]. Furthermore, three distinct domains were identified in the luminal domain of ATF6 that bind to BiP and can induce release of BiP from ATF6[357]. In yeast Ire1p, a ten amino acid deletion at position 226 interfered with release of BiP from Ire1p in response to ER stress and conferred tunicamycin sensitivity to Δ226-Ire1p carrying yeast cells[326]. Tunicamycin resistance was restored when the BiP binding site in Δ226-Ire1p was deleted[326].

The second model for activation of IRE1 by unfolded proteins is based on the recently published 3.0 Å crystal structure for the core ER luminal domain of IRE1[388]. One IRE1 ER luminal domain forms a half side of a groove that resembles the peptide binding groove in the major histocompatibility complex (MHC). *In vitro*, the ER luminal domain of IRE1α forms dimers with high affinity[324,325], suggesting the peptide binding groove is present in unstressed cells and awaiting interaction with unfolded or unstructured regions of proteins. Mutation of amino acid residues involved in formation of IRE1 oligomers and of residues that line the bottom of the groove interfered with signal transduction by Ire1p[388]. Protein levels for these mutant proteins were identical to wild type Ire1p[388], suggesting that these mutants fold efficiently. Thus, IRE1, and PERK, may directly sense the folding status of the protein cargo in the ER. Precedence for directly sensing the folding status of proteins exists in the activation of the degradative protease S (DegS) through interaction with *C*-termini of unfolded outer membrane porins in the envelope stress response in gram-negative bacteria[389]. Finally, the crystal structure of the core ER luminal domain also supports the idea that Ire1p forms fibrillar homooligomers in the ER membrane.

In contrast to ATF6, where the *N*-linked glycosylation status may be involved in activation of ATF6, this seems not to be the case for IRE1. Of the *N*-glycosylation sites found on human IRE1, only one is conserved in yeast. Mutation of this site did not interfere with UPR signaling by Ire1p[104]. However, functional redundancy between the conserved *N*-glycosylation site and non-conserved *N*-glycosylation sites may also explain this observation.

3.2.2   *Activation of the RNase domain*

Di-, oligomerization, or a conformational change induced by binding of unfolded proteins to the ER luminal domain of Ire1p activates Ire1p. Initially, two cytosolic domains come into close contact, resulting in autophosphorylation of conserved L-serine residues in the activation loop of the kinase domain *in trans*[378,380,390,391]. Phosphorylation of these L-serine residues fully opens the ATP binding pocket of Ire1p. These steps are analogous to activation of growth factor tyrosine kinases[378,390]. Then, surprisingly, occupancy of the ATP binding pocket with ADP or ATP is sufficient to induce a conformational change in the RNase domain that activates the RNase domain[391]. In this respect, activation of the kinase domain of Ire1p is similar to activation of RNase L, which, like Ire1p, has a non-functional kinase like domain and a *C*-terminal RNase domain that is allosterically activated by 2′,5′-oligoadenylate in response to viral infections[392,393].

3.2.3   The mechanism of mRNA splicing employed by IRE1

The only known substrate for the activated *IRE1* RNase domain in yeast is the mRNA encoding the bZIP transcription factor homologous to ATF6/CREB 1 protein (Hac1p)[394,395]. *HAC1* mRNA is constitutively transcribed in an inactive form, *HAC1*[u] (u for uninduced)[396]. Activated Ire1p initiates non-spliceosomal splicing of *HAC1*[u] mRNA by cleaving the consensus sequence BCNG↓MNGHR (B = A, C, G; H = A, C, U; M = A, C; N = A, C, G, U; R = A, G; ↓ = cleavage position) in hairpin loops[397]. *HAC1*[u] mRNA consists of two exons separated by a 3′ intron of 252 bp. Activated Ire1p cleaves both 5′ and 3′ exon intron junctions[379] and generates 5′-OH and 2′,3′-cyclic phosphate ends at the splice junctions[397] (Figure 7). tRNA ligase (Trl1p)/RNA ligase (Rlg1p) then joins the *HAC1* exons[398], yielding spliced or induced (i) *HAC1* mRNA (*HAC1*[i]). A 2′-phosphate group derived from the splice junction phosphate is left on the initial ligation product by the ligase and removed by the NAD[+] dependent tRNA 2′-phosphotransferase 1 protein (Tpt1p)[397]. In this reaction NAD[+] serves as phosphate acceptor in a reaction that yields nicotinamide and ADP-ribose 1″,2″-cyclic phosphate (ADP-ribose>P)[399]. ADP-ribose>P may play a role in signal transduction during the UPR. ADP-ribose>P is dephosphorylated to ADP-ribose by sequential action of the phosphatases cyclic nucleotide phosphodiesterase 1 protein (Cpd1p)[400] and phosphatase of ADP-ribose 1″-phosphate 1 protein (Poa1p)[401]. Thus, *HAC1*[u] mRNA is an mRNA that is spliced by a tRNA splicing mechanism. *In vitro*, the two *HAC1* exons form a hybrid, thus explaining the preference for Trl1p to ligate the exons[397].

The metazoan counterpart for *HAC1* mRNA is the mRNA for the bZIP transcription factor X box binding protein 1 (XBP-1)[96,402–404]. IRE1 cleaves both exon intron junctions in *XBP-1* mRNA[96,402–404]. It is not known if a 2′,3′-cyclic phosphate end is formed in *XBP-1* splicing intermediates. RNA ligase activities with similar properties as Trl1p have been described in wheat germ[405–408], *Chlamydomonas*[409,410], and mammalian cells[411]. However, only the plant tRNA ligase was recently cloned[412]. An additional tRNA ligase activity, which incorporates the junction phosphate into the spliced tRNA, has been described in mammalian cells (Figure 7)[413]. This ligase has a molecular weight of ∼160 kDa and ligates several RNAs with 5′-OH and 2′,3′-cyclic phosphate ends[414,415]. Which, if any, of these ligases, is the *XBP-1* ligase, is unknown. Of the phosphatases, mammalian tRNA 2′-phosphotransferase (TRPT1) complements a defect in Tpt1p in yeast[416], and homologs for Poa1p can be identified in BLAST searches. However, a homolog for Cpd1p has not been identified.

HAC1 mRNA is likely to be spliced in the cytosol, because splicing was observed in polysomes[417], but can also occur outside polysomes[418]. The majority of *HAC1*[u] mRNA is located in the cytosol[419]. This cytosolic pool is a substrate for activated Ire1p[417]. tRNAs are spliced in the nucleus[420], and tRNA ligase localizes to the nucleus in yeast[421]. However, the mutation H148Y in tRNA ligase specifically

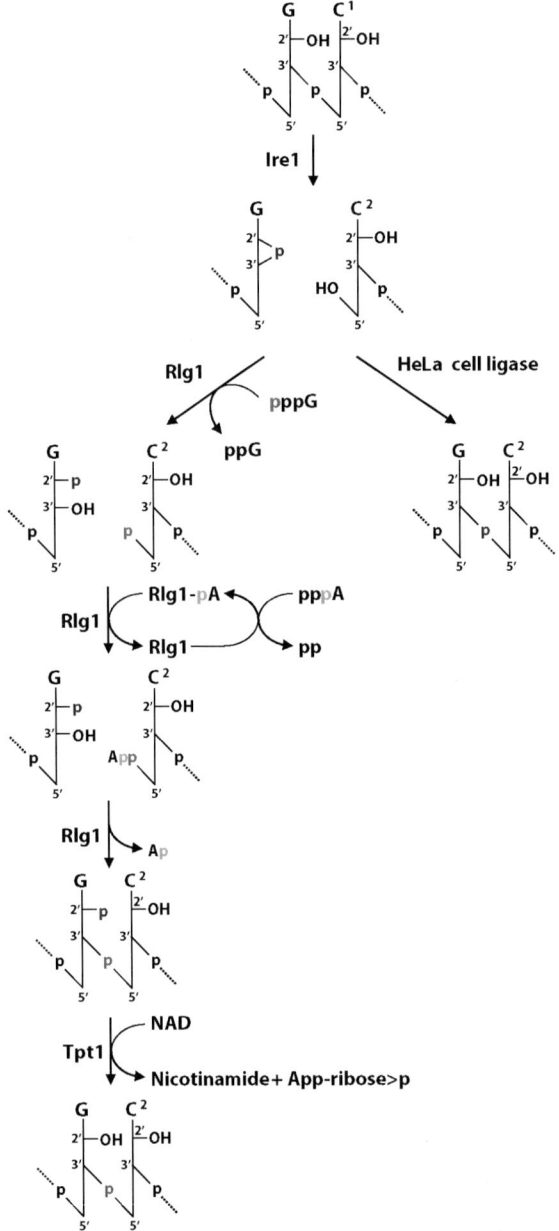

Figure 7. Mechanism of *HAC1*[u] mRNA splicing in yeast and of tRNA splicing in mammals. Reprinted, with permission, from the *Annual Review of Biochemistry*, Volume 74 ©2005 by Annual Reviews www.annualreviews.org

abolishes the UPR, but has no effect on tRNA splicing[398], indicating that this L-histidine residue is either involved in substrate recognition by the ligase, or involved in shuttling the ligase out of the nucleus during activation of the UPR. In contrast, in mammals IRE1α was convincingly localized to the inner nuclear envelope, showing that *XBP-1* splicing is a nuclear event[404].

3.2.4 Translational regulation of Hac1^{i}p and XBP-1

The *HAC1*[u] intron attenuates translation of *HAC1*[u] mRNA[417,419]. By hybridizing to a complementary 19 nucleotide (nt) sequence in the 5′-untranslated region (UTR), the intron stalls *HAC1*[u] mRNA on polyribosomes, thus preventing its translation[417,419,422]. To explain how ribosomes are loaded onto backfolded *HAC1*[u] mRNA, it was proposed that ribosomes are loaded onto *HAC1*[u] mRNA immediately when the mRNA exits the nucleus[417]. The hybrid between the 5′-UTR and the intron is only formed after the 3′ sequence has emerged from the nucleus, thus trapping some ribosomes already engaged with *HAC1*[u] mRNA on the mRNA[417]. Although the contribution of hybrid formation between the 5′-UTR and the intron to translational attenuation is relatively small, it increases significantly during times of ER stress[417]. Constitutive, high-level activation of the UPR is detrimental to cell growth[377,378,422]. Thus, it is imperative that synthesis of Hac1p is tightly controlled. A further control is added in that the DNA binding domain and the transcriptional activation domain are located on two separate exons that are only joined when the mRNA is spliced. Theoretically, this allows for the possibility that the unspliced Hac1[u]p functions in processes unrelated to the UPR[423].

In contrast to *HAC1*[u] mRNA, no translational attenuator is found in the *XBP-1*[u] intron, and differential regulation of translation of *XBP-1*[u] and *XBP-1*[s] (s for spliced) mRNA is not observed[403,404]. Instead, mammalian cells rely on competition for binding sites and dimerization partners between XBP-1[u] and XBP-1[s] to suppress transcriptional activation by XBP-1[s][424]. Efficient UPR activation in mammalian cells requires rapid proteasomal degradation of XBP-1[u] in order that this competition is relieved[424,425].

3.2.5 Transcriptional regulation by Hac1^{i}p and XBP-1^s

In Hac1p, the splicing reaction replaces the ten 3′ codons of the first exon with 18 codons encoded by the second exon[379,396]. The alternative *C*-terminus in Hac1[i]p is more acidic than the *C*-terminus of Hac1[u]p and is a ~10 fold stronger transcriptional activation domain[426]. Similarly, splicing of *XBP-1*[u] mRNA introduces a new *C*-terminus into XBP-1[s] with increased transcriptional activation potential[96,403,404]. Hac1[i]p translocates to the nucleus where it binds to specific sequence motifs in several promoters. These include the UPR elements (UPRE) UPRE-1 (consensus sequence CAGCGTG)[386,427] and UPRE-2 (consensus sequence TACGTG)[428], which are found in the promoter regions of genes encoding ER chaperones including BiP[386,427,429–431], Lhs1p[430], and protein foldases, i.e. Pdi1p[430], ER protein unnecessary for growth 1 protein (Eug1p)[430], and Fkb2p (Figure 8)[430,432]. Hac1[i]p binds to UPRE-1 and -2 as a heterodimer with the bZIP transcription

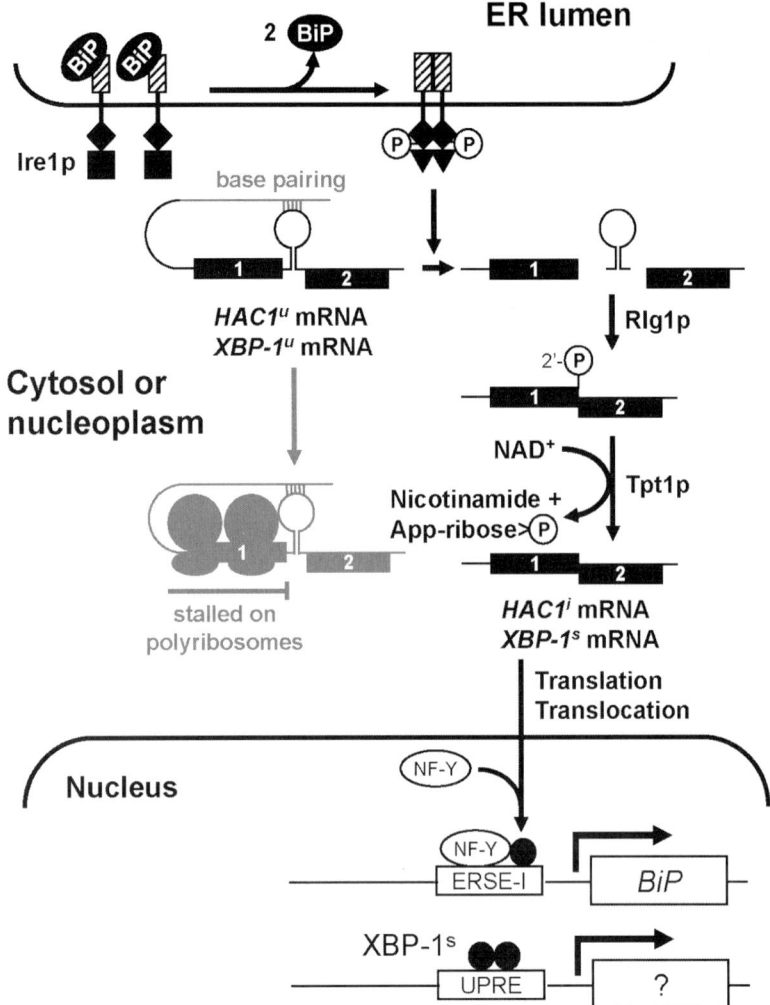

Figure 8. Activation of Hac1ip and XBP-1s through spliceosome independent splicing of their mRNAs by IRE1, tRNA ligase, and the 2′ NAD+-dependent tRNA 2′-phosphotransferase (Tpt1p/TRPT). The conformational change in the IRE1 RNase domain is indicated by a change from a square to a triangle. The translational frame switch between the 1st and 2nd exon is indicated by showing the 2nd exon staggered to the 1st exon. Events specific for *HAC1* mRNA are shown in grey. Abbreviations: App-ribose>P – ADP-ribose 1″, 2″-cyclic phosphate and NAD+ - nicotine adenine dinucleotide. Reprinted from *Current Molecular Medicine*, copyright 2006, with permission from Bentham Science Publishers

factor general control non-derepressible 4 protein (Gcn4p)[428]. A yeast two-hybrid screen isolated the transcriptional co-activator Gcn5p/transcriptional adaptor 4 protein (Ada4p) as an interaction partner for Ire1p[433]. Transcriptional co-activators mediate interaction between transcriptional activators and TATA box occupying general transcription factors[434]. Gcn5p is part of the Spt-Ada-Gcn5 acetyltransferase (SAGA) histone acetyltransferase complex (HAT)[435–439], a multimeric complex consisting of a number of transcriptional co-activators including Ada1p, Ada2p, Ada3p, Ada5p/Spt20p, Spt3p, Spt7p, and Spt8p[435–439], and of the SAGA related complex SAGA-like (SLIK)[440]/SAGA altered, Spt8 absent (SALSA)[441]. Deletion of individual components within the SAGA HAT diminished the UPR, while deletion of *ADA5* prevented UPR signaling altogether[439], although these components were dispensable for the heat shock response[433]. Transcriptional upregulation of UPRE containing genes by Hac1^{i}p may involve recruitment of SAGA to the UPRE by Hac1^{i}p and subsequent acetylation of the *N*-terminal core nucleosomal histone tails to facilitate transcription[442]. In addition, Hac1^{i}p regulates transcription on the upstream repressing sequence 1 (URS1, consensus sequence TCGGCGGCT)[443] and the subtelomeric ATF/CREB GTA variant element (SACE) (ATGGTATCAT)[444]. Interestingly, Hac1^{i}p is a negative regulator on URS1, making it the only known yeast bZIP factor to act as both a transcriptional activator and repressor. URS1 controls the expression of early meiotic genes (EMGs)[445] as well as genes encoding proteins involved in metabolism of nitrogen and carbon sources[446]. Transcriptional repression mediated through URS1 requires binding of the transcriptional regulator unscheduled meiosis 6 protein (Ume6p) to URS1[447]. Ume6p recruits two transcriptional repression complexes, the imitation of switch 2 (ISW2) chromatin remodeling complex[448] and the reduced potassium dependency 3 protein (Rpd3p) – switch independent 3 protein (Sin3p) histone deacetylase complex (HDAC)[449]. Hac1^{i}p dependent repression at URS1 required the presence of a catalytically functional Rpd3p-Sin3p HDAC on URS1, but the chromatin remodeling complex was dispensable for transcriptional repression by Hac1^{i}p on URS1[443]. The Rpd3p-Sin3p HDAC exists in two complexes in cells, a large, promoter bound complex that regulates transcription initiation, and a small complex tethered to open reading frames (ORF) to suppress spurious initiation of transcription while the chromatin structure of the ORF is opened when the elongating RNA polymerase II is passing through the ORF[450,451]. Both complexes are characterized by specific subunits[450,451]. Deletion of components specific to the large Rpd3p complex abolished transcriptional repression by Hac1^{i}p[443]. This observation is consistent with the observation that Hac1^{i}p is repressing transcription on the promoter element URS1. Coimmunoprecipitation experiments demonstrated a direct interaction between Hac1^{i}p and the Rpd3p-Sin3p HDAC[443]. Despite this observation, *hac1*Δ yeast retained some, but not all, repression by the Rpd3-Sin3 HDAC, suggesting that Hac1^{i}p is not integral to the function of the HDAC[443]. Instead, we proposed that HDAC mediated repression is enhanced by its association with Hac1^{i}p[443]. These data suggest that potentially Hac1^{i}p and the UPR repress all of the Ume6p controlled genes within yeast, which make up ∼20% of all yeast genes[446], many of which are required for carbon and nitrogen metabolism as well as induction of meiosis[446]. Thus, it is possible that downregulation of metabolism is a protective arm of the UPR in yeast that attenuates the influx of nascent, unfolded polypeptide chains into the ER during ER stress. This arm of the UPR may substitute for transcriptional[72–74] and translational[333–335] attenuation of secretory protein synthesis observed during ER stress in higher eukaryotes.

SACE is located within the promoter regions of ER resident chaperones and of *COS* genes of unknown function[444]. The role of SACE as a Hac1^{i}p target has yet to be substantiated as data has only been obtained indirectly using SACE *lacZ* reporters in yeast strains deleted for a number of bZIP transcription factors[444].

3.2.6   Regulation of signal transduction by IRE1

Both Hac1^{i}p[452] and XBP-1^{s404} autostimulate their transcription. High levels of *HAC1* mRNA were not maintained in cells lacking the *HAC1* UPRE and as such, these cells were sensitive to prolonged ER stress[452]. For *XBP-1*, impairment of this positive feedback mechanism is a genetic risk factor for bipolar disorder in humans[453]. In yeast, Ire1p is dephosphorylated by the protein phosphatase phosphatase two C 2 protein (Ptc2p)[454]. Whether this dephosphorylation is constitutive or responsive to ER stress is not known. Furthermore, as activation of the RNase domain by occupancy of the ATP binding pocket of Ire1p can bypass phosphorylation of the activation loop in Ire1p, an initial role for Ptc2p in deactivation of Ire1p seems unlikely. In mammalian cells, the sarcoma formation (Src) homology 2/3 domain containing protein NCK-1, has been shown to interact with IRE1α and to inhibit signal transduction by IRE1α, e.g. activation of extracellular signal regulated kinase (ERK)[455]. A yeast two hybrid screen identified ju-nana (Jun) activation domain binding protein 1 (JAB1) as an IRE1α interaction partner in unstressed cells, where it downregulates the UPR[454,456]. Upon ER stress JAB1 dissociates from IRE1α. Thus, JAB1 may be a negative regulator of IRE1α, i.e. sterically interfere with oligomerization or *trans* autophosphorylation of IRE1α.

3.3   PERK

The discovery of higher eukaryotic IRE1 homologs prompted a search for proteins with similar functional roles and resulted in the discovery of another type 1 transmembrane protein kinase, PERK/PEK in *C. elegans* and mammals[333–335]. Structurally, PERK is similar to IRE1, consisting of an ER luminal stress sensing domain and a cytosolic protein kinase domain. However, PERK does not contain the RNase domain found within IRE1, instead the *C*-terminal kinase domain shares substantial sequence homology with the eIF2α kinases, PKR, GCN2 and heme regulated translational inhibitor (HRI)[333–335]. The ER luminal domains of IRE1 and PERK display limited sequence homology, however, despite the absence of a PERK gene within *Saccharomyces cerevisiae*, chimeras of the ER luminal domain of PERK and the transmembrane domain and cytosolic portion of Ire1p are functional[104]. Activation of PERK results in phosphorylation and subsequent activation of its two substrates, eukaryotic translation initiation factor 2α (eIF2α) and the bZIP cap'n'collar transcription factor nuclear factor erythroid 2 (NF-E2) related factor 2 (NRF2) (Figure 9)[457].

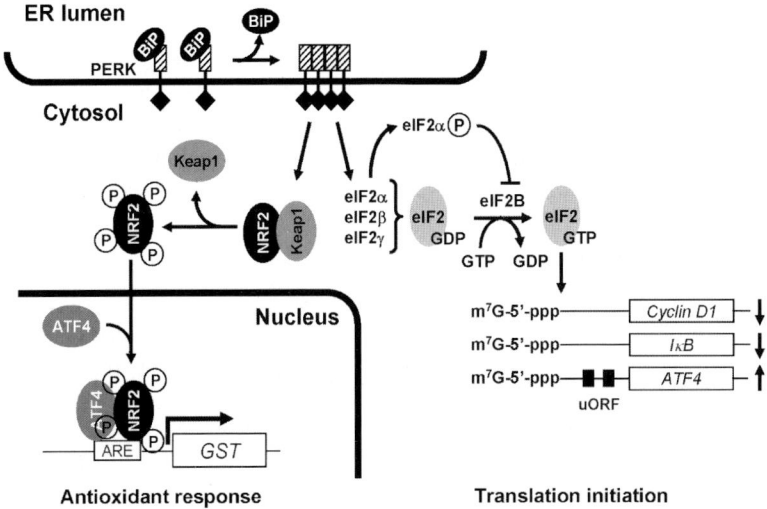

Figure 9. Signal transduction by PERK. Abbreviation: m^7G – 7-methylguanosine. Reprinted from *Current Molecular Medicine*, copyright 2006, with permission from Bentham Science Publishers.

3.3.1   Consequences of eIF2α phosphorylation

eIF2α is the α subunit of the heterotrimeric eukaryotic translation initiation complex eIF2[458] (Figure 10). Phosphorylation of eIF2α by PERK has three consequences (Figure 9). First, general translation is attenuated. Second, short lived proteins are cleared from the cell, when general translation is attenuated. Important physiological consequences are growth arrest in the G_1 phase of the cell cycle and activation of an immune and inflammatory response by the transcription factor NF-κB. Finally, translation of a subset of mRNAs, e.g. the mRNA for the bZIP transcription factor ATF4, is stimulated. How these seemingly opposite outcomes are achieved by the same regulatory event is discussed in the following sections.

3.3.2   Attenuation of general translation

eIF2 forms a heterotrimeric complex with GTP and methionine methionyl initiator tRNA (Met-tRNA$_i^{Met}$)[458]. This complex binds the small, 40S ribosomal subunit, forming the 43 S preinitiation complex (Figure 10)[458]. The eIF2·GTP·Met-tRNA$_i^{Met}$ complex recognizes the 5′ cap structure at the 5′ end of a mRNA, and initiates scanning of the mRNA in a 5′ to 3′ direction until it reaches the first start codon[458], usually AUG, in a context favorable for initiation of translation, the Kozak sequence[459] (A/GNNAUGG). Upon recognition of the start codon, GTP hydrolysis is stimulated by the GTPase activating protein (GAP) eIF5, and GDP·eIF2 is released. The large, 60 S ribosomal subunit associates with the small subunit and the complete, 80 S ribosome initiates translation. To replenish the heterotrimeric eIF2·GTP·Met-tRNA$_i^{Met}$ complex, GDP is exchanged on eIF2 for GTP. This reaction is catalyzed by the GTP exchange factor eIF2B[458]. Phosphorylation of eIF2α at

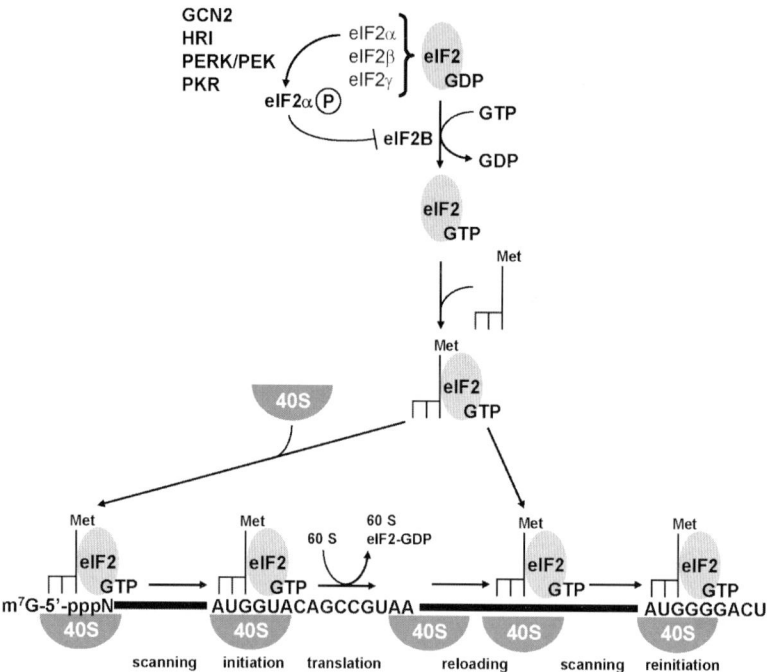

Figure 10. Control of mRNA translation through regulation of reinitiation. For simplicity only events relevant to reinitiation through modulation of Met-tRNAiMet·eIF2·GTP levels are shown. Reprinted, with permission, from the *Annual Review of Biochemistry*, Volume 74 ©2005 by Annual Reviews www.annualreviews.org.

Ser51 by PERK traps eIF2 in its GDP bound form by inhibiting the GTP exchange activity of eIF2B through formation of a stable complex between eIF2B and eIF2α. Thus, the cellular concentration of the 43 S preinitiation complex decreases, and cap dependent translation is inhibited. eIF2 is present in cells in a 10–20 fold excess over eIF2B[460]. Therefore, small changes in eIF2α phosphorylation have a profound effect on translation. Attenuation of general translation decreases the influx of nascent, unfolded polypeptide chains into the ER, and thus, protects against ER stress. Indeed, *perk*[−/−] cells are sensitive to ER stress, and, importantly, can be partially rescued by the translation inhibitor cycloheximide[333].

3.3.3   Clearance of short lived proteins from the cell

Attenuation of general translation has one important consequence: Short lived proteins are cleared from the cell, arresting growth of mammalian cells in the G_1 phase of the cell cycle and activation of an immune and inflammatory response by RelA (NF-κB) transcription factors.

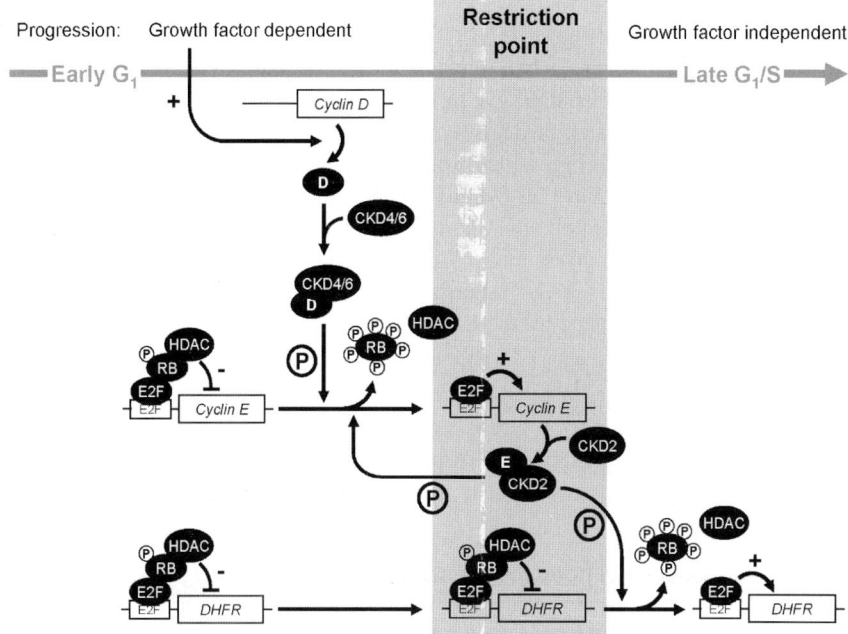

Figure 11. Restriction point control. Boxes represent genes, ellipses proteins, and a circled P a phosphorylation event.

3.3.4   G₁ cell cycle arrest

Late in G_1 phase a cell makes the decision to divide or not to divide[461,462]. This point in the cell cycle is called the restriction point (Figure 11). Passage through the restriction point is controlled by cyclin dependent kinases (CDK). CDKs are activated by association with cyclins, i.e. cyclin D activates CDK4 and CDK6, and cyclin E activates CDK2. These cyclin · CDK complexes phosphorylate the transcriptional co-repressor retinoblastoma protein (RB)[463–465]. In the hypophosphorylated state RB interacts with the E2 promoter binding factor (E2F)/differentiation-regulated transcription factor (DRTF1) transcription factor family[466–470]. The E2F family consists of six family members, E2F-1, E2F-2, E2F-3, E2F-4, E2F-5, and E2F-6, which form heterodimers with the transcription factors DRTF1 polypeptide 1 (DP-1) and DP-2[471]. The E2F transcription factors regulate transcription of genes whose products are important for entry into S phase, e.g. the dihydrofolate reductase (DHFR), thymidine kinase (TK), thymidylate synthase (TS), DNA polymerase-α (POL), CDC2, and cyclin E genes[461]. Through interaction with E2F hypophosphorylated RB recruits the transcriptional repressors HDAC1[472–474] and HDAC2[472,474] to E2F targets and represses their transcription. Hyperphosphorylation of RB by the cyclin D · CDK4/6 complex disrupts association of RB with HDAC1/2 and E2F and activates expression of cyclin E. Phosphorylation of RB and other, unknown targets by cyclin E · CDK2 further increases cyclin E expression and induces expression of genes activated by E2F (Figure 11)[461,462]. Expression of cyclin D is stimulated by growth factors and mitogens, whereas cyclin E expression is not regulated by growth factors and mitogens. Thus, until cyclin E is expressed, and then autostimulates its own expression, progression through G_1 is dependent on the presence of mitogens. At the restriction point, progression through G_1 and S phase becomes independent of mitogens. D type cyclins, i.e. cyclins D1, D2, and D3, are short lived proteins and are depleted from the cell when translation is attenuated by phosphorylation of eIF2α[475-477], resulting in arrest in G_1. Consistent with this model is that ER stress does not arrest $rb^{-/-}$ cells in G_1[477]. Furthermore, overexpression of cyclin E should bypass the G_1 arrest induced by the UPR.

3.3.5   Activation of RelA family transcription factors

The RelA transcription factor family comprises five members[478-483]: c-Rel, NF-κB1/p105/50, NF-κB2/p100/p52, RelA/p65, and RelB (Figure 12). These proteins share a Rel homology domain that is involved in homo- and heterodimerization,

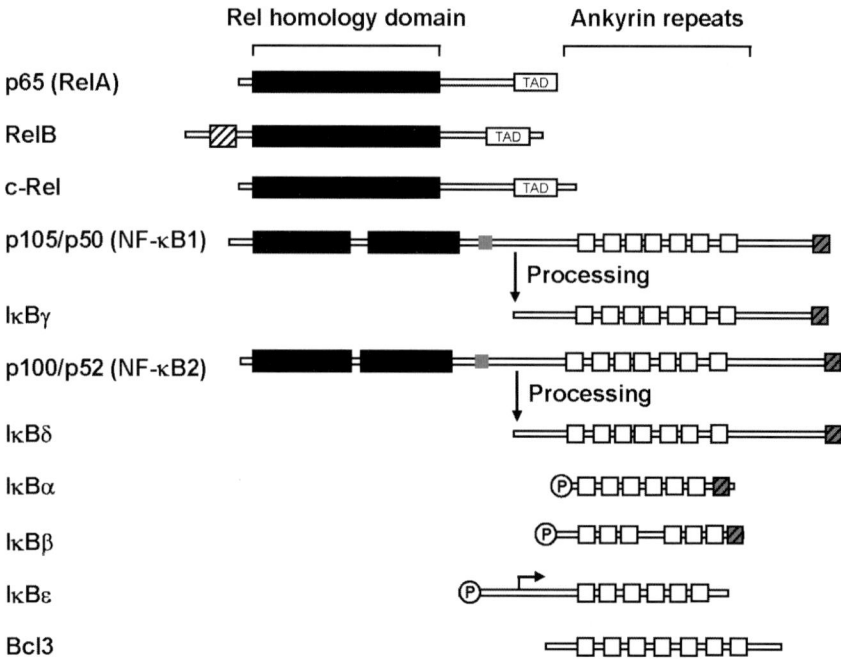

Figure 12. Rel family transcription factors. Grey boxes represent glycine rich regions, boxes with diagonal lines on a white background a leucine zipper, and boxes with diagonal lines on a grey background PEST (P – L-proline, E – L-glutamic acid, S – L-serine, and T – L-threonine) regions. Phosphorylation sites are indicated by a circled P and an alternative translation initiation site in IκBϵ by a horizontal arrow. Drawings are not to scale. Reprinted from *Current Molecular Medicine*, copyright 2006, with permission from Bentham Science Publishers

DNA binding, and interaction with IκBs. NF-κB dimers are retained in the cytosol through interaction with IκBs in unstimulated cells. Several IκBs are known: IκBα, IκBβ, IκBγ, IκBε, IκBζ, and Bcl-3 (Figure 12). IkBs contain several ankyrin repeats that mediate their interaction with NF-κB dimers. In this respect, the *C*-terminal ankyrin repeats found in NF-κB1 and NF-κB2 function as intramolecular IκBs. In stimulated cells, NF-κB dimers become phosphorylated, ubiquitinated, and after degradation of an interacting IκB, translocate to the nucleus to regulate transcription. NF-κB1 and NF-κB2 are activated by proteasomal degradation of their *C*-terminal ankyrin repeats. Thus, the activity of NF-κB, usually the prototypic heterodimer between RelA and p50, is regulated by the balance between synthesis and degradation of IκBs. Phosphorylation of eIF2α and attenuation of general translation shifts this balance towards degradation of IκBs and activation of NF-κB[484–487]. NF-κB primarily induces genes playing a role in immune and inflammatory responses, cell growth, and pro- and antiapoptotic genes[482,488,489]. Activation of an immune and inflammatory response in response to ER stress may seem surprising at first sight. However, activation of these responses may be necessary to efficiently combat viral infections, which, by inducing massive glycoprotein synthesis in the ER, activate the UPR.

3.3.6   Escape mechanisms from translational attenuation

Mounting an efficient ER stress response requires protein synthesis. How is this achieved when general translation is attenuated? Cellular mRNAs have three known mechanisms at their disposal to escape translational attenuation by eIF2α phosphorylation (Figure 13). These are referred to as reinitiation after a short upstream ORF (uORF), leaky scanning, and internal ribosomal entry sites (IRES). These mechanisms are reviewed in the following sections.

3.3.7   Reinitiation after a short uORF

Following translation of an ORF, the ribosome dissociates into its 40 S and 60 S subunits[460,490]. The 60 S subunit and the major fraction of the small subunits dissociate from mRNAs. The minor fraction of 40 S subunits that remain associated with mRNAs resume scanning along the mRNA in 5′ to 3′ direction. If the 40 S subunit is reloaded with the ternary eIF2 · GTP · Met-tRNA$_i^{Met}$ complex before it reaches the next start codon in a strong Kozak context, translation reinitiates at a downstream ORF. Whether reinitiation happens at a downstream start codon is influenced by four factors, (*i*) the concentration of the ternary eIF2 · GTP · Met-tRNA$_i^{Met}$ complex, (*ii*) the distance the 40 S subunit has to scan through between two ORFs, (*iii*) secondary structure elements between the ORFs that slow down movement of the 40 S subunit or serve as efficient exit sites for the 40 S subunit and (*iv*) whether the secondary structure of the mRNA is remodeled during scanning. Usually, two ORFs have to be close together to allow for efficient reinitiation. Thus, depending on the eIF2·GTP·Met-tRNA$_i^{Met}$ concentration, the 40 S ribosomal subunit can scan through specific secondary structure elements before it, after reloading with eIF2·GTP·Met-tRNA$_i^{Met}$, initiates translation at a downstream ORF. An example of an mRNA

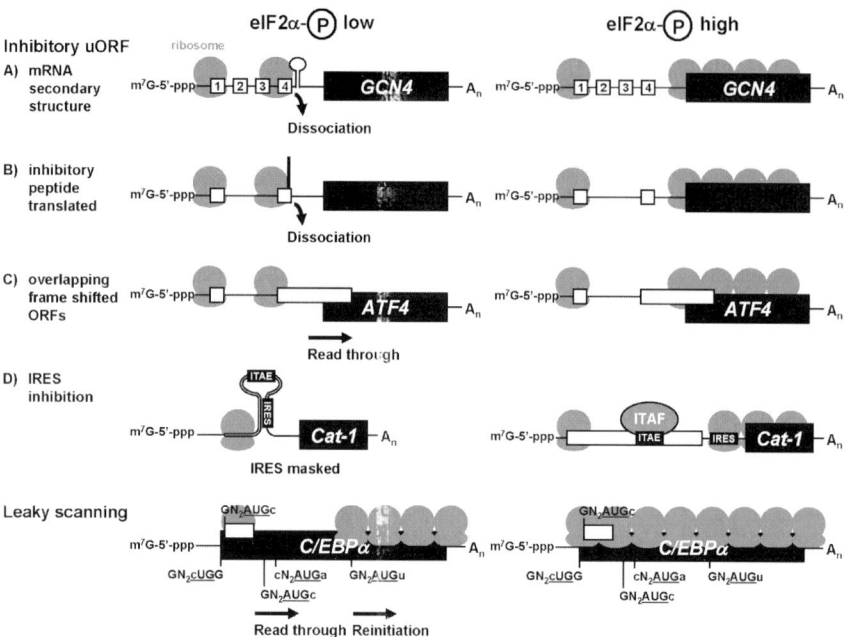

Figure 13. Escape mechanisms for cellular mRNAs from translational repression mediated by eIF2α phosphorylation. uORFs are in white, the major ORF is black, and start codons in *C/EBPα* mRNA are underlined. Nucleotides in upper case correspond to the consensus Kozak sequence (RNNAUGG, R = A, G), nucleotides in lower case do not. Staggered ORFs are in two different reading frames. Abbreviations: A_n – polyadenyl tail, ITAE – IRES *trans* acting element, ITAF – IRES *trans* acting factor, IRES – internal ribosomal entry sequence. Reprinted from *Current Molecular Medicine*, copyright 2006, with permission from Bentham Science Publishers

whose translation is regulated by reinitiation is the mRNA for the yeast bZIP transcription factor Gcn4p. The 5′-UTR of *GCN4* contains four short uORFs preceding the *GCN4* ORF (Figure 13A). Translation always initiates at the first uORF. When eIF2·GTP·Met-tRNA$_i^{Met}$ is abundant, translation is reinitiated at uORF4. Base pairing around the termination codon of uORF4 causes the 40 S subunit to efficiently dissociate from the mRNA after termination of translation of uORF4[491,492]. Thus, translation the *GCN4* ORF is repressed when eIF2·GTP·Met-tRNA$_i^{Met}$ is abundant. However, when the amount of eIF2·GTP·Met-tRNA$_i^{Met}$ decreases, the 40 S subunit scans further along the mRNA and through uORF4, resulting in reinitiation at the *GCN4* start codon. Alternatively, the inhibitory upstream ORF may encode a peptide that promotes exit of the 40 S subunit from the mRNA (Figure 13B), or the uORF and the major ORF overlap (Figure 13C). Thus, phosphorylation of eIF2α, which inhibits GDP GTP exchange on eIF2 and decreases the cellular eIF2·GTP·Met-tRNA$_i^{Met}$ concentration activates translation of *GCN4*. In the mammalian UPR, translation of the mRNA encoding the bZIP transcription factor ATF4 is regulated in a similar way[493–495].

3.3.8   Leaky scanning

Leaky scanning is the second known major mechanism to escape translational attenuation[490]. Leaky scanning relies on the use of alternative start codons, i.e. ACG, CUG, and GUG in a strong Kozak context and of the regular start codon AUG in a weak Kozak context. If the scanning 43 S preinitiation complex encounters an AUG start codon in a strong Kozak context, nearly all ribosomes will initiate at this site. However, if the start codon is weak, only a fraction of the scanning 43 S preinitiation complex will recognize this start codon and initiate translation. Those 43 S preinitiation complexes that did not recognize this start codon will continue scanning along the mRNA. Thus, leaky scanning allows initiation at downstream start codons if upstream start codons are in a weak context. Translation of mRNAs encoding the bZIP transcription factors C/EBPα and C/EBPβ is regulated by leaky scanning (Figure 13)[460,490,496–498]. The phosphorylation status of eIF2α influences recognition of start codons in these mRNAs[497,498]. High levels of the ternary complex eIF2·GTP·Met-tRNA$_i^{Met}$ favor initiation at the small out of frame uORF, subsequent scanning through the in frame start codons GN$_2$AUGc and CN$_2$AUGa, and initiation at the distal start codon GN$_2$AUGu. Thus, a truncated bZIP transcription factor lacking its transactivation domain, liver-specific transcriptional inhibitor protein (LIP), is synthesized, that functions as a repressor of other C/EBPs[496]. When eIF2·GTP·Met-tRNA$_i^{Met}$ levels decrease, translation preferentially initiates at the two in frame start codons GN$_2$AUGc and CN$_2$AUGa. Liver-specific transcriptional activator protein (LAP) that in addition to the DNA binding domain and leucine zipper of LIP also contains a transactivation domain is synthesized under these conditions[499]. Mice homozygous for knock-in of the Ser51Ala mutation in eIF2α that abolishes phosphorylation of eIF2α by GCN2, HRI, PERK, and PKR, die shortly after birth because of fatal hypoglycaemia caused by defective induction of gluconeogenic enzymes[494]. This phenotype is similar to the phenotype of c/ebpα[−/−] mice[500], arguing that regulation of translation start site selection in C/EBPα mRNA is required for proper development.

3.3.9   IRES

Finally, IRES represent the third known mechanism to escape translational attenuation induced by eIF2α phosphorylation (Figure 13). In essence, the presence of an IRES in a mRNA renders its translation independent of its cap structure. IRES are commonly found in viral mRNAs and allow viral mRNAs to escape shut-off of host cell protein synthesis by phosphorylation of eIF2α by PKR. IRES are also found in some cellular mRNAs[501], including BiP mRNA[502,503]. BiP translation is induced by heat shock, but reporter constructs whose expression is controlled by the BiP IRES were unresponsive to ER stress[504]. In CAT-1 mRNA, the 5′-UTR is remodeled in response to eIF2α phosphorylation, unmasking an IRES when eIF2α phosphorylation is induced[505]. Whether ER stress induces translation of CAT-1 mRNA is unknown.

3.3.10   Pathways downstream of ATF4

ATF4 homodimerizes and forms heterodimers with other bZIP transcription factors. Through heterodimerization with NRF2, ATF4 induces an antioxidant response[506]. C/EBP·ATF4 heterodimers induce the bZIP transcription factors ATF3[507], CHOP[508], the membrane protein hyperhomocysteinemia induced ER stress responsive protein (HERP)/methyl methanesulfonate inducible fragment 1 (MIF1)[509], phosphoenolpyruvate carboxykinase (PEPCK)[510], and the neuronal cell death-inducible putative kinase (NIPK)/tribbles-related protein 3 (TRB3)[511]. ATF3 homodimers repress their own transcription[512], and transcription of CHOP[513,514], PEPCK, and the fructose bisphosphatase gene[515].

The bZIP transcription factor CHOP heterodimerizes with the bZIP proteins ATF3[516], ATF4[516], and the C/EBP proteins[517]. CHOP mainly functions as a repressor of these transcription factors, because two conserved residues in the basic region are mutated to proline in CHOP[517]. However, some CHOP heterodimers can also activate transcription[64,511,518]. Expression of CHOP is activated by all major pathways of the UPR[519], with the PERK eIF2α ATF4 pathway being the most important[508]. CHOP, and ATF4, are negatively regulated by TRB3[511,518]. The major function of CHOP in the UPR is to induce apoptosis by repressing transcription of Bcl-2[520]. $chop^{-/-}$ [520,521], $c/ebp\beta^{-/-}$ [521], but also TRB knockdown[518] cells, are resistant to ER stress induced apoptosis.

3.3.11   Consequences of NRF2 phosphorylation

NRF2 is located within the cytosol of unstressed cells in a complex with the cytoskeletal anchor protein kelch-like Ech-associated protein 1 (KEAP1)[457]. NRF2 is phosphorylated by PERK during ER stress, causing it to dissociate from Keap1. Subsequently, phosphorylated NRF2 translocates to the nucleus where it, as a heterodimer with other bZIP transcription factors, such as ATF4[522], c-Jun, Jun-B and Jun-D[506] activates transcription of genes containing an antioxidant response element (ARE) in their promoter[506,523,524]. Heterodimerization of NRF2 with the small bZIP protein MafK inhibits NRF2[506,524]. Enzymes involved in phase II metabolism of xenobiotics to reduce cellular stress typically contain an ARE. These include the A1 and A2 subunits of glutathione S-transferase, NAD(P)H:quinone oxidoreductase, γ-glutamylcysteine synthetase, heme oxygenase 1 and UDP-glucuronosyl transferase[506]. Reactive oxygen species accumulate in response to ER stress in $perk^{-/-}$ cells[77]. $nrf2^{-/-}$ cells show decreased viability following ER stress compared to wild type cells[457,525]. Thus, it appears that ER stress induces an imbalance in cellular redox metabolism. In support of this idea, ER stress activates and antioxidants deactivate the redox regulated transcription factor NF-κB[526]. Two sources for generation of reactive oxygen species during ER stress have been proposed, increased oxidative phosphorylation in mitochondria to meet the elevated energy demand for additional folding attempts of unfolded proteins[527] and oxidative protein folding in the ER itself[64]. Accumulation of reactive oxygen species in yeast cells defective in upregulation of ER chaperone genes in response to ER stress was attenuated[527], supporting these ideas.

3.3.12   Regulation of PERK signaling

The inhibitory effect of PERK on translation is transient to allow for recovery from prolonged ER stress. Protein phosphatase 1 (PP1) counters phosphorylation of eIF2α by PERK[528–530]. Five tissue specific isoforms of PP1 are known, PP1α_1, PP1α_2, PP1β, PP1γ_1, and PP1γ_2[531]. ~50 regulatory subunits regulate the catalytic subunit of PP1, PP1c[531]. Four regulatory subunits target PP1c to dephosphorylate eIF2α: Avian sarcoma virus NL-S[532], herpes simplex virus protein $\gamma_1$34.5/infected cell protein 34.5 (ICP34.5)[533–536], constitutive repressor of eIF2α phosphorylation (*CReP*)/protein phosphatase 1 regulator 15N (PPP1R15B)[529], and GADD34/ myeloid differentiation primary response gene 116 (MyD116)[528,532,537]. *CReP* is a constitutive activator of PP1[529], NL-S and $\gamma_1$34.5 are a viral defense mechanism against translational shut-down of host cell protein synthesis by phosphorylation of eIF2α. GADD34 is not detected in unstressed cells[529], but induced by ATF4 late in the UPR[528,530]. GADD34 itself is inhibited by several other proteins. These include the DnaJ co-chaperone B cell leukemia/lymphoma 2 (Bcl-2) associated athanogene protein 1 (BAG-1)[538], inhibitor-1[532], the chromatin remodeling factor human sucrose non-fermenting 5 (hSNF5)/integrase-interacting protein 1 (INI1)[539], the chimeric leukemic human trithorax (HRX) fusion protein[540], and the Src-related tyrosine kinase lymphocyte protein tyrosine kinase (Lck)/yes-related tyrosine kinase (Lyn)[541]. How this additional level of regulation affects translational attenuation by PERK in response to ER stress is currently unknown. In addition to PP1, a calyculin A sensitive phosphatase is recruited onto eIF2 by the SH2/3 homology domain containing adaptor protein NCK-1[542]. The first and third SH3 domains of NCK-1 interact with the *C*-terminal portion of the β subunit of eIF2[543].

PERK is also inhibited by interaction with the tetratricopeptide repeat (TPR) containing cytosolic DnaJ cochaperone 58 kDa PKR inhibitor (P58^{IPK})[544–546]. XBP-1^{s}[545] and ATF6[545,546] induce P58^{IPK}. The TPR domain of P58^{IPK}, but not its DnaJ domain, is required for inhibition of PKR[547]. The 52 kDa regulator of P58^{IPK} (P52^{rIPK}) interacts with and inhibits P58^{IPK}[548,549]. P52^{rIPK} contains a charged domain with homology to HSP90[548,549]. Induction of P58^{IPK} in the UPR serves the same purpose as activation of GADD34, that is, to limit the time window of translational attenuation in response to ER stress to ensure that an efficient response to prolonged ER stress can be mounted. Exactly how this is achieved is poorly understood.

4.   ADAPTIVE SIGNAL TRANSDUCTION MECHANISMS EMPLOYED BY THE UPR

Initially, the UPR is a homeostatic response to ER stress and induces adaptive responses to this stress situation. Only if this fails, or in response to prolonged ER stress, apoptosis is induced. However, evidence for a biphasic, i.e. an initial adaptive response that then turns into an apoptotic response, has not yet been reported. Adaptive responses regulated by the UPR either increase the protein folding capacity of the ER or decrease the protein folding demand imposed on the ER (Figure 1). These responses include the transcriptional activation of genes encoding ER chaperones and foldases, stimulation of phospholipid and membrane biosynthesis to increase the size of the ER and to decrease the unfolded protein concentration in the ER, attenuation of synthesis of secretory proteins by inhibition of transcription and translation, stimulation of ERAD, and activation of an antioxi-dant response. Translational attenuation and the antioxidant response were discussed before. Repression of transcription of genes encoding secretory proteins, called repression under secretion stress (RESS), has been observed in yeast, filamentous fungi, and plants. RESS is still poorly understood. Here we will discuss how the UPR induces chaperone and foldase expression, stimulates ERAD and phospholipid synthesis.

4.1   Chaperone and Foldase Expression

A number of transcriptional profiling studies have identified genes which are upreg-ulated in response to UPR signaling in both yeast[550] and mammalian cells[77,494,551,552]. These include ER chaperones, foldases, and genes which are specifically required as intermediates in the secretory pathway. In particular, ATF6[551] and XBP-1[552] were identified as important in upregulation of gene transcription in response to unfolded proteins. Cells treated with *ATF6α* or *ATF6β* siRNA did not reveal any specific transcriptional targets for these regulators, although silencing of *ATF6α* in *xbp-1⁻/⁻* cells significantly impaired induction of certain UPR targets including GRP94[552], but not others such as CHOP and GRP78. This suggests that while the genes induced by ATF6 and XBP-1 are likely to overlap, the genes up-regulated by PERK and ATF6, with the exception of *CHOP*, are likely to be separate. These signaling pathways are likely to play distinct roles in the mammalian UPR[551].

4.2   ERAD

In addition to upregulation of ER resident chaperone proteins, the UPR upreg-ulates proteins required for ERAD[423,550,553-555], thereby increasing the clearance of misfolded proteins and decreasing the protein load within the ER. Proteins that are recognized as terminally misfolded are directed to the cytosol for protea-somal degradation in a multistep process called ERAD[556-558]. ERAD starts when a polypeptide chain is recognized as unfolded by BiP, PDI, or EDEM[559]. The protein is then transported to the ER luminal side of the retrotranslocation pore, which in many, but not all, cases is the Sec61p translocon[217]. This already is no trivial task, because enzymes involved in quality control can be located in post-ER, pre-Golgi compartments, e.g. UGGT[560], and some soluble ERAD substrates have to be retrieved from the *cis*-Golgi[561-563]. After retrotranslocation, the protein needs to be recognized as a proteasomal substrate again, polyubiquitinated, deglycosy-lated, and targeted towards the proteasome. As the protein emerges from the ER, it is modified by the addition of ubiquitin (Ub), a small 76 amino acid protein found in all eukaryotic cells, which targets the protein for degradation by the 26S proteasome. Modification of a protein with Ub is a three step process. The first step, activation of ubiquitin, is catalyzed by a class of enzymes known as the Ub activating enzymes/E1s which bind and adenylate the *C*-terminus of Ub. This is followed by Ub conjugation and ligation, where the *C*-terminus of activated Ub is attached to the target protein, catalyzed by two groups of enzymes, the Ub conjugating enzymes (Ubc)/E2s and Ub ligases (Ubl)/E3s (Figure 14A)[564,565]. In yeast, two membrane bound ubiquitin ligase complexes have been described (Figure 14B). The 3-hydroxy-3-methylglutaryl-coenzyme A (HMG-CoA) reductase degradation 1 protein (Hrd1p)/degradation in the ER 3 protein (Der3p) multimembrane spanning really interesting new gene (RING) finger ubiquitin ligase[566,567] cooperates with Ubc7p and Ubc1p[554,566], and a putative targeting protein, Hrd3p. The second RING finger ubiquitin ligase degradation of α2 10 protein (Doa10p) utilizes Ubc6p and a complex of coupling of ubiquitin conjugation to ER degradation 1 protein (Cue1p) and Ubc7p as ubiquitin conjugating enzymes[568]. Mammalian counterparts for these two RING finger ubiquitin ligases exist[569–571]. In mammals, additional soluble ubiquitin ligases participate in ERAD, e.g. CHIP[572,573], Parkin[574–576], and the suppressor of kinetochore protein mutant 1 protein (Skp1)·cullin 1·F box (SCF) complex. Interestingly, two F box proteins, F box protein that recognizes sugar

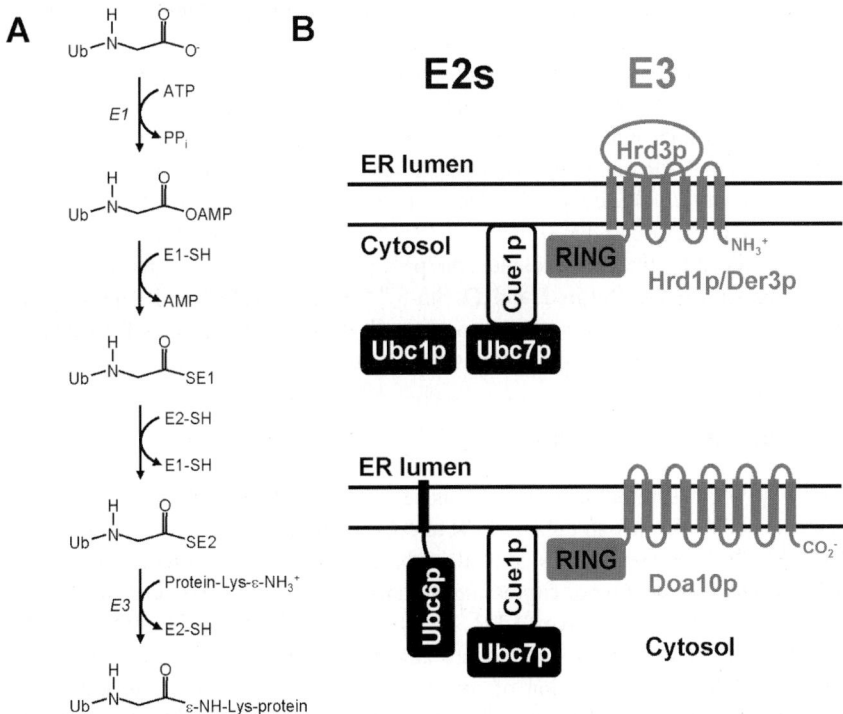

Figure 14. ERAD. A Ubiquitination mechanism. B Ubiquitination complexes found in the ER membrane in yeast. There is only one gene encoding an ubiquitin activating enzyme in the yeast genome

chains (FBS1)/F box protein X 2 (FBX2)/neural F box protein of 42 kDa (NFB42) and FBS2/FBX6/F box gene 2 (FBG2) recognize N-glycosylated proteins, indicating that SCF^{FBS} ubiquitinates retrotranslocated N-glycoproteins[577,578]. After retrotranslocation, the AAA ATPase Cdc48p/p97/valosin-containing protein (VCP) disassembles protein aggregates prior to their degradation by the proteasome[579–583]. In yeast, Cdc48p associates with Hrd4p/nuclear protein localization 4 protein (Npl4p)[584,585] and ubiquitin fusion degradation 1 protein (Ufd1p)[586], two ubiquitin binding proteins. Finally, proteasomes have to localize close to the ER membrane. In yeast, the vicinity of the ER membrane and of the nuclear envelope is their default localization[587]. In mammalian cells they are dispersed throughout the nucleoplasm and the cytosol[588]. A specialized proteasome, the immunoproteasome preferentially localizes to the ER membrane[589]. ER resident transmembrane proteins containing ubiquitin like domains such as HERP may serve as anchors for proteasomes in the ER membrane[423].

Currently, three aspects of ERAD are known to be regulated by the UPR, recognition of unfolded proteins, ubiquitin ligase activities, and recruitment of proteasomes to the ER membrane in mammalian cells. EDEM, the α-mannosidase-like protein that extracts persistently unfolded proteins from the CNX/CRT cycle, is induced by XBP-1^s, but not ATF6[559]. Kinetically, XBP-1 is activated after ATF6 in the UPR, leading to the proposal of two phases within the UPR. The first of these, mediated by ATF6, induces chaperones and foldase expression to promote folding of slowly folding proteins. The second phase is bi-directional, where both protein folding and ERAD proceed simultaneously. This phase is initiated by splicing of *XBP-1*, which then induces EDEM[559]. From synthetic lethalities between defects in ERAD and UPR signaling in yeast, it can be concluded that ERAD and the UPR are constitutively active[554,590]. Furthermore, strains in which ERAD is compromised display elevated UPR signaling[396,424,591]. Ubiquitin conjugating enzymes and ligases induced by the UPR include Ubc7p[554], Cue1p[554], Hrd1p[554], and mammalian homologs of Der1p, Derlin-2 and Derlin-3[592]. Finally, ATF6 and ATF4 induce HERP[423,555]. *herp*$^{-/-}$ embryonic carcinoma cells displayed increased UPR signaling and were more vulnerable to ER stress than wild type cells[591].

4.3   Regulation of ER Size

Cells with high secretory activity such as hepatocytes, plasma cells[593] or pancreatic exocrine cells[65] display an enlarged ER to cope with the increase in protein folding demand. The increase in ER size is also observed in normal cells that are subject to ER stress[95] or which are expressing mutant or folding incompetent proteins[594]. In addition to being sensitive to ER stress, *ire1*Δ, *hac1*Δ, and *rlg1-100* (H148Y Trl1p) yeast cells are also *myo*-inositol auxotrophs[329,331,398,595]. This phenotype is linked to defects in expression of genes related to phospholipid metabolism, in particular the gene encoding *myo*-inositol 3-phosphate synthase (*INO1*)[596]. *HAC1* and *IRE1* mediate induction of *INO1* in response to ER stress[550,597]. A functional UPR was required for induction of membrane proliferation when certain membrane proteins were overexpressed[597,598], but this requirement for the UPR in membrane proliferation is not general[599]. Yeast grown in the absence of *myo*-inositol require *HAC1* and *IRE1* for transcription of *INO1*[597]. In line with these findings, a direct role for the UPR in controling proliferation of ER membrane synthesis through directly regulating *INO1* expression was proposed[597]. Expression of *INO1* is repressed by the overproduction of *myo*-inositol 1 protein (Opi1p)[600], a bZIP protein. Through formation of a heterodimer with Opi1p Hac1[i]p may interfere with repression of *INO1*, providing an explanation for the requirement of *IRE1*, *HAC1* and a functional tRNA ligase for activation of *INO1* by the UPR. However, experimental evidence for this hypothesis is not available.

More recent data suggests that the UPR controls *INO1* and phospholipid metabolism in yeast in a more complex manner. Overexpression of acetyl-CoA synthetase regulation 1 protein (Acr1p), an inner mitochondrial membrane protein, did not cause ER membrane proliferation, but nevertheless was lethal in *ire1*Δ cells[598]. This lethality was rescued when the cells were grown on oleate[598]. Further, overexpression of the peroxisomal membrane protein peroxisome related 15 protein (Pex15p) induced ER membrane proliferation and was lethal in *ire1*Δ cells, but again, this lethality was rescued when cells were grown on oleate[598]. Taken together, these, albeit circumstantial, observations suggest that the UPR does not regulate ER membrane proliferation and that the lethalities seen in *ire1*Δ cells are caused by a more general disturbance in lipid metabolism in these cells. Henry and coworkers[601] provided more direct evidence for a general disturbance of lipid metabolism in UPR deficient yeast cells. *sec14[ts]cki1* cells display an overproduction of *myo*-inositol (Opi⁻) phenotype caused by elevated phosphatidic acid levels (Figure 15). In Opi⁻ cells *IRE1* and *HAC1* were not required for expression or regulation of *INO1* in response to changes in lipid metabolism, suggesting that increased phosphatidic acid levels can bypass the requirement for *IRE1* and *HAC1* in regulation of *INO1*. In *ire1*Δ and *hac1*Δ cells cytidine diphosphate (CDP) diacylglycerol levels were increased, and levels for phosphatidic acid and phosphatidylinositol decreased[601]. Interestingly, Opi1p is tethered to the ER membrane and held there in an inactive state by binding to phosphatidic acid and the ER membrane protein suppressor of choline sensitivity 2 protein (Scs2p)[602]. *scs2*Δ cells are *myo*-inositol auxotrophs and overexpression of Scs2p suppresses the *myo*-inositol auxotrophy of *ire1*Δ cells[603]. These data explain how Opi1p senses *myo*-inositol levels and how *IRE1* and *HAC1* regulate *INO1* expression. When *myo*-inositol levels rise, phosphatidic acid is converted to phosphatidylinositol, releasing Opi1p from the ER membrane. Opi1p then translocates into the nucleus and represses *INO1*. Activation of *INO1* correlates with location of the *INO1* gene to the nuclear periphery[604]. In *ire1*Δ and *hac1*Δ cells phosphatidic acid content is decreased, resulting in activation of Opi1p and repression of *INO1*.

In mammalian cells ATF6 (cf. above) and XBP-1ˢ have been linked to regulation of ER size and lipid metabolism. Overexpression of XBP-1ˢ in mature B cells increased cell size, and organelle content[605]. Knockdown of XBP-1 with microRNAs slightly diminished the ER[605]. In addition, overexpression of XBP-1ˢ increased

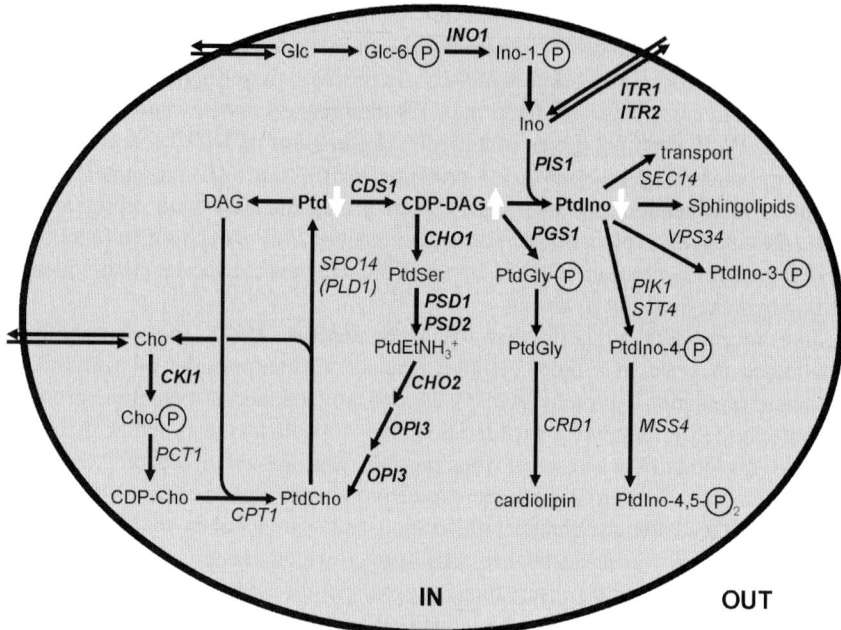

Figure 15. Phospholipid biosynthesis. Phospholipids whose levels are altered in *ire1*Δ and *hac1*Δ strains during *myo*-inositol starvation are in white, and arrows (↓ or ↑) indicate if their levels are decreased or increased[601]. Abbreviations: Cho – choline, DAG – diacylglycerol, EtNH$_3^+$ - ethanolamine, Glc – D-glucose, Gly – glycerol, Ino – myo-inositol, Ptd - phosphatidic acid or phosphatidyl, Ser – serine. Reprinted from *Mutation Research*, *569*, Martin Schröder and Randal J. Kaufman, ER stress and the unfolded protein response, Copyright (2005), with permission from Elsevier

protein content, mitochondrial mass and function, ribosome number and total protein synthesis. Thus, because of these pleiotrophic effects of XBP-1^s on cell physiology, it is not clear if the effects of XBP-1^s on organelle content reflect a function directly attributable to XBP-1^s. In NIH-3T3 cells XBP-1^s induced synthesis of the primary phospholipid of the ER membrane in mammalian cells, phosphatidylcholine[606]. These cells also displayed an increased surface area and volume of the rough ER and increased activity of the CDP-choline pathway and phosphatidylcholine biosynthesis.

5. REGULATION OF APOPTOSIS BY THE UPR

The initial response to ER stress appears to be protective. Overexpression of BiP, GRP94, calreticulin, or PDI reduce cell death in response to oxidative stress, disturbances of calcium homeostasis and hypoxia[607,608]. Persistent accumulation of misfolded proteins elicits apoptotic pathways resulting in cell death[75,609]. However,

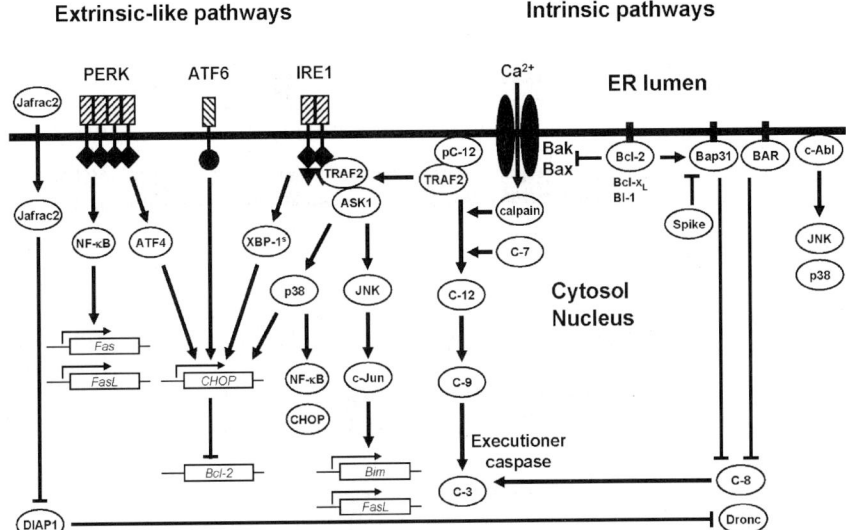

Figure 16. Apoptotic mechanisms triggered by ER stress. Abbreviations: C – caspase, pC – procaspase. Reprinted from *Current Molecular Medicine*, copyright 2006, with permission from Bentham Science Publishers

causative evidence for a biphasic UPR, i.e. an initial protective UPR that changes to an apoptotic UPR, is not available. Formally, the adaptive response to ER stress may solely have the purpose to give cells the time required to induce apoptosis effectively.

Apoptosis is controlled by two main pathways, an intrinsic and an extrinsic pathway (Figure 16)[63,64]. The extrinsic pathway responds to extracellular stimuli and is activated by self association of cell surface receptors and initiation of a caspase cascade. Caspases are a family of cysteine dependent L-aspartate specific proteases produced as inactive zymogens that require cleavage for activation. Topologically, the ER is equivalent to the extracellular space. Thus, in the UPR, apoptosis induced through IRE1, PERK, and ATF6 can be considered to be the equivalent of the extrinsic pathway, the extrinsic-like pathway. The intrinsic pathway responds to intracellular insults such as DNA damage and is regulated by both pro- and anti-apoptotic Bcl-2 protein family members (Figure 17). Activation of proapoptotic family members, i.e. Bcl-x_L/Bcl-2 associated death promoter (Bad), Bcl-2 homologous antagonist/killer (Bak), and Bcl-2 associated X protein (Bax), results in their oligomerization and insertion into the outer mitochondrial membrane. This induces the formation of a pore and subsequent cytochrome *c* release from mitochondria. The presence of cytochrome *c* in the cytosol enables a complex to form between Apaf-1 and procaspase 9 leading to initiation of the caspase cascade and activation of the executioner caspase caspase 3[610]. The intrinsic pathway also promotes apoptosis in response to ER stress.

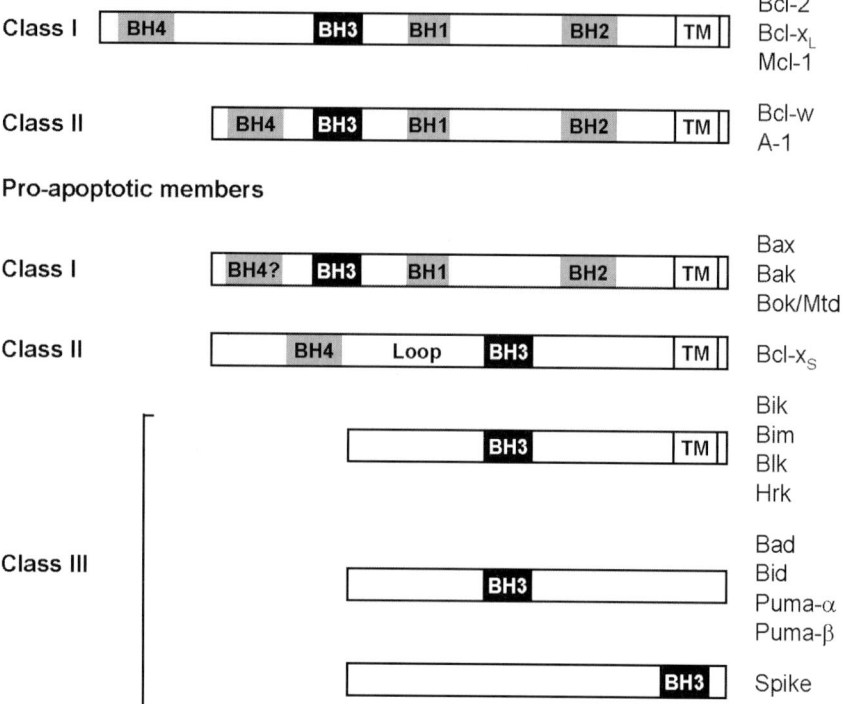

Figure 17. Primary structure of Bcl-2 family members. Reprinted from *Current Molecular Medicine*, copyright 2006, with permission from Bentham Science Publishers

5.1 The Extrinsic-like Pathway

The extrinsic-like pathway that regulates apoptosis in response to ER stress can be considered to consist of three upstream elements, ATF6, IRE1, and PERK (Figure 16). How these transmembrane proteins promote apoptosis in response to ER stress is discussed in the following sections. In addition, release of proteins from the ER into the cytosol or the nucleus, in analogy to release of cytochrome *c* from mitochondria into the cytosol, may activate apoptosis in response to ER stress. The only known example for this mechanism comes from *Drosophila*. Here, the thioredoxin peroxidase JAFRAC2 is released into the cytosol in response to DNA damage or inhibition of protein folding in the ER, binds to the *Drosophila* inhibitor of apoptosis protein 1 (DIAP1), and disrupts a complex between DIAP1 and the caspase *Drosophila* Nedd2 like caspase (Dronc), resulting in activation of Dronc[611].

5.1.1   ATF6

ATF6 promotes apoptosis through induction of the proapoptotic transcription factor CHOP[519]. However, the contribution of ATF6 to induction of CHOP appears to be minor[519]. Thus, ATF6 largely has a protective function in the UPR. CHOP inhibits expression of antiapoptotic Bcl-2[520]. $chop^{-/-}$ cells are resistant to ER stress induced apoptosis[520].

5.1.2   IRE1

IRE1 promotes apoptosis through three pathways: Splicing of $XBP-1^u$ mRNA, activation of mitogen activated protein kinase (MAPK) signaling, and activation of caspases bound to the ER membrane.

5.1.3   Activation of apoptosis by XBP-1

Through induction of CHOP, XBP-1 promotes apoptosis[519]. Overexpression of RNase defective IRE1 mutants suppressed cell death induced by homocysteine[612], suggesting that the proapoptotic function of XBP-1 is of physiological relevance.

5.1.4   Activation of MAPK signaling

IRE1α forms a proapoptotic, heterotrimeric complex with tumor necrosis factor receptor associated factor 2 (TRAF2)[613] and apoptosis signal-regulating kinase 1 (ASK1)[99]. Kinase defective IRE1α mutants did not interact with TRAF2, suggesting that TRAF2 specifically interacts with oligomeric, phosphorylated, and active IRE1α[613]. $ask1^{-/-}$ primary neurons are protected from apoptosis[99]. Dominant negative mutants of TRAF2 show decreased cell death in response to ER stress induced by homocysteine[612]. The IRE1α·TRAF2·ASK1 complex activates a MAPK cascade to promote apoptosis (Figure 18). The MAPK kinase kinase (MAPKKK) ASK1 activates two MAPKK modules, consisting of MAP or ERK kinase 2 (MEK3)/MAP kinase kinase 3 (MKK3), MEK6/MKK6, and MEK4/MKK4/stress activated protein kinase (SAPK) or ERK kinase-1 (SEK1)/MEK7. These MAPKKs then activate the MAPKs p38 and c-Jun N-terminal kinase (JNK). MEK4 is a JNK kinase, whereas MEK3 and MEK6 preferentially activate p38[614]. Four isoforms of p38 are known, p38α, p38β, p38γ, and p38δ[614]. MEK6 shows no selectivity for these isoforms, but MEK3 preferentially accepts p38α and p38β as substrates[614]. Signaling specificity may be encoded by the activation loop in p38 through selective interaction of MEK3 with different activation loops in p38 isoforms[615,616]. In TNFR signaling ubiquitination of TRAF2 by the ubiquitin conjugating enzyme UBC13 is required for activation of JNKs, but not for activation of p38 or NF-κB[617]. A similar event may regulate signaling specificity of the IRE1·TRAF2·ASK1 complex in the UPR. p38 potentiates the function of several transcription factors, i.e. ATF1, ATF2, CHOP, Elk-1, E-twenty six specific (Ets-1), muscle-specific enhancer 2A (MEF2A), NF-κB, p53, and sphingolipid activator protein 1 (Sap-1) (Figure 18)[618,619]. In addition, p38 activates several mitogen kinases[614,619]. JNKs also phosphorylate and potentiate the function of several transcription factors, i.e. the bZIP transcription

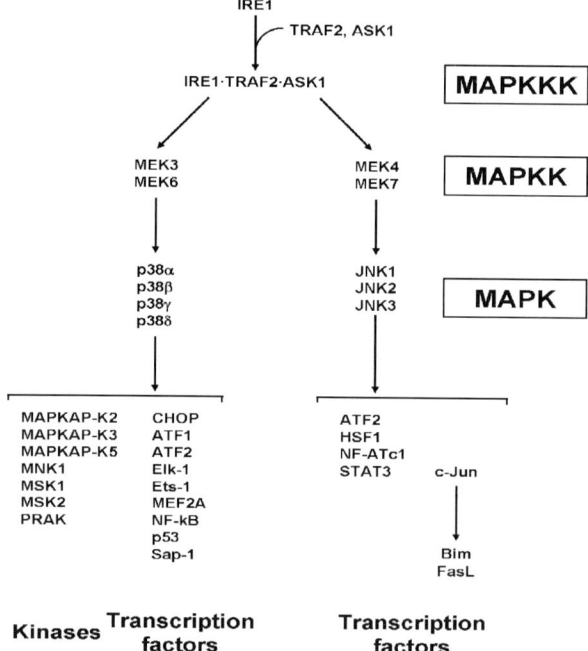

Figure 18. The MAP kinase signaling pathway downstream of IRE1. Abbreviations: MAPKAP-K2/3 – MAPK activated protein kinase 2/3, MNK – MAP interacting kinase, MSK – mitogen and stress activated protein kinase, NF-ATc1 – nuclear factor in activated T cells c1, PRAK – p38 regulated/activated kinase, STAT – signal transducer and activator of transcription. Reprinted from *Current Molecular Medicine*, copyright 2006, with permission from Bentham Science Publishers

factor c-Jun[620]. Phosphorylation of proapoptotic Bcl-2 family proteins, i.e. Bcl-2 interacting mediator of cell death (BIM), by JNK promotes apoptosis[621,622]. *ask1*[−/−] cells are defective in activation of JNKs, showing that ASK1 is the major activator of JNKs in response to ER stress. Again, both p38 and JNKs can promote pro- and antiapoptotic responses. How signaling specificity of these pathways is established during the UPR is currently unknown.

5.1.5   *Activation of ER associated caspases*

Procaspase 12 is associated with the ER membrane[623] and interacts with TRAF2[624]. Upon activation and autophosphorylation, IRE1α sequesters TRAF2 from procaspase 12, resulting in clustering and activation of this caspase[624]. ER stress induces procaspase 12[625]. A second negative regulator of caspase 12 is the microsome associated melanoma associated antigen 3 (MAGE-3) which inhibits activation of procaspase 12 by ER stress[626]. The proenzyme is activated by overexpression of caspase 12[626]. A second mechanism for activation of procaspase 12 is its cleavage by caspase 7 after translocation of caspase 7 from the cytosol to the ER[625]. Activated caspase 12 then proteolytically activates procaspase 9, and activated caspase 9 activates the executioner caspase caspase 3/cysteine protease of 32 kDa (CPP32)[625,626]. Caspase 3 cleaves the antiapoptotic protein Bcl-2, making cell death an irreversible process[627]. This apoptotic pathway was reported to be independent of apoptotic protease activating factor-1 (APAF-1) and mitochondrial cytochrome *c* release[626]. Activation of procaspase 12 by ER stress seems to be relevant for induction of apoptosis in response to ER stress in mice.

In human cells the *CASPASE 12* gene is disrupted by a nonsense mutation and inactive[628]. Furthermore, inactivating mutations are present in the small percentage of humans that carry an otherwise intact *CASPASE 12* ORF[629]. In human cells, caspase 4 may substitute for caspase 12. Procaspase 4 is associated with the ER membrane and cleaved when apoptosis was induced by ER stress, but not other apoptotic stimuli[630]. Knock-down of caspase 4 protected cells from apoptosis induced by ER stress or expression of amyloid β (Aβ)[629].

5.1.6 PERK

Perk promotes apoptosis through induction of ATF4 and possibly activation of NF-κB. ATF4 is the major activator of *CHOP* expression in the UPR[494]. ATF4·NFR2 heterodimers induce an antioxidant response and thus protect from reactive oxygen species[457,506,525]. As reactive oxygen species induce apoptosis[75], ATF4 fulfills both pro- and antiapoptotic functions in the UPR. The same may be true for NF-κB. NF-κB is generally considered an antiapoptotic transcription factor[482]. However, NF-κB also regulates expression of proapoptotic genes, i.e. *Fas* and *Fas ligand* (*FasL*)[482]. In contrast to ATF4, the contribution of NF-κB to regulation of apoptosis by the UPR has not yet been studied in detail.

5.2 Bcl-2 Protein Family Members in the Intrinsic Pathway

An antiapoptotic and a proapoptotic subfamily make up the Bcl-2 protein family (Figure 17). These proteins contain Bcl-2 homology (BH) domains of which four types (BH1 – BH4) are distinguished. Antiapoptotic Bcl-2 proteins are Bcl-2, Bcl-x_L, Bcl-w, myeloid cell leukemia 1 (Mcl-1), and A-1. Proapoptotic subfamily members are Bcl-2 associated X protein (Bax), Bcl-2 homologous antagonist/killer (Bak), matador (Mtd)/Bcl-2 related ovarian killer (Bok), Bcl-x_S, Bcl-2 interacting killer (Bik)/New Bcl-2 interacting killer (Nbk), hara-kiri (Hrk), Bik-like killer (Blk), Bim, Bcl-x_L/Bcl-2 associated death promoter (Bad), and Bid (Figure 17)[631]. The activity of these proteins is determined through formation of heterodimers between individual family members. For example, proapoptotic members inhibit antiapoptotic members through formation of heterodimeric complexes and *vice versa*. BH1 and BH2 domains found in Bcl-2, Bcl-x_L, Bax, and Bak form channels. Pro- and antiapoptotic channels can be distinguished by their ion selectivity. Proapoptotic channels are Cl$^-$ selective, whereas antiapoptotic channels are K$^+$ selective[632]. Apoptotic stimuli induce a conformational change in Bax and Bak that exposes their *N*-termini. Oligomeric Bax and Bak insert themselves into the mitochondrial membrane, resulting in cytochrome c release from mitochondria[633]. Some Bcl-2 proteins have been localized to the ER membrane, i.e. Bcl-2, Bak, and Bax[634,635]. Bak and Bax oligomerize at and insert themselves into the ER membrane and cause release of Ca^{2+} from the ER[636,637]. Bcl-2 artificially tethered to the ER membrane rendered cells resistant to apoptosis induced by ER stress[638,639], by serum starvation in some, but not all, cell types[640,641], and by certain types of DNA damage[642]. The BH3 only protein Bcl-2 binding component 3 (Bbc3)/p53 upregulated modulator of apoptosis (Puma) was induced by ER stress[643]. $puma^{-/-}$ cells were resistant to ER stress induced apoptosis[643]. Another ER membrane protein interacting with Bcl-2 family members, Bax inhibitor-1 (BI-1), protects mice from kidney damage induced by tunicamycin and from stroke injury[644]. Taken together, these data suggest the existence of a direct link between ER stress and regulation of the activity of Bcl-2 protein family members.

The missing link between ER stress and regulation of Bcl-2 protein family member activity may be membrane bound chaperones. B cell receptor (BCR) associated protein of 31 kDa (BAP31) and bifunctional apoptosis regulator (BAR) interact with Bcl-2 and Bcl-x_L. Both localize to the ER membrane[645,646]. BAP31 forms homo- and heterooligomers with closely related BAP29[645,647]. Through their death effector domain-like (DED-L) region, BAP31 and BAR interact with DED containing caspases, e.g. procaspase 8/Fas associated protein with a novel death domain (FADD) like interleukin (IL) 1β converting enzyme (ICE) (FLICE)/mammalian cell death abnormality 3 (CED-3) homologue 5 (Mch5), and form a complex with each caspase, Bcl-2, and Bcl-x_L[645,646]. Full-length BAP31 suppresses apoptosis induced by stimulation of Fas. The caspases caspase 1/ICE and caspase 8 cleave BAP31 at two caspase recognition sites and liberae a proapoptotic 20 kDa transmembrane fragment[648]. Association of Bcl-2 and Bcl-x_L with BAP31 inhibits activation of BAP31 by caspases[645]. BAP31 is also a cargo receptor for ER to Golgi transport[647,649-653]. Inhibition of protein folding by inhibiting N-linked glycosylation with the drug tunicamycin induced apoptosis and cleavage of BAP31[654]. Accumulation of folding incompetent α1-antitrypsin Z in the ER induced cleavage of BAP31[655]. BAP31 also interacts with calnexin. Calnexin deficient cells displayed a decrease in cleavage of BAP31 and were resistant to ER stress induced apoptosis[656]. Thus, BAP31 may function as a relay between the ER chaperone machinery, Bcl-2 proteins, and caspases. Caspase cleavage of BAP31 may also serve to make the cell death program irreversible as discussed above.

Increases in the cytosolic Ca^{2+} concentration activate calpain[657], which proteolytically activates procaspase 12. Calpain, and to a lesser degree caspase 3, may also cleave GRP94 after its translocation to the ER membrane during apoptosis[658]. Again, destruction of cytoprotective proteins such as GRP94 and Bcl-2 during apoptosis may explain why the cell death program is irreversible. Increases in cytosolic Ca^{2+} in response to ER stress also initiate apoptosis via a mitochondrial mediated pathway. Ca^{2+} is readily taken up by mitochondria which may result in collapse of the inner membrane potential. Influx of Ca^{2+} into mitochondria results in opening of the permeability transition pore (PTP), which is formed between the multiprotein voltage dependent anion channel, the adenine nucleotide translocase and cyclophilin D[659]. Opening of this pore releases cytochrome c into the cytosol and activates apoptotic mediators, i.e. procaspase 3. The PTP also recruits Bax to the outer mitochondrial membrane[659]. Mitochondrial membrane depolarization in response to ER stress can be blocked by overexpression of Bcl-x_L[660]. Further evidence for the role of calcium signaling in apoptosis is provided by the discovery that inhibition of the type I inositol 1,4,5-triphosphate receptor (IP$_3$R) Ca^{2+} release channel in the ER results in an increased resistance to many apoptotic stimuli[661]. Furthermore, calnexin interacts with both the sarcoplasmic reticulum ATPase (SERCA) transporter to prevent uptake of cytosolic Ca^{2+} into the ER[662] and the IP$_3$R to regulate Ca^{2+} release[663].

Finally, the tyrosine kinase cellular Abelson murine leukemia virus transforming protein (c-Abl) translocated from the ER to mitochondria in response to ER stress induced by depleting ER luminal Ca^{2+} stores or inhibiting N-linked glycosylation, resulting in release of cytochrome c[664]. c-Abl also contributes to activation of the MAPK JNK and p38. c-$abl^{-/-}$ cells were protected from ER stress induced apoptosis[664].

6. THE UPR IN NORMAL CELLULAR PHYSIOLOGY

The UPR is most active under times of pathophysiological or pharmacologically induced ER stress, but also plays important roles in normal cellular processes. Three such processes have been identified to date: Regulation of phospholipid metabolism and organelle biogenesis (cf. above), control of starvation and differentiation responses, i.e. pseudohyphal growth and meiosis in the budding yeast *Saccharomyces cerevisiae*[76,443] and differentiation of secretory cells, such as β islet cells of the pancreas and terminal differentiation of B cells into antibody secreting plasma cells. Direct evidence for low level constitutive UPR signaling in unstressed cells is found in the basal levels of *HAC1* mRNA splicing that occurs in exponentially growing yeast (between 3 and 5%)[76,418]. This low level activity may match ER protein folding capacity with its actual unfolded protein load. Even in cells considered as being unstressed, unfolded proteins are present in the ER lumen, because in yeast abrogation of the UPR and ERAD is synthetic lethal[553,554,590].

6.1 Regulation of Differentiation Programs by the UPR

The UPR controls differentiation programs in yeast and mammalian cells. These are discussed in the following sections.

6.1.1 Differentiation programs induced by starvation in budding yeast

The UPR has been shown to be important for regulating differentiation and starvation responses to severe nitrogen limitation in yeast[76]. Diploid yeast initiate one of two differentiation programs in response to severe nitrogen starvation, pseudohyphal growth[665] or meiosis[666]. Compared to the normal growth form, pseudohyphal growth is characterized by an elongated cell shape, cells sticking together after cell division has been completed, a change from a bi- to an unipolar budding pattern, and synchronization of cell division in mother and daughter cells[667,668]. These changes allow immobile yeast cells to form chains of cells, which are a form of movement to forage for nutrients. In contrast, meiosis or sporulation ultimately yields an ascus containing four haploid, stress resistant spores. Pseudohyphal growth was repressed by overexpression of Hac1^{i}p and was derepressed in $hac1\Delta/hac1\Delta$ and $ire1\Delta/ire1\Delta$ strains[76]. Furthermore, pseudohyphal growth was inhibited by UPR activators such as 2-deoxy-D-glucose and tunicamycin at sublethal concentrations in an *IRE1* and *HAC1* dependent manner[76]. Cells commit themselves to completion of meiosis after initiation of the first nuclear division[669]. This early phase is dominated by expression of the early meiotic genes (EMGs), a set of genes induced early in meiosis. Induction of EMGs was blunted in yeast overexpressing *HAC1^i* and derepressed in $hac1\Delta$ strains[76,443]. These data show that the UPR, through regulating synthesis of Hac1^{i}p, represses these two differentiation programs in yeast. Furthermore, the level to which the UPR is activated in healthy, exponentially growing cells is dependent on the metabolic state of the cell. For example, when grown on D-glucose or D-fructose, the preferred, fermentable carbon sources of yeast, levels of *HAC1* splicing were low, but increased when yeast were grown on increasingly poor carbon sources, such as the disaccharide D-maltose, and non-fermentable carbon sources such as acetate or ethanol[76,418]. Splicing of *HAC1* mRNA that occurred under normal conditions was abolished less than 5 min after induction of nitrogen starvation. The addition of ammonium salts to nitrogen starved cells reactivated *HAC1* splicing with kinetics comparable to those observed before nitrogen starvation and to levels comparable prior to induction of nitrogen starvation. With a half life of 1–2 minutes[422], Hac1^{i}p is rapidly removed from the cell following nitrogen starvation. Thus, Ire1p mediated splicing of *HAC1* mRNA and Hac1^{i}p provide a mechanism for sensing nitrogen in exponentially growing unstressed cells and control differentiation programs induced in response to this starvation condition.

6.1.2 Terminal B cell differentiation

Terminal differentiation of B cells into antibody secreting plasma cells is associated with a 5–10 fold increase in ER size[593]. During this differentiation program splicing of *XBP-1^u* mRNA is observed[96,670,671], showing that these cells experience ER stress. Correlative evidence suggests that B cells initiate a two phase response with respect to the UPR. Activation of UPR genes and UPR target genes precedes production of immunoglobulins (Ig)[97,670,672,673] and may predispose the ER to future increases in secretory activity. For example, IL-4 induces *XBP-1^u* mRNA[97], and signaling through the B cell receptor induces *BiP*, *GRP94*, the calreticulin gene, *EDEM-1*, and *CHOP*[673].

The second phase is characterized by increased unfolded protein levels in the ER, and activation of the classical UPR pathways. The UPR is activated and required for B cell differentiation as evidenced by the requirement for XBP-1[675] for formation of plasma cells and observation of *XBP-1^u* splicing when B cell differentiation is

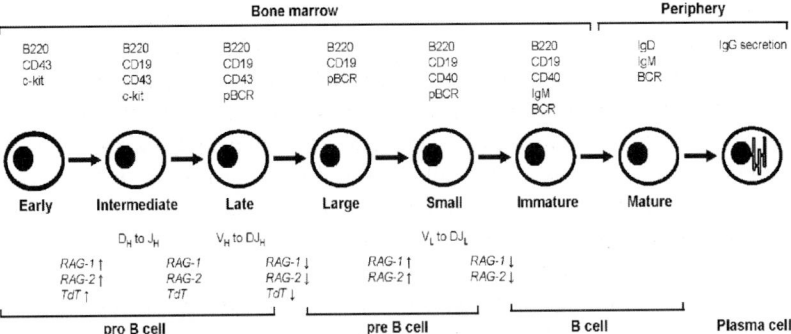

Figure 19. Simplified scheme of B cell differentiation. Above the cells cell surface markers are summarized, below the cells recombination events, and changes in expression level of *RAG-1*, *RAG-2* and *TdT*[674]. Abbreviations: B220 - B cell-specific member of the T200 glycoprotein family, CD – cluster of differentiation, and pBCR – pre B cell receptor. Reprinted from *Current Molecular Medicine*, copyright 2006, with permission from Bentham Science Publishers

induced, i.e. by stimulation with lipopolysaccharides[96,97,671]. Expansion of the ER during production of plasma cells requires increased production of phospholipids, which may be induced by XBP-1^{s}[605,606]. IRE1α is required for two steps in B cell differentiation. Through splicing of *XBP-1*^{u} mRNA IRE1α is required for differentiation of mature B cells into plasma cells. However, IRE1α is also required in an early step of B cell differentiation (Figure 19), because *ire1α*$^{-/-}$ fetal hematopoietic cells did not express B cell receptors and did not undergo early Ig heavy chain locus rearrangements[671], because of decreased expression of recombination activating gene 1 (*RAG-1*), *RAG-2*, and terminal deoxynucleotidyl transferase (*TdT*)[671]. This function of IRE1α is independent of its kinase and RNase activity, but required its cytosolic domain[671]. IRE1α may serve as a scaffold to assemble a signaling complex or a transcriptional activation complex to promote expression of these genes. Terminal differentiation of B lymphocytes requires Blimp-1 dependent repression of *c-myc*[676] *via* recruitment of histone deacetylase to the *c-myc* promoter[115]. In addition, the kinetics of UPR activation as evidenced by *XBP-1*^{u} mRNA splicing, repression of *c-myc* and activation of Blimp-1[97] are similar. These data suggest a role for the UPR in driving differentiation or in maintenance of the differentiated state in mammalian cells.

6.2   The UPR as a Nutrient Sensor in Higher Eukaryotic Cells

Despite a substantial amount of evidence for the role of the UPR in control of yeast starvation and differentiation programs, there is less conclusive evidence for such a role in mammalian cells. D-glucose starvation and anaerobiosis induces expression of the key ER resident molecular chaperones, including BiP/GRP78 and GRP94[126,127,677], suggesting that the ability of the UPR to respond to nutritional status is conserved in higher eukaryotic organisms. eIF2α phosphorylation and PERK are essential for *in vivo* D-glucose homeostasis[494,678,679]. Mice homozygous for knock-in of the Ser51Ala mutation in eIF2α that abolishes its phosphorylation died within 18 hours of birth because of fatal hypoglycemia caused by defective induction of gluconeogenic enzymes and a deficiency in pancreatic β cells[494]. Consistent with these observations is a model where the UPR senses the D-glucose concentration and responds by mediating proinsulin translation in pancreatic β cells. When intracellular D-glucose levels fall, protein folding within the ER becomes slow, thus activating PERK and leading to translational inhibition *via* eIF2α phosphorylation, and a drop in translation of proinsulin mRNA. The rate of protein folding is increased when D-glucose levels are restored, increasing protein glycosylation and folding, thus decreasing the UPR and restoring proinsulin production[494]. The decreased rate of protein folding may be caused by the requirement of D-glucose for *N*-glycosylation and of ATP for protein folding in the ER. Elevated levels of IRE1α[103] and PERK[333] in β cells may predispose these cells to glucose sensing, allowing these cells to detect even small changes in the free BiP pool by IRE1α and PERK. These observations can also be explained by β cell dysfunction, where a large number of β cells have been lost due to UPR induced apoptosis[680]. In support of this model, the rate of proinsulin mRNA translation in islet cells isolated from *perk*[−/−] and wild type mice were comparable[681].

A second physiological response to D-glucose deprivation that may be regulated by the UPR is repression of cholesterol synthesis[682,683] through activation of ATF6, formation of an heterotrimeric ATF6·SREBP-2·HDAC1 complex and repression of cholesterogenic targets of SREBP-2[78]. Gluconeogenesis, activated during D-glucose starvation, and cholesterol synthesis compete for citric acid cycle intermediates. Therefore, repression of cholesterol synthesis may preserve citric acid cycle intermediates for gluconeogenesis.

Finally, ATF4 activates transcription of amino acid biosynthetic genes on the amino acid response element[684] and is induced by the UPR. This activity of ATF4 may constitute a positive feedback loop to maintain high levels of amino acids and protein synthesis to sustain elevated secretory activity.

7.   ENGINEERING OF THE UPR TO INCREASE CELL SPECIFIC PRODUCTIVITIES

Several engineering approaches can be used to improve recombinant protein production by manipulating protein folding in the ER. First, metabolic engineering to increase the flux through metabolic pathways will provide more ATP which is required for protein folding and has been shown to increase cell specific production rates[30,32,33]. Experimental evidence supports the idea that high secretory activity is an energy burden for the cell[53]. Cell engineering has mostly focused on engineering chaperone systems, to create apoptosis resistant cells through overexpression of anti-apoptotic proteins, such as Bcl-2[685–692] and Bcl-x$_L$[689,691], or knockdown of proapoptotic genes, i.e. caspase 3[693], c-Jun[694], and to engineer cell lines with enhanced growth properties, i.e. through overexpression of cyclin E[695], or E2F-1[696]. These approaches, most likely because activation of the UPR is linked to cell cycle control and regulation of apoptosis, can also increase cell specific productivities[686]. A more direct approach to increase cell specific productivities is to engineer chaperone systems. These have been engineered by overexpressing individual chaperones and foldases. Usually, the effect of overexpression of the chaperone on production of an individual recombinant protein has been studied. However, these approaches have not been consistently successful[21,697,698]. For example, overexpression of BiP[699,700] and PDI[701] reduced secretion of recombinant proteins. One explanation for these conflicting observations is that there may be some specificity to chaperone client protein interactions, which may be imposed on a general chaperone such as BiP through interaction with its cochaperones. Second, overexpression of an ATP dependent HSP70 chaperone such as BiP without co-overexpressing its co-chaperones, providing additional ATP sources, or cooperating chaperones such as Lhs1p, may not result in an increase in functional BiP in the ER. Even in cases where overexpression of a chaperone or a foldase has been shown to improve secretion of a recombinant protein, it cannot be ruled out that enhancement of cell specific productivity is caused by secondary or pleiotrophic effects of protein or chaperone overexpression and that action of the overexpressed chaperone or foldase itself on the recombinant protein is virtually without effect. Finally, chaperones and foldases are a double edged sword. They assist nascent polypeptide chains in folding, but also monitor the progress a folding substrate makes, and if that progress is too slow, redirect their substrate for proteasomal destruction.

Engineering of the UPR holds promise to address the concern that to increase the capacity of the protein folding machinery in the ER it is necessary to elevate the level of many, and not just one chaperone. At first sight, this could easily be achieved by overexpressing transcription factors activated by the UPR, i.e. ATF6, XBP-1^s, or ATF4. This approach has been used to engineer the UPR in yeast and filamentous fungi. Secretion of *Bacillus amyloliquefaciens* α-amylase by *S. cerevisiae* increased when Hac1^{i}p or its homolog from *Trichoderma reesei* were expressed in *S. cerevisiae*[702]. Likewise, secretion of *Trametes versicolor* laccase and of bovine preprochymosin increased when *Aspergillus niger* HACA was overexpressed in *A. niger*[703]. In contrast, secretion of *T. reesei* endoglucanase EGI by yeast was unaffected by overexpression of Hac1^{i}p[702]. Constitutive activation of the UPR, i.e. through overexpression of Ire1p[378] or Hac1^{i}p[426], alters the morphological appearance of the secretory pathway and retards growth. Further, through repression of metabolic genes Hac1^{i}p may attenuate metabolism[443]. Thus, even in unicellular eukaryotes it may be necessary to separate properties of Hac1^{i}p beneficial for cell engineering from unwanted properties. This problem is potentiated in higher eukaryotic cells. It should be obvious that one objective to successfully engineer the UPR to increase cell specific productivities has to be to try to separate adaptive and proapoptotic functions of molecules involved in the higher eukaryotic UPR. Further, careful consideration has to be given to the cell type and the organism from which the cell line was derived. For example, caspase 12 plays an important role in apoptosis in response to ER stress in mice, but not in humans[623,628,629]. Further, the UPR is transduced through activation of bZIP transcription factors that extensively heterodimerize with other bZIP proteins, of which there are ~50 in mammalian cells. Cell and tissue type specific expression of these proteins, as seen for OASIS, will have tremendous impact on successfully engineering of the UPR. In summary, whereas our current molecular understanding of the UPR, especially when we look at signaling specificity in the UPR, is incomplete, engineering of the UPR and gaining further insight into molecular mechanisms of the UPR hold great promise to improve cell specific production rates and to develop the next generation of production cell lines.

8.   ACKNOWLEDGEMENTS

This work was supported by funding from the Royal Society, the Biotechnology and Biological Sciences Research Council (BB/C513418/1), and the European Commission (MC IRG 016411) to M.S.

9.   REFERENCES

1. L. Chu and D. K. Robinson, Industrial choices for protein production by large-scale cell culture, *Curr. Opin. Biotechnol.* **12**(2), 180–187 (2001).
2. P. D. Minor, Ensuring safety and consistency in cell culture production processes: viral screening and inactivation, *Trends Biotechnol.* **12**(7), 257–261 (1994).
3. M. J. Jorgensen, A. B. Cantor, B. C. Furie and B. Furie, Expression of completely gamma-carboxylated recombinant human prothrombin, *J. Biol. Chem.* **262**(14), 6729–6734 (1987).
4. A. W. Stephens, A. Siddiqui and C. H. Hirs, Expression of functionally active human antithrombin III, *Proc. Natl. Acad. Sci. U S A* **84**(11), 3886–3890 (1987).
5. G. Zettlmeissl, H. Ragg and H. E. Karges, Expression of biologically active human antithrombin III in Chinese hamster ovary cells, *Biotechnology* **5**(7), 720–725 (1987).
6. J. D. Walls, D. T. Berg, S. B. Yan and B. W. Grinnell, Amplification of multicistronic plasmids in the human 293 cell line and secretion of correctly processed recombinant human protein C, *Gene* **81**(1), 139–149 (1989).
7. L. S. Gillespie, K. K. Hillesland and D. J. Knauer, Expression of biologically active human antithrombin III by recombinant baculovirus in Spodoptera frugiperda cells, *J. Biol. Chem.* **266**(6), 3995–4001 (1991).
8. G. Russo, K. Mertens, L. De Magistris, H. Lange, R. Cortese and C. Pietropaolo, Biologically active recombinant prothrombin and antithrombin III expressed in a human hepatoma/vaccinia virus system, *Biotechnol. Appl. Biochem.* **14**(2), 222–233 (1991).
9. M. M. Guarna, C. H. Fann, S. J. Busby, K. M. Walker, D. G. Kilburn and J. M. Piret, Effect of cDNA copy number on secretion rate of activated protein C, *Biotechnol. Bioeng.* **46**(1), 22–27 (1995).
10. H. Schellekens, The immunogenicity of biopharmaceuticals, *Neurology* **61**(Suppl 5), S11–12 (2003).
11. S. W. Pipe, J. M. Saint-Remy and C. E. Walsh, New high-technology products for the treatment of haemophilia, *Haemophilia* **10**(Suppl 4), 55–63 (2004).
12. W. Y. Hwang and J. Foote, Immunogenicity of engineered antibodies, *Methods* **36**(1), 3–10 (2005).
13. S. W. Pipe, The promise and challenges of bioengineered recombinant clotting factors, *J. Thromb. Haemost.* **3**(8), 1692–1701 (2005).
14. W. Jelkmann, The enigma of the metabolic fate of circulating erythropoietin (Epo) in view of the pharmacokinetics of the recombinant drugs rhEpo and NESP, *Eur. J. Haematol.* **69**(5–6), 265–274 (2002).

15. I. C. Macdougall, Optimizing the use of erythropoietic agents – pharmacokinetic and pharmaco-dynamic considerations, *Nephrol. Dial. Transplant.* **17**(Suppl 5), 66–70 (2002).
16. E. L. Saenko, N. M. Ananyeva, M. Shima, C. A. Hauser and S. W. Pipe, The future of recombinant coagulation factors, *J. Thromb. Haemost.* **1**(5), 922–930 (2003).
17. S. A. Brooks, Appropriate glycosylation of recombinant proteins for human use: implications of choice of expression system, *Mol. Biotechnol.* **28**(3), 241–255 (2004).
18. S. Geisse, H. Gram, B. Kleuser and H. P. Kocher, Eukaryotic expression systems: a comparison, *Protein Expr. Purif.* **8**(3), 271–282 (1996).
19. L. M. Barnes, C. M. Bentley and A. J. Dickson, Advances in animal cell recombinant protein production: GS-NS0 expression system, *Cytotechnology* **32**(2), 109–123 (2000).
20. P. Thomas and T. G. Smart, HEK293 cell line: a vehicle for the expression of recombinant proteins, *J. Pharmacol. Toxicol. Methods* **51**(3), 187–200 (2005).
21. D. M. Dinnis and D. C. James, Engineering mammalian cell factories for improved recombinant monoclonal antibody production: lessons from nature?, *Biotechnol. Bioeng.* **91**(2), 180–189 (2005).
22. P. W. Sauer, J. E. Burky, M. C. Wesson, H. D. Sternard and L. Qu, A high-yielding, generic fed-batch cell culture process for production of recombinant antibodies, *Biotechnol. Bioeng.* **67**(5), 585–597 (2000).
23. C.-H. Liu, I.-M. Chu and S.-M. Hwang, Factorial designs combined with the steepest ascent method to optimize serum-free media for CHO cells, *Enzyme Microb. Technol.* **28**(4–5), 314–321 (2001).
24. C. Chun, K. Heineken, D. Szeto, T. Ryll, S. Chamow and J. D. Chung, Application of factorial design to accelerate identification of CHO growth factor requirements, *Biotechnol. Prog.* **19**(1), 52–57 (2003).
25. J. Dempsey, S. Ruddock, M. Osborne, A. Ridley, S. Sturt and R. Field, Improved fermentation processes for NS0 cell lines expressing human antibodies and glutamine synthetase, *Biotechnol. Prog.* **19**(1), 175–178 (2003).
26. J. Vives, S. Juanola, J. J. Cairó and F. Gòdia, Metabolic engineering of apoptosis in cultured animal cells: implications for the biotechnology industry, *Metab. Eng.* **5**(2), 124–132 (2003).
27. N. Arden and M. J. Betenbaugh, Life and death in mammalian cell culture: strategies for apoptosis inhibition, *Trends Biotechnol.* **22**(4), 174–180 (2004).
28. S. L. Bell, C. Bebbington, M. F. Scott, J. N. Wardell, R. E. Spier, M. E. Bushell and P. G. Sanders, Genetic engineering of hybridoma glutamine metabolism, *Enzyme Microb. Technol.* **17**(2), 98–106 (1995).
29. N. Irani, M. Wirth, J. van Den Heuvel and R. Wagner, Improvement of the primary metabolism of cell cultures by introducing a new cytoplasmic pyruvate carboxylase reaction, *Biotechnol. Bioeng.* **66**(4), 238–246 (1999).
30. H. J. Cruz, J. L. Moreira and M. J. Carrondo, Metabolically optimised BHK cell fed-batch cultures, *J. Biotechnol.* **80**(2), 109–118 (2000).
31. H. Park, I.-H. Kim, I.-Y. Kim, K.-H. Kim and H.-J. Kim, Expression of carbamoyl phosphate synthetase I and ornithine transcarbamoylase genes in Chinese hamster ovary *dhfr*-cells decreases accumulation of ammonium ion in culture media, *J. Biotechnol.* **81**(2–3), 129–140 (2000).
32. K. Chen, Q. Liu, L. Xie, P. A. Sharp and D. I. Wang, Engineering of a mammalian cell line for reduction of lactate formation and high monoclonal antibody production, *Biotechnol. Bioeng.* **72**(1), 55–61 (2001).
33. N. Irani, A. J. Beccaria and R. Wagner, Expression of recombinant cytoplasmic yeast pyruvate carboxylase for the improvement of the production of human erythropoietin by recombinant BHK-21 cells, *J. Biotechnol.* **93**(3), 269–282 (2002).
34. C. B. Elias, C. Carpentier, Y. Durocher, L. Bisson, R. Wagner and A. Kamen, Improving glucose and glutamine metabolism of human HEK 293 and *Trichoplusia ni* insect cells engineered to express a cytosolic pyruvate carboxylase enzyme, *Biotechnol. Prog.* **19**(1), 90–97 (2003).
35. F. W. Alt, R. E. Kellems, J. R. Bertino and R. T. Schimke, Selective multiplication of dihydrofolate reductase genes in methotrexate-resistant variants of cultured murine cells, *J. Biol. Chem.* **253**(5), 1357–1370 (1978).

36. R. J. Kaufman, P. C. Brown and R. T. Schimke, Amplified dihydrofolate reductase genes in unstably methotrexate-resistant cells are associated with double minute chromosomes, *Proc. Natl. Acad. Sci. U S A* **76**(11), 5669–5673 (1979).

37. M. J. Page and M. A. Sydenham, High level expression of the humanized monoclonal antibody Campath-1H in Chinese hamster ovary cells, *Biotechnology (N Y)* **9**(1), 64–68 (1991).

38. R. J. Kaufman and P. A. Sharp, Amplification and expression of sequences cotransfected with a modular dihydrofolate reductase complementary DNA gene, *J. Mol. Biol.* **159**(4), 601–621 (1982).

39. J. A. Lewis, J. P. Davide and P. W. Melera, Selective amplification of polymorphic dihydrofolate reductase gene loci in Chinese hamster lung cells, *Proc. Natl. Acad. Sci. U S A* **79**(22), 6961–6965 (1982).

40. A. P. Young and G. M. Ringold, Mouse 3T6 cells that overproduce glutamine synthetase, *J. Biol. Chem.* **258**(18), 11260–11266 (1983).

41. P. G. Sanders and R. H. Wilson, Amplification and cloning of the Chinese hamster glutamine synthetase gene, *EMBO J.* **3**(1), 65–71 (1984).

42. C. R. Bebbington, G. Renner, S. Thomson, D. King, D. Abrams and G. T. Yarranton, High-level expression of a recombinant antibody from myeloma cells using a glutamine synthetase gene as an amplifiable selectable marker, *Biotechnology (N Y)* **10**(2), 169–175 (1992).

43. J. Running Deer and D. S. Allison, High-level expression of proteins in mammalian cells using transcription regulatory sequences from the Chinese hamster EF-1α gene, *Biotechnol. Prog.* **20**(3), 880–889 (2004).

44. L. A. Dickinson, T. Joh, Y. Kohwi and T. Kohwi-Shigematsu, A tissue-specific MAR/SAR DNA-binding protein with unusual binding site recognition, *Cell* **70**(4), 631–645 (1992).

45. A. N. Antoniou, S. Ford, J. D. Taurog, G. W. Butcher and S. J. Powis, Formation of HLA-B27 homodimers and their relationship to assembly kinetics, *J. Biol. Chem.* **279**(10), 8895–8902 (2004).

46. R. K. Koduri, J. T. Miller and P. Thammana, An efficient homologous recombination vector pTV(I) contains a hot spot for increased recombinant protein expression in Chinese hamster ovary cells, *Gene* **280**(1–2), 87–95 (2001).

47. R. J. Gouka, P. J. Punt and C. A. M. J. J. van den Hondel, Efficient production of secreted proteins by *Aspergillus*: progress, limitations and prospects, *Appl. Microbiol. Biotechnol.* **47**(1), 1–11 (1997).

48. N. Kraševec, C. A. M. J. J. van den Hondel and R. Komel, Expression of human lymphotoxin α in *Aspergillus niger*, *Pflügers Arch.* **440**(5 Suppl), R83–85 (2000).

49. F. J. Moralejo, R. E. Cardoza, S. Gutierrez, H. Sisniega, I. Faus and J. F. Martin, Overexpression and lack of degradation of thaumatin in an aspergillopepsin A-defective mutant of *Aspergillus awamori* containing an insertion in the *pepA* gene, *Appl. Microbiol. Biotechnol.* **54**(6), 772–777 (2000).

50. N. S. Outchkourov, W. J. Stiekema and M. A. Jongsma, Optimization of the expression of equistatin in *Pichia pastoris, Protein Expr. Purif.* **24**(1), 18–24 (2002).

51. R. E. Cardoza, S. Gutiérrez, N. Ortega, A. Colina, J. Casqueiro and J. F. Martín, Expression of a synthetic copy of the bovine chymosin gene in *Aspergillus awamori* from constitutive and pH-regulated promoters and secretion using two different pre-pro sequences, *Biotechnol. Bioeng.* **83**(3), 249–259 (2003).

52. L. A. Fouser, S. L. Swanberg, B. Y. Lin, M. Benedict, K. Kelleher, D. A. Cumming and G. E. Riedel, High level expression on a chimeric anti-ganglioside GD2 antibody: genomic kappa sequences improve expression in COS and CHO cells, *Biotechnology (N Y)* **10**(10), 1121–1127 (1992).

53. M. Schröder, R. Schäfer and P. Friedl, Induction of protein aggregation in an early secretory compartment by elevation of expression level, *Biotechnol. Bioeng.* **78**(2), 131–140 (2002).

54. R. P. Singh, M. Alrubeai, C. D. Gregory and A. N. Emery, Cell death in bioreactors – a role for apoptosis, *Biotechnol. Bioeng.* **44**(6), 720–726 (1994).

55. M. Schröder and P. Friedl, Overexpression of recombinant human antithrombin III in Chinese hamster ovary cells results in malformation and decreased secretion of the recombinant protein, *Biotechnol. Bioeng.* **53**(6), 547–559 (1997).

56. R. Parekh, K. Forrester and D. Wittrup, Multicopy overexpression of bovine pancreatic trypsin inhibitor saturates the protein folding and secretory capacity of *Saccharomyces cerevisiae, Protein Expr. Purif.* **6**(4), 537–545 (1995).

57. G. J. Pendse, S. Karkare and J. E. Bailey, Effect of cloned gene dosage on cell growth and hepatitis B surface antigen synthesis and secretion in recombinant CHO cells, *Biotechnol. Bioeng.* **40**(1), 119–129 (1992).

58. M. Schröder, C. Körner and P. Friedl, Quantitative analysis of transcription and translation in gene amplified Chinese hamster ovary cells on the basis of a kinetic model, *Cytotechnology* **29**(2), 93–102 (1999).

59. L. M. Barnes, C. M. Bentley and A. J. Dickson, Molecular definition of predictive indicators of stable protein expression in recombinant NS0 myeloma cells, *Biotechnol. Bioeng.* **85**(2), 115–121 (2004).

60. H. Hohenblum, N. Borth and D. Mattanovich, Assessing viability and cell-associated product of recombinant protein producing *Pichia pastoris* with flow cytometry, *J. Biotechnol.* **102**(3), 281–290 (2003).

61. K. J. Kauffman, E. M. Pridgen, F. J. Doyle, 3rd, P. S. Dhurjati and A. S. Robinson, Decreased protein expression and intermittent recoveries in BiP levels result from cellular stress during heterologous protein expression in *Saccharomyces cerevisiae, Biotechnol. Prog.* **18**(5), 942–950 (2002).

62. A. J. Dorner, D. G. Bole and R. J. Kaufman, The relationship of N-linked glycosylation and heavy chain-binding protein association with the secretion of glycoproteins, *J. Cell Biol.* **105**(6 Pt 1), 2665–2674 (1987).

63. M. Schröder and R. J. Kaufman, ER stress and the unfolded protein response, *Mutat. Res.* **569**(1–2), 29–63 (2005).

64. M. Schröder and R. J. Kaufman, The mammalian unfolded protein response, *Annu. Rev. Biochem.* **74**, 739–789 (2005).

65. G. Palade, Intracellular aspects of the process of protein synthesis, *Science* **189**(4200), 347–358 (1975).

66. T. A. Rapoport, B. Jungnickel and U. Kutay, Protein transport across the eukaryotic endoplasmic reticulum and bacterial inner membranes, *Annu. Rev. Biochem.* **65**, 271–303 (1996).

67. C. Hwang, A. J. Sinskey and H. F. Lodish, Oxidized redox state of glutathione in the endoplasmic reticulum, *Science* **257**(5076), 1496–1502 (1992).

68. S. Orrenius, B. Zhivotovsky and P. Nicotera, Regulation of cell death: the calcium-apoptosis link, *Nat. Rev. Mol. Cell. Biol.* **4**(7), 552–565 (2003).

69. F. J. Stevens and Y. Argon, Protein folding in the ER, *Semin. Cell Dev. Biol.* **10**(5), 443–454 (1999).

70. C. Hammond and A. Helenius, Quality control in the secretory pathway: retention of a misfolded viral membrane glycoprotein involves cycling between the ER, intermediate compartment, and Golgi apparatus, *J. Cell Biol.* **126**(1), 41–52 (1994).

71. E. D. Werner, J. L. Brodsky and A. A. McCracken, Proteasome-dependent endoplasmic reticulum-associated protein degradation: an unconventional route to a familiar fate, *Proc. Natl. Acad. Sci. U S A* **93**(24), 13797–13801 (1996).

72. I. M. Martínez and M. J. Chrispeels, Genomic analysis of the unfolded protein response in Arabidopsis shows its connection to important cellular processes, *Plant Cell* **15**(2), 561–576 (2003).

73. T. M. Pakula, M. Laxell, A. Huuskonen, J. Uusitalo, M. Saloheimo and M. Penttilä, The effects of drugs inhibiting protein secretion in the filamentous fungus *Trichoderma reesei*. Evidence for down-regulation of genes that encode secreted proteins in the stressed cells, *J. Biol. Chem.* **278**(45), 45011–45020 (2003).

74. Y. Kimata, Y. Ishiwata-Kimata, S. Yamada and K. Kohno, Yeast unfolded protein response pathway regulates expression of genes for anti-oxidative stress and for cell surface proteins, *Genes Cells* **11**(1), 59–69 (2006).

75. C. Xu, B. Bailly-Maitre and J. C. Reed, Endoplasmic reticulum stress: cell life and death decisions, *J. Clin. Invest.* **115**(10), 2656–2664 (2005).

76. M. Schröder, J. S. Chang and R. J. Kaufman, The unfolded protein response represses nitrogen-starvation induced developmental differentiation in yeast, *Genes Dev.* **14**(23), 2962–2975 (2000).

77. H. P. Harding, Y. Zhang, H. Zeng, I. Novoa, P. D. Lu, M. Calfon, N. Sadri, C. Yun, B. Popko, R. Paules, D. F. Stojdl, J. C. Bell, T. Hettmann, J. M. Leiden and D. Ron, An integrated stress response regulates amino acid metabolism and resistance to oxidative stress, *Mol. Cell* **11**(3), 619–633 (2003).

78. L. Zeng, M. Lu, K. Mori, S. Luo, A. S. Lee, Y. Zhu and J. Y. Shyy, ATF6 modulates SREBP2-mediated lipogenesis, *EMBO J.* **23**(4), 950–958 (2004).

79. R. J. Kaufman, D. Scheuner, M. Schröder, X. Shen, K. Lee, C. Y. Liu and S. M. Arnold, The unfolded protein response in nutrient sensing and differentiation, *Nat. Rev. Mol. Cell. Biol.* **3**(6), 411–421 (2002).

80. A. Takatsuki and G. Tamura, Effect of tunicamycin on the synthesis of macromolecules in cultures of chick embryo fibroblasts infected with Newcastle disease virus, *J. Antibiot. (Tokyo)* **24**(11), 785–794 (1971).

81. L. Lehle and W. Tanner, The specific site of tunicamycin inhibition in the formation of dolichol-bound N-acetylglucosamine derivatives, *FEBS Lett.* **71**(1), 167–170 (1976).

82. A. Heifetz, R. W. Keenan and A. D. Elbein, Mechanism of action of tunicamycin on the UDP-GlcNAc:dolichyl-phosphate Glc-NAc-1-phosphate transferase, *Biochemistry* **18**(11), 2186–2192 (1979).

83. R. K. Keller, D. Y. Boon and F. C. Crum, *N*-Acetylglucosamine-1-phosphate transferase from hen oviduct: solubilization, characterization, and inhibition by tunicamycin, *Biochemistry* **18**(18), 3946–3952 (1979).

84. G. Barnes, W. J. Hansen, C. L. Holcomb and J. Rine, Asparagine-linked glycosylation in *Saccharomyces cerevisiae*: genetic analysis of an early step, *Mol. Cell. Biol.* **4**(11), 2381–2388 (1984).

85. M. F. Schmidt, R. T. Schwarz and C. Scholtissek, Interference of nucleoside diphosphate derivatives of 2-deoxy-D-glucose with the glycosylation of virus-specific glycoproteins in vivo, *Eur. J. Biochem.* **70**(1), 55–62 (1976).

86. R. T. Schwarz, M. F. Schmidt and L. Lehle, Glycosylation in vitro of Semliki-Forest-virus and influenza-virus glycoproteins and its suppression by nucleotide-2-deoxy-hexose, *Eur. J. Biochem.* **85**(1), 163–172 (1978).

87. R. Datema, R. P. Lezica, P. W. Robbins and R. T. Schwarz, Deoxyglucose inhibition of protein glycosylation: effects of nucleotide deoxysugars on the formation of glucosylated lipid intermediates, *Arch. Biochem. Biophys.* **206**(1), 65–71 (1981).

88. S. H. Back, M. Schröder, K. Lee, K. Zhang and R. J. Kaufman, ER stress signaling by regulated splicing: IRE1/HAC1/XBP1, *Methods* **35**(4), 395–416 (2005).

89. C. K. Suzuki, J. S. Bonifacino, A. Y. Lin, M. M. Davis and R. D. Klausner, Regulating the retention of T-cell receptor α chain variants within the endoplasmic reticulum: Ca^{2+}-dependent association with BiP, *J. Cell Biol.* **114**(2), 189–205 (1991).

90. W. J. Ou, J. J. Bergeron, Y. Li, C. Y. Kang and D. Y. Thomas, Conformational changes induced in the endoplasmic reticulum luminal domain of calnexin by Mg-ATP and Ca^{2+}, *J. Biol. Chem.* **270**(30), 18051–18059 (1995).

91. E. F. Corbett, K. Oikawa, P. Francois, D. C. Tessier, C. Kay, J. J. Bergeron, D. Y. Thomas, K. H. Krause and M. Michalak, Ca^{2+} regulation of interactions between endoplasmic reticulum chaperones, *J. Biol. Chem.* **274**(10), 6203–6211 (1999).

92. D. R. Wetmore and K. D. Hardman, Roles of the propeptide and metal ions in the folding and stability of the catalytic domain of stromelysin (matrix metalloproteinase 3), *Biochemistry* **35**(21), 6549–6558 (1996).

93. Y. Kozutsumi, M. Segal, K. Normington, M.-J. Gething and J. Sambrook, The presence of malfolded proteins in the endoplasmic reticulum signals the induction of glucose-regulated proteins, *Nature* **332**(6163), 462–464 (1988).

94. R. J. Kaufman, L. C. Wasley and A. J. Dorner, Synthesis, processing, and secretion of recombinant human factor VIII expressed in mammalian cells, *J. Biol. Chem.* **263**(13), 6352–6362 (1988).

95. A. J. Dorner, L. C. Wasley and R. J. Kaufman, Increased synthesis of secreted proteins induces expression of glucose-regulated proteins in butyrate-treated Chinese hamster ovary cells, *J. Biol. Chem.* **264**(34), 20602–20607 (1989).

96. M. Calfon, H. Zeng, F. Urano, J. H. Till, S. R. Hubbard, H. P. Harding, S. G. Clark and D. Ron, IRE1 couples endoplasmic reticulum load to secretory capacity by processing the *XBP-1* mRNA, *Nature* **415**(6867), 92–96 (2002).

97. N. N. Iwakoshi, A.-H. Lee, P. Vallabhajosyula, K. L. Otipoby, K. Rajewsky and L. H. Glimcher, Plasma cell differentiation and the unfolded protein response intersect at the transcription factor XBP-1, *Nat. Immunol.* **4**(4), 321–329 (2003).

98. E. P. Jelitto-Van Dooren, S. Vidal and J. Denecke, Anticipating endoplasmic reticulum stress. A novel early response before pathogenesis-related gene induction, *Plant Cell* **11**(10), 1935–1944 (1999).

99. H. Nishitoh, A. Matsuzawa, K. Tobiume, K. Saegusa, K. Takeda, K. Inoue, S. Hori, A. Kakizuka and H. Ichijo, ASK1 is essential for endoplasmic reticulum stress-induced neuronal cell death triggered by expanded polyglutamine repeats, *Genes Dev.* **16**(11), 1345–1355 (2002).

100. I. G. Haas and M. Wabl, Immunoglobulin heavy chain binding protein, *Nature* **306**(5941), 387–389 (1983).

101. S. Munro and H. R. Pelham, An Hsp70-like protein in the ER: identity with the 78 kd glucose-regulated protein and immunoglobulin heavy chain binding protein, *Cell* **46**(2), 291–300 (1986).

102. S. L. Sanders, K. M. Whitfield, J. P. Vogel, M. D. Rose and R. W. Schekman, Sec61p and BiP directly facilitate polypeptide translocation into the ER, *Cell* **69**(2), 353–365 (1992).

103. A. Bertolotti, Y. Zhang, L. M. Hendershot, H. P. Harding and D. Ron, Dynamic interaction of BiP and ER stress transducers in the unfolded-protein response, *Nat. Cell Biol.* **2**(6), 326–332 (2000).

104. C. Y. Liu, M. Schröder and R. J. Kaufman, Ligand-independent dimerization activates the stress response kinases IRE1 and PERK in the lumen of the endoplasmic reticulum, *J. Biol. Chem.* **275**(32), 24881–24885 (2000).

105. K. Okamura, Y. Kimata, H. Higashio, A. Tsuru and K. Kohno, Dissociation of Kar2p/BiP from an ER sensory molecule, Ire1p, triggers the unfolded protein response in yeast, *Biochem. Biophys. Res. Commun.* **279**(2), 445–450 (2000).

106. H. Y. Lin, P. Masso-Welch, Y. P. Di, J. W. Cai, J. W. Shen and J. R. Subjeck, The 170-kDa glucose-regulated stress protein is an endoplasmic reticulum protein that binds immunoglobulin, *Mol. Biol. Cell* **4**(11), 1109–1119 (1993).

107. B. K. Baxter, P. James, T. Evans and E. A. Craig, *SSI1* encodes a novel Hsp70 of the *Saccharomyces cerevisiae* endoplasmic reticulum, *Mol. Cell. Biol.* **16**(11), 6444–6456 (1996).

108. R. A. Craven, M. Egerton and C. J. Stirling, A novel Hsp70 of the yeast ER lumen is required for the efficient translocation of a number of protein precursors, *EMBO J.* **15**(11), 2640–2650 (1996).

109. N. Saris and M. Makarow, Transient ER retention as stress response: conformational repair of heat-damaged proteins to secretion-competent structures, *J. Cell Sci.* **111**(11), 1575–1582 (1998).

110. T. G. Hamilton, T. B. Norris, P. R. Tsuruda and G. C. Flynn, Cer1p functions as a molecular chaperone in the endoplasmic reticulum of *Saccharomyces cerevisiae*, *Mol. Cell. Biol.* **19**(8), 5298–5307 (1999).

111. M. Chevalier, H. Rhee, E. C. Elguindi and S. Y. Blond, Interaction of murine BiP/GRP78 with the DnaJ homologue MTJ1, *J. Biol. Chem.* **275**(26), 19620–19627 (2000).

112. G. Schlenstedt, S. Harris, B. Risse, R. Lill and P. A. Silver, A yeast DnaJ homologue, Scj1p, can function in the endoplasmic reticulum with BiP/Kar2p via a conserved domain that specifies interactions with Hsp70s, *J. Cell Biol.* **129**(4), 979–988 (1995).

113. S. Silberstein, G. Schlenstedt, P. A. Silver and R. Gilmore, A role for the DnaJ homologue Scj1p in protein folding in the yeast endoplasmic reticulum, *J. Cell Biol.* **143**(4), 921–933 (1998).

114. C. Bies, S. Guth, K. Janoschek, W. Nastainczyk, J. Volkmer and R. Zimmermann, A Scj1p homolog and folding catalysts present in dog pancreas microsomes, *Biol. Chem.* **380**(10), 1175–1182 (1999).

115. J. Yu, C. Angelin-Duclos, J. Greenwood, J. Liao and K. Calame, Transcriptional repression by blimp-1 (PRDI-BF1) involves recruitment of histone deacetylase, *Mol. Cell. Biol.* **20**(7), 2592–2603 (2000).

116. Y. Shen, L. Meunier and L. M. Hendershot, Identification and characterization of a novel endo-plasmic reticulum (ER) DnaJ homologue, which stimulates ATPase activity of BiP *in vitro* and is induced by ER stress, *J. Biol. Chem.* **277**(18), 15947–15956 (2002).

117. P. M. Cunnea, A. Miranda-Vizuete, G. Bertoli, T. Simmen, A. E. Damdimopoulos, S. Hermann, S. Leinonen, M. P. Huikko, J.-Å. Gustafsson, R. Sitia and G. Spyrou, ERdj5, an endoplasmic reticulum (ER)-resident protein containing DnaJ and thioredoxin domains, is expressed in secretory cells or following ER stress, *J. Biol. Chem.* **278**(2), 1059–1066 (2003).

118. S. Nishikawa and T. Endo, The yeast JEM1p is a DnaJ-like protein of the endoplasmic reticulum membrane required for nuclear fusion, *J. Biol. Chem.* **272**(20), 12889–12892 (1997).

119. I. Sadler, A. Chiang, T. Kurihara, J. Rothblatt, J. Way and P. Silver, A yeast gene important for protein assembly into the endoplasmic reticulum and the nucleus has homology to DnaJ, an *Escherichia coli* heat shock protein, *J. Cell Biol.* **109**(6 Pt 1), 2665–2675 (1989).

120. D. Feldheim, J. Rothblatt and R. Schekman, Topology and functional domains of Sec63p, an endoplasmic reticulum membrane protein required for secretory protein translocation, *Mol. Cell. Biol.* **12**(7), 3288–3296 (1992).

121. M. A. Scidmore, H. H. Okamura and M. D. Rose, Genetic interactions between *KAR2* and *SEC63*, encoding eukaryotic homologues of DnaK and DnaJ in the endoplasmic reticulum, *Mol. Biol. Cell* **4**(11), 1145–1159 (1993).

122. K. T. Chung, Y. Shen and L. M. Hendershot, BAP, a mammalian BiP-associated protein, is a nucleotide exchange factor that regulates the ATPase activity of BiP, *J. Biol. Chem.* **277**(49), 47557–47563 (2002).

123. A. Boisramé, M. Kabani, J. M. Beckerich, E. Hartmann and C. Gaillardin, Interaction of Kar2p and Sls1p is required for efficient co-translational translocation of secreted proteins in the yeast *Yarrowia lipolytica*, *J. Biol. Chem.* **273**(47), 30903–30908 (1998).

124. M. Kabani, J. M. Beckerich and C. Gaillardin, Sls1p stimulates Sec63p-mediated activation of Kar2p in a conformation-dependent manner in the yeast endoplasmic reticulum, *Mol. Cell. Biol.* **20**(18), 6923–6934 (2000).

125. J. R. Tyson and C. J. Stirling, *LHS1* and *SIL1* provide a lumenal function that is essential for protein translocation into the endoplasmic reticulum, *EMBO J.* **19**(23), 6440–6452 (2000).

126. J. Pouysségur, R. P. Shiu and I. Pastan, Induction of two transformation-sensitive membrane polypeptides in normal fibroblasts by a block in glycoprotein synthesis or glucose deprivation, *Cell* **11**(4), 941–947 (1977).

127. R. P. Shiu, J. Pouyssegur and I. Pastan, Glucose depletion accounts for the induction of two transformation-sensitive membrane proteinsin Rous sarcoma virus-transformed chick embryo fibroblasts, *Proc. Natl. Acad. Sci. U S A* **74**(9), 3840–3844 (1977).

128. G. Koch, M. Smith, D. Macer, P. Webster and R. Mortara, Endoplasmic reticulum contains a common, abundant calcium-binding glycoprotein, endoplasmin, *J. Cell Sci.* **86**(1), 217–232 (1986).

129. R. G. Maki, L. J. Old and P. K. Srivastava, Human homologue of murine tumor rejection antigen gp96: 5'-regulatory and coding regions and relationship to stress-induced proteins, *Proc. Natl. Acad. Sci. U S A* **87**(15), 5658–5662 (1990).

130. D. Watanabe, K. Yamada, Y. Nishina, Y. Tajima, U. Koshimizu, A. Nagata and Y. Nishimune, Molecular cloning of a novel Ca^{2+}-binding protein (calmegin) specifically expressed during male meiotic germ cell development, *J. Biol. Chem.* **269**(10), 7744–7749 (1994).

131. N. Ahluwalia, J. J. Bergeron, I. Wada, E. Degen and D. B. Williams, The p88 molecular chaperone is identical to the endoplasmic reticulum membrane protein, calnexin, *J. Biol. Chem.* **267**(15), 10914–10918 (1992).

132. K. Galvin, S. Krishna, F. Ponchel, M. Frohlich, D. E. Cummings, R. Carlson, J. R. Wands, K. J. Isselbacher, S. Pillai and M. Ozturk, The major histocompatibility complex class I antigen-binding protein p88 is the product of the calnexin gene, *Proc. Natl. Acad. Sci. U S A* **89**(18), 8452–8456 (1992).

133. D. M. Waisman, B. P. Salimath and M. J. Anderson, Isolation and characterization of CAB-63, a novel calcium-binding protein, *J. Biol. Chem.* **260**(3), 1652–1660 (1985).

134. L. Fliegel, K. Burns, D. H. MacLennan, R. A. Reithmeier and M. Michalak, Molecular cloning of the high affinity calcium-binding protein (calreticulin) of skeletal muscle sarcoplasmic reticulum, *J. Biol. Chem.* **264**(36), 21522–21528 (1989).

135. C. A. Jakob, D. Bodmer, U. Spirig, P. Bättig, A. Marcil, D. Dignard, J. J. Bergeron, D. Y. Thomas and M. Aebi, Htm1p, a mannosidase-like protein, is involved in glycoprotein degradation in yeast, *EMBO Rep.* **2**(5), 423–430 (2001).

136. N. Hosokawa, I. Wada, K. Hasegawa, T. Yorihuzi, L. O. Tremblay, A. Herscovics and K. Nagata, A novel ER α-mannosidase-like protein accelerates ER-associated degradation, *EMBO Rep.* **2**(5), 415–422 (2001).

137. K. Nakatsukasa, S.-i. Nishikawa, N. Hosokawa, K. Nagata and T. Endo, Mnl1p, an α-mannosidase-like protein in yeast *Saccharomyces cerevisiae*, is required for endoplasmic reticulum-associated degradation of glycoproteins, *J. Biol. Chem.* **276**(12), 8635–8638 (2001).

138. S. W. Mast, K. Diekman, K. Karaveg, A. Davis, R. N. Sifers and K. W. Moremen, Human EDEM2, a novel homolog of family 47 glycosidases, is involved in ER-associated degradation of glycoproteins, *Glycobiology* **15**(4), 421–436 (2005).

139. S. Olivari, C. Galli, H. Alanen, L. Ruddock and M. Molinari, A novel stress-induced EDEM variant regulating endoplasmic reticulum-associated glycoprotein degradation, *J. Biol. Chem.* **280**(4), 2424–2428 (2005).

140. M. C. Sousa, M. A. Ferrero-Garcia and A. J. Parodi, Recognition of the oligosaccharide and protein moieties of glycoproteins by the UDP-Glc:glycoprotein glucosyltransferase, *Biochemistry* **31**(1), 97–105 (1992).

141. S. E. Trombetta and A. J. Parodi, Purification to apparent homogeneity and partial characterization of rat liver UDP-glucose:glycoprotein glucosyltransferase, *J. Biol. Chem.* **267**(13), 9236–9240 (1992).

142. H. Hettkamp, G. Legler and E. Bause, Purification by affinity chromatography of glucosidase I, an endoplasmic reticulum hydrolase involved in the processing of asparagine-linked oligosaccharides, *Eur. J. Biochem.* **142**(1), 85–90 (1984).

143. D. Brada and U. C. Dubach, Isolation of a homogeneous glucosidase II from pig kidney microsomes, *Eur. J. Biochem.* **141**(1), 149–156 (1984).

144. S. Saxena, K. Shailubhai, B. Dong-Yu and I. K. Vijay, Purification and characterization of glucosidase II involved in *N*-linked glycoprotein processing in bovine mammary gland, *Biochem. J.* **247**(3), 563–570 (1987).

145. J. Bischoff and R. Kornfeld, Evidence for an α-mannosidase in endoplasmic reticulum of rat liver, *J. Biol. Chem.* **258**(13), 7907–7910 (1983).

146. J. Bischoff, L. Liscum and R. Kornfeld, The use of 1-deoxymannojirimycin to evaluate the role of various α-mannosidases in oligosaccharide processing in intact cells, *J. Biol. Chem.* **261**(10), 4766–4774 (1986).

147. S. Weng and R. G. Spiro, Demonstration that a kifunensine-resistant α-mannosidase with a unique processing action on *N*-linked oligosaccharides occurs in rat liver endoplasmic reticulum and various cultured cells, *J. Biol. Chem.* **268**(34), 25656–25663 (1993).

148. S. Weng and R. G. Spiro, Endoplasmic reticulum kifunensine-resistant α-mannosidase is enzymatically and immunologically related to the cytosolic α-mannosidase, *Arch. Biochem. Biophys.* **325**(1), 113–123 (1996).

149. A. R. Frand and C. A. Kaiser, The *ERO1* gene of yeast is required for oxidation of protein dithiols in the endoplasmic reticulum, *Mol. Cell* **1**(2), 161–170 (1998).

150. M. G. Pollard, K. J. Travers and J. S. Weissman, Ero1p: a novel and ubiquitous protein with an essential role in oxidative protein folding in the endoplasmic reticulum, *Mol. Cell* **1**(2), 171–182 (1998).

151. A. Cabibbo, M. Pagani, M. Fabbri, M. Rocchi, M. R. Farmery, N. J. Bulleid and R. Sitia, ERO1-L, a human protein that favors disulfide bond formation in the endoplasmic reticulum, *J. Biol. Chem.* **275**(7), 4827–4833 (2000).

152. M. Pagani, M. Fabbri, C. Benedetti, A. Fassio, S. Pilati, N. J. Bulleid, A. Cabibbo and R. Sitia, Endoplasmic reticulum oxidoreductin 1-Lβ (ERO1-Lβ), a human gene induced in the course of the unfolded protein response, *J. Biol. Chem.* **275**(31), 23685–23692 (2000).

153. J. Gerber, U. Mühlenhoff, G. Hofhaus, R. Lill and T. Lisowsky, Yeast Erv2p is the first micro-
     somal FAD-linked sulfhydryl oxidase of the Erv1p/Alrp protein family, *J. Biol. Chem.* **276**(26),
     23486–23491 (2001).
154. C. S. Sevier, J. W. Cuozzo, A. Vala, F. Åslund and C. A. Kaiser, A flavoprotein oxidase defines
     a new endoplasmic reticulum pathway for biosynthetic disulphide bond formation, *Nat. Cell Biol.*
     3(10), 874–882 (2001).
155. J. C. Edman, L. Ellis, R. W. Blacher, R. A. Roth and W. J. Rutter, Sequence of protein disulphide
     isomerase and implications of its relationship to thioredoxin, *Nature* **317**(6034), 267–270 (1985).
156. N. J. Bulleid and R. B. Freedman, Defective co-translational formation of disulphide bonds in
     protein disulphide-isomerase-deficient microsomes, *Nature* **335**(6191), 649–651 (1988).
157. J. A. Oberdorf, D. Lebeche, J. F. Head and B. Kaminer, Identification of a calsequestrin-like
     protein from sea urchin eggs, *J. Biol. Chem.* **263**(14), 6806–6809 (1988).
158. R. A. Mazzarella, M. Srinivasan, S. M. Haugejorden and M. Green, ERp72, an abundant luminal
     endoplasmic reticulum protein, contains three copies of the active site sequences of protein disulfide
     isomerase, *J. Biol. Chem.* **265**(2), 1094–1101 (1990).
159. M. LaMantia, T. Miura, H. Tachikawa, H. A. Kaplan, W. J. Lennarz and T. Mizunaga, Glyco-
     sylation site binding protein and protein disulfide isomerase are identical and essential for cell
     viability in yeast, *Proc. Natl. Acad. Sci. U S A* **88**(10), 4453–4457 (1991).
160. H. A. Lucero, D. Lebeche and B. Kaminer, ERcalcistorin/protein disulfide isomerase (PDI).
     Sequence determination and expression of a cDNA clone encoding a calcium storage protein
     with PDI activity from endoplasmic reticulum of the sea urchin egg, *J. Biol. Chem.* **269**(37),
     23112–23119 (1994).
161. M. G. Desilva, J. Lu, G. Donadel, W. S. Modi, H. Xie, A. L. Notkins and M. S. Lan, Characteri-
     zation and chromosomal localization of a new protein disulfide isomerase, PDIp, highly expressed
     in human pancreas, *DNA Cell Biol.* **15**(1), 9–16 (1996).
162. M. G. Desilva, A. L. Notkins and M. S. Lan, Molecular characterization of a pancreas-specific
     protein disulfide isomerase, PDIp, *DNA Cell Biol.* **16**(3), 269–274 (1997).
163. M. van Lith, N. Hartigan, J. Hatch and A. M. Benham, PDILT, a divergent testis-specific protein
     disulfide isomerase with a non-classical SXXC motif that engages in disulfide-dependent interac-
     tions in the endoplasmic reticulum, *J. Biol. Chem.* **280**(2), 1376–1383 (2005).
164. P. N. Van, F. Peter and H. D. Söling, Four intracisternal calcium-binding glycoproteins from
     rat liver microsomes with high affinity for calcium. No indication for calsequestrin-like proteins
     in inositol 1,4,5-trisphosphate-sensitive calcium sequestering rat liver vesicles, *J. Biol. Chem.*
     264(29), 17494–17501 (1989).
165. J. Füllekrug, B. Sönnichsen, U. Wünsch, K. Arseven, P. Nguyen Van, H. D. Söling and G. Mieskes,
     CaBP1, a calcium binding protein of the thioredoxin family, is a resident KDEL protein of the ER
     and not of the intermediate compartment, *J. Cell Sci.* **107**(10), 2719–2727 (1994).
166. J. Lundström-Ljung, U. Birnbach, K. Rupp, H. D. Söling and A. Holmgren, Two resident ER-
     proteins, CaBP1 and CaBP2, with thioredoxin domains, are substrates for thioredoxin reductase:
     comparison with protein disulfide isomerase, *FEBS Lett.* **357**(3), 305–308 (1995).
167. H. Tachikawa, Y. Takeuchi, W. Funahashi, T. Miura, X. D. Gao, D. Fujimoto, T. Mizunaga
     and K. Onodera, Isolation and characterization of a yeast gene, *MPD1*, the overexpression of
     which suppresses inviability caused by protein disulfide isomerase depletion, *FEBS Lett.* **369**(2–3),
     212–216 (1995).
168. M. J. Lewis, R. A. Mazzarella and M. Green, Structure and assembly of the endoplasmic reticulum.
     The synthesis of three major endoplasmic reticulum proteins during lipopolysaccharide-induced
     differentiation of murine lymphocytes, *J. Biol. Chem.* **260**(5), 3050–3057 (1985).
169. P. N. Van, K. Rupp, A. Lampen and H. D. Söling, CaBP2 is a rat homolog of ERp72 with
     proteindisulfide isomerase activity, *Eur. J. Biochem.* **213**(2), 789–795 (1993).
170. M. J. Lewis, R. A. Mazzarella and M. Green, Structure and assembly of the endoplasmic reticulum:
     biosynthesis and intracellular sorting of ERp61, ERp59, and ERp49, three protein components of
     murine endoplasmic reticulum, *Arch. Biochem. Biophys.* **245**(2), 389–403 (1986).

171. R. Urade, M. Nasu, T. Moriyama, K. Wada and M. Kito, Protein degradation by the phosphoinositide-specific phospholipase C-α family from rat liver endoplasmic reticulum, *J. Biol. Chem.* **267**(21), 15152–15159 (1992).

172. C. Tachibana and T. H. Stevens, The yeast *EUG1* gene encodes an endoplasmic reticulum protein that is functionally related to protein disulfide isomerase, *Mol. Cell. Biol.* **12**(10), 4601–4611 (1992).

173. P. Koivunen, T. Helaakoski, P. Annunen, J. Veijola, S. Raisanen, T. Pihlajaniemi and K. I. Kivirikko, ERp60 does not substitute for protein disulphide isomerase as the β-subunit of prolyl 4-hydroxylase, *Biochem. J.* **316**(2), 599–605 (1996).

174. N. Hirano, F. Shibasaki, R. Sakai, T. Tanaka, J. Nishida, Y. Yazaki, T. Takenawa and H. Hirai, Molecular cloning of the human glucose-regulated protein ERp57/GRP58, a thiol-dependent reductase. Identification of its secretory form and inducible expression by the oncogenic transformation, *Eur. J. Biochem.* **234**(1), 336–342 (1995).

175. T. Hayano and M. Kikuchi, Molecular cloning of the cDNA encoding a novel protein disulfide isomerase-related protein (PDIR), *FEBS Lett.* **372**(2–3), 210–214 (1995).

176. B. Knoblach, B. O. Keller, J. Groenendyk, S. Aldred, J. Zheng, B. D. Lemire, L. Li and M. Michalak, ERp19 and ERp46, new members of the thioredoxin family of endoplasmic reticulum proteins, *Mol. Cell. Proteomics* **2**(10), 1104–1119 (2003).

177. T. Anelli, M. Alessio, A. Bachi, L. Bergamelli, G. Bertoli, S. Camerini, A. Mezghrani, E. Ruffato, T. Simmen and R. Sitia, Thiol-mediated protein retention in the endoplasmic reticulum: the role of ERp44, *EMBO J.* **22**(19), 5015–5022 (2003).

178. N. Akiyama, Y. Matsuo, H. Sai, M. Noda and S. Kizaka-Kondoh, Identification of a series of transforming growth factor β-responsive genes by retrovirus-mediated gene trap screening, *Mol. Cell. Biol.* **20**(9), 3266–3273 (2000).

179. Y. Matsuo, N. Akiyama, H. Nakamura, J. Yodoi, M. Noda and S. Kizaka-Kondoh, Identification of a novel thioredoxin-related transmembrane protein, *J. Biol. Chem.* **276**(13), 10032–10038 (2001).

180. Y. Matsuo, Y. Nishinaka, S. Suzuki, M. Kojima, S. Kizaka-Kondoh, N. Kondo, A. Son, J. Sakakura-Nishiyama, Y. Yamaguchi, H. Masutani, Y. Ishii and J. Yodoi, TMX, a human transmembrane oxidoreductase of the thioredoxin family: the possible role in disulfide-linked protein folding in the endoplasmic reticulum, *Arch. Biochem. Biophys.* **423**(1), 81–87 (2004).

181. H. Tachikawa, W. Funahashi, Y. Takeuchi, H. Nakanishi, R. Nishihara, S. Katoh, X. D. Gao, T. Mizunaga and D. Fujimoto, Overproduction of Mpd2p suppresses the lethality of protein disulfide isomerase depletion in a CXXC sequence dependent manner, *Biochem. Biophys. Res. Commun.* **239**(3), 710–714 (1997).

182. Q. Wang and A. Chang, Eps1, a novel PDI-related protein involved in ER quality control in yeast, *EMBO J.* **18**(21), 5972–5982 (1999).

183. Q. Wang and A. Chang, Substrate recognition in ER-associated degradation mediated by Eps1, a member of the protein disulfide isomerase family, *EMBO J.* **22**(15), 3792–3802 (2003).

184. D. M. Ferrari, P. Nguyen Van, H. D. Kratzin and H. D. Söling, ERp28, a human endoplasmic-reticulum-lumenal protein, is a member of the protein disulfide isomerase family but lacks a CXXC thioredoxin-box motif, *Eur. J. Biochem.* **255**(3), 570–579 (1998).

185. M. Konsolaki and T. Schüpbach, *windbeutel*, a gene required for dorsoventral patterning in *Drosophila*, encodes a protein that has homologies to vertebrate proteins of the endoplasmic reticulum, *Genes Dev.* **12**(1), 120–131 (1998).

186. E. Liepinsh, M. Baryshev, A. Sharipo, M. Ingelman-Sundberg, G. Otting and S. Mkrtchian, Thioredoxin fold as homodimerization module in the putative chaperone ERp29: NMR structures of the domains and experimental model of the 51 kDa dimer, *Structure (Camb)* **9**(6), 457–471 (2001).

187. E. Sargsyan, M. Baryshev, L. Szekely, A. Sharipo and S. Mkrtchian, Identification of ERp29, an endoplasmic reticulum lumenal protein, as a new member of the thyroglobulin folding complex, *J. Biol. Chem.* **277**(19), 17009–17015 (2002).

188. J. K. Suh, L. L. Poulsen, D. M. Ziegler and J. D. Robertus, Yeast flavin-containing monooxygenase generates oxidizing equivalents that control protein folding in the endoplasmic reticulum, *Proc. Natl. Acad. Sci. U S A* **96**(6), 2687–2691 (1999).

189. P. Caroni, A. Rothenfluh, E. McGlynn and C. Schneider, S-cyclophilin. New member of the cyclophilin family associated with the secretory pathway, *J. Biol. Chem.* **266**(17), 10739–10742 (1991).

190. G. Spik, B. Haendler, O. Delmas, C. Mariller, M. Chamoux, P. Maes, A. Tartar, J. Montreuil, K. Stedman, H. P. Kocher and et al., A novel secreted cyclophilin-like protein (SCYLP), *J. Biol. Chem.* **266**(17), 10735–10738 (1991).

191. N. Iwai and T. Inagami, Molecular cloning of a complementary DNA to rat cyclophilin-like protein mRNA, *Kidney Int.* **37**(6), 1460–1465 (1990).

192. P. L. Koser, D. Sylvester, G. P. Livi and D. J. Bergsma, A second cyclophilin-related gene in *Saccharomyces cerevisiae*, *Nucleic Acids Res.* **18**(6), 1643 (1990).

193. E. R. Price, L. D. Zydowsky, M. J. Jin, C. H. Baker, F. D. McKeon and C. T. Walsh, Human cyclophilin B: a second cyclophilin gene encodes a peptidyl-prolyl isomerase with a signal sequence, *Proc. Natl. Acad. Sci. U S A* **88**(5), 1903–1907 (1991).

194. B. H. Shieh, M. A. Stamnes, S. Seavello, G. L. Harris and C. S. Zuker, The *ninaA* gene required for visual transduction in *Drosophila* encodes a homologue of cyclosporin A-binding protein, *Nature* **338**(6210), 67–70 (1989).

195. N. J. Colley, E. K. Baker, M. A. Stamnes and C. S. Zuker, The cyclophilin homolog ninaA is required in the secretory pathway, *Cell* **67**(2), 255–263 (1991).

196. M. A. Stamnes, B. H. Shieh, L. Chuman, G. L. Harris and C. S. Zuker, The cyclophilin homolog ninaA is a tissue-specific integral membrane protein required for the proper synthesis of a subset of Drosophila rhodopsins, *Cell* **65**(2), 219–227 (1991).

197. E. K. Baker, N. J. Colley and C. S. Zuker, The cyclophilin homolog NinaA functions as a chaperone, forming a stable complex *in vivo* with its protein target rhodopsin, *EMBO J.* **13**(20), 4886–4895 (1994).

198. S. K. Nigam, Y. J. Jin, M. J. Jin, K. T. Bush, B. E. Bierer and S. J. Burakoff, Localization of the FK506-binding protein, FKBP 13, to the lumen of the endoplasmic reticulum, *Biochem. J.* **294**(2), 511–515 (1993).

199. K. T. Bush, B. A. Hendrickson and S. K. Nigam, Induction of the FK506-binding protein, FKBP13, under conditions which misfold proteins in the endoplasmic reticulum, *Biochem. J.* **303**(3), 705–708 (1994).

200. M. C. Coss, D. Winterstein, R. C. Sowder, II, and S. L. Simek, Molecular cloning, DNA sequence analysis, and biochemical characterization of a novel 65-kDa FK506-binding protein (FKBP65), *J. Biol. Chem.* **270**(49), 29336–29341 (1995).

201. L. G. Josefsson and L. L. Randall, Processing *in vivo* of precursor maltose-binding protein in *Escherichia coli* occurs post-translationally as well as co-translationally, *J. Biol. Chem.* **256**(5), 2504–2507 (1981).

202. L. G. Josefsson and L. L. Randall, Different exported proteins in E. coli show differences in the temporal mode of processing in vivo, *Cell* **25**(1), 151–157 (1981).

203. D. T. Rutkowski, C. M. Ott, J. R. Polansky and V. R. Lingappa, Signal sequences initiate the pathway of maturation in the endoplasmic reticulum lumen, *J. Biol. Chem.* **278**(32), 30365–30372 (2003).

204. Y. Li, J. J. Bergeron, L. Luo, W. J. Ou, D. Y. Thomas and C. Y. Kang, Effects of inefficient cleavage of the signal sequence of HIV-1 gp 120 on its association with calnexin, folding, and intracellular transport, *Proc. Natl. Acad. Sci. U S A* **93**(18), 9606–9611 (1996).

205. E. A. Craig, The heat shock response, *CRC Crit. Rev. Biochem.* **18**(3), 239–280 (1985).

206. S. Lindquist, The heat-shock response, *Annu. Rev. Biochem.* **55**, 1151–1191 (1986).

207. P. K. Sorger, Heat shock factor and the heat shock response, *Cell* **65**(3), 363–366 (1991).

208. A. I. Gragerov, E. S. Martin, M. A. Krupenko, M. V. Kashlev and V. G. Nikiforov, Protein aggregation and inclusion body formation in *Escherichia coli* rpoH mutant defective in heat shock protein induction, *FEBS Lett.* **291**(2), 222–224 (1991).

209. G. C. Flynn, J. Pohl, M. T. Flocco and J. E. Rothman, Peptide-binding specificity of the molecular chaperone BiP, *Nature* **353**(6346), 726–730 (1991).

210. S. Blond-Elguindi, S. E. Cwirla, W. J. Dower, R. J. Lipshutz, S. R. Sprang, J. F. Sambrook and M.-J. Gething, Affinity panning of a library of peptides displayed on bacteriophages reveals the binding specificity of BiP, *Cell* **75**(4), 717–728 (1993).
211. J. L. Brodsky, J. Goeckeler and R. Schekman, BiP and Sec63p are required for both co- and posttranslational protein translocation into the yeast endoplasmic reticulum, *Proc. Natl. Acad. Sci. U S A* **92**(21), 9643–9646 (1995).
212. S. K. Lyman and R. Schekman, Interaction between BiP and Sec63p is required for the completion of protein translocation into the ER of *Saccharomyces cerevisiae*, *J. Cell Biol.* **131**(5), 1163–1171 (1995).
213. S. K. Lyman and R. Schekman, Binding of secretory precursor polypeptides to a translocon subcomplex is regulated by BiP, *Cell* **88**(1), 85–96 (1997).
214. B. D. Hamman, L. M. Hendershot and A. E. Johnson, BiP maintains the permeability barrier of the ER membrane by sealing the lumenal end of the translocon pore before and early in translocation, *Cell* **92**(6), 747–758 (1998).
215. B. P. Young, R. A. Craven, P. J. Reid, M. Willer and C. J. Stirling, Sec63p and Kar2p are required for the translocation of SRP-dependent precursors into the yeast endoplasmic reticulum *in vivo*, *EMBO J.* **20**(1–2), 262–271 (2001).
216. N. N. Alder, Y. Shen, J. L. Brodsky, L. M. Hendershot and A. E. Johnson, The molecular mechanisms underlying BiP-mediated gating of the Sec61 translocon of the endoplasmic reticulum, *J. Cell Biol.* **168**(3), 389–399 (2005).
217. J. L. Brodsky, E. D. Werner, M. E. Dubas, J. L. Goeckeler, K. B. Kruse and A. A. McCracken, The requirement for molecular chaperones during endoplasmic reticulum-associated protein degradation demonstrates that protein export and import are mechanistically distinct, *J. Biol. Chem.* **274**(6), 3453–3460 (1999).
218. S.-i. Nishikawa, S. W. Fewell, Y. Kato, J. L. Brodsky and T. Endo, Molecular chaperones in the yeast endoplasmic reticulum maintain the solubility of proteins for retrotranslocation and degradation, *J. Cell Biol.* **153**(5), 1061–1070 (2001).
219. G. J. Steel, D. M. Fullerton, J. R. Tyson and C. J. Stirling, Coordinated activation of Hsp70 chaperones, *Science* **303**(5654), 98–101 (2004).
220. B. Bukau and A. L. Horwich, The Hsp70 and Hsp60 chaperone machines, *Cell* **92**(3), 351–366 (1998).
221. J. C. Young, I. Moarefi and F. U. Hartl, Hsp90: a specialized but essential protein-folding tool, *J. Cell Biol.* **154**(2), 267–273 (2001).
222. X. Zhu, X. Zhao, W. F. Burkholder, A. Gragerov, C. M. Ogata, M. E. Gottesman and W. A. Hendrickson, Structural analysis of substrate binding by the molecular chaperone DnaK, *Science* **272**(5268), 1606–1614 (1996).
223. M.-J. Gething, Role and regulation of the ER chaperone BiP, *Semin. Cell Dev. Biol.* **10**(5), 465–472 (1999).
224. P. Spee, J. Subjeck and J. Neefjes, Identification of novel peptide binding proteins in the endoplasmic reticulum: ERp72, calnexin, and grp170, *Biochemistry* **38**(32), 10559–10566 (1999).
225. G. C. Flynn, T. G. Chappell and J. E. Rothman, Peptide binding and release by proteins implicated as catalysts of protein assembly, *Science* **245**(4916), 385–390 (1989).
226. S. Blond-Elguindi, A. M. Fourie, J. F. Sambrook and M. J. Gething, Peptide-dependent stimulation of the ATPase activity of the molecular chaperone BiP is the result of conversion of oligomers to active monomers, *J. Biol. Chem.* **268**(17), 12730–12735 (1993).
227. A. M. Fourie, J. F. Sambrook and M.-J. Gething, Common and divergent peptide binding specificities of hsp70 molecular chaperones, *J. Biol. Chem.* **269**(48), 30470–30478 (1994).
228. G. Knarr, M.-J. Gething, S. Modrow and J. Buchner, BiP binding sequences in antibodies, *J. Biol. Chem.* **270**(46), 27589–27594 (1995).
229. J. D. Jamieson and G. E. Palade, Intracellular transport of secretory proteins in the pancreatic exocrine cell. IV. Metabolic requirements, *J. Cell Biol.* **39**(3), 589–603 (1968).
230. A. J. Dorner, L. C. Wasley and R. J. Kaufman, Protein dissociation from GRP78 and secretion are blocked by depletion of cellular ATP levels, *Proc. Natl. Acad. Sci. U S A* **87**(19), 7429–7432 (1990).

231. I. Braakman, J. Helenius and A. Helenius, Role of ATP and disulphide bonds during protein folding in the endoplasmic reticulum, *Nature* **356**(6366), 260–262 (1992).
232. A. J. Dorner and R. J. Kaufman, The levels of endoplasmic reticulum proteins and ATP affect folding and secretion of selective proteins, *Biologicals* **22**(2), 103–112 (1994).
233. B. Misselwitz, O. Staeck and T. A. Rapoport, J proteins catalytically activate Hsp70 molecules to trap a wide range of peptide sequences. *Mol. Cell* **2**(5), 593–603 (1998).
234. B. Misselwitz, O. Staeck, K. E. Matlack and T. A. Rapoport, Interaction of BiP with the J-domain of the Sec63p component of the endoplasmic reticulum protein translocation complex, *J. Biol. Chem.* **274**(29), 20110–20115 (1999).
235. S. E. Brightman, G. L. Blatch and B. R. Zetter, Isolation of a mouse cDNA encoding MTJ1, a new murine member of the DnaJ family of proteins, *Gene* **153**(2), 249–254 (1995).
236. C. Bies, R. Blum, J. Dudek, W. Nastainczyk, S. Oberhauser, M. Jung and R. Zimmermann, Characterization of pancreatic ERj3p, a homolog of yeast DnaJ-like protein Scj1p, *Biol. Chem.* **385**(5), 389–395 (2004).
237. M. Yu, R. H. Haslam and D. B. Haslam, HEDJ, an Hsp40 co-chaperone localized to the endoplasmic reticulum of human cells, *J. Biol. Chem.* **275**(32), 24984–24992 (2000).
238. F. Pröls, M. P. Mayer, O. Renner, P. G. Czarnecki, M. Ast, C. Gässler, J. Wilting, H. Kurz and B. Christ, Upregulation of the cochaperone Mdg1 in endothelial cells is induced by stress and during in vitro angiogenesis, *Exp. Cell Res.* **269**(1), 42–53 (2001).
239. M. H. Skowronek, M. Rotter and I. G. Haas, Molecular characterization of a novel mammalian DnaJ-like Sec63p homolog, *Biol. Chem.* **380**(9), 1133–1138 (1999).
240. L. Zhao, C. Longo-Guess, B. S. Harris, J. W. Lee and S. L. Ackerman, Protein accumulation and neurodegeneration in the woozy mutant mouse is caused by disruption of SIL1, a cochaperone of BiP, *Nat. Genet.* **37**(9), 974–979 (2005).
241. A. Boisramé, J. M. Beckerich and C. Gaillardin, Sls1p, an endoplasmic reticulum component, is involved in the protein translocation process in the yeast *Yarrowia lipolytica, J. Biol. Chem.* **271**(20), 11668–11675 (1996).
242. L. E. Hightower, S. E. Sadis and I. M. Takenaka, in: *The Biology of Heat Shock Proteins and Molecular Chaperones*, edited by R. I. Morimoto, A. Tissières and C. Georgopoulos (Cold Spring Harbor Laboratory Press, Cold Spring Harbor, 1994), pp. 197–207.
243. C. B. Hirschberg, P. W. Robbins and C. Abeijon, Transporters of nucleotide sugars, ATP, and nucleotide sulfate in the endoplasmic reticulum and Golgi apparatus, *Annu. Rev. Biochem.* **67**, 49–69 (1998).
244. L. M. Hendershot, J. Ting and A. S. Lee, Identity of the immunoglobulin heavy-chain-binding protein with the 78,000-dalton glucose-regulated protein and the role of posttranslational modifications in its binding function, *Mol. Cell. Biol.* **8**(10), 4250–4256 (1988).
245. A. Carlino, H. Toledo, D. Skaleris, R. DeLisio, H. Weissbach and N. Brot, Interactions of liver Grp78 and *Escherichia coli* recombinant Grp78 with ATP: multiple species and disaggregation, *Proc. Natl. Acad. Sci. U S A* **89**(6), 2081–2085 (1992).
246. P. J. Freiden, J. R. Gaut and L. M. Hendershot, Interconversion of three differentially modified and assembled forms of BiP, *EMBO J.* **11**(1), 63–70 (1992).
247. W. J. Welch, J. I. Garrels, G. P. Thomas, J. J. Lin and J. R. Feramisco, Biochemical characterization of the mammalian stress proteins and identification of two stress proteins as glucose- and Ca^{2+}-ionophore-regulated proteins, *J. Biol. Chem.* **258**(11), 7102–7111 (1983).
248. T. Leustek, H. Toledo, N. Brot and H. Weissbach, Calcium-dependent autophosphorylation of the glucose-regulated protein, Grp78, *Arch. Biochem. Biophys.* **289**(2), 256–261 (1991).
249. T. Leustek, D. Amir-Shapira, H. Toledo, N. Brot and H. Weissbach, Autophosphorylation of 70 kDa heat shock proteins, *Cell. Mol. Biol.* **38**(1), 1–10 (1992).
250. J. R. Gaut, In vivo threonine phosphorylation of immunoglobulin binding protein (BiP) maps to its protein binding domain, *Cell Stress Chaperones* **2**(4), 252–262 (1997).
251. L. Carlsson and E. Lazarides, ADP-ribosylation of the M_r 83,000 stress-inducible and glucose-regulated protein in avian and mammalian cells: modulation by heat shock and glucose starvation, *Proc. Natl. Acad. Sci. U S A* **80**(15), 4664–4668 (1983).

252. G. H. Leno and B. E. Ledford, Reversible ADP-ribosylation of the 78 kDa glucose-regulated protein, *FEBS Lett.* **276**(1–2), 29–33 (1990).

253. B. E. Ledford and G. H. Leno, ADP-ribosylation of the molecular chaperone GRP78/BiP, *Mol. Cell. Biochem.* **138**(1–2), 141–148 (1994).

254. A. L. Laitusis, M. A. Brostrom and C. O. Brostrom, The dynamic role of GRP78/BiP in the coordination of mRNA translation with protein processing, *J. Biol. Chem.* **274**(1), 486–493 (1999).

255. K. S. Crowley, S. Liao, V. E. Worrell, G. D. Reinhart and A. E. Johnson, Secretory proteins move through the endoplasmic reticulum membrane via an aqueous, gated pore, *Cell* **78**(3), 461–471 (1994).

256. A. S. Lee, J. Bell and J. Ting, Biochemical characterization of the 94- and 78-kilodalton glucose-regulated proteins in hamster fibroblasts, *J. Biol. Chem.* **259**(7), 4616–4621 (1984).

257. Y. Argon and B. B. Simen, GRP94, an ER chaperone with protein and peptide binding properties, *Semin. Cell Dev. Biol.* **10**(5), 495–505 (1999).

258. K. A. Hutchison, B. Nevins, F. Perini and I. H. Fox, Soluble and membrane-associated human low-affinity adenosine binding protein (adenotin): properties and homology with mammalian and avian stress proteins, *Biochemistry* **29**(21), 5138–5144 (1990).

259. T. Fein, E. Schulze, J. Bär and U. Schwabe, Purification and characterization of an adenotin-like adenosine binding protein from human platelets, *Naunyn Schmiedebergs Arch Pharmacol* **349**(4), 374–380 (1994).

260. P. K. Srivastava, Y.-T. Chen and L. J. Old, 5'-structural analysis of genes encoding polymorphic antigens of chemically induced tumors, *Proc. Natl. Acad. Sci. U S A* **84**(11), 3807–3811 (1987).

261. R. G. Maki, R. L. Eddy, Jr., M. Byers, T. B. Shows and P. K. Srivastava, Mapping of the genes for human endoplasmic reticular heat shock protein gp96/grp94, *Somat. Cell Mol. Genet.* **19**(1), 73–81 (1993).

262. R. A. Mazzarella and M. Green, ERp99, an abundant, conserved glycoprotein of the endoplasmic reticulum, is homologous to the 90-kDa heat shock protein (hsp90) and the 94-kDa glucose regulated protein (GRP94), *J. Biol. Chem.* **262**(18), 8875–8883 (1987).

263. M. S. Kulomaa, N. L. Weigel, D. A. Kleinsek, W. G. Beattie, O. M. Conneely, C. March, T. Zarucki-Schulz, W. T. Schrader and B. W. O'Malley, Amino acid sequence of a chicken heat shock protein derived from the complementary DNA nucleotide sequence, *Biochemistry* **25**(20), 6244–6251 (1986).

264. U. Dechert, M. Weber, M. Weber-Schaeuffelen and E. Wollny, Isolation and partial characterization of an 80,000-dalton protein kinase from the microvessels of the porcine brain, *J. Neurochem.* **53**(4), 1268–1275 (1989).

265. S.-i. Yamada, T. Ono, A. Mizuno and T. K. Nemoto, A hydrophobic segment within the C-terminal domain is essential for both client-binding and dimer formation of the HSP90-family molecular chaperone, *Eur. J. Biochem.* **270**(1), 146–154 (2003).

266. J. Melnick, J. L. Dul and Y. Argon, Sequential interaction of the chaperones BiP and GRP94 with immunoglobulin chains in the endoplasmic reticulum, *Nature* **370**(6488), 373–375 (1994).

267. C. A. Clairmont, A. De Maio and C. B. Hirschberg, Translocation of ATP into the lumen of rough endoplasmic reticulum-derived vesicles and its binding to luminal proteins including BiP (GRP 78) and GRP 94, *J. Biol. Chem.* **267**(6), 3983–3990 (1992).

268. T. Dierks, J. Volkmer, G. Schlenstedt, C. Jung, U. Sandholzer, K. Zachmann, P. Schlotterhose, K. Neifer, B. Schmidt and R. Zimmermann, A microsomal ATP-binding protein involved in efficient protein transport into the mammalian endoplasmic reticulum, *EMBO J.* **15**(24), 6931–6942 (1996).

269. Z. Li and P. K. Srivastava, Tumor rejection antigen gp96/grp94 is an ATPase: implications for protein folding and antigen presentation, *EMBO J.* **12**(8), 3143–3151 (1993).

270. W. M. Obermann, H. Sondermann, A. A. Russo, N. P. Pavletich and F. U. Hartl, In vivo function of Hsp90 is dependent on ATP binding and ATP hydrolysis, *J. Cell Biol.* **143**(4), 901–910 (1998).

271. B. Panaretou, C. Prodromou, S. M. Roe, R. O'Brien, J. E. Ladbury, P. W. Piper and L. H. Pearl, ATP binding and hydrolysis are essential to the function of the Hsp90 molecular chaperone in vivo, *EMBO J.* **17**(16), 4829–4836 (1998).

272. K. L. Soldano, A. Jivan, C. V. Nicchitta and D. T. Gewirth, Structure of the N-terminal domain of GRP94. Basis for ligand specificity and regulation, *J. Biol. Chem.* **278**(48), 48330–48338 (2003).

273. M. F. Rosser, B. M. Trotta, M. R. Marshall, B. Berwin and C. V. Nicchitta, Adenosine nucleotides and the regulation of GRP94-client protein interactions, *Biochemistry* **43**(27), 8835–8845 (2004).

274. M. F. Rosser and C. V. Nicchitta, Ligand interactions in the adenosine nucleotide-binding domain of the Hsp90 chaperone, GRP94. I. Evidence for allosteric regulation of ligand binding, *J. Biol. Chem.* **275**(30), 22798–22805 (2000).

275. A. J. Caplan, Hsp90's secrets unfold: new insights from structural and functional studies, *Trends Cell Biol.* **9**(7), 262–268 (1999).

276. J. Buchner, Hsp90 & Co. - a holding for folding, *Trends. Biochem. Sci.* **24**(4), 136–141 (1999).

277. D. Picard, Heat-shock protein 90, a chaperone for folding and regulation, *Cell. Mol. Life Sci.* **59**(10), 1640–1648 (2002).

278. J. C. Young, J. M. Barral and F. Ulrich Hartl, More than folding: localized functions of cytosolic chaperones, *Trends. Biochem. Sci.* **28**(10), 541–547 (2003).

279. W. T. Schaiff, K. A. Hruska, Jr., D. W. McCourt, M. Green and B. D. Schwartz, HLA-DR associates with specific stress proteins and is retained in the endoplasmic reticulum in invariant chain negative cells, *J. Exp. Med.* **176**(3), 657–666 (1992).

280. L. R. Ferreira, K. Norris, T. Smith, C. Hebert and J. J. Sauk, Association of Hsp47, Grp78, and Grp94 with procollagen supports the successive or coupled action of molecular chaperones, *J. Cell. Biochem.* **56**(4), 518–526 (1994).

281. G. Kuznetsov, L. B. Chen and S. K. Nigam, Multiple molecular chaperones complex with misfolded large oligomeric glycoproteins in the endoplasmic reticulum, *J. Biol. Chem.* **272**(5), 3057–3063 (1997).

282. L. Meunier, Y. K. Usherwood, K. T. Chung and L. M. Hendershot, A subset of chaperones and folding enzymes form multiprotein complexes in endoplasmic reticulum to bind nascent proteins, *Mol. Biol. Cell* **13**(12), 4456–4469 (2002).

283. S. C. Hubbard and R. J. Ivatt, Synthesis and processing of asparagine-linked oligosaccharides, *Annu. Rev. Biochem.* **50**, 555–583 (1981).

284. R. Kornfeld and S. Kornfeld, Assembly of asparagine-linked oligosaccharides, *Annu. Rev. Biochem.* **54**, 631–664 (1985).

285. M. R. Wormald and R. A. Dwek, Glycoproteins: glycan presentation and protein-fold stability, *Structure (Camb)* **7**(7), R155–160 (1999).

286. L. Ellgaard, M. Molinari and A. Helenius, Setting the standards: quality control in the secretory pathway, *Science* **286**(5446), 1882–1888 (1999).

287. L. Ellgaard and A. Helenius, Quality control in the endoplasmic reticulum, *Nat. Rev. Mol. Cell. Biol.* **4**(3), 181–191 (2003).

288. E. Degen and D. B. Williams, Participation of a novel 88–kD protein in the biogenesis of murine class I histocompatibility molecules, *J. Cell Biol.* **112**(6), 1099–1115 (1991).

289. I. Wada, D. Rindress, P. H. Cameron, W. J. Ou, J. J. Doherty II, D. Louvard, A. W. Bell, D. Dignard, D. Y. Thomas and J. J. Bergeron, SSRα and associated calnexin are major calcium binding proteins of the endoplasmic reticulum membrane, *J. Biol. Chem.* **266**(29), 19599–19610 (1991).

290. M. J. Smith and G. L. E. Koch, Multiple zones in the sequence of calreticulin (CRP55, calregulin, HACBP), a major calcium binding ER/SR protein, *EMBO J.* **8**(12), 3581–3586 (1989).

291. N. C. Khanna, M. Tokuda and D. M. Waisman, Conformational changes induced by binding of divalent cations to calregulin, *J. Biol. Chem.* **261**(19), 8883–8887 (1986).

292. D. R. Macer and G. L. Koch, Identification of a set of calcium-binding proteins in reticuloplasm, the luminal content of the endoplasmic reticulum, *J. Cell Sci.* **91**(1), 61–70 (1988).

293. T. J. Ostwald and D. H. MacLennan, Isolation of a high affinity calcium-binding protein from sarcoplasmic reticulum, *J. Biol. Chem.* **249**(3), 974–979 (1974).

294. W.-J. Ou, P. H. Cameron, D. Y. Thomas and J. J. Bergeron, Association of folding intermediates of glycoproteins with calnexin during protein maturation, *Nature* **364**(6440), 771–776 (1993).

295. J. R. Peterson, A. Ora, P. N. Van and A. Helenius, Transient, lectin-like association of calreticulin with folding intermediates of cellular and viral glycoproteins, *Mol. Biol. Cell* **6**(9), 1173–1184 (1995).

296. J. D. Schrag, J. J. M. Bergeron, Y. Li, S. Borisova, M. Hahn, D. Y. Thomas and M. Cygler, The structure of calnexin, an ER chaperone involved in quality control of protein folding, *Mol. Cell* **8**(3), 633–644 (2001).

297. A. J. Parodi, Protein glucosylation and its role in protein folding, *Annu. Rev. Biochem.* **69**, 69–93 (2000).

298. C. Hammond, I. Braakman and A. Helenius, Role of N-linked oligosaccharide recognition, glucose trimming, and calnexin in glycoprotein folding and quality control, *Proc. Natl. Acad. Sci. U S A* **91**(3), 913–917 (1994).

299. F. Fernández, C. D'Alessio, S. Fanchiotti and A. J. Parodi, A misfolded protein conformation is not a sufficient condition for *in vivo* glucosylation by the UDP-Glc:glycoprotein glucosyltransferase, *EMBO J.* **17**(20), 5877–5886 (1998).

300. M. Molinari, C. Galli, O. Vanoni, S. M. Arnold and R. J. Kaufman, Persistent glycoprotein misfolding activates the glucosidase II/UGT1-driven calnexin cycle to delay aggregation and loss of folding competence, *Mol. Cell* **20**(4), 503–512 (2005).

301. M. Sousa and A. J. Parodi, The molecular basis for the recognition of misfolded glycoproteins by the UDP-Glc:glycoprotein glucosyltransferase, *EMBO J.* **14**(17), 4196–4203 (1995).

302. C. Ritter and A. Helenius, Recognition of local glycoprotein misfolding by the ER folding sensor UDP-glucose:glycoprotein glucosyltransferase, *Nat. Struct. Biol.* **7**(4), 278–280 (2000).

303. S. C. Taylor, A. D. Ferguson, J. J. Bergeron and D. Y. Thomas, The ER protein folding sensor UDP-glucose glycoprotein-glucosyltransferase modifies substrates distant to local changes in glycoprotein conformation, *Nat. Struct. Mol. Biol.* **11**(2), 128–134 (2004).

304. C. Ritter, K. Quirin, M. Kowarik and A. Helenius, Minor folding defects trigger local modification of glycoproteins by the ER folding sensor GT, *EMBO J.* **24**(9), 1730–1738 (2005).

305. C. Appenzeller, H. Andersson, F. Kappeler and H.-P. Hauri, The lectin ERGIC-53 is a cargo transport receptor for glycoproteins, *Nat. Cell Biol.* **1**(6), 330–334 (1999).

306. M. E. Egan, J. Glöckner-Pagel, C. Ambrose, P. A. Cahill, L. Pappoe, N. Balamuth, E. Cho, S. Canny, C. A. Wagner, J. Geibel and M. J. Caplan, Calcium-pump inhibitors induce functional surface expression of ΔF508-CFTR protein in cystic fibrosis epithelial cells, *Nat. Med.* **8**(5), 485–492 (2002).

307. L. S. Grinna and P. W. Robbins, Substrate specificities of rat liver microsomal glucosidases which process glycoproteins, *J. Biol. Chem.* **255**(6), 2255–2258 (1980).

308. R. G. Spiro, Q. Zhu, V. Bhoyroo and H. D. Söling, Definition of the lectin-like properties of the molecular chaperone, calreticulin, and demonstration of its copurification with endomannosidase from rat liver Golgi, *J. Biol. Chem.* **271**(19), 11588–11594 (1996).

309. A. Vassilakos, M. Michalak, M. A. Lehrman and D. B. Williams, Oligosaccharide binding characterics of the molecular chaperones calnexin and calreticulin, *Biochemistry* **37**(10), 3480–3490 (1998).

310. G. Z. Lederkremer and M. H. Glickman, A window of opportunity: timing protein degradation by trimming of sugars and ubiquitins, *Trends. Biochem. Sci.* **30**(6), 297–303 (2005).

311. M. Molinari, V. Calanca, C. Galli, P. Lucca and P. Paganetti, Role of EDEM in the release of misfolded glycoproteins from the calnexin cycle, *Science* **299**(5611), 1397–1400 (2003).

312. Y. Oda, N. Hosokawa, I. Wada and K. Nagata, EDEM as an acceptor of terminally misfolded glycoproteins released from calnexin, *Science* **299**(5611), 1394–1397 (2003).

313. C. E. Jessop, S. Chakravarthi, R. H. Watkins and N. J. Bulleid, Oxidative protein folding in the mammalian endoplasmic reticulum, *Biochem. Soc. Trans.* **32**(5), 655–658 (2004).

314. D. M. Ferrari and H. D. Söling, The protein disulphide-isomerase family: unravelling a string of folds, *Biochem. J.* **339**(1), 1–10 (1999).

315. M. M. Lyles and H. F. Gilbert, Catalysis of the oxidative folding of ribonuclease A by protein disulfide isomerase: pre-steady-state kinetics and the utilization of the oxidizing equivalents of the isomerase, *Biochemistry* **30**(3), 619–625 (1991).

316. B. P. Tu and J. S. Weissman, The FAD- and O_2-dependent reaction cycle of Ero1-mediated oxidative protein folding in the endoplasmic reticulum, *Mol. Cell* **10**(5), 983–994 (2002).

317. A. Kerem, C. Kronman, S. Bar-Nun, A. Shafferman and B. Velan, Interrelations between assembly and secretion of recombinant human acetylcholinesterase, *J. Biol. Chem.* **268**(1), 180–184 (1993).

318. R. Sitia, M. Neuberger, C. Alberini, P. Bet, A. Fra, C. Valetti, G. Williams and C. Milstein, Developmental regulation of IgM secretion: the role of the carboxy-terminal cysteine, *Cell* **60**(5), 781–790 (1990).

319. E. Anken and I. Braakman, Endoplasmic reticulum stress and the making of a professional secretory cell, *Crit. Rev. Biochem. Mol. Biol.* **40**(5), 269–283 (2005).

320. P. Gillece, J. M. Luz, W. J. Lennarz, F. J. de La Cruz and K. Römisch, Export of a cysteine-free misfolded secretory protein from the endoplasmic reticulum for degradation requires interaction with protein disulfide isomerase, *J. Cell Biol.* **147**(7), 1443–1456 (1999).

321. S. Schneuwly, R. D. Shortridge, D. C. Larrivee, T. Ono, M. Ozaki and W. L. Pak, *Drosophila ninaA* gene encodes an eye-specific cyclophilin (cyclosporine A binding protein), *Proc. Natl. Acad. Sci. U S A* **86**(14), 5390–5394 (1989).

322. Y.-J. Jin, M. W. Albers, W. S. Lane, B. E. Bierer, S. L. Schreiber and S. J. Burakoff, Molecular cloning of a membrane-associated human FK506- and rapamycin-binding protein, FKBP-13, *Proc. Natl. Acad. Sci. U S A* **88**(15), 6677–6681 (1991).

323. K. Dolinski, S. Muir, M. Cardenas and J. Heitman, All cyclophilins and FK506 binding proteins are, individually and collectively, dispensable for viability in *Saccharomyces cerevisiae, Proc. Natl. Acad. Sci. U S A* **94**(24), 13093–13098 (1997).

324. C. Y. Liu, Z. Xu and R. J. Kaufman, Structure and intermolecular interactions of the luminal dimerization domain of human IRE1α, *J. Biol. Chem.* **278**(20), 17680–17687 (2003).

325. C. Y. Liu, H. N. Wong, J. A. Schauerte and R. J. Kaufman, The protein kinase/endoribonuclease IRE1α that signals the unfolded protein response has a luminal *N*-terminal ligand-independent dimerization domain, *J. Biol. Chem.* **277**(21), 18346–18356 (2002).

326. Y. Kimata, D. Oikawa, Y. Shimizu, Y. Ishiwata-Kimata and K. Kohno, A role for BiP as an adjustor for the endoplasmic reticulum stress-sensing protein Ire1, *J. Cell Biol.* **167**(3), 445–456 (2004).

327. K. Haze, H. Yoshida, H. Yanagi, T. Yura and K. Mori, Mammalian transcription factor ATF6 is synthesized as a transmembrane protein and activated by proteolysis in response to endoplasmic reticulum stress, *Mol. Biol. Cell* **10**(11), 3787–3799 (1999).

328. M. Li, P. Baumeister, B. Roy, T. Phan, D. Foti, S. Luo and A. S. Lee, ATF6 as a transcription activator of the endoplasmic reticulum stress element: thapsigargin stress-induced changes and synergistic interactions with NF-Y and YY1, *Mol. Cell. Biol.* **20**(14), 5096–5106 (2000).

329. J. Nikawa and S. Yamashita, IRE1 encodes a putative protein kinase containing a membrane-spanning domain and is required for inositol phototrophy in Saccharomyces cerevisiae, *Mol. Microbiol.* **6**(11), 1441–1446 (1992).

330. P. Bork and C. Sander, A hybrid protein kinase-RNase in an interferon-induced pathway?, *FEBS Lett.* **334**(2), 149–152 (1993).

331. J. S. Cox, C. E. Shamu and P. Walter, Transcriptional induction of genes encoding endoplasmic reticulum resident proteins requires a transmembrane protein kinase, *Cell* **73**(6), 1197–1206 (1993).

332. K. Mori, W. Ma, M.-J. Gething and J. Sambrook, A transmembrane protein with a cdc2+/CDC28-related kinase activity is required for signaling from the ER to the nucleus, *Cell* **74**(4), 743–756 (1993).

333. H. P. Harding, Y. Zhang and D. Ron, Protein translation and folding are coupled by an endoplasmic-reticulum-resident kinase, *Nature* **397**(6716), 271–274 (1999).

334. Y. Shi, K. M. Vattem, R. Sood, J. An, J. Liang, L. Stramm and R. C. Wek, Identification and characterization of pancreatic eukaryotic initiation factor 2α-subunit kinase, PEK, involved in translational control, *Mol. Cell. Biol.* **18**(12), 7499–7509 (1998).

335. Y. Shi, J. An, J. Liang, S. E. Hayes, G. E. Sandusky, L. E. Stramm and N. N. Yang, Characterization of a mutant pancreatic eIF-2α kinase, PEK, and co-localization with somatostatin in islet delta cells, *J. Biol. Chem.* **274**(9), 5723–5730 (1999).

336. T. W. Hai, F. Liu, W. J. Coukos and M. R. Green, Transcription factor ATF cDNA clones: an extensive family of leucine zipper proteins able to selectively form DNA-binding heterodimers, *Genes Dev.* **3**(12B), 2083–2090 (1989).

337. J. Min, H. Shukla, H. Kozono, S. K. Bronson, S. M. Weissman and D. D. Chaplin, A novel Creb family gene telomeric of HLA-DRA in the HLA complex, *Genomics* **30**(2), 149–156 (1995).

338. A. Khanna and R. D. Campbell, The gene *G13* in the class III region of the human MHC encodes a potential DNA-binding protein, *Biochem. J.* **319**(1), 81–89 (1996).

339. K. Haze, T. Okada, H. Yoshida, H. Yanagi, T. Yura, M. Negishi and K. Mori, Identification of the G13 (cAMP-response-element-binding protein-related protein) gene product related to activating transcription factor 6 as a transcriptional activator of the mammalian unfolded protein response, *Biochem. J.* **355**(1), 19–28 (2001).

340. S. Kondo, T. Murakami, K. Tatsumi, M. Ogata, S. Kanemoto, K. Otori, K. Iseki, A. Wanaka and K. Imaizumi, OASIS, a CREB/ATF-family member, modulates UPR signalling in astrocytes, *Nat. Cell Biol.* **7**(2), 186–194 (2005).

341. T. Murakami, S. Kondo, M. Ogata, S. Kanemoto, A. Saito, A. Wanaka and K. Imaizumi, Cleavage of the membrane-bound transcription factor OASIS in response to endoplasmic reticulum stress, *J. Neurochem.* **96**(4), 1090–1100 (2006).

342. R. Lu, P. Yang, P. O'Hare and V. Misra, Luman, a new member of the CREB/ATF family, binds to herpes simplex virus VP16-associated host cellular factor, *Mol. Cell. Biol.* **17**(9), 5117–5126 (1997).

343. C. Raggo, N. Rapin, J. Stirling, P. Gobeil, E. Smith-Windsor, P. O'Hare and V. Misra, Luman, the cellular counterpart of herpes simplex virus VP16, is processed by regulated intramembrane proteolysis, *Mol. Cell. Biol.* **22**(16), 5639–5649 (2002).

344. L. M. DenBoer, P. W. Hardy-Smith, M. R. Hogan, G. P. Cockram, T. E. Audas and R. Lu, Luman is capable of binding and activating transcription from the unfolded protein response element, *Biochem. Biophys. Res. Commun.* **331**(1), 113–119 (2005).

345. P. D. Burbelo, G. C. Gabriel, M. C. Kibbey, Y. Yamada, H. K. Kleinman and B. S. Weeks, LZIP-1 and LZIP-2: two novel members of the bZIP family, *Gene* **139**(2), 241–245 (1994).

346. J. Stirling and P. O'Hare, CREB4, a transmembrane bZip transcription factor and potential new substrate for regulation and cleavage by S1P, *Mol. Biol. Cell* **17**(1), 413–426 (2005).

347. C. T. Storlazzi, F. Mertens, A. Nascimento, M. Isaksson, J. Wejde, O. Brosjö, N. Mandahl and I. Panagopoulos, Fusion of the *FUS* and *BBF2H7* genes in low grade fibromyxoid sarcoma, *Hum. Mol. Genet.* **12**(18), 2349–2358 (2003).

348. X. Chen, J. Shen and R. Prywes, The luminal domain of ATF6 senses endoplasmic reticulum (ER) stress and causes translocation of ATF6 from the ER to the Golgi, *J. Biol. Chem.* **277**(15), 13045–13052 (2002).

349. J. Ye, R. B. Rawson, R. Komuro, X. Chen, U. P. Dave, R. Prywes, M. S. Brown and J. L. Goldstein, ER stress induces cleavage of membrane-bound ATF6 by the same proteases that process SREBPs, *Mol. Cell* **6**(6), 1355–1364 (2000).

350. J. Shen and R. Prywes, Dependence of site-2 protease cleavage of ATF6 on prior site-1 protease digestion is determined by the size of the luminal domain of ATF6, *J. Biol. Chem.* **279**(41), 43046–43051 (2004).

351. Y. Wang, J. Shen, N. Arenzana, W. Tirasophon, R. J. Kaufman and R. Prywes, Activation of ATF6 and an ATF6 DNA binding site by the endoplasmic reticulum stress response, *J. Biol. Chem.* **275**(35), 27013–27020 (2000).

352. H. Yoshida, K. Haze, H. Yanagi, T. Yura and K. Mori, Identification of the *cis*-acting endoplasmic reticulum stress response element responsible for transcriptional induction of mammalian glucose-regulated proteins. Involvement of basic leucine zipper transcription factors, *J. Biol. Chem.* **273**(50), 33741–33749 (1998).

353. K. Kokame, H. Kato and T. Miyata, Identification of ERSE-II, a new *cis*-acting element responsible for the ATF6-dependent mammalian unfolded protein response, *J. Biol. Chem.* **276**(12), 9199–9205 (2001).

354. H. Yoshida, T. Okada, K. Haze, H. Yanagi, T. Yura, M. Negishi and K. Mori, ATF6 activated by proteolysis binds in the presence of NF-Y (CBF) directly to the *cis*-acting element responsible for the mammalian unfolded protein response, *Mol. Cell. Biol.* **20**(18), 6755–6767 (2000).

355. D. J. Thuerauf, L. Morrison and C. C. Glembotski, Opposing roles for ATF6α and ATF6β in ER stress response gene induction, *J. Biol. Chem.* **279**(20), 21078–21084 (2004).

356. J. Shen, X. Chen, L. Hendershot and R. Prywes, ER stress regulation of ATF6 localization by dissociation of BiP/GRP78 binding and unmasking of Golgi localization signals, *Dev. Cell* **3**(1), 99–111 (2002).

357. J. S. Shen, E. L. Snapp, J. Lippincott-Schwartz and R. Prywes, Stable binding of ATF6 to BiP in the endoplasmic reticulum stress response, *Mol. Cell. Biol.* **25**(3), 921–932 (2005).

358. M. Hong, S. Luo, P. Baumeister, J.-M. Huang, R. K. Gogia, M. Li and A. S. Lee, Underglyco-sylation of ATF6 as a novel sensing mechanism for activation of the unfolded protein response, *J. Biol. Chem.* **279**(12), 11354–11363 (2004).

359. M. Hong, M. Li, C. Mao and A. S. Lee, Endoplasmic reticulum stress triggers an acute proteasome-dependent degradation of ATF6, *J. Cell. Biochem.* **92**(4), 723–732 (2004).

360. Y. Honma, K. Kanazawa, T. Mori, Y. Tanno, M. Tojo, H. Kiyosawa, J. Takeda, T. Nikaido, T. Tsukamoto, S. Yokoya and A. Wanaka, Identification of a novel gene, OASIS, which encodes for a putative CREB/ATF family transcription factor in the long-term cultured astrocytes and gliotic tissue, *Brain Res. Mol. Brain Res.* **69**(1), 93–103 (1999).

361. Y. Omori, J. Imai, Y. Suzuki, S. Watanabe, A. Tanigami and S. Sugano, OASIS is a transcriptional activator of CREB/ATF family with a transmembrane domain, *Biochem. Biophys. Res. Commun.* **293**(1), 470–477 (2002).

362. T. Nikaido, S. Yokoya, T. Mori, S. Hagino, K. Iseki, Y. Zhang, M. Takeuchi, H. Takaki, S. Kikuchi and A. Wanaka, Expression of the novel transcription factor OASIS, which belongs to the CREB/ATF family, in mouse embryo with special reference to bone development, *Histochem. Cell Biol.* **116**(2), 141–148 (2001).

363. J. D. Horton, J. L. Goldstein and M. S. Brown, SREBPs: activators of the complete program of cholesterol and fatty acid synthesis in the liver, *J. Clin. Invest.* **109**(9), 1125–1131 (2002).

364. M. S. Brown and J. L. Goldstein, A proteolytic pathway that controls the cholesterol content of membranes, cells, and blood, *Proc. Natl. Acad. Sci. U S A* **96**(20), 11041–11048 (1999).

365. Y. Hirano, S. Murata, K. Tanaka, M. Shimizu and R. Sato, Sterol regulatory element-binding proteins are negatively regulated through SUMO-1 modification independent of the ubiquitin/26 S proteasome pathway, *J. Biol. Chem.* **278**(19), 16809–16819 (2003).

366. X. Hua, A. Nohturfft, J. L. Goldstein and M. S. Brown, Sterol resistance in CHO cells traced to point mutation in SREBP cleavage-activating protein, *Cell* **87**(3), 415–426 (1996).

367. C. J. Loewen and T. P. Levine, Cholesterol homeostasis: not until the SCAP lady INSIGs, *Curr. Biol.* **12**(22), R779–R781 (2002).

368. D. Yabe, M. S. Brown and J. L. Goldstein, Insig-2, a second endoplasmic reticulum protein that binds SCAP and blocks export of sterol regulatory element-binding proteins, *Proc. Natl. Acad. Sci. U S A* **99**(20), 12753–12758 (2002).

369. T. Yang, P. J. Espenshade, M. E. Wright, D. Yabe, Y. Gong, R. Aebersold, J. L. Goldstein and M. S. Brown, Crucial step in cholesterol homeostasis: sterols promote binding of SCAP to INSIG-1, a membrane protein that facilitates retention of SREBPs in ER, *Cell* **110**(4), 489–500 (2002).

370. A. J. Brown, L. Sun, J. D. Feramisco, M. S. Brown and J. L. Goldstein, Cholesterol addition to ER membranes alters conformation of SCAP, the SREBP escort protein that regulates cholesterol metabolism, *Mol. Cell* **10**(2), 237–245 (2002).

371. B. Feng, P. M. Yao, Y. Li, C. M. Devlin, D. Zhang, H. P. Harding, M. Sweeney, J. X. Rong, G. Kuriakose, E. A. Fisher, A. R. Marks, D. Ron and I. Tabas, The endoplasmic reticulum is the site of cholesterol-induced cytotoxicity in macrophages, *Nat. Cell Biol.* **5**(9), 781–792 (2003).

372. R. E. Soccio, R. M. Adams, K. N. Maxwell and J. L. Breslow, Differential gene regulation of StarD4 and StarD5 cholesterol transfer proteins. Activation of StarD4 by sterol regulatory element-binding protein-2 and StarD5 by endoplasmic reticulum stress, *J. Biol. Chem.* **280**(19), 19410–19418 (2005).

373. S. Zhang, T. J. Burkett, I. Yamashita and D. J. Garfinkel, Genetic redundancy between *SPT23* and *MGA2*: regulators of Ty-induced mutations and Ty1 transcription in *Saccharomyces cerevisiae*, *Mol. Cell. Biol.* **17**(8), 4718–4729 (1997).

374. S. Zhang, Y. Skalsky and D. J. Garfinkel, *MGA2* or *SPT23* is required for transcription of the Δ9 fatty acid desaturase gene, *OLE1*, and nuclear membrane integrity in *Saccharomyces cerevisiae, Genetics* **151**(2), 473–483 (1999).

375. T. Hoppe, K. Matuschewski, M. Rape, S. Schlenker, H. D. Ulrich and S. Jentsch, Activation of a membrane-bound transcription factor by regulated ubiquitin/proteasome-dependent processing, *Cell* **102**(5), 577–586 (2000).

376. W. Tirasophon, A. A. Welihinda and R. J. Kaufman, A stress response pathway from the endoplasmic reticulum to the nucleus requires a novel bifunctional protein kinase/endoribonuclease (Ire1p) in mammalian cells, *Genes Dev.* **12**(12), 1812–1824 (1998).

377. X. Z. Wang, H. P. Harding, Y. Zhang, E. M. Jolicoeur, M. Kuroda and D. Ron, Cloning of mammalian Ire1 reveals diversity in the ER stress responses, *EMBO J.* **17**(19), 5708–5717 (1998).

378. C. E. Shamu and P. Walter, Oligomerization and phosphorylation of the Ire1p kinase during intracellular signaling from the endoplasmic reticulum to the nucleus, *EMBO J.* **15**(12), 3028–3039 (1996).

379. C. Sidrauski and P. Walter, The transmembrane kinase Ire1p is a site-specific endonuclease that initiates mRNA splicing in the unfolded protein response, *Cell* **90**(6), 1031–1039 (1997).

380. A. A. Welihinda and R. J. Kaufman, The unfolded protein response pathway in Saccharomyces cerevisiae. Oligomerization and trans-phosphorylation of Ire1p (Ern1p) are required for kinase activation, *J. Biol. Chem.* **271**(30), 18181–18187 (1996).

381. W. Tirasophon, K. Lee, B. Callaghan, A. Welihinda and R. J. Kaufman, The endoribonuclease activity of mammalian IRE1 autoregulates its mRNA and is required for the unfolded protein response, *Genes Dev.* **14**(21), 2725–2736 (2000).

382. Y. Kimata, Y. I. Kimata, Y. Shimizu, H. Abe, I. C. Farcasanu, M. Takeuchi, M. D. Rose and K. Kohno, Genetic evidence for a role of BiP/Kar2 that regulates Ire1 in response to accumulation of unfolded proteins, *Mol. Biol. Cell* **14**(6), 2559–2569 (2003).

383. C. Herman, D. Thévenet, R. D'Ari and P. Bouloc, Degradation of σ^{32}, the heat shock regulator in *Escherichia coli*, is governed by HflB, *Proc. Natl. Acad. Sci. U S A* **92**(8), 3516–3520 (1995).

384. T. Tomoyasu, J. Gamer, B. Bukau, M. Kanemori, H. Mori, A. J. Rutman, A. B. Oppenheim, T. Yura, K. Yamanaka, H. Niki, S. Hiraga and T. Ogura, *Escherichia coli* FtsH is a membrane-bound, ATP-dependent protease which degrades the heat-shock transcription factor σ^{32}, *EMBO J.* **14**(11), 2551–2560 (1995).

385. J. Zou, Y. Guo, T. Guettouche, D. F. Smith and R. Voellmy, Repression of heat shock transcription factor HSF1 activation by HSP90 (HSP90 complex) that forms a stress-sensitive complex with HSF1, *Cell* **94**(4), 471–480 (1998).

386. K. Kohno, K. Normington, J. Sambrook, M. J. Gething and K. Mori, The promoter region of the yeast *KAR2* (BiP) gene contains a regulatory domain that responds to the presence of unfolded proteins in the endoplasmic reticulum, *Mol. Cell. Biol.* **13**(2), 877–890 (1993).

387. K. Ma, K. M. Vattem and R. C. Wek, Dimerization and release of molecular chaperone inhibition facilitate activation of eukaryotic initiation factor-2 kinase in response to endoplasmic reticulum stress, *J. Biol. Chem.* **277**(21), 18728–18735 (2002).

388. J. J. Credle, J. S. Finer-Moore, F. R. Papa, R. M. Stroud and P. Walter, Inaugural Article: On the mechanism of sensing unfolded protein in the endoplasmic reticulum, *Proc. Natl. Acad. Sci. U S A* **102**(52), 18773–18784 (2005).

389. B. M. Alba and C. A. Gross, Regulation of the *Escherichia coli* σ^E-dependent envelope stress response, *Mol. Microbiol.* **52**(3), 613–619 (2004).

390. A. Weiss and J. Schlessinger, Switching signals on or off by receptor dimerization, *Cell* **94**(3), 277–280 (1998).

391. F. R. Papa, C. Zhang, K. Shokat and P. Walter, Bypassing a kinase activity with an ATP-competitive drug, *Science* **302**(5650), 1533–1537 (2003).

392. B. Dong, M. Niwa, P. Walter and R. H. Silverman, Basis for regulated RNA cleavage by functional analysis of RNase L and Ire1p, *RNA* **7**(3), 361–373 (2001).

393. S. Naik, J. M. Paranjape and R. H. Silverman, RNase L dimerization in a mammalian two-hybrid system in response to 2′, 5′-oligoadenylates, *Nucleic Acids Res.* **26**(6), 1522–1527 (1998).

394. H. Nojima, S. H. Leem, H. Araki, A. Sakai, N. Nakashima, Y. Kanaoka and Y. Ono, Hac1: a novel yeast bZIP protein binding to the CRE motif is a multicopy suppressor for cdc10 mutant of *Schizosaccharomyces pombe, Nucleic Acids Res.* **22**(24), 5279–5288 (1994).

395. M. Niwa, C. K. Patil, J. DeRisi and P. Walter, Genome-scale approaches for discovering novel nonconventional splicing substrates of the Ire1 nuclease, *Genome Biol.* **6**(1), R3 (2005).

396. J. S. Cox and P. Walter, A novel mechanism for regulating activity of a transcription factor that controls the unfolded protein response, *Cell* **87**(3), 391–404 (1996).

397. T. N. Gonzalez, C. Sidrauski, S. Dörfler and P. Walter, Mechanism of non-spliceosomal mRNA splicing in the unfolded protein response pathway, *EMBO J.* **18**(11), 3119–3132 (1999).

398. C. Sidrauski, J. S. Cox and P. Walter, tRNA ligase is required for regulated mRNA splicing in the unfolded protein response, *Cell* **87**(3), 405–413 (1996).

399. S. M. McCraith and E. M. Phizicky, A highly specific phosphatase from *Saccharomyces cerevisiae* implicated in tRNA splicing, *Mol. Cell. Biol.* **10**(3), 1049–1055 (1990).

400. G. M. Culver, S. A. Consaul, K. T. Tycowski, W. Filipowicz and E. M. Phizicky, tRNA splicing in yeast and wheat germ. A cyclic phosphodiesterase implicated in the metabolism of ADP-ribose $1''$, $2''$-cyclic phosphate, *J. Biol. Chem.* **269**(40), 24928–24934 (1994).

401. N. P. Shull, S. L. Spinelli and E. M. Phizicky, A highly specific phosphatase that acts on ADP-ribose $1''$-phosphate, a metabolite of tRNA splicing in Saccharomyces cerevisiae, *Nucleic Acids Res.* **33**(2), 650–660 (2005).

402. X. Shen, R. E. Ellis, K. Lee, C.-Y. Liu, K. Yang, A. Solomon, H. Yoshida, R. Morimoto, D. M. Kurnit, K. Mori and R. J. Kaufman, Complementary signaling pathways regulate the unfolded protein response and are required for *C. elegans* development, *Cell* **107**(7), 893–903 (2001).

403. H. Yoshida, T. Matsui, A. Yamamoto, T. Okada and K. Mori, XBP1 mRNA is induced by ATF6 and spliced by IRE1 in response to ER stress to produce a highly active transcription factor, *Cell* **107**(7), 881–891 (2001).

404. K. Lee, W. Tirasophon, X. Shen, M. Michalak, R. Prywes, T. Okada, H. Yoshida, K. Mori and R. J. Kaufman, IRE1-mediated unconventional mRNA splicing and S2P-mediated ATF6 cleavage merge to regulate XBP1 in signaling the unfolded protein response, *Genes Dev.* **16**(4), 452–466 (2002).

405. M. Konarska, W. Filipowicz, H. Domdey and H. J. Gross, Formation of a $2'$-phosphomonoester, $3'$, $5'$-phosphodiester linkage by a novel RNA ligase in wheat germ, *Nature* **293**(5828), 112–116 (1981).

406. M. Konarska, W. Filipowicz and H. J. Gross, RNA ligation via $2'$-phosphomonoester, $3'5'$-phosphodiester linkage: requirement of $2'$, $3'$-cyclic phosphate termini and involvement of a $5'$-hydroxyl polynucleotide kinase, *Proc. Natl. Acad. Sci. U S A* **79**(5), 1474–1478 (1982).

407. L. Pick, H. Furneaux and J. Hurwitz, Purification of wheat germ RNA ligase. II. Mechanism of action of wheat germ RNA ligase, *J. Biol. Chem.* **261**(15), 6694–6704 (1986).

408. L. Pick and J. Hurwitz, Purification of wheat germ RNA ligase. I. Characterization of a ligase-associated $5'$-hydroxyl polynucleotide kinase activity, *J. Biol. Chem.* **261**(15), 6684–6693 (1986).

409. Y. Kikuchi, K. Tyc, W. Filipowicz, H. L. Sänger and H. J. Gross, Circularization of linear viroid RNA via $2'$-phosphomonoester, $3'$, $5'$-phosphodiester bonds by a novel type of RNA ligase from wheat germ and *Chlamydomonas, Nucleic Acids Res.* **10**(23), 7521–7529 (1982).

410. K. Tyc, Y. Kikuchi, M. Konarska, W. Filipowicz and H. J. Gross, Ligation of endogenous tRNA half molecules to their corresponding $5'$ halves *via* $2'$-phosphmonoester, $3'$, $5'$-phosphodiester bonds in extracts from *Chlamydomonas, EMBO J.* **2**(4), 605–610 (1983).

411. M. Zillmann, M. A. Gorovsky and E. M. Phizicky, Conserved mechanism of tRNA splicing in eukaryotes, *Mol. Cell. Biol.* **11**(11), 5410–5416 (1991).

412. M. Englert and H. Beier, Plant tRNA ligases are multifunctional enzymes that have diverged in sequence and substrate specificity from RNA ligases of other phylogenetic origins, *Nucleic Acids Res.* **33**(1), 388–399 (2005).

413. K. Nishikura and E. M. De Robertis, RNA processing in microinjected Xenopus oocytes. Sequential addition of base modifications in the spliced transfer RNA, *J. Mol. Biol.* **145**(2), 405–420 (1981).

414. W. Filipowicz, M. Konarska, H. J. Gross and A. J. Shatkin, RNA 3′-terminal phosphate cyclase activity and RNA ligation in HeLa cell extract, *Nucleic Acids Res.* **11**(5), 1405–1418 (1983).

415. K. K. Perkins, H. Furneaux and J. Hurwitz, Isolation and characterization of an RNA ligase from HeLa cells, *Proc. Natl. Acad. Sci. U S A* **82**(3), 684–688 (1985).

416. Q. D. Hu, H. Lu, K. Huo, K. Ying, J. Li, Y. Xie, Y. Mao and Y. Y. Li, A human homolog of the yeast gene encoding tRNA 2′-phosphotransferase: cloning, characterization and complementation analysis, *Cell. Mol. Life Sci.* **60**(8), 1725–1732 (2003).

417. U. Rüegsegger, J. H. Leber and P. Walter, Block of HAC1 mRNA translation by long-range base pairing is released by cytoplasmic splicing upon induction of the unfolded protein response, *Cell* **107**(1), 103–114 (2001).

418. K. M. Kuhn, J. L. DeRisi, P. O. Brown and P. Sarnow, Global and specific translational regulation in the genomic response of *Saccharomyces cerevisiae* to a rapid transfer from a fermentable to a nonfermentable carbon source, *Mol. Cell. Biol.* **21**(3), 916–927 (2001).

419. R. E. Chapman and P. Walter, Translational attenuation mediated by an mRNA intron, *Curr. Biol.* **7**(11), 850–859 (1997).

420. I. Winicov and J. D. Button, Nuclear ligation of RNA 5′-OH kinase products in tRNA, *Mol. Cell. Biol.* **2**(3), 241–249 (1982).

421. M. W. Clark and J. Abelson, The subnuclear localization of tRNA ligase in yeast, *J. Cell Biol.* **105**(4), 1515–1526 (1987).

422. T. Kawahara, H. Yanagi, T. Yura and K. Mori, Endoplasmic reticulum stress-induced mRNA splicing permits synthesis of transcription factor Hac1p/Ern4p that activates the unfolded protein response, *Mol. Biol. Cell* **8**(10), 1845–1862 (1997).

423. T. van Laar, A. J. van der Eb and C. Terleth, Mif1: a missing link between the unfolded protein response pathway and ER-associated protein degradation?, *Curr. Protein Pept. Sci.* **2**(2), 169–190 (2001).

424. A. H. Lee, N. N. Iwakoshi, K. C. Anderson and L. H. Glimcher, Proteasome inhibitors disrupt the unfolded protein response in myeloma cells, *Proc. Natl. Acad. Sci. U S A* **100**(17), 9946–9951 (2003).

425. H. Yoshida, M. Oku, M. Suzuki and K. Mori, pXBP1(U) encoded in XBP1 pre-mRNA negatively regulates unfolded protein response activator pXBP1(S) in mammalian ER stress response, *J. Cell Biol.* **172**(4), 565–575 (2006).

426. K. Mori, N. Ogawa, T. Kawahara, H. Yanagi and T. Yura, mRNA splicing-mediated C-terminal replacement of transcription factor Hac1p is required for efficient activation of the unfolded protein response, *Proc. Natl. Acad. Sci. U S A* **97**(9), 4660–4665 (2000).

427. K. Mori, A. Sant, K. Kohno, K. Normington, M. J. Gething and J. F. Sambrook, A 22 bp cis-acting element is necessary and sufficient for the induction of the yeast *KAR2* (BiP) gene by unfolded proteins, *EMBO J.* **11**(7), 2583–2593 (1992).

428. C. K. Patil, H. Li and P. Walter, Gcn4p and novel upstream activating sequences regulate targets of the unfolded protein response, *PLoS Biol.* **2**(8), 1208–1223 (2004).

429. K. Mori, T. Kawahara, H. Yoshida, H. Yanagi and T. Yura, Signalling from endoplasmic reticulum to nucleus: transcription factor with a basic-leucine zipper motif is required for the unfolded protein-response pathway, *Genes Cells* **1**(9), 803–817 (1996).

430. K. Mori, N. Ogawa, T. Kawahara, H. Yanagi and T. Yura, Palindrome with spacer of one nucleotide is characteristic of the cis-acting unfolded protein response element in *Saccharomyces cerevisiae*, *J. Biol. Chem.* **273**(16), 9912–9920 (1998).

431. T. Zimmer, A. Ogura, A. Ohta and M. Takagi, Misfolded membrane-bound cytochrome P450 activates *KAR2* induction through two distinct mechanisms, *J. Biochem. (Tokyo)* **126**(6), 1080–1089 (1999).

432. J. A. Partaledis and V. Berlin, The *FKB2* gene of *Saccharomyces cerevisiae*, encoding the immunosuppressant-binding protein FKBP-13, is regulated in response to accumulation of unfolded proteins in the endoplasmic reticulum, *Proc. Natl. Acad. Sci. U S A* **90**(12), 5450–5454 (1993).

433. A. A. Welihinda, W. Tirasophon, S. R. Green and R. J. Kaufman, Gene induction in response to unfolded protein in the endoplasmic reticulum is mediated through Ire1p kinase interaction

with a transcriptional coactivator complex containing Ada5p, *Proc. Natl. Acad. Sci. U S A* **94**(9), 4289–4294 (1997).

434. R. J. Kaufman, Stress signaling from the lumen of the endoplasmic reticulum: coordination of gene transcriptional and translational controls, *Genes Dev.* **13**(10), 1211–1233 (1999).

435. T. Georgakopoulos and G. Thireos, Two distinct yeast transcriptional activators require the function of the GCN5 protein to promote normal levels of transcription, *EMBO J.* **11**(11), 4145–4152 (1992).

436. G. A. Marcus, N. Silverman, S. L. Berger, J. Horiuchi and L. Guarente, Functional similarity and physical association between GCN5 and ADA2: putative transcriptional adaptors, *EMBO J.* **13**(20), 4807–4815 (1994).

437. G. A. Marcus, J. Horiuchi, N. Silverman and L. Guarente, *ADA5/SPT20* links the *ADA* and *SPT* genes, which are involved in yeast transcription, *Mol. Cell. Biol.* **16**(6), 3197–3205 (1996).

438. P. A. Grant, L. Duggan, J. Côté, S. M. Roberts, J. E. Brownell, R. Candau, R. Ohba, T. Owen-Hughes, C. D. Allis, F. Winston, S. L. Berger and J. L. Workman, Yeast Gcn5 functions in two multisubunit complexes to acetylate nucleosomal histones: characterization of an Ada complex and the SAGA (Spt/Ada) complex, *Genes Dev.* **11**(13), 1640–1650 (1997).

439. A. A. Welihinda, W. Tirasophon and R. J. Kaufman, The transcriptional co-activator *ADA5* is required for *HAC1* mRNA processing *in vivo, J. Biol. Chem.* **275**(5), 3377–3381 (2000).

440. M. G. Pray-Grant, D. Schieltz, S. J. McMahon, J. M. Wood, E. L. Kennedy, R. G. Cook, J. L. Workman, J. R. Yates, III and P. A. Grant, The novel SLIK histone acetyltransferase complex functions in the yeast retrograde response pathway, *Mol. Cell. Biol.* **22**(24), 8774–8786 (2002).

441. D. E. Sterner, R. Belotserkovskaya and S. L. Berger, SALSA, a variant of yeast SAGA, contains truncated Spt7, which correlates with activated transcription, *Proc. Natl. Acad. Sci. U S A* **99**(18), 11622–11627 (2002).

442. S. K. Kurdistani and M. Grunstein, Histone acetylation and deacetylation in yeast, *Nat. Rev. Mol. Cell. Biol.* **4**(4), 276–284 (2003).

443. M. Schröder, R. Clark, C. Y. Liu and R. J. Kaufman, The unfolded protein response represses differentiation through the *RPD3-SIN3* histone deacetylase, *EMBO J.* **23**(11), 2281–2292 (2004).

444. I. Spode, D. Maiwald, C. P. Hollenberg and M. Suckow, ATF/CREB sites present in sub-telomeric regions of *Saccharomyces cerevisiae* chromosomes are part of promoters and act as UAS/URS of highly conserved *COS* genes, *J. Mol. Biol.* **319**(2), 407–420 (2002).

445. A. P. Mitchell, Control of meiotic gene expression in *Saccharomyces cerevisiae, Microbiol. Rev.* **58**(1), 56–70 (1994).

446. R. M. Williams, M. Primig, B. K. Washburn, E. A. Winzeler, M. Bellis, C. Sarrauste de Menthiere, R. W. Davis and R. E. Esposito, The Ume6 regulon coordinates metabolic and meiotic gene expression in yeast, *Proc. Natl. Acad. Sci. U S A* **99**(21), 13431–13436 (2002).

447. R. Strich, R. T. Surosky, C. Steber, E. Dubois, F. Messenguy and R. E. Esposito, UME6 is a key regulator of nitrogen repression and meiotic development, *Genes Dev.* **8**(7), 796–810 (1994).

448. J. P. Goldmark, T. G. Fazzio, P. W. Estep, G. M. Church and T. Tsukiyama, The Isw2 chromatin remodeling complex represses early meiotic genes upon recruitment by Ume6p, *Cell* **103**(3), 423–433 (2000).

449. D. Kadosh and K. Struhl, Repression by Ume6 involves recruitment of a complex containing Sin3 corepressor and Rpd3 histone deacetylase to target promoters, *Cell* **89**(3), 365–371 (1997).

450. M. J. Carrozza, L. Florens, S. K. Swanson, W.-J. Shia, S. Anderson, J. Yates, M. P. Washburn and J. L. Workman, Stable incorporation of sequence specific repressors Ash1 and Ume6 into the Rpd3L complex, *Biochim. Biophys. Acta.* **1731**(2), 77–87 (2005).

451. M. J. Carrozza, B. Li, L. Florens, T. Suganuma, S. K. Swanson, K. K. Lee, W. J. Shia, S. Anderson, J. Yates, M. P. Washburn and J. L. Workman, Histone H3 methylation by Set2 directs deacetylation of coding regions by Rpd3S to suppress spurious intragenic transcription, *Cell* **123**(4), 581–592 (2005).

452. N. Ogawa and K. Mori, Autoregulation of the *HAC1* gene is required for sustained activation of the yeast unfolded protein response, *Genes Cells* **9**(2), 95–104 (2004).

453. C. Kakiuchi, K. Iwamoto, M. Ishiwata, M. Bundo, T. Kasahara, I. Kusumi, T. Tsujita, Y. Okazaki, S. Nanko, H. Kunugi, T. Sasaki and T. Kato, Impaired feedback regulation of XBP1 as a genetic risk factor for bipolar disorder, *Nat. Genet.* **35**(2), 171–175 (2003).

454. A. A. Welihinda, W. Tirasophon, S. R. Green and R. J. Kaufman, Protein serine/threonine phosphatase Ptc2p negatively regulates the unfolded-protein response by dephosphorylating Ire1p kinase, *Mol. Cell. Biol.* **18**(4), 1967–1977 (1998).

455. D. T. Nguyên, S. Kebache, A. Fazel, H. N. Wong, S. Jenna, A. Emadali, E. H. Lee, J. J. Bergeron, R. J. Kaufman, L. Larose and E. Chevet, Nck-dependent activation of extracellular signal regulated kinase-1 and regulation of cell survival during endoplasmic reticulum stress, *Mol. Biol. Cell* **15**(9), 4248–4260 (2004).

456. K. Oono, T. Yoneda, T. Manabe, S. Yamagishi, S. Matsuda, J. Hitomi, S. Miyata, T. Mizuno, K. Imaizumi, T. Katayama and M. Tohyama, JAB1 participates in unfolded protein responses by association and dissociation with IRE1, *Neurochem. Int.* **45**(5), 765–772 (2004).

457. S. B. Cullinan, D. Zhang, M. Hannink, E. Arvisais, R. J. Kaufman and J. A. Diehl, Nrf2 is a direct PERK substrate and effector of PERK-dependent cell survival, *Mol. Cell. Biol.* **23**(20), 7198–7209 (2003).

458. J. W. B. Hershey, Translational control in mammalian cells, *Annu. Rev. Biochem.* **60**, 717–755 (1991).

459. M. Kozak, Structural features in eukaryotic mRNAs that modulate the initiation of translation, *J. Biol. Chem.* **266**(30), 19867–19870 (1991).

460. R. J. Kaufman, Regulation of mRNA translation by protein folding in the endoplasmic reticulum, *Trends. Biochem. Sci.* **29**(3), 152–158 (2004).

461. C. J. Sherr, Cancer cell cycles, *Science* **274**(5293), 1672–1677 (1996).

462. A. Ho and S. F. Dowdy, Regulation of G_1 cell-cycle progression by oncogenes and tumor suppressor genes, *Curr. Opin. Genet. Dev.* **12**(1), 47–52 (2002).

463. S. F. Dowdy, P. W. Hinds, K. Louie, S. I. Reed, A. Arnold and R. A. Weinberg, Physical interaction of the retinoblastoma protein with human D cyclins, *Cell* **73**(3), 499–511 (1993).

464. M. E. Ewen, H. K. Sluss, C. J. Sherr, H. Matsushime, J. Kato and D. M. Livingston, Functional interactions of the retinoblastoma protein with mammalian D-type cyclins, *Cell* **73**(3), 487–497 (1993).

465. J. Kato, H. Matsushime, S. W. Hiebert, M. E. Ewen and C. J. Sherr, Direct binding of cyclin D to the retinoblastoma gene product (pRb) and pRb phosphorylation by the cyclin D-dependent kinase CDK4, *Genes Dev.* **7**(3), 331–342 (1993).

466. P. A. Hamel, R. M. Gill, R. A. Phillips and B. L. Gallie, Transcriptional repression of the E2-containing promoters EIIaE, c-*myc*, and *RB1* by the product of the *RB1* gene, *Mol. Cell. Biol.* **12**(8), 3431–3438 (1992).

467. S. J. Weintraub, C. A. Prater and D. C. Dean, Retinoblastoma protein switches the E2F site from positive to negative element, *Nature* **358**(6383), 259–261 (1992).

468. E. K. Flemington, S. H. Speck and W. G. Kaelin, Jr., E2F-1-mediated transactivation is inhibited by complex formation with the retinoblastoma susceptibility gene product, *Proc. Natl. Acad. Sci. U S A* **90**(15), 6914–6918 (1993).

469. E. W. Lam and R. J. Watson, An E2F-binding site mediates cell-cycle regulated repression of mouse B-*myb* transcription, *EMBO J.* **12**(7), 2705–2713 (1993).

470. S. J. Weintraub, K. N. Chow, R. X. Luo, S. H. Zhang, S. He and D. C. Dean, Mechanism of active transcriptional repression by the retinoblastoma protein, *Nature* **375**(6534), 812–815 (1995).

471. N. Dyson, The regulation of E2F by pRB-family proteins, *Genes Dev.* **12**(15), 2245–2262 (1998).

472. A. Brehm, E. A. Miska, D. J. McCance, J. L. Reid, A. J. Bannister and T. Kouzarides, Retinoblastoma protein recruits histone deacetylase to repress transcription, *Nature* **391**(6667), 597–601 (1998).

473. R. X. Luo, A. A. Postigo and D. C. Dean, Rb interacts with histone deacetylase to repress transcription, *Cell* **92**(4), 463–473 (1998).

474. L. Magnaghi-Jaulin, R. Groisman, I. Naguibneva, P. Robin, S. Lorain, J. P. Le Villain, F. Troalen, D. Trouche and A. Harel-Bellan, Retinoblastoma protein represses transcription by recruiting a histone deacetylase, *Nature* **391**(6667), 601–605 (1998).

475. A. Tomida, H. Suzuki, H. D. Kim and T. Tsuruo, Glucose-regulated stresses cause decreased expression of cyclin D1 and hypophosphorylation of retinoblastoma protein in human cancer cells, *Oncogene* **13**(12), 2699–2705 (1996).

476. J. W. Brewer, L. M. Hendershot, C. J. Sherr and J. A. Diehl, Mammalian unfolded protein response inhibits cyclin D1 translation and cell-cycle progression, *Proc. Natl. Acad. Sci. U S A* **96**(15), 8505–8510 (1999).

477. J. W. Brewer and J. A. Diehl, PERK mediates cell-cycle exit during the mammalian unfolded protein response, *Proc. Natl. Acad. Sci. U S A* **97**(23), 12625–12630 (2000).

478. A. S. Baldwin, Jr., The NF-κB and IκB proteins: new discoveries and insights, *Annu. Rev. Immunol.* **14**, 649–683 (1996).

479. S. Ghosh, M. J. May and E. B. Kopp, NF-κB and Rel proteins: evolutionarily conserved mediators of immune responses, *Annu. Rev. Immunol.* **16**, 225–260 (1998).

480. M. Barkett and T. D. Gilmore, Control of apoptosis by Rel/NF-κB transcription factors, *Oncogene* **18**(49), 6910–6924 (1999).

481. F. E. Chen and G. Ghosh, Regulation of DNA binding by Rel/NF-κB transcription factors: structural views, *Oncogene* **18**(49), 6845–6852 (1999).

482. J. Caamano and C. A. Hunter, NF-κB family of transcription factors: central regulators of innate and adaptive immune functions, *Clin. Microbiol. Rev.* **15**(3), 414–429 (2002).

483. Q. Li and I. M. Verma, NF-κB regulation in the immune system, *Nat. Rev. Immunol.* **2**(10), 725–734 (2002).

484. H. Y. Jiang, S. A. Wek, B. C. McGrath, D. Scheuner, R. J. Kaufman, D. R. Cavener and R. C. Wek, Phosphorylation of the α subunit of eukaryotic initiation factor 2 is required for activation of NF-κB in response to diverse cellular stresses, *Mol. Cell. Biol.* **23**(16), 5651–5663 (2003).

485. J. Deng, P. D. Lu, Y. Zhang, D. Scheuner, R. J. Kaufman, N. Sonenberg, H. P. Harding and D. Ron, Translational repression mediates activation of nuclear factor kappa B by phosphorylated translation initiation factor 2, *Mol. Cell. Biol.* **24**(23), 10161–10168 (2004).

486. S. Wu, M. Tan, Y. Hu, J. L. Wang, D. Scheuner and R. J. Kaufman, Ultraviolet light activates NFκB through translational inhibition of IκBα synthesis, *J. Biol. Chem.* **279**(33), 34898–34902 (2004).

487. H. Y. Jiang and R. C. Wek, GCN2 phosphorylation of eIF2 α activates NF-κB in response to UV irradiation, *Biochem. J.* **385**(2), 371–380 (2005).

488. P. A. Baeuerle and T. Henkel, Function and activation of NF-κB in the immune system, *Annu. Rev. Immunol.* **12**, 141–179 (1994).

489. U. Siebenlist, G. Franzoso and K. Brown, Structure, regulation and function of NF-κB, *Annu. Rev. Cell Biol.* **10**, 405–455 (1994).

490. M. Kozak, Pushing the limits of the scanning mechanism for initiation of translation, *Gene* **299**(1–2), 1–34 (2002).

491. P. F. Miller and A. G. Hinnebusch, Sequences that surround the stop codons of upstream open reading frames in GCN4 mRNA determine their distinct functions in translational control, *Genes Dev.* **3**(8), 1217–1225 (1989).

492. C. M. Grant and A. G. Hinnebusch, Effect of sequence context at stop codons on efficiency of reinitiation in *GCN4* translational control, *Mol. Cell. Biol.* **14**(1), 606–618 (1994).

493. H. P. Harding, I. Novoa, Y. Zhang, H. Zeng, R. Wek, M. Schapira and D. Ron, Regulated translation initiation controls stress-induced gene expression in mammalian cells, *Mol. Cell* **6**(5), 1099–1108 (2000).

494. D. Scheuner, B. Song, E. McEwen, C. Liu, R. Laybutt, P. Gillespie, T. Saunders, S. Bonner-Weir and R. J. Kaufman, Translational control is required for the unfolded protein response and in vivo glucose homeostasis, *Mol. Cell* **7**(6), 1165–1176 (2001).

495. P. D. Lu, H. P. Harding and D. Ron, Translation reinitiation at alternative open reading frames regulates gene expression in an integrated stress response, *J. Cell Biol.* **167**(1), 27–33 (2004).

496. P. Cornelius, O. A. MacDougald and M. D. Lane, Regulation of adipocyte development, *Annu. Rev. Nutr.* **14**, 99–129 (1994).

497. C. F. Calkhoven, C. Müller and A. Leutz, Translational control of C/EBPα and C/EBPβ isoform expression, *Genes Dev.* **14**(15), 1920–1932 (2000).

498. C. F. Calkhoven, C. Müller and A. Leutz, Translational control of gene expression and disease, *Trends Mol. Med.* **8**(12), 577–583 (2002).

499. P. Descombes and U. Schibler, A liver-enriched transcriptional activator protein, LAP, and a transcriptional inhibitory protein, LIP, are translated from the same mRNA, *Cell* **67**(3), 569–579 (1991).

500. N. D. Wang, M. J. Finegold, A. Bradley, C. N. Ou, S. V. Abdelsayed, M. D. Wilde, L. R. Taylor, D. R. Wilson and G. J. Darlington, Impaired energy homeostasis in C/EBPα knockout mice, *Science* **269**(5227), 1108–1112 (1995).

501. C. U. Hellen and P. Sarnow, Internal ribosome entry sites in eukaryotic mRNA molecules, *Genes Dev.* **15**(13), 1593–1612 (2001).

502. D. G. Macejak and P. Sarnow, Translational regulation of the immunoglobulin heavy-chain binding protein mRNA, *Enzyme* **44**(1–4), 310–319 (1990).

503. D. G. Macejak and P. Sarnow, Internal initiation of translation mediated by the 5′ leader of a cellular mRNA, *Nature* **353**(6339), 90–94 (1991).

504. J. Fernandez, B. Bode, A. Koromilas, J. A. Diehl, I. Krukovets, M. D. Snider and M. Hatzoglou, Translation mediated by the internal ribosome entry site of the *cat-1* mRNA is regulated by glucose availability in a PERK kinase-dependent manner, *J. Biol. Chem.* **277**(14), 11780–11787 (2002).

505. I. Yaman, J. Fernandez, H. Liu, M. Caprara, A. A. Komar, A. E. Koromilas, L. Zhou, M. D. Snider, D. Scheuner, R. J. Kaufman and M. Hatzoglou, The zipper model of translational control: a small upstream ORF is the switch that controls structural remodeling of an mRNA leader, *Cell* **113**(4), 519–531 (2003).

506. T. Nguyen, P. J. Sherratt and C. B. Pickett, Regulatory mechanisms controlling gene expression mediated by the antioxidant response element, *Annu. Rev. Pharmacol. Toxicol.* **43**, 233–260 (2003).

507. H. Y. Jiang, S. A. Wek, B. C. McGrath, D. Lu, T. Hai, H. P. Harding, X. Wang, D. Ron, D. R. Cavener and R. C. Wek, Activating transcription factor 3 is integral to the eukaryotic initiation factor 2 kinase stress response, *Mol. Cell. Biol.* **24**(3), 1365–1377 (2004).

508. Y. Ma, J. W. Brewer, J. A. Diehl and L. M. Hendershot, Two distinct stress signaling pathways converge upon the CHOP promoter during the mammalian unfolded protein response, *J. Mol. Biol.* **318**(5), 1351–1365 (2002).

509. Y. Ma and L. M. Hendershot, Herp is dually regulated by both the endoplasmic reticulum stress-specific branch of the unfolded protein response and a branch that is shared with other cellular stress pathways, *J. Biol. Chem.* **279**(14), 13792–13799 (2004).

510. M. Vallejo, D. Ron, C. P. Miller and J. F. Habener, C/ATF, a member of the activating transcription factor family of DNA-binding proteins, dimerizes with CAAT/enhancer-binding proteins and directs their binding to cAMP response elements, *Proc. Natl. Acad. Sci. U S A* **90**(10), 4679–4683 (1993).

511. D. Örd and T. Örd, Characterization of human NIPK (*TRB3, SKIP3*) gene activation in stressful conditions, *Biochem. Biophys. Res. Commun.* **330**(1), 210–218 (2005).

512. C. D. Wolfgang, G. Liang, Y. Okamoto, A. E. Allen and T. Hai, Transcriptional autorepression of the stress-inducible gene *ATF3*, *J. Biol. Chem.* **275**(22), 16865–16870 (2000).

513. C. D. Wolfgang, B. P. Chen, J. L. Martindale, N. J. Holbrook and T. Hai, *gadd153/Chop10*, a potential target gene of the transcriptional repressor ATF3, *Mol. Cell. Biol.* **17**(11), 6700–6707 (1997).

514. T. W. Fawcett, J. L. Martindale, K. Z. Guyton, T. Hai and N. J. Holbrook, Complexes containing activating transcription factor (ATF)/cAMP-responsive-element-binding protein (CREB) interact with the CCAAT/enhancer-binding protein (C/EBP)-ATF composite site to regulate *Gadd153* expression during the stress response, *Biochem. J.* **339**(1), 135–141 (1999).

515. A. E. Allen-Jennings, M. G. Hartman, G. J. Kociba and T. Hai, The roles of ATF3 in glucose homeostasis. A transgenic mouse model with liver dysfunction and defects in endocrine pancreas, *J. Biol. Chem.* **276**(31), 29507–29514 (2001).

516. B. P. Chen, C. D. Wolfgang and T. Hai, Analysis of ATF3, a transcription factor induced by physiological stresses and modulated by gadd153/Chop10, *Mol. Cell. Biol.* **16**(3), 1157–1168 (1996).

517. D. Ron and J. F. Habener, CHOP, a novel developmentally regulated nuclear protein that dimerizes with transcription factors C/EBP and LAP and functions as a dominant-negative inhibitor of gene transcription, *Genes Dev.* **6**(3), 439–453 (1992).

518. N. Ohoka, S. Yoshii, T. Hattori, K. Onozaki and H. Hayashi, *TRB3*, a novel ER stress-inducible gene, is induced via ATF4-CHOP pathway and is involved in cell death, *EMBO J.* **24**(6), 1243–1255 (2005).

519. S. Oyadomari and M. Mori, Roles of CHOP/GADD153 in endoplasmic reticulum stress, *Cell Death Differ.* **11**(4), 381–389 (2004).

520. K. D. McCullough, J. L. Martindale, L. O. Klotz, T. Y. Aw and N. J. Holbrook, Gadd153 sensitizes cells to endoplasmic reticulum stress by down-regulating Bcl2 and perturbing the cellular redox state, *Mol. Cell. Biol.* **21**(4), 1249–1259 (2001).

521. H. Zinszner, M. Kuroda, X. Wang, N. Batchvarova, R. T. Lightfoot, H. Remotti, J. L. Stevens and D. Ron, CHOP is implicated in programmed cell death in response to impaired function of the endoplasmic reticulum, *Genes Dev.* **12**(7), 982–995 (1998).

522. C. H. He, P. Gong, B. Hu, D. Stewart, M. E. Choi, A. M. Choi and J. Alam, Identification of activating transcription factor 4 (ATF4) as an Nrf2-interacting protein. Implication for heme oxygenase-1 gene regulation, *J. Biol. Chem.* **276**(24), 20858–20865 (2001).

523. R. Venugopal and A. K. Jaiswal, Nrf2 and Nrf1 in association with Jun proteins regulate antioxidant response element-mediated expression and coordinated induction of genes encoding detoxifying enzymes, *Oncogene* **17**(24), 3145–3156 (1998).

524. T. Nguyen, H. C. Huang and C. B. Pickett, Transcriptional regulation of the antioxidant response element. Activation by Nrf2 and repression by MafK, *J. Biol. Chem.* **275**(20), 15466–15473 (2000).

525. S. B. Cullinan and J. A. Diehl, PERK-dependent activation of Nrf2 contributes to redox homeostasis and cell survival following endoplasmic reticulum stress, *J. Biol. Chem.* **279**(19), 20108–20117 (2004).

526. H. L. Pahl and P. A. Baeuerle, A novel signal transduction pathway from the endoplasmic reticulum to the nucleus is mediated by transcription factor NF-κB, *EMBO J.* **14**(11), 2580–2588 (1995).

527. C. M. Haynes, E. A. Titus and A. A. Cooper, Degradation of misfolded proteins prevents ER-derived oxidative stress and cell death, *Mol. Cell* **15**(5), 767–776 (2004).

528. I. Novoa, H. Zeng, H. P. Harding and D. Ron, Feedback inhibition of the unfolded protein response by *GADD34*-mediated dephosphorylation of eIF2α, *J. Cell Biol.* **153**(5), 1011–1022 (2001).

529. C. Jousse, S. Oyadomari, I. Novoa, P. Lu, Y. Zhang, H. P. Harding and D. Ron, Inhibition of a constitutive translation initiation factor 2α phosphatase, *CReP*, promotes survival of stressed cells, *J. Cell Biol.* **163**(4), 767–775 (2003).

530. Y. Ma and L. M. Hendershot, Delineation of a negative feedback regulatory loop that controls protein translation during endoplasmic reticulum stress, *J. Biol. Chem.* **278**(37), 34864–34873 (2003).

531. P. T. Cohen, Protein phosphatase 1–targeted in many directions, *J. Cell Sci.* **115**(2), 241–256 (2002).

532. J. H. Connor, D. C. Weiser, S. Li, J. M. Hallenbeck and S. Shenolikar, Growth arrest and DNA damage-inducible protein GADD34 assembles a novel signaling complex containing protein phosphatase 1 and inhibitor 1, *Mol. Cell. Biol.* **21**(20), 6841–6850 (2001).

533. J. Chou and B. Roizman, Herpes simplex virus 1 γ₁34.5 gene function, which blocks the host response to infection, maps in the homologous domain of the genes expressed during growth arrest and DNA damage, *Proc. Natl. Acad. Sci. U S A* **91**(12), 5247–5251 (1994).

534. S. M. Brown, A. R. MacLean, E. A. McKie and J. Harland, The herpes simplex virus virulence factor ICP34.5 and the cellular protein MyD116 complex with proliferating cell nuclear antigen through the 63-amino-acid domain conserved in ICP34.5, MyD116, and GADD34, *J. Virol.* **71**(12), 9442–9449 (1997).

535. B. He, M. Gross and B. Roizman, The γ₁34.5 protein of herpes simplex virus 1 complexes with protein phosphatase 1α to dephosphorylate the α subunit of the eukaryotic translation initiation factor 2 and preclude the shutoff of protein synthesis by double-stranded RNA-activated protein kinase, *Proc. Natl. Acad. Sci. U S A* **94**(3), 843–848 (1997).

536. B. He, M. Gross and B. Roizman, The $\gamma_1$34.5 protein of herpes simplex virus 1 has the structural and functional attributes of a protein phosphatase 1 regulatory subunit and is present in a high molecular weight complex with the enzyme in infected cells, *J. Biol. Chem.* **273**(33), 20737–20743 (1998).

537. B. He, J. Chou, D. A. Liebermann, B. Hoffman and B. Roizman, The carboxyl terminus of the murine MyD116 gene substitutes for the corresponding domain of the $\gamma_1$34.5 gene of herpes simplex virus to preclude the premature shutoff of total protein synthesis in infected human cells, *J. Virol.* **70**(1), 84–90 (1996).

538. W. J. Hung, R. S. Roberson, J. Taft and D. Y. Wu, Human BAG-1 proteins bind to the cellular stress response protein GADD34 and interfere with GADD34 functions, *Mol. Cell. Biol.* **23**(10), 3477–3486 (2003).

539. D. Y. Wu, D. C. Tkachuck, R. S. Roberson and W. H. Schubach, The human SNF5/INI1 protein facilitates the function of the growth arrest and DNA damage-inducible protein (GADD34) and modulates GADD34-bound protein phosphatase-1 activity, *J. Biol. Chem.* **277**(31), 27706–27715 (2002).

540. H. T. Adler, R. Chinery, D. Y. Wu, S. J. Kussick, J. M. Payne, A. J. Fornace, Jr. and D. C. Tkachuk, Leukemic HRX fusion proteins inhibit GADD34-induced apoptosis and associate with the GADD34 and hSNF5/INI1 proteins, *Mol. Cell. Biol.* **19**(10), 7050–7060 (1999).

541. A. V. Grishin, O. Azhipa, I. Semenov and S. J. Corey, Interaction between growth arrest-DNA damage protein 34 and Src kinase Lyn negatively regulates genotoxic apoptosis, *Proc. Natl. Acad. Sci. U S A* **98**(18), 10172–10177 (2001).

542. S. Kebache, E. Cardin, D. T. Nguyên, E. Chevet and L. Larose, Nck-1 antagonizes the endoplasmic reticulum stress-induced inhibition of translation, *J. Biol. Chem.* **279**(10), 9662–9671 (2004).

543. S. Kebache, D. Zuo, E. Chevet and L. Larose, Modulation of protein translation by Nck-1, *Proc. Natl. Acad. Sci. U S A* **99**(8), 5406–5411 (2002).

544. M. Gale, Jr., S. L. Tan, M. Wambach and M. G. Katze, Interaction of the interferon-induced PKR protein kinase with inhibitory proteins P58^{IPK} and vaccinia virus K3L is mediated by unique domains: implications for kinase regulation, *Mol. Cell. Biol.* **16**(8), 4172–4181 (1996).

545. W. Yan, C. L. Frank, M. J. Korth, B. L. Sopher, I. Novoa, D. Ron and M. G. Katze, Control of PERK eIF2α kinase activity by the endoplasmic reticulum stress-induced molecular chaperone P58^{IPK}, *Proc. Natl. Acad. Sci. U S A* **99**(25), 15920–15925 (2002).

546. R. van Huizen, J. L. Martindale, M. Gorospe and N. J. Holbrook, P58^{IPK}, a novel endoplasmic reticulum stress-inducible protein and potential negative regulator of eIF2α signaling, *J. Biol. Chem.* **278**(18), 15558–15564 (2003).

547. W. Yan, M. J. Gale, Jr., S. L. Tan and M. G. Katze, Inactivation of the PKR protein kinase and stimulation of mRNA translation by the cellular co-chaperone P58^{IPK} does not require J domain function, *Biochemistry* **41**(15), 4938–4945 (2002).

548. M. Gale, Jr., C. M. Blakely, D. A. Hopkins, M. W. Melville, M. Wambach, P. R. Romano and M. G. Katze, Regulation of interferon-induced protein kinase PKR: modulation of P58^{IPK} inhibitory function by a novel protein, P52^{rIPK}, *Mol. Cell. Biol.* **18**(2), 859–871 (1998).

549. M. Gale, Jr., C. M. Blakely, A. Darveau, P. R. Romano, M. J. Korth and M. G. Katze, P52^{rIPK} regulates the molecular cochaperone P58^{IPK} to mediate control of the RNA-dependent protein kinase in response to cytoplasmic stress, *Biochemistry* **41**(39), 11878–11887 (2002).

550. K. J. Travers, C. K. Patil, L. Wodicka, D. J. Lockhart, J. S. Weissman and P. Walter, Functional and genomic analyses reveal an essential coordination between the unfolded protein response and ER-associated degradation, *Cell* **101**(3), 249–258 (2000).

551. T. Okada, H. Yoshida, R. Akazawa, M. Negishi and K. Mori, Distinct roles of activating transcription factor 6 (ATF6) and double-stranded RNA-activated protein kinase-like endoplasmic reticulum kinase (PERK) in transcription during the mammalian unfolded protein response, *Biochem. J.* **366**(2), 585–594 (2002).

552. A. H. Lee, N. N. Iwakoshi and L. H. Glimcher, XBP-1 regulates a subset of endoplasmic reticulum resident chaperone genes in the unfolded protein response, *Mol. Cell. Biol.* **23**(21), 7448–7459 (2003).

553. R. Casagrande, P. Stern, M. Diehn, C. Shamu, M. Osario, M. Zúñiga, P. O. Brown and H. Ploegh, Degradation of proteins from the ER of S. cerevisiae requires an intact unfolded protein response pathway, *Mol. Cell* **5**(4), 729–735 (2000).

554. R. Friedlander, E. Jarosch, J. Urban, C. Volkwein and T. Sommer, A regulatory link between ER-associated protein degradation and the unfolded-protein response, *Nat. Cell Biol.* **2**(7), 379–384 (2000).

555. T. van Laar, T. Schouten, E. Hoogervorst, M. van Eck, A. J. van der Eb and C. Terleth, The novel MMS-inducible gene *Mif1/KIAA0025* is a target of the unfolded protein response pathway, *FEBS Lett.* **469**(1), 123–131 (2000).

556. R. Y. Hampton, ER-associated degradation in protein quality control and cellular regulation, *Curr. Opin. Cell Biol.* **14**(4), 476–482 (2002).

557. A. A. McCracken and J. L. Brodsky, Evolving questions and paradigm shifts in endoplasmic-reticulum-associated degradation (ERAD), *Bioessays* **25**(9), 868–877 (2003).

558. B. Meusser, C. Hirsch, E. Jarosch and T. Sommer, ERAD: the long road to destruction, *Nat. Cell Biol.* **7**(8), 766–772 (2005).

559. H. Yoshida, T. Matsui, N. Hosokawa, R. J. Kaufman, K. Nagata and K. Mori, A time-dependent phase shift in the mammalian unfolded protein response, *Dev. Cell* **4**(2), 265–271 (2003).

560. C. Zuber, J.-y. Fan, B. Guhl, A. Parodi, J. H. Fessler, C. Parker and J. Roth, Immunolocalization of UDP-glucose:glycoprotein glucosyltransferase indicates involvement of pre-Golgi intermediates in protein quality control, *Proc. Natl. Acad. Sci. U S A* **98**(19), 10710–10715 (2001).

561. C. Zuber, J. Y. Fan, B. Guhl and J. Roth, Misfolded proinsulin accumulates in expanded pre-Golgi intermediates and endoplasmic reticulum subdomains in pancreatic beta cells of Akita mice, *FASEB J.* **18**(3), U341–U360 (2004).

562. S. R. Caldwell, K. J. Hill and A. A. Cooper, Degradation of endoplasmic reticulum (ER) quality control substrates requires transport between the ER and Golgi, *J. Biol. Chem.* **276**(26), 23296–23303 (2001).

563. S. Vashist, W. Kim, W. J. Belden, E. D. Spear, C. Barlowe and D. T. Ng, Distinct retrieval and retention mechanisms are required for the quality control of endoplasmic reticulum protein folding, *J. Cell Biol.* **155**(3), 355–368 (2001).

564. A. Hershko and A. Ciechanover, The ubiquitin system, *Annu. Rev. Biochem.* **67**, 425–479 (1998).

565. R. N. Freiman and R. Tjian, Regulating the regulators: lysine modifications make their mark, *Cell* **112**(1), 11–17 (2003).

566. N. W. Bays, R. G. Gardner, L. P. Seelig, C. A. Joazeiro and R. Y. Hampton, Hrd1p/Der3p is a membrane-anchored ubiquitin ligase required for ER-associated degradation, *Nat. Cell Biol.* **3**(1), 24–29 (2001).

567. P. M. Deak and D. H. Wolf, Membrane topology and function of Der3/Hrd1p as a ubiquitin-protein ligase (*E*3) involved in endoplasmic reticulum degradation, *J. Biol. Chem.* **276**(14), 10663–10669 (2001).

568. R. Swanson, M. Locher and M. Hochstrasser, A conserved ubiquitin ligase of the nuclear enve-lope/endoplasmic reticulum that functions in both ER-associated and Matα2 repressor degradation, *Genes Dev.* **15**(20), 2660–2674 (2001).

569. S. Fang, M. Ferrone, C. Yang, J. P. Jensen, S. Tiwari and A. M. Weissman, The tumor autocrine motility factor receptor, gp78, is a ubiquitin protein ligase implicated in degradation from the endoplasmic reticulum, *Proc. Natl. Acad. Sci. U S A* **98**(25), 14422–14427 (2001).

570. U. Lenk, H. Yu, J. Walter, M. S. Gelman, E. Hartmann, R. R. Kopito and T. Sommer, A role for mammalian Ubc6 homologues in ER-associated protein degradation, *J. Cell Sci.* **115**(14), 3007–3014 (2002).

571. M. Kikkert, R. Doolman, M. Dai, R. Avner, G. Hassink, S. van Voorden, S. Thanedar, J. Roitelman, V. Chau and E. Wiertz, Human HRD1 is an E3 ubiquitin ligase involved in degradation of proteins from the endoplasmic reticulum, *J. Biol. Chem.* **279**(5), 3525–3534 (2004).

572. P. Connell, C. A. Ballinger, J. Jiang, Y. Wu, L. J. Thompson, J. Hohfeld and C. Patterson, The co-chaperone CHIP regulates protein triage decisions mediated by heat-shock proteins, *Nat. Cell Biol.* **3**(1), 93–96 (2001).

573. G. C. Meacham, C. Patterson, W. Zhang, J. M. Younger and D. M. Cyr, The Hsc70 co-chaperone CHIP targets immature CFTR for proteasomal degradation, *Nat. Cell Biol.* **3**(1), 100–105 (2001).

574. Y. Imai, M. Soda and R. Takahashi, Parkin suppresses unfolded protein stress-induced cell death through its E3 ubiquitin-protein ligase activity, *J. Biol. Chem.* **275**(46), 35661–35664 (2000).

575. Y. Imai, M. Soda, H. Inoue, N. Hattori, Y. Mizuno and R. Takahashi, An unfolded putative transmembrane polypeptide, which can lead to endoplasmic reticulum stress, is a substrate of Parkin, *Cell* **105**(7), 891–902 (2001).

576. Y. Imai, M. Soda, S. Hatakeyama, T. Akagi, T. Hashikawa, K.-I. Nakayama and R. Takahashi, CHIP is associated with Parkin, a gene responsible for familial Parkinson's disease, and enhances its ubiquitin ligase activity, *Mol. Cell* **10**(1), 55–67 (2002).

577. Y. Yoshida, T. Chiba, F. Tokunaga, H. Kawasaki, K. Iwai, T. Suzuki, Y. Ito, K. Matsuoka, M. Yoshida, K. Tanaka and T. Tai, E3 ubiquitin ligase that recognizes sugar chains, *Nature* **418**(6896), 438–442 (2002).

578. Y. Yoshida, F. Tokunaga, T. Chiba, K. Iwai, K. Tanaka and T. Tai, Fbs2 is a new member of the E3 ubiquitin ligase family that recognizes sugar chains, *J. Biol. Chem.* **278**(44), 43877–43884 (2003).

579. M. Rape, T. Hoppe, I. Gorr, M. Kalocay, H. Richly and S. Jentsch, Mobilization of processed, membrane-tethered SPT23 transcription factor by CDC48$^{UFD1/NPL4}$, a ubiquitin-selective chaperone, *Cell* **107**(5), 667–677 (2001).

580. Y. Ye, H. H. Meyer and T. A. Rapoport, The AAA ATPase Cdc48/p97 and its partners transport proteins from the ER into the cytosol, *Nature* **414**(6864), 652–656 (2001).

581. S. Braun, K. Matuschewski, M. Rape, S. Thoms and S. Jentsch, Role of the ubiquitin-selective CDC48$^{UFD1/NPL4}$ chaperone (segregase) in ERAD of OLE1 and other substrates, *EMBO J.* **21**(4), 615–621 (2002).

582. E. Jarosch, C. Taxis, C. Volkwein, J. Bordallo, D. Finley, D. H. Wolf and T. Sommer, Protein dislocation from the ER requires polyubiquitination and the AAA-ATPase Cdc48, *Nat. Cell Biol.* **4**(2), 134–139 (2002).

583. E. Rabinovich, A. Kerem, K. U. Fröhlich, N. Diamant and S. Bar-Nun, AAA-ATPase p97/Cdc48p, a cytosolic chaperone required for endoplasmic reticulum-associated protein degradation, *Mol. Cell. Biol.* **22**(2), 626–634 (2002).

584. N. W. Bays, S. K. Wilhovsky, A. Goradia, K. Hodgkiss-Harlow and R. Y. Hampton, *HRD4/NPL4* is required for the proteasomal processing of ubiquitinated ER proteins, *Mol. Biol. Cell* **12**(12), 4114–4128 (2001).

585. A. L. Hitchcock, H. Krebber, S. Frietze, A. Lin, M. Latterich and P. A. Silver, The conserved Npl4 protein complex mediates proteasome-dependent membrane-bound transcription factor activation, *Mol. Biol. Cell* **12**(10), 3226–3241 (2001).

586. H. H. Meyer, Y. Wang and G. Warren, Direct binding of ubiquitin conjugates by the mammalian p97 adaptor complexes, p47 and Ufd1-Npl4, *EMBO J.* **21**(21), 5645–5652 (2002).

587. C. Enenkel, A. Lehmann and P.-M. Kloetzel, Subcellular distribution of proteasomes implicates a major location of protein degradation in the nuclear envelope-ER network in yeast, *EMBO J.* **17**(21), 6144–6154 (1998).

588. P. Brooks, G. Fuertes, R. Z. Murray, S. Bose, E. Knecht, M. C. Rechsteiner, K. B. Hendil, K. Tanaka, J. Dyson and J. Rivett, Subcellular localization of proteasomes and their regulatory complexes in mammalian cells, *Biochem. J.* **346**(1), 155–161 (2000).

589. P. Brooks, R. Z. Murray, G. G. Mason, K. B. Hendil and A. J. Rivett, Association of immunoproteasomes with the endoplasmic reticulum, *Biochem. J.* **352**(3), 611–615 (2000).

590. K. J. Travers, C. K. Patil and J. S. Weissman, Functional genomic approaches to understanding molecular chaperones and stress responses, *Adv. Protein Chem.* **59**, 345–390 (2001).

591. O. Hori, F. Ichinoda, A. Yamaguchi, T. Tamatani, M. Taniguchi, Y. Koyama, T. Katayama, M. Tohyama, D. M. Stern, K. Ozawa, Y. Kitao and S. Ogawa, Role of Herp in the endoplasmic reticulum stress response, *Genes Cells* **9**(5), 457–469 (2004).

592. Y. Oda, T. Okada, H. Yoshida, R. J. Kaufman, K. Nagata and K. Mori, Derlin-2 and Derlin-3 are regulated by the mammalian unfolded protein response and are required for ER-associated degradation, *J. Cell Biol.* **172**(3), 383–393 (2006).

593. D. L. Wiest, J. K. Burkhardt, S. Hester, M. Hortsch, D. I. Meyer and Y. Argon, Membrane biogenesis during B cell differentiation: most endoplasmic reticulum proteins are expressed coordinately, *J. Cell Biol.* **110**(5), 1501–1511 (1990).

594. A. Masuda, M. Kuwano and T. Shimada, Ultrastructural changes during the enhancement of cellular 3-hydroxy-3-methyl-glutaryl-coenzyme A reductase in a Chinese hamster cell mutant resistant to compactin (ML 236B), *Cell Struct. Funct.* **8**(3), 309–312 (1983).

595. J. Nikawa, M. Akiyoshi, S. Hirata and T. Fukuda, Saccharomyces cerevisiae *IRE2/HAC1* is involved in *IRE1*-mediated *KAR2* expression, *Nucleic Acids Res.* **24**(21), 4222–4226 (1996).

596. H. J. Chang, S. A. Jesch, M. L. Gaspar and S. A. Henry, Role of the unfolded protein response pathway in secretory stress and regulation of *INO1* expression in *Saccharomyces cerevisiae*, *Genetics* **168**(4), 1899–1913 (2004).

597. J. S. Cox, R. E. Chapman and P. Walter, The unfolded protein response coordinates the production of endoplasmic reticulum protein and endoplasmic reticulum membrane, *Mol. Biol. Cell* **8**(9), 1805–1814 (1997).

598. A. K. Stroobants, E. H. Hettema, M. van den Berg and H. F. Tabak, Enlargement of the endoplasmic reticulum membrane in *Saccharomyces cerevisiae* is not necessarily linked to the unfolded protein response via Ire1p, *FEBS Lett.* **453**(1–2), 210–214 (1999).

599. M. Hyde, L. Block-Alper, J. Felix, P. Webster and D. I. Meyer, Induction of secretory pathway components in yeast is associated with increased stability of their mRNA, *J. Cell Biol.* **156**(6), 993–1001 (2002).

600. J. P. Hirsch and S. A. Henry, Expression of the *Saccharomyces cerevisiae* inositol-1-phosphate synthase (*INO1*) gene is regulated by factors that affect phospholipid synthesis, *Mol. Cell. Biol.* **6**(10), 3320–3328 (1986).

601. H. J. Chang, E. W. Jones and S. A. Henry, Role of the unfolded protein response pathway in regulation of *INO1* and in the *sec14* bypass mechanism in *Saccharomyces cerevisiae*, *Genetics* **162**(1), 29–43 (2002).

602. C. J. Loewen, M. L. Gaspar, S. A. Jesch, C. Delon, N. T. Ktistakis, S. A. Henry and T. P. Levine, Phospholipid metabolism regulated by a transcription factor sensing phosphatidic acid, *Science* **304**(5677), 1644–1647 (2004).

603. C. J. Loewen, A. Roy and T. P. Levine, A conserved ER targeting motif in three families of lipid binding proteins and in Opi1p binds VAP, *EMBO J.* **22**(9), 2025–2035 (2003).

604. J. H. Brickner and P. Walter, Gene recruitment of the activated *INO1* locus to the nuclear membrane, *PLoS Biol.* **2**(11), 1–11 (2004).

605. A. L. Shaffer, M. Shapiro-Shelef, N. N. Iwakoshi, A.-H. Lee, S. B. Qian, H. Zhao, X. Yu, L. Yang, B. K. Tan, A. Rosenwald, E. M. Hurt, E. Petroulakis, N. Sonenberg, J. W. Yewdell, K. Calame, L. H. Glimcher and L. M. Staudt, XBP1, downstream of Blimp-1, expands the secretory apparatus and other organelles, and increases protein synthesis in plasma cell differentiation, *Immunity* **21**(1), 81–93 (2004).

606. R. Sriburi, S. Jackowski, K. Mori and J. W. Brewer, XBP1: a link between the unfolded protein response, lipid biosynthesis, and biogenesis of the endoplasmic reticulum, *J. Cell Biol.* **167**(1), 35–41 (2004).

607. H. Liu, R. C. Bowes, III, B. van de Water, C. Sillence, J. F. Nagelkerke and J. L. Stevens, Endoplasmic reticulum chaperones GRP78 and calreticulin prevent oxidative stress, Ca^{2+} disturbances, and cell death in renal epithelial cells, *J. Biol. Chem.* **272**(35), 21751–21759 (1997).

608. S. Tanaka, T. Uehara and Y. Nomura, Up-regulation of protein-disulfide isomerase in response to hypoxia/brain ischemia and its protective effect against apoptotic cell death, *J. Biol. Chem.* **275**(14), 10388–10393 (2000).

609. M. Schröder and R. J. Kaufman, Divergent roles of Ire1α and PERK in the unfolded protein response, *Curr. Mol. Med.* **6**(1), 5–36 (2006).

610. D. T. Rutkowski and R. J. Kaufman, A trip to the ER: coping with stress, *Trends Cell Biol.* **14**(1), 20–28 (2004).

611. T. Tenev, A. Zachariou, R. Wilson, A. Paul and P. Meier, Jafrac2 is an IAP antagonist that promotes cell death by liberating Dronc from DIAP1, *EMBO J.* **21**(19), 5118–5129 (2002).

612. C. Zhang, Y. Cai, M. T. Adachi, S. Oshiro, T. Aso, R. J. Kaufman and S. Kitajima, Homocysteine induces programmed cell death in human vascular endothelial cells through activation of the unfolded protein response, *J. Biol. Chem.* **276**(38), 35867–35874 (2001).

613. F. Urano, X. Wang, A. Bertolotti, Y. Zhang, P. Chung, H. P. Harding and D. Ron, Coupling of stress in the ER to activation of JNK protein kinases by transmembrane protein kinase IRE1, *Science* **287**(5453), 664–666 (2000).

614. P. P. Roux and J. Blenis, ERK and p38 MAPK-activated protein kinases: a family of protein kinases with diverse biological functions, *Microbiol. Mol. Biol. Rev.* **68**(2), 320–344 (2004).

615. H. Enslen, D. M. Brancho and R. J. Davis, Molecular determinants that mediate selective activation of p38 MAP kinase isoforms, *EMBO J.* **19**(6), 1301–1311 (2000).

616. D. Brancho, N. Tanaka, A. Jaeschke, J. J. Ventura, N. Kelkar, Y. Tanaka, M. Kyuuma, T. Takeshita, R. A. Flavell and R. J. Davis, Mechanism of p38 MAP kinase activation in vivo, *Genes Dev.* **17**(16), 1969–1978 (2003).

617. H. Habelhah, S. Takahashi, S. G. Cho, T. Kadoya, T. Watanabe and Z. Ronai, Ubiquitination and translocation of TRAF2 is required for activation of JNK but not of p38 or NF-κB, *EMBO J.* **23**(2), 322–332 (2004).

618. X. Z. Wang and D. Ron, Stress-induced phosphorylation and activation of the transcription factor CHOP (GADD153) by p38 MAP Kinase, *Science* **272**(5266), 1347–1349 (1996).

619. J. M. Kyriakis and J. Avruch, Mammalian mitogen-activated protein kinase signal transduction pathways activated by stress and inflammation, *Physiol. Rev.* **81**(2), 807–869 (2001).

620. C. Dunn, C. Wiltshire, A. MacLaren and D. A. Gillespie, Molecular mechanism and biological functions of c-Jun N-terminal kinase signalling via the c-Jun transcription factor, *Cell. Signal.* **14**(7), 585–593 (2002).

621. K. Lei and R. J. Davis, JNK phosphorylation of Bim-related members of the Bcl2 family induces Bax-dependent apoptosis, *Proc. Natl. Acad. Sci. U S A* **100**(5), 2432–2437 (2003).

622. G. V. Putcha, S. Le, S. Frank, C. G. Besirli, K. Clark, B. Chu, S. Alix, R. J. Youle, A. LaMarche, A. C. Maroney and E. M. Johnson, Jr., JNK-mediated BIM phosphorylation potentiates BAX-dependent apoptosis, *Neuron* **38**(6), 899–914 (2003).

623. T. Nakagawa, H. Zhu, N. Morishima, E. Li, J. Xu, B. A. Yankner and J. Yuan, Caspase-12 mediates endoplasmic-reticulum-specific apoptosis and cytotoxicity by amyloid-β, *Nature* **403**(6765), 98–103 (2000).

624. T. Yoneda, K. Imaizumi, K. Oono, D. Yui, F. Gomi, T. Katayama and M. Tohyama, Activation of caspase-12, an endoplastic reticulum (ER) resident caspase, through tumor necrosis factor receptor-associated factor 2-dependent mechanism in response to the ER stress, *J. Biol. Chem.* **276**(17), 13935–13940 (2001).

625. R. V. Rao, E. Hermel, S. Castro-Obregon, G. del Rio, L. M. Ellerby, H. M. Ellerby and D. E. Bredesen, Coupling endoplasmic reticulum stress to the cell death program. Mechanism of caspase activation, *J. Biol. Chem.* **276**(36), 33869–33874 (2001).

626. N. Morishima, K. Nakanishi, H. Takenouchi, T. Shibata and Y. Yasuhiko, An endoplasmic reticulum stress-specific caspase cascade in apoptosis. Cytochrome *c*-independent activation of caspase-9 by caspase-12, *J. Biol. Chem.* **277**(37), 34287–34294 (2002).

627. E. H.-Y. Cheng, D. G. Kirsch, R. J. Clem, R. Ravi, M. B. Kastan, A. Bedi, K. Ueno and J. M. Hardwick, Conversion of Bcl-2 to a Bax-like death effector by caspases, *Science* **278**(5345), 1966–1968 (1997).

628. H. Fischer, U. Koenig, L. Eckhart and E. Tschachler, Human caspase 12 has acquired deleterious mutations, *Biochem. Biophys. Res. Commun.* **293**(2), 722–726 (2002).

629. M. Saleh, J. P. Vaillancourt, R. K. Graham, M. Huyck, S. M. Srinivasula, E. S. Alnemri, M. H. Steinberg, V. Nolan, C. T. Baldwin, R. S. Hotchkiss, T. G. Buchman, B. A. Zehnbauer, M. R. Hayden, L. A. Farrer, S. Roy and D. W. Nicholson, Differential modulation of endotoxin responsiveness by human caspase-12 polymorphisms, *Nature* **429**(6987), 75–79 (2004).

630. J. Hitomi, T. Katayama, Y. Eguchi, T. Kudo, M. Taniguchi, Y. Koyama, T. Manabe, S. Yamagishi, Y. Bando, K. Imaizumi, Y. Tsujimoto and M. Tohyama, Involvement of caspase-4 in endoplasmic reticulum stress-induced apoptosis and Aβ-induced cell death, *J. Cell Biol.* **165**(3), 347–356 (2004).

631. B. Antonsson and J.-C. Martinou, The Bcl-2 protein family, *Exp. Cell Res.* **256**(1), 50–57 (2000).
632. D. T. Chao and S. J. Korsmeyer, BCL-2 family: regulators of cell death, *Annu. Rev. Immunol.* **16**, 395–419 (1998).
633. M. C. Wei, W. X. Zong, E. H. Cheng, T. Lindsten, V. Panoutsakopoulou, A. J. Ross, K. A. Roth, G. R. MacGregor, C. B. Thompson and S. J. Korsmeyer, Proapoptotic BAX and BAK: a requisite gateway to mitochondrial dysfunction and death, *Science* **292**(5517), 727–730 (2001).
634. Y. Akao, Y. Otsuki, S. Kataoka, Y. Ito and Y. Tsujimoto, Multiple subcellular localization of bcl-2: detection in nuclear outer membrane, endoplasmic reticulum membrane, and mitochondrial membranes, *Cancer Res.* **54**(9), 2468–2471 (1994).
635. T. Lithgow, R. van Driel, J. F. Bertram and A. Strasser, The protein product of the oncogene *bcl-2* is a component of the nuclear envelope, the endoplasmic reticulum, and the outer mitochondrial membrane, *Cell Growth Differ.* **5**(4), 411–417 (1994).
636. L. Scorrano, S. A. Oakes, J. T. Opferman, E. H. Cheng, M. D. Sorcinelli, T. Pozzan and S. J. Korsmeyer, BAX and BAK regulation of endoplasmic reticulum Ca^{2+}: a control point for apoptosis, *Science* **300**(5616), 135–139 (2003).
637. W. X. Zong, C. Li, G. Hatzivassiliou, T. Lindsten, Q.-C. Yu, J. Yuan and C. B. Thompson, Bax and Bak can localize to the endoplasmic reticulum to initiate apoptosis, *J. Cell Biol.* **162**(1), 59–69 (2003).
638. J. Häcki, L. Egger, L. Monney, S. Conus, T. Rossé, I. Fellay and C. Borner, Apoptotic crosstalk between the endoplasmic reticulum and mitochondria controlled by Bcl-2, *Oncogene* **19**(19), 2286–2295 (2000).
639. N. S. Wang, M. T. Unkila, E. Z. Reineks and C. W. Distelhorst, Transient expression of wild-type or mitochondrially targeted Bcl-2 induces apoptosis, whereas transient expression of endoplasmic reticulum-targeted Bcl-2 is protective against Bax-induced cell death, *J. Biol. Chem.* **276**(47), 44117–44128 (2001).
640. W. Zhu, A. Cowie, G. W. Wasfy, L. Z. Penn, B. Leber and D. W. Andrews, Bcl-2 mutants with restricted subcellular location reveal spatially distinct pathways for apoptosis in different cell types, *EMBO J.* **15**(16), 4130–4141 (1996).
641. M. G. Annis, N. Zamzami, W. Zhu, L. Z. Penn, G. Kroemer, B. Leber and D. W. Andrews, Endoplasmic reticulum localized Bcl-2 prevents apoptosis when redistribution of cytochrome c is a late event, *Oncogene* **20**(16), 1939–1952 (2001).
642. J. Rudner, A. Lepple-Wienhues, W. Budach, J. Berschauer, B. Friedrich, S. Wesselborg, K. Schulze-Osthoff and C. Belka, Wild-type, mitochondrial and ER-restricted Bcl-2 inhibit DNA damage-induced apoptosis but do not affect death receptor-induced apoptosis, *J. Cell Sci.* **114**(23), 4161–4172 (2001).
643. C. Reimertz, D. Kögel, A. Rami, T. Chittenden and J. H. Prehn, Gene expression during ER stress-induced apoptosis in neurons: induction of the BH3-only protein Bbc3/PUMA and activation of the mitochondrial apoptosis pathway, *J. Cell Biol.* **162**(4), 587–597 (2003).
644. H. J. Chae, H. R. Kim, C. Xu, B. Bailly-Maitre, M. Krajewska, S. Krajewski, S. Banares, J. Cui, M. Digicaylioglu, N. Ke, S. Kitada, E. Monosov, M. Thomas, C. L. Kress, J. R. Babendure, R. Y. Tsien, S. A. Lipton and J. C. Reed, BI-1 regulates an apoptosis pathway linked to endoplasmic reticulum stress, *Mol. Cell* **15**(3), 355–366 (2004).
645. F. W. Ng, M. Nguyen, T. Kwan, P. E. Branton, D. W. Nicholson, J. A. Cromlish and G. C. Shore, p28 Bap31, a Bcl-2/Bcl-X_L- and procaspase-8-associated protein in the endoplasmic reticulum, *J. Cell Biol.* **139**(2), 327–338 (1997).
646. H. Zhang, Q. Xu, S. Krajewski, M. Krajewska, Z. Xie, S. Fuess, S. Kitada, K. Pawlowski, A. Godzik and J. C. Reed, BAR: An apoptosis regulator at the intersection of caspases and Bcl-2 family proteins, *Proc. Natl. Acad. Sci. U S A* **97**(6), 2597–2602 (2000).
647. T. Adachi, W. W. Schamel, K. M. Kim, T. Watanabe, B. Becker, P. J. Nielsen and M. Reth, The specificity of association of the IgD molecule with the accessory proteins BAP31/BAP29 lies in the IgD transmembrane sequence, *EMBO J.* **15**(7), 1534–1541 (1996).
648. M. Nguyen, D. G. Breckenridge, A. Ducret and G. C. Shore, Caspase-resistant BAP31 inhibits fas-mediated apoptotic membrane fragmentation and release of cytochrome c from mitochondria, *Mol. Cell. Biol.* **20**(18), 6731–6740 (2000).

649. W. G. Annaert, B. Becker, U. Kistner, M. Reth and R. Jahn, Export of cellubrevin from the endoplasmic reticulum is controlled by BAP31, *J. Cell Biol.* **139**(6), 1397–1410 (1997).
650. E. T. Spiliotis, H. Manley, M. Osorio, M. C. Zúñiga and M. Edidin, Selective export of MHC class I molecules from the ER after their dissociation from TAP, *Immunity* **13**(6), 841–851 (2000).
651. G. Lambert, B. Becker, R. Schreiber, A. Boucherot, M. Reth and K. Kunzelmann, Control of cystic fibrosis transmembrane conductance regulator expression by BAP31, *J. Biol. Chem.* **276**(23), 20340–20345 (2001).
652. W. W. Schamel, S. Kuppig, B. Becker, K. Gimborn, H. P. Hauri and M. Reth, A high-molecular-weight complex of membrane proteins BAP29/BAP31 is involved in the retention of membrane-bound IgD in the endoplasmic reticulum, *Proc. Natl. Acad. Sci. U S A* **100**(17), 9861–9866 (2003).
653. M. E. Paquet, M. Cohen-Doyle, G. C. Shore and D. B. Williams, Bap29/31 influences the intracellular traffic of MHC class I molecules, *J. Immunol.* **172**(12), 7548–7555 (2004).
654. J. Määttä, O. Hallikas, S. Welti, P. Hildén, J. Schröder and E. Kuismanen, Limited caspase cleavage of human BAP31, *FEBS Lett.* **484**(3), 202–206 (2000).
655. T. Hidvegi, B. Z. Schmidt, P. Hale and D. H. Perlmutter, Accumulation of mutant α1 antitrypsin Z in the ER activates caspases-4 and -12, NFκB and BAP31 but not the unfolded protein response, *J. Biol. Chem.* **280**(47), 39002–39015 (2005).
656. A. Zuppini, J. Groenendyk, L. A. Cormack, G. Shore, M. Opas, R. C. Bleackley and M. Michalak, Calnexin deficiency and endoplasmic reticulum stress-induced apoptosis, *Biochemistry* **41**(8), 2850–2858 (2002).
657. T. Nakagawa and J. Yuan, Cross-talk between two cysteine protease families. Activation of caspase-12 by calpain in apoptosis, *J. Cell Biol.* **150**(4), 887–894 (2000).
658. R. K. Reddy, J. Lu and A. S. Lee, The endoplasmic reticulum chaperone glycoprotein GRP94 with Ca^{2+}-binding and antiapoptotic properties is a novel proteolytic target of calpain during etoposide-induced apoptosis, *J. Biol. Chem.* **274**(40), 28476–28483 (1999).
659. M. Crompton, The mitochondrial permeability transition pore and its role in cell death, *Biochem. J.* **341**(2), 233–249 (1999).
660. P. Boya, I. Cohen, N. Zamzami, H. L. Vieira and G. Kroemer, Endoplasmic reticulum stress-induced cell death requires mitochondrial membrane permeabilization, *Cell Death Differ.* **9**(4), 465–467 (2002).
661. D. Boehning, R. L. Patterson and S. H. Snyder, Apoptosis and calcium: new roles for cytochrome c and inositol 1,4,5-trisphosphate, *Cell Cycle* **3**(3), 252–254 (2004).
662. G. Münch, B. Bölck, P. Karczewski and R. H. Schwinger, Evidence for calcineurin-mediated regulation of SERCA 2a activity in human myocardium, *J. Mol. Cell. Cardiol.* **34**(3), 321–334 (2002).
663. S. K. Joseph, D. Boehning, S. Bokkala, R. Watkins and J. Widjaja, Biosynthesis of inositol trisphosphate receptors: selective association with the molecular chaperone calnexin, *Biochem. J.* **342**(1), 153–161 (1999).
664. Y. Ito, P. Pandey, N. Mishra, S. Kumar, N. Narula, S. Kharbanda, S. Saxena and D. Kufe, Targeting of the c-Abl tyrosine kinase to mitochondria in endoplasmic reticulum stress-induced apoptosis, *Mol. Cell. Biol.* **21**(18), 6233–6242 (2001).
665. C. J. Gimeno, P. O. Ljungdahl, C. A. Styles and G. R. Fink, Unipolar cell divisions in the yeast S. cerevisiae lead to filamentous growth: regulation by starvation and RAS, *Cell* **68**(6), 1077–1090 (1992).
666. S. M. Honigberg and K. Purnapatre, Signal pathway integration in the switch from the mitotic cell cycle to meiosis in yeast, *J. Cell Sci.* **116**(11), 2137–2147 (2003).
667. S. J. Kron and N. A. Gow, Budding yeast morphogenesis: signalling, cytoskeleton and cell cycle, *Curr. Opin. Cell Biol.* **7**(6), 845–855 (1995).
668. D. Rua, B. T. Tobe and S. J. Kron, Cell cycle control of yeast filamentous growth, *Curr. Opin. Microbiol.* **4**(6), 720–727 (2001).
669. S. M. Honigberg and R. E. Esposito, Reversal of cell determination in yeast meiosis: postcommitment arrest allows return to mitotic growth, *Proc. Natl. Acad. Sci. U S A* **91**(14), 6559–6563 (1994).

670. J. N. Gass, N. M. Gifford and J. W. Brewer, Activation of an unfolded protein response during differentiation of antibody-secreting B cells, *J. Biol. Chem.* **277**(50), 49047–49054 (2002).

671. K. Zhang, H. N. Wong, B. Song, C. N. Miller, D. Scheuner and R. J. Kaufman, The unfolded protein response sensor IRE1α is required at 2 distinct steps in B cell lymphopoiesis, *J. Clin. Invest.* **115**(2), 268–281 (2005).

672. E. van Anken, E. P. Romijn, C. Maggioni, A. Mezghrani, R. Sitia, I. Braakman and A. J. Heck, Sequential waves of functionally related proteins are expressed when B cells prepare for antibody secretion, *Immunity* **18**(2), 243–253 (2003).

673. A. H. Skalet, J. A. Isler, L. B. King, H. P. Harding, D. Ron and J. G. Monroe, Rapid BCR-induced unfolded protein response in non-secretory B cells correlates with pro- versus anti-apoptotic cell fate, *J. Biol. Chem.* **280**(48), 39762–39771 (2005).

674. A. Henderson and K. Calame, Transcriptional regulation during B cell development, *Annu. Rev. Immunol.* **16**, 163–200 (1998).

675. A. M. Reimold, N. N. Iwakoshi, J. Manis, P. Vallabhajosyula, E. Szomolanyi-Tsuda, E. M. Gravallese, D. Friend, M. J. Grusby, F. Alt and L. H. Glimcher, Plasma cell differentiation requires the transcription factor XBP-1, *Nature* **412**(6844), 300–307 (2001).

676. K. I. Lin, Y. Lin and K. Calame, Repression of c-*myc* is necessary but not sufficient for terminal differentiation of B lymphocytes in vitro, *Mol. Cell. Biol.* **20**(23), 8684–8695 (2000).

677. J. J. Sciandra, J. R. Subjeck and C. S. Hughes, Induction of glucose-regulated proteins during anaerobic exposure and of heat-shock proteins after reoxygenation, *Proc. Natl. Acad. Sci. U S A* **81**(15), 4843–4847 (1984).

678. H. P. Harding, H. Zeng, Y. Zhang, R. Jungries, P. Chung, H. Plesken, D. D. Sabatini and D. Ron, Diabetes mellitus and exocrine pancreatic dysfunction in *Perk-/-* mice reveals a role for translational control in secretory cell survival, *Mol. Cell* **7**(6), 1153–1163 (2001).

679. W. C. Ladiges, S. E. Knoblaugh, J. F. Morton, M. J. Korth, B. L. Sopher, C. R. Baskin, A. MacAuley, A. G. Goodman, R. C. LeBoeuf and M. G. Katze, Pancreatic β-cell failure and diabetes in mice with a deletion mutation of the endoplasmic reticulum molecular chaperone gene P58^{IPK}, *Diabetes* **54**(4), 1074–1081 (2005).

680. P. Zhang, B. McGrath, S. Li, A. Frank, F. Zambito, J. Reinert, M. Gannon, K. Ma, K. McNaughton and D. R. Cavener, The PERK eukaryotic initiation factor 2α kinase is required for the development of the skeletal system, postnatal growth, and the function and viability of the pancreas, *Mol. Cell. Biol.* **22**(11), 3864–3874 (2002).

681. H. P. Harding, Y. Zhang, A. Bertolotti, H. Zeng and D. Ron, *Perk* is essential for translational regulation and cell survival during the unfolded protein response, *Mol. Cell* **5**(5), 897–904 (2000).

682. R. A. Easom and V. A. Zammit, Acute effects of starvation and treatment of rats with anti-insulin serum, glucagon and catecholamines on the state of phosphorylation of hepatic 3-hydroxy-3-methylglutaryl-CoA reductase *in vivo*, *Biochem. J.* **241**(1), 183–188 (1987).

683. M. Okuyama, M. Tsunogai, N. Watanabe, Y. Asakura and A. Shigematsu, Study of the de novo synthesis of cholesterol in the rat liver: a newly developed radiotracer technique, "TLC-autoradioluminography", *Biol. Pharm. Bull.* **18**(11), 1467–1471 (1995).

684. M. S. Kilberg, Y.-X. Pan, H. Chen and V. Leung-Pineda, Nutritional control of gene expression: How mammalian cells respond to amino acid limitation, *Annu. Rev. Nutr.* **25**, 59–85 (2005).

685. E. S. Alnemri, N. M. Robertson, T. F. Fernandes, C. M. Croce and G. Litwack, Overexpressed full-length human BCL2 extends the survival of baculovirus-infected Sf9 insect cells, *Proc. Natl. Acad. Sci. U S A* **89**(16), 7295–7299 (1992).

686. Y. Itoh, H. Ueda and E. Suzuki, Overexpression of bcl-2, apoptosis suppressing gene – prolonged viable culture period of hybridoma and enhanced antibody-production, *Biotechnol. Bioeng.* **48**(2), 118–122 (1995).

687. A. J. Mastrangelo, J. M. Hardwick and M. J. Betenbaugh, Bcl-2 inhibits apoptosis and extends recombinant protein production in cells infected with Sindbis viral vectors, *Cytotechnology* **22**(1–3), 169–178 (1996).

688. R. P. Singh, A. N. Emery and M. Al-Rubeai, Enhancement of survivability of mammalian cells by overexpression of the apoptosis-suppressor gene bcl-2, *Biotechnol. Bioeng.* **52**(1), 166–175 (1996).

689. A. J. Mastrangelo, J. M. Hardwick, S. Zou and M. J. Betenbaugh, Part II. Overexpression of *bcl-2* family members enhances survival of mammalian cells in response to various culture insults, *Biotechnol. Bioeng.* **67**(5), 555–564 (2000).

690. M. Fussenegger, D. Fassnacht, R. Schwartz, J. A. Zanghi, M. Graf, J. E. Bailey and R. Portner, Regulated overexpression of the survival factor *bcl-2* in CHO cells increases viable cell density in batch culture and decreases DNA release in extended fixed-bed cultivation, *Cytotechnology* **32**(1), 45–61 (2000).

691. H. Meents, B. Enenkel, H. M. Eppenberger, R. G. Werner and M. Fussenegger, Impact of coexpression and coamplification of sICAM and antiapoptosis determinants $bcl-2/bcl-x_L$ on productivity, cell survival, and mitochondria number in CHO-DG44 grown in suspension and serum-free media, *Biotechnol. Bioeng.* **80**(6), 706–716 (2002).

692. B. T. Tey, R. P. Singh, L. Piredda, M. Piacentini and M. Al-Rubeai, Influence of Bcl-2 on cell death during the cultivation of a Chinese hamster ovary cell line expressing a chimeric antibody, *Biotechnol. Bioeng.* **68**(1), 31–43 (2000).

693. N. S. Kim and G. M. Lee, Inhibition of sodium butyrate-induced apoptosis in recombinant Chinese hamster ovary cells by constitutively expressing antisense RNA of caspase-3, *Biotechnol. Bioeng.* **78**(2), 217–228 (2002).

694. Y. H. Kim, T. Iida, T. Fujita, S. Terada, A. Kitayama, H. Ueda, E. V. Prochownik and E. Suzuki, Establishment of an apoptosis-resistant and growth-controllable cell line by transfecting with inducible antisense c-Jun gene, *Biotechnol. Bioeng.* **58**(1), 65–72 (1998).

695. W. A. Renner, K. H. Lee, V. Hatzimanikatis, J. E. Bailey and H. M. Eppenberger, Recombinant cyclin E expression activates proliferation and obviates surface attachment of chinese-hamster ovary (CHO) cells in protein-free medium, *Biotechnol. Bioeng.* **47**(4), 476–482 (1995).

696. K. H. Lee, A. Sburlati, W. A. Renner and J. E. Bailey, Deregulated expression of cloned transcription factor E2F-1 in Chinese hamster ovary cells shifts protein patterns and activates growth in protein-free medium, *Biotechnol. Bioeng.* **50**(3), 273–279 (1996).

697. M. J. Betenbaugh, E. Ailor, E. Whiteley, P. Hinderliter and T. A. Hsu, Chaperone and foldase coexpression in the baculovirus-insect cell expression system, *Cytotechnology* **20**(1–3), 149–159 (1996).

698. R. E. Cudna and A. J. Dickson, Endoplasmic reticulum signaling as a determinant of recombinant protein expression, *Biotechnol. Bioeng.* **81**(1), 56–65 (2003).

699. A. J. Dorner, L. C. Wasley and R. J. Kaufman, Overexpression of GRP78 mitigates stress induction of glucose regulated proteins and blocks secretion of selective proteins in Chinese hamster ovary cells, *EMBO J.* **11**(4), 1563–1571 (1992).

700. A. J. Dorner, M. G. Krane and R. J. Kaufman, Reduction of endogenous GRP78 levels improves secretion of a heterologous protein in CHO cells, *Mol. Cell. Biol.* **8**(10), 4063–4070 (1988).

701. K. Kitchin and M. C. Flickinger, Alteration of hybridoma viability and antibody secretion in transfectomas with inducible overexpression of protein disulfide isomerase, *Biotechnol. Prog.* **11**(5), 565–574 (1995).

702. M. Valkonen, M. Penttilä and M. Saloheimo, Effects of inactivation and constitutive expression of the unfolded-protein response pathway on protein production in the yeast *Saccharomyces cerevisiae*, *Appl. Environ. Microbiol.* **69**(4), 2065–2072 (2003).

703. M. Valkonen, M. Ward, H. Wang, M. Penttilä and M. Saloheimo, Improvement of foreign-protein production in *Aspergillus niger* var. *awamori* by constitutive induction of the unfolded-protein response, *Appl. Environ. Microbiol.* **69**(12), 6979–6986 (2003).

CHAPTER 5

ENGINEERING OF CELL PROLIFERATION
VIA MYC MODULATION

VASILIKI IFANDI[1] AND MOHAMED AL-RUBEAI[2]

[1] *Clinical Sciences Research Institute, Warwick Medical School, Walsgrave Hospital, Coventry, CV2 2DX, UK and Biomedical Research Institute, Dept. of Biological Sciences, University of Warwick, Coventry, CV4 7AL, UK*
[2] *School of Chemical and Bioprocess Engineering, and Centre for Synthesis and Chemical Biology, University College Dublin, Belfield, Dublin 4, Ireland*

Abstract: The use of metabolic engineering for the enhancement and control of cell proliferation is a rapidly developing field in biotechnology. Great research interest has
been directed towards the development of proliferation and apoptosis controlled cell
lines with high cell density, regulated proliferation, apoptosis resistance, and easy
adaptation into serum free cultures. These are some of the desirable characteristics
for the cost effective production of biopharmaceuticals, mainly because genetically
modified cell lines can afford greater efficiency and control. Some of the strategies
employed by metabolic engineering for the management of cell proliferation include
the control of external factors in the culture environment, suppression of growth
inhibitors, and over-expression of important regulators of proliferation and apoptosis pathways, such as growth factors and genes. *c-myc* is such a prime candidate
to achieve these objectives. In its role as a transcription factor Myc can regulate
an extensive array of biological activities by modulating the expression of a large
number of genes that in turn regulate multiple downstream events. More importantly, Myc can regulate cell proliferation and transform cells in such a manner as to
consider the advantages of utilising its unique characteristics into the development
of cell lines with major significance in animal cell culture biotechnology

Keywords: c-Myc, proliferation, metabolic engineering, apoptosis

1.   INTRODUCTION

The onset of biotechnology in the 1970s has revealed new areas of possibilities
for the development of new therapeutic and diagnostic tools. An extensive number
of cellular systems are employed for the expression of human proteins that were

157

M. Al-Rubeai and M. Fussenegger (eds.), Systems Biology, 157–183.

normally obtainable in exceedingly small quantities, and one of these systems incorporates the culture of animal cells. In particular, proteins that are complicated in structure and function, and unsuitable to be produced in bacteria and yeast, can be produced in mammalian cells, with the benefit that the end product is biologically active. Animal cells are fastidious and costly to grow, but they produce correctly modified, fully active proteins. Consequently, much effort in the biotechnology industry has been dedicated to utilizing bioreactor systems for large-scale culture of animal cells; the aim is to establish an optimal physical and chemical environment, resulting in an increase in efficiency and intensification of the production process. There are several parameters that play a significant role in determining the outcome of cell culture, namely, nutrient supply, pH, shear forces, oxygenation and accumulation of waste products. Furthermore, the appropriate balance of survival and growth factors supplemented by serum is also fundamental in maintaining high cell viability. However, in large-scale operations it is imperative that culture conditions and components meet safety standards for public health. The undefined nature of serum proves to be a problem, which considered along with its direct cost, generates the necessity to develop new approaches for industrial animal cell culture. However, many cell lines undergo apoptosis upon withdrawal of serum, and although serum-free media are available, they are expensive, and in most cases have to be tailor made for each cell line of interest.

Apoptosis was first described in the 19th century by a German scientist, Walther Flemming, who was studying cell death, and named this process 'chromatolysis'[1]. But it was not until 1972 when Kerr, Wyllie, and Currie published the first detailed paper that clearly described the morphology of the cells undergoing this form of cell death, they named apoptosis (derived from Greek, meaning 'falling off' leaves from trees)[2]. Yet it was not until a couple of decades later that the idea of an integral and controlled cell death programme was widely accepted. Nowadays the terms apoptosis, programmed cell death (PCD) and cellular suicide, have been used interchangeably to describe this mechanism of cell death.

Apoptosis is a physiological process that plays a fundamental role in development, and regulation, of tissue homeostasis in multicellular organisms. Hence, it is under very tight control mechanisms, and its disrupted regulation is implicated in a number of disorders, ranging from cancers to autoimmune diseases, to degenerative syndromes. It is additionally used by the cells as a self-defense mechanism, by destroying unwanted and harmful cells, such as self-reactive lymphocytes, viral infected cells and tumour cells. In apoptosis the cell in its entirety is disassembled and its contents are enclosed inside membrane bound vesicles that prohibit the leakage of the intracellular components resulted by the death of the cell, that would effectively induce an immune response. Furthermore the resulting apoptotic fragments are rapidly phagocytosed by macrophages or neighbouring cells to prevent leakage of the apoptotic cell contents. In direct contrast, in necrosis, another form of cell death, the cell in its entirety ruptures and its contents are released into the neighbouring cells and surrounding tissue, resulting in an inflammatory response.

The cells that die from necrosis usually are forced to follow this path due to profound physical, chemical or genotoxic insult.

The serious implications of apoptosis and its deregulation became a very important area of biomedical research. However, it was not until Al-Rubeai et. al.[3] first described that apoptosis was a form of cell death observed in bioreactors, that the importance of this process was acknowledged in the field of biotechnology. The main aim of industrial processes for animal cell culture is to develop an optimal physical and chemical environment of the culture, in such a way as to advance efficiency, and enhance productivity. However, cells undergo apoptosis due to nutrient limitations, toxic metabolites, and hydrodynamic stresses, resulting into reduced productivity of the cell culture by reducing the number of viable cell factories which are also continuously proliferating leading to increased biomass.

Since 1858 when Rudolf Virchow proposed that *"Omnis cellula e cellula"* ("every cell originates from another cell")[4] the fact that cells replicate themselves and this replication is essential for the continuation of life has been one of the most fundamental doctrines in biology. All living organisms, from the simple single cell to the complex multicellular species are the result of continuous and repeated cell replication and proliferation events. Understandably, these events that form the cell cycle, are some of the most controlled and orderly events in the life of a cell.

Brielfy, the cell cycle is a tightly regulated process by which cells replicate themselves, with the end result being two identical daughter cells. Depending on various internal and external stimuli cells can enter a resting non-proliferative phase/state which can be protracted or even permanent-termed G_0. Eukaryotic cells have evolved a very complex network of regulatory proteins that form part of the cell cycle control system that is responsible for the progression through the cell cycle. This network governs each phase of the cell cycle and at the heart of this are the cyclin-dependent kinases (CDKs). The cyclins are divided in groups depending on which stage of the cell cycle they are involved; The G1-S cyclins (cyclin D_1, D_2, and D_3); S cyclins (cyclin A); M cyclins (cyclin B); and lastly G_1 cyclins that are involved in the progression through start or the restriction point in late $G1^4$.

In order for the cell cycle to be completed, which necessitates the successful completion of various events such as DNA replication, duplication and partitioning of chromosomes there are a number of checkpoints that monitor the cell cycle progress. These checkpoints are regulated by cyclin dependent kinase inhibitors (CKIs) which can stop the cell cycle progression if there are any problems. Damaged DNA, normally induces cell cycle arrest, therefore failure of these checkpoints leads to the propagation of the mutated DNA to daughter cells, which in turn can be the beginning of cancer. Mutations in genes that play roles in the cell cycle (oncogenes) or checkpoints that arrest the cycle or induce apoptosis (tumour suppressors) are responsible for the development of cancer. Some immortalized cells (and hybridomas) are examples of cycling cells for which their cell cycles lack the checkpoints. This situation is clear in many continuous cell lines in culture when depletion of nutrients results in reducing protein synthesis, thereby inducing apoptosis.

c-Myc plays a very important role in the regulation of cell proliferation, which will be discussed in details in the following sections.

1.1   Regulation of Cell Proliferation and Apoptosis

The control of cell proliferation and apoptosis is a vital function for the development and regulation of tissue homeostasis in multicellular organisms, and allows the tight control of cell numbers and tissue size. The proliferating cell that imposes a risk of neoplasia to the intact multicellular organism is obliged to trigger its own cell death, and its survival depends on factors that suppress apoptosis. Cell proliferation and apoptosis are closely related, and may even be considered as sides of the same coin. Interestingly, there are factors that promote not only cell proliferation and entry into the cell cycle, but also apoptosis. Therefore, cells that enter the cell cycle will continue to proliferate, unless they receive signals that will induce cell death, and the same factors such as cytokines or gene products that will influence cell survival, will also act as the apoptotic signals. This is all part of the very tight regulated homeostasis mechanism imposed on the cells, in order to regulate cell proliferation within specific tissues and in response to trauma.

In the cell culture environment the importance of regulating cell proliferation and apoptosis is further highlighted due to the lack of the intrinsic mechanisms that the protective environment of the tissue can provide. While the culture medium and various conditions are also tightly regulated and defined in order to provide an optimum environment, what in essence is asked by the cells is to overcome those exact internal mechanisms that govern cell life and death. The mechanisms that control this process are too complex and beyond the scope of this review. This review will focus instead on the capability of certain oncogenes to act as cell proliferation promoters, and apoptosis inducers, and in particular the proto-oncogene *c-myc* one of the most studied genes with this dichotomous ability to induce both apoptosis and proliferation[5].

2.   THE C-MYC PROTEIN

The *myc* gene was first identified as the transforming sequence of avian myelocy-tomatosis virus MC29 (*v-myc*)[6], and subsequently, the cellular homologue *c-myc* was also identified[7]. Since the discovery of the *myc* genes, more than 20 years ago, there has been an extensive amount of work that provides evidence as to their role in tumour formation, in a number of models systems, as well as in human neoplasias. In particular, some of the most documented examples are the translocations of the *c-myc* locus in Burkitt's lymphoma, and the amplifications of other members of the *myc* family of genes, in neuroblastomas, and small cell carcinomas [8,9]. Mutations in signaling pathways regulating *c-myc* expression, or lesions of the *myc* locus, have resulted in growth factor independent expression of Myc in those tumours[10,11]. The initiation and formation of tumour is obviously not the normal function of *c-myc* but it was nonetheless an indication of its ability to influence cell cycle progression. The *c-myc* gene and protein have been the subject of vast number of studies; their complex regulation, and biological activities have been explored by a host of research groups. The following sections are a review of the structure and function of the c-Myc protein based on a small sample of the vast available literature.

The *c-myc* gene encodes a short-lived (20–30 min.) nuclear phosphoprotein, and its expression generally correlates with cell proliferation. Sequence analysis of *c-myc* showed that it possesses a DNA binding/dimerization domain and a transactivation domain (TAD)[10,12], therefore it was considered as a *bonafide* transcription factor[11]. The DNA binding/dimerization domain and the TAD domain are present in the C-terminal and N-terminal of the protein respectively.

The Myc carboxy-terminal domain (CTD) comprises a sequence-specific DNA binding and dimerization domain of the basic helix-loop-helix zipper (b/HLH/Z) motif; the bHLHZ domain is characteristic of a class of transcription factors that bind to specific DNA recognition –CACGTG- E box motifs. The function of this domain is to mediate oligomerisation through the HLH/Zip region, and DNA interaction through the basic region. It is through this region that Myc is able to form heterodimers with another protein, Max ('Myc associated X') (Figure 1), a small bHLHZ protein[13], that in contrast to Myc does not contain a TAD domain[14]. Max proteins can form homodimers, which appear to be transcriptionally inactive[15] while on the other hand, Myc proteins do not homodimerise in cells, instead their DNA binding and biological activity depends on heterodimerisation with Max[15-18].

The c-Myc amino-terminal domain (NTD) comprises the transcriptional activation domain (TAD), and also includes two 20 amino acid segments namely the myc boxes 1 and 2 (MB1, MB2), which are conserved in all myc family proteins, and are required for the transactivation function of Myc. The exact mechanisms by which NTD activates transcription are unclear, although it is thought that the activity of NTD might be influenced by a number of proteins such as TRRAP, an ATM-related protein, p107, a retinoblastoma related protein, α-tubulin, PAM, MM-1, and AMY-1[19,20].

c-Myc is the main member of a small family of Myc proteins, which also includes N-myc, L-myc, S-myc and B-myc. The first *c-myc* related gene was discovered in neuroblastomas and was designated N-*myc*. Subsequent studies identified several other *c-myc* related genes. A gene designate L-*myc* was isolated

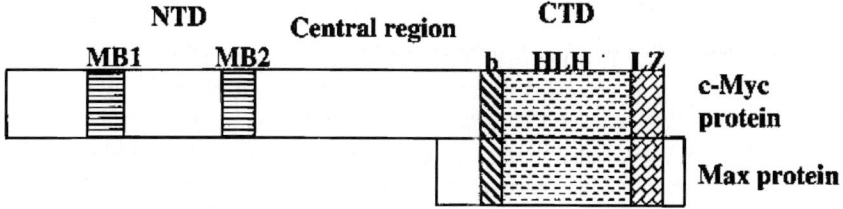

Figure 1. The structure of the c-Myc protein

from small lung carcinomas, followed by B-*myc* a mouse specific gene and S-*myc* a rat specific gene. The Myc family of proteins are highly conserved through evolution, and share the greatest degree of homology in their N-terminal domain (NTD) Myc box regions (MB1 and MB2), and at the C-terminal domain bHLHZ motif, while the amino acids in the NTD excluding the MB1 and MB2 regions, and the central amino acids are less conserved[7,21]. The c-, N-, and L-Myc family members have transforming activities and neoplastic potential, while B-and S-Myc are inhibitory.

2.1   The Biological Activities of c-Myc

In response to cellular signals Myc can regulate a wide array of diverse biological activities. It has been shown to induce cell cycle progression and proliferation, cell growth, potentiate apoptosis, block differentiation, drive transformation, activate genomic instability and stimulate angiogenesis (Figure 2). Obviously, Myc is a multifunctional protein that plays a very important role in the regulation of several cellular processes, essential for the control of cell proliferation. The ability of Myc to direct such a diverse set of activities is because Myc is functioning as a regulator of gene transcription, and specifically, that Myc regulates a set of target genes that results in downstream signaling cascades, leading to the implementation of each distinct biological activity. The following sections will present an overview of the Myc regulation of those biological activities, and the role of Myc as a transcription factor will also be discussed.

2.1.1   Cell proliferation

c-Myc is a member of a set of cellular messengers commonly known as 'immediate early response' genes, because their expression is activated by a variety of mitogenic stimuli (such as platelet derived growth factors-PDGF; colony stimulating factor –CSF; epidermal growth factor; interleukin-IL7 and IL2) independent of *de novo* protein synthesis, early during the Go to G1 transition of cells during the cell cycle[7,21,22]. In quiescent cells *in vitro*, c-Myc expression is practically unnoticeable. Yet following stimulation by the presence of mitogens or serum, c-Myc mRNA and protein levels quickly increase and cells enter the G1 phase of the cell cycle. Subsequently, the mRNA and protein levels decrease to low, but detectable, steady state levels usually observed in proliferating cells. If serum or mitogens are completely removed, c-Myc levels decrease again to undetectable levels and cells arrest.

Early on, evidence that the expression of the *c-myc* gene is closely related to cell proliferation was provided by a number of studies. Rabbits et al.[23] showed that c-Myc levels increased 2–4 hours following addition of mitogens, and were subsequently reduced, but remained constitutively expressed, throughout the cell cycle unlike other immediate early response genes (e.g. *fos* and *jun*) while Mateyak et al. [24] showed that deletion of the c-myc gene by homologous recombination lead to cell cycle lengthening. c-Myc expression is maintained throughout the cell cycle,

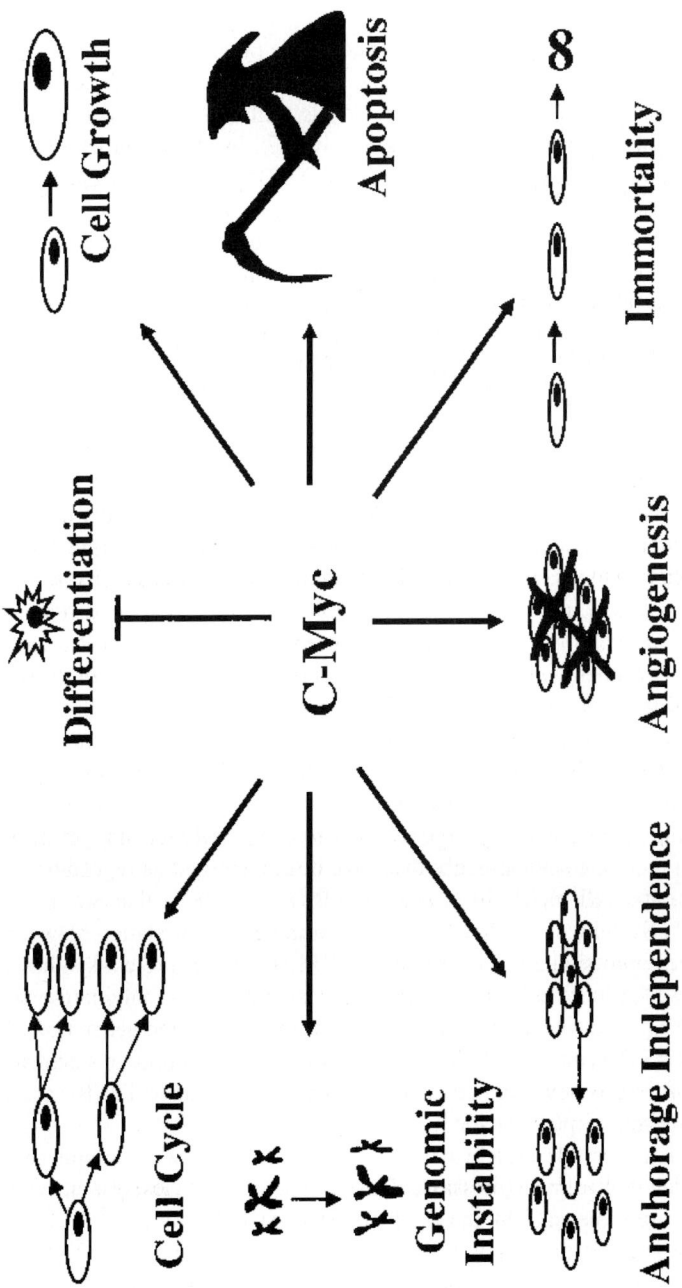

Figure 2. The biological functions of the c-Myc protein

and it plays an important role in the regulation of both entry in the cell cycle, and maintenance of cell proliferation[10,25–30].

At a molecular level, Myc appears to regulate cell cycle progression by promoting changes in the expression of a variety of genes. c-Myc expression leads to direct or indirect activation of *cyclin D1, D2, E,* and *A,* as well as *CDK4* and *Cdc25A.* The progression of G1-S in eukaryotic cells is governed by the activities of the cyclin dependent kinase (CDK) complexes, cyclin D-CDK4 and cyclin E-CDK2. Myc also suppresses cell cycle checkpoint genes, such as *gadd45* and *gadd153,* as well as cyclin kinase inhibitors *p15^{INK4b}, p21^{Cip1},* and *p27^{Kip1}*[18,31]. Additionally, Myc can stimulate the expression of the Id2 protein that inhibits the function of the gene product of the retinoblastoma gene, pRb, which in turn is a restriction point for G1 cyclin/cdk complexes. Finally, Myc also induces E2F1 and E2F2, which in turn regulate an independent parallel pathway that plays an important synergistic control with Myc in cell cycle control[22].

2.1.2   Cell growth

When cells proliferate, apart from doubling their DNA content, they also double their mass. Cell division is coupled to cell growth, which is defined as the accumulation of cell mass or cell size. Experimental evidence in a number of systems suggests that Myc is also involved in the regulation of cell growth. In particular, studies of the *Drosophila myc* ortholog, *dmyc,* showed that an increase or decrease in cell size was dependent on over-expression or reduced expression of dMyc respectively[32]. Evidence to support the role of c-Myc in the regulation of cell growth in mammalian cells has also been reported in studies with B cells. Schuhmacher et al. [33] have shown that B cells that could not enter the cell cycle could still be induced to grow in size upon Myc activation.

The exact mechanism engaged by Myc to regulate cell growth has not yet been defined, although there are several genes that have been identified as regulators of protein synthesis, and cell metabolism, such as *eIF4E* and *eIF2α* that are direct c-Myc targets[34–36]. Schmidt et al[37] have reported that there is a correlation between a decrease in Myc expression and a decrease in *eIF4E* and *eIF2α* protein synthesis and DNA synthesis. Additionally there is also evidence that RNA polymerase III (pol III) might also play a role. In particular, pol III is involved in the generation of transfer RNA and 5S ribosomal RNA that are further necessary for protein synthesis in growing cells. Pol III is however activated by c-Myc via binding to TFIIIB, a pol III specific general transcription factor[38].

Although this is a possible system where c-myc can interact with components of the growth-regulating pathway, further evaluation is required to provide insights into the growth control mechanism and the function of Myc.

2.1.3   Apoptosis

A detailed analysis of the role that Myc plays in apoptosis is beyond the scope of this chapter. Nevertheless because it is such an important function and due to the vast amount of literature that is available, a brief overview will be provided in the following section.

One of the most studied biological activities of c-Myc is that of induction of apoptosis. Deregulated expression of Myc is sufficient to induce or sensitise cells to apoptosis. This was first demonstrated in IL3 (interleukin 3) dependent myeloid 32D cells, and was later studied in a variety of systems, such as Rat-1 fibroblasts, primary rat embryo fibroblasts (REFs), various mouse cell lines and mouse embryo fibroblasts (MEFs), hepatocytes, osteosarcomas, epithelial cells and lymphoid cells[39]. Initially, it was thought that c-Myc was required for apoptosis, induced by death receptors, but recently it has been shown that c-Myc expression can sensitise the cells to apoptosis following a wide range of stimuli, such as, serum or growth factor deprivation, transcription and translation inhibitors, hypoxia, glucose deprivation, heat shock, chemotoxins, DNA damage, virus infection, and cancer chemotherapeutics[40]. The molecular mechanism by which Myc exerts this effect has not yet been fully elucidated, although there are several models that have been described[27,40–42]. Insight into this issue has been provided using three main approaches: Structure function studies to identify the regions of the Myc protein important for apoptosis; identification of Myc target genes; and determination of rate limiting steps in the apoptotic pathways activated downstream of Myc.

Several studies that involved analysis of Myc deletions mutants have concluded there is an overlap in the terminal regions of Myc required for apoptosis and transformation, with the Myc Box containing NTD and bHLHZ CTD being essential for the biological activities of Myc. In particular, the MB2 regions of Myc, as well as phosphorylation sites, are involved in the apoptotic potential of Myc, which also requires interaction with Max[43]. Moreover, those regions are also important in the regulation of Myc stability and its protein interaction abilities, and therefore would present an impact factor on the downstream regulation of the Myc apoptogenic program[17,40].

The identification of putative Myc target genes involved in apoptosis has been the subject of intense research. The introduction of microarray technology has lead to a plethora of reports, describing a vast number of potential Myc targets that are either activated, or repressed, in response to Myc [31]. Although there is evidence that some of the known target genes play an important role in Myc regulated apoptosis, no gene up to date has been identified that is essential for this process. Included among the target genes activated by Myc are *ornithine decarboxylase (ODC)* (involved in polyamine synthesis and required for cell proliferation); *cyclin A* (involved in the cell cycle, associates with CDK2 to promote cell cycle progression into S phase); *cdc25A* (a phosphatase that activates both CDK2 and CDK4, and its activity is required for the G1/S transition); *lactate dehydrogenase A* (a metabolic enzyme, part of the normal anaerobic glycolysis pathway); p19[ARF] (plays a very important role in cell proliferation)[31,39,40].

Myc is also associated with inducing apoptosis through activation of FADD, which is the downstream effector of the Fas mediated apoptosis. The exact mechanism has not been established, but it is thought that c-Myc is capable of activating this pathway by repressing the expression of FLIP, which is a negative regulator of FADD or DISC and mediator of inhibition of caspase activation[44].

Functional studies have provided important information regarding the involvement of Myc during apoptosis, by identifying the rate-limiting step required by Myc to drive the apoptotic process. Several studies indicate that Myc might be involved in the mitochondrial apoptotic pathway, downstream of a variety of apoptotic triggers[45,46]. First indication as to the role of Myc in the mitochondria associated pathway of apoptosis came from the studies by Juin and colleagues[45], that showed Myc induced release of cytochrome c from the mitochondria during apoptosis, and that ectopic addition of cytochrome c resulted in apoptosis in a similar manner as with the addition of Myc. More recently, Soucie and colleagues[46] showed that a rate-limiting step during apoptosis in the presence of Myc might be present upstream cytochrome c release, and involves the functional activation of Bax. Briefly, Bax is a member of the Bcl-2 family of proteins and during apoptotic stimuli, Bax sustains conformational changes and translocates to the outer membrane of the mitochondria. Bax translocation has effects on the membrane integrity, which allows cytochrome c release. Although the Bax associated apoptotic pathway has been studied extensively, the Myc-mediated Bax activation remains unclear. One possibility is that Myc itself translocates to the mitochondria in response to apoptotic stimuli and directly activates Bax. Alternatively, Myc might activate Bax indirectly, by regulating Bax upstream activators such as Bid, or caspase 8.

There are various theories as to how Myc can induce apoptosis. One thing that is definite though is that there is no uniform pathway, and that Myc induced apoptosis is governed by the cell type, along with the tissue location. Furthermore, it is also modulated by the existence of further mutations in other genes that are either pro- or anti-apoptotic[47].

2.1.4   Block of Differentiation

The exact mechanism by which Myc suppresses differentiation has yet to be fully understood, and one major obstacle is the inability to define suitable systems that allow experimental manipulation. Some insight has been provided from observations of the effects of deregulation of*myc* expression in a number of cells or organisms. Firstly, each myc family member has a unique pattern of gene expression during development, which suggests non-redundant contribution *in vivo*. Secondly, down regulation of c-myc expression is typically observed in cells triggered to undergo differentiation; and thirdly, ectopic expression of Myc effectively blocks differentiation in a wide variety of cells in both in vivo and in vitro.

The differentiation program is so sensitive to Myc levels that downregulation of Myc is sufficient to trigger differentiation and suppression of Myc expression is an essential component to a full genetic program of differentiation. More-over, this suppression is either directly or indirectly linked to the 'master regulator' of differentiation of each cell type, for example, C/EBPa (plays a role in differentiation of myoblasts to granulocytes and hepatocytes and adipocytes); MyoD, myogenin and Myf-5 (all involved in differentiating myogenic cell lines). It is unclear whether Myc is involved directly in repressing transcription of the so-called master regulators, or whether this effect is an indirect consequence of Myc driven cell proliferation. Other studies have highlighted the importance of the Myc/Max/Mad network in regulating cell proliferation and differentiation[18].

Whether it is a direct or indirect effect of Myc, inhibition of cell differentiation is a powerful biological activity that directly contributes to transformation[7,16,21,48].

2.1.5 Transformation

The ability of Myc to induce cellular transformations has been established both *in vitro* and *in vivo*. There is a plethora of literature regarding the role of Myc in transformation and this function of Myc will be mentioned only briefly in this chapter.

In the cell culture environment, it has been shown that constitutive over-expression of Myc is able to immortalize fibroblasts and prevent withdrawal from the cell cycle[49] accompanied by decreased requirement in survival factors[50] and shortening of the G1 phase[51]. Myc also co-operates with other oncogenes such as *ras*, to immortalize and transform cells in culture; constitutive over-expression of *c-myc* along with *ras* results in transformation of primary rat fibroblasts and haematopoietic cells[52,53]. The role of Myc as a transcription factor is quite evident in cellular transformation. Although the genes that are regulated downstream of Myc, and play an important role in transformation, have not been fully elucidated, ectopic expression of known target genes such as *odc*, *HMG-I/Y* and *rcl* have been known to transform Rat1a cells and as such have being linked to the ability of Myc to transform cells *in vitro*. Additionally, Myc has been also shown to upregulate telomerase activity by transcriptional activation of *hTert*, and telomerase activity is linked with cellular immortalisation and might also play a role in tumourigenesis[54].

The transformation potential of Myc in tissue culture is measured by both focus formation, and anchorage independent growth of the cells. Cells over-expressing Myc have been shown to form focci and grow in soft agar in an anchorage independent manner, with loss of contact inhibition and increase in growth rates[8,55]. Myc contributes to this type of cell transformation by repressing the genes that encode several proteins involved in cell adhesion, cytoskeletal structure, and extracellular matrix. Such genes include the lymphocyte function-associated antigen-1 (*LFA-1*) which is important for cell–cell contacts; the $\alpha 3\beta 1$ integrin, downregulation of which contributes to loss of contact inhibition; and the collagens $\alpha 1$ (I), $\alpha 2$ (I), $\alpha 3$ (VI), and $\alpha 1$ (III)[31], the down regulation of which affects cell anchorage[18,55].

Studies in mouse models show that the oncogenic potential of Myc is sufficient to instigate the transformation process *in vivo*. Studying transformation in mouse models has the advantage of identifying relevant aspects of the oncogene-induced phenotype that are not detected in cell culture, such as angiogenesis, as a result of the oncogene activation, or metastatic potential of individual tumours. The ability of c-Myc to promote angiogenesis has been demonstrated in a number of studies [56–60]. Myc, by repressing the angiogenesis inhibitor Thrombosporin-1 and by inducing

VEGF (vascular epithelial growth factor) and HIF1α[61] has being shown to be angiogenic in a number of cell types. The relationship between Myc and angiogenesis has not been fully understood as of date; however it is clear that myc influences tumourigenesis and the ability of tumours to induce an angiogenic response.

Structure function analysis has revealed that the MB2 region of Myc is essential in driving cellular transformation and while a mutation or deletion in the MB2 region affects the ability of Myc to promote transformation, deletions in MB1 have less impact[55].

2.1.6   Genomic instability

Genomic instability can be defined as an accumulation of genetic abnormalities such as chromosomal alterations, gene amplifications, and nucleotide changes[62]. The role of Myc in promoting genomic instability has been observed in both cultured cells and *in vivo* animal models, and it has been shown that transient or constitutive over-expression of Myc results in chromosomal and extrachromosomal gene amplification and rearrangement[63–66]. Although, firm conclusions regarding the molecular mechanism by which Myc is able to promote genetic instability have far from been reached, several suggestions exist. Firstly, Myc over-expression might lead to bypassing a checkpoint upon spindle disruption in the G2/M phase of the cell cycle leading to endoreplication and apoptosis. At this stage secondary mutations e.g. loss of p53 or over-expression of Bcl-X$_L$, can co-operate with Myc and result in polyploidy in a variety of cell lines[67]. Additionally, Myc might also trigger shortening of the G1 cell cycle phase, which results to premature S-phase entry, and has been shown to lead to genomic instability due to DNA damage or replication in the presence of insufficient nucleotide pools[68]. Finally, Myc itself might be involved in driving DNA replication directly[69].

2.1.7   Myc as a transcription factor

As it has been previously mentioned, Myc mediates its biological activities, by regulating the expression of specific target genes. A number of those genes have been identified as potential Myc targets and they are involved in a wide variety of cellular responses. The transcriptional role of Myc can be either upregulation (activation) or downregulation (repression) of the target gene. In addition, there are direct and indirect target genes, the former being a gene whose expression is altered by direct interaction with the Myc protein, while an indirect target gene is expressed as a result of upstream Myc regulation[31,39–40,70,71]. For a more detailed review and extensive lists of c-Myc targets, there is a web site dedicated for this exact purpose: www.myc-cancer-gene.org.

While the exact mechanisms of Myc activation and repression of target genes are beyond the scope of this review, a brief overview of Myc activation has been illustrated in Figure 3. The mechanisms of gene repression are not very well understood, but probably there is some functional interference with transcriptional activators.

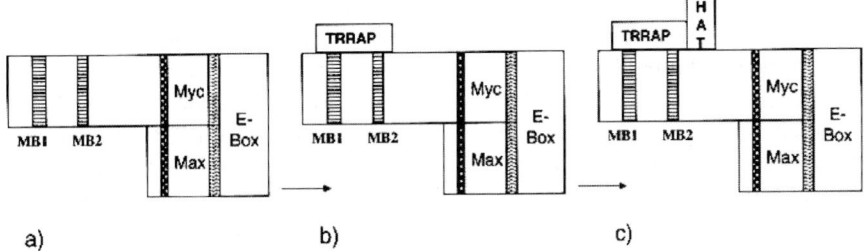

Figure 3. Myc as a transcription factor; a) the Myc-Max complex binds to an E-box sequence of a target gene b) TRRAP (transformation/transcription domain-associated protein) binds to MBII C) TRRAP a member of a complex with HAT (histone acetyltransferase) activity acetylates nucleosomal histones which results in chromatin remodeling and allows access to Myc-Max complexes to target DNA; the result is the expression of the target gene

The transcriptional activity of Myc requires two Myc regions: the bHLHZ region of CTD, and the TAD of the NT domain. These regions are important for Myc heterodimerisation with Max in order to permit sequence specific DNA binding followed by Myc-dependent activation of transcription[18].

3.   EXPLOITING METABOLIC ENGINEERING IN CELL CULTURE TECHNOLOGY

Large-scale animal cultures have been used effectively in the biotechnology and biopharmaceutical industry for the production of therapeutic biologics. Understanding of the physiological processes governing cell proliferation, growth and metabolism in those environments, has been the main tool of overcoming the problems associated with culture systems, and has been engaged extensively for the development and practical design of higher performance bioproduction processes.

At the centre of every strategy implemented for optimisation of cell culture, is one main goal: increasing viability and survival of the cells utilised, i.e. keeping the cells content in their culture environment. In animal cell culture, viability is the main single aspect that decides the success of each system, and in turn viability can be influenced by a number of factors; such as cell line instability, rate-limiting steps introduced by nutrient limitation, chemical and physical factors e.g. pH, shear forces, oxygenation, and accumulation of waste products. Many strategies have been implemented to provide solutions regarding controlled proliferation, enhanced survival, and improved cellular responses to the external environment. Such solutions include, design of new media and formulation protocols, controlled feeding techniques, and utilisation of high-density perfusion cultures, all of which are aimed at one main problem: cell death in bioreactors.

Cells survive and proliferate in culture when they are supplied with adequate amounts of nutrients and a suitable physical environment. The extracellular environment in the bioreactor must meet the intracellular requirements of the cells for them to survive and continue proliferating. When optimal conditions cease to exist the cells begin to die, and availability of growth and survival factors influence the probability of cell death. It is widely acceptable that cell proliferation and death are sides of the same coin, and are both controlled by an integral mechanism that is influenced by the physical environment in cell culture among other things. The presence of various growth factors such as insulin-like growth factor (IGF-1), interleukin-2 (IL-2) and platelet derived growth factors (PDGF) have all been shown to inhibit cell death in the form of apoptosis. The latter has been proven to be a major problem in large-scale cultures[72], and most strategies regarding process optimisation aim to overcome this problem. Supplementation with serum which provides growth factors and other nutrients to the cells, addition of extra growth factors, or inhibitors of apoptosis have also been considered for the regulation of apoptosis. However, in large-scale operations where the end product is to be used for therapeutic purposes, it is vital that culture conditions and components meet safety standards for public health. This proves to be a problem due to the undefined nature of serum, its direct cost, and the fact that many cell lines undergo apoptosis upon withdrawal of serum. One of the fields that has developed significantly in the recent years to overcome this problem, is the design of serum-free and protein-free media. The former still contain certain proteins such as albumin, IGF-1 and transferrin, which are derived from animal sources, and require monitoring and testing for microbial agents such as mycoplasma and viruses. On the other hand, protein-free media may prove to be the solution to the introduction of animal microbial agents since the proteins are replaced with non-protein substitutes e.g. albumin can be replaced by a-cyclodextrin and transferrin by iron gluconate[73]. Developing and designing media in order to advance production, improve product quality and ensure the elimination of potential public health concerns (e.g. BSE) in the production of pharmaceuticals, is a rapidly growing field in industry. Nevertheless, serum and protein–free media are expensive and need to be tailored-made for each cell line of interest. Therefore, a new approach, implementing the use of metabolic engineering might provide new insights and solutions in the race of optimisation of bioprocesses.

The most widespread method of overcoming apoptosis in animal cell cultures has been the introduction of oncogenes and in particular, the anti-apoptotic oncogene *bcl-2*. The ability of this gene to suppress apoptosis has been extensively research by various groups[74] and it has been widely proven that implementation of this gene in animal cell cultures protects the cells form a range of diverse apoptotic stimuli, including, nutrient limitation (glucose; glutamine; amino acids; survival/growth factors), ammonia toxicity, hydrodynamic stress, hyperosmolality, hypoxia and hyperoxia and sub-optimal pH conditions. Control of apoptosis, however, is not sufficient to optimise large-scale cultures. Increased proliferation and improved growth characteristics are also very important. The critical role played by *c-myc* in the proliferation and death pathways, indicates that its direct manipulation and use in cells might result in significant improvements in cell culture. Indeed, transformed cells with *c-myc*[75] have been shown to exhibit a substantial improvement in growth rates and antibody productivity. However, as discussed previously, cells overexpressing Myc undergo apoptosis in the absence of growth factors present in serum; and while serum appears to be very important for cell survival, from an economic and technical point of view it is an unknown quantity that proves problematic in large-scale culture. Addition of serum in cultures might result in introduction of contaminants and hinder the isolation of target protein[76], therefore new strategies for the development of serum-free media, or a cell line that will be able to grow in the absence of serum is an important objective for the biotechnology industry.

Chinese Hamster Ovary (CHO) is a cell line that has emerged as a favourite candidate for large-scale production of recombinant proteins. CHO cells are very robust in a number of culture conditions and have been used for the last 40 years therefore are very well characterised. A large number of proteins have been produced on CHO-based systems such as tPA, Factor VIII, interferon-γ[77], and there are many different strategies currently employed that involve the use of metabolic engineering in order to design a system that will incorporate the robust CHO nature with increased cell proliferation and the use of serum and/or protein-free media. Such systems include the development of 'super-CHO' that express IGF-1 and transferrin and has the ability to grow in totally protein-free media[78]; 'veggie-CHO' a cell line that grows in serum-free media in the absence of transferrin and IGF-1 that produces the recombinant protein Flt-3L[79]; the multicistronic CHO cell lines that incorporate the p21-, p-27- or p53-containing tetracycline repressible expression vectors or the p27-bcl$_{XL}$-tricistronic CHO system that all aim to increase cell proliferation, productivity and decrease cell death[80–82]. Other work on CHO includes the uncoupling of cell growth and proliferation by the over-expression of p21 which resulted in enhancement of productivity and provided new insight into how improved productivity can be achieved in a cell line commonly used for large-scale production of pharmaceutical proteins[83] as well as the work which has shown a decreased apoptosis, greater attachment tendency, higher cell density and reduced serum dependency due to the over-expression of hTERT[84].

With the use of metabolic engineering, a new strategy has been developed which implemented the proliferative abilities of c-Myc, along with the robustness of CHO cells, in order to create a cell line that exhibits enhanced survivability and proliferation. The following sections will provide an overview of this strategy and how c-Myc was implemented to control cell proliferation *in vitro* and to further transform CHO-K1 cells into a cell line with a great potential for use in biotechnology.

3.1   The Effect of c-Myc Over-Expression on CHO Cells

In order to study the effect of c-Myc over-expression on cell proliferation and in transformation in general, a new cell line was created as previously described[85] namely cmyc-cho and its corresponding control neo-cho. The cmyc-cho cell line over-expressed c-Myc as was determined by means of western blotting and flow cytometry, with results indicating a 30 fold increase in the c-Myc expression. Following transfection the cell line was subjected to various culture conditions and the effect of c-Myc over-expression was assayed in relation to the control cell line.

3.2   Effect of c-Myc Over-Expression on Cell Proliferation of CHO

Cell proliferation is a process that is highly regulated by intracellular factors and depends on appropriate and specific extracellular signals. A number of studies have shown that c-myc is such an intracellular factor that affects cell proliferation in normal, non tumourigenic cell lines, with unregulated expression of the protein resulting in tumourigenic phenotypes[23-25,30,39]. Furthermore, several cell systems have been developed where c-Myc is over-expressed either by transfection, infection or microinjection that affected proliferation rates[55]. However, evidence up to date suggested that c-myc although itself an important regulator of cell proliferation is still regulated by extracellular signals in normal cells[39,40].

The results from the study described above have shown that constitutive over-expression of c-Myc results in an increase of proliferation rate with an additional, but relative, independence to external factors affecting and regulating proliferation, such as cell to cell contact and growth factors present in serum. Cmyc-cho cells reached higher cell numbers and exhibited higher growth rates with less glucose utilisation when compared to the control cell line. This pattern emerged from all the different types of culture conditions, from static batch, to static fed batch, to suspension cultures, and the results will be discussed in the following sections.

3.2.1   Static batch cultures

Maximum cell number has been shown to increase by 54% in the cmyc-cho cultures as compared with neo-cho control cultures. Viability on the other hand remained above 80% for both the cmyc-cho and the neo-cho cultures over the first 6 days. Table 1, shows a summary of the results in static cultures; cmyc-cho cells reached higher cell numbers and exhibited higher growth rates with less glucose utilisation when compared to the control cell line

The cell cycle distribution during the batch culture of cmyc-cho and neo-cho can be seen in Figure 4. The percentage of cmyc-cho cells in the S phase was about 46% and it started to decline after 2 to 3 days although the cell did not reach confluence until 4 days of culture. The reduction of S phase cells corresponded with an increase in the percentage in G1 cells, while the percentage of cells in G2/M ranged between 35% and 25% throughout the duration of the culture. Control cells showed different cell cycle distributions. There was definite increase in G1 cells throughout the culture and a steady decline in S phase cells. At the beginning of

Table 1. Maximum viable cell density, Growth and Glucose utilisation rates in static batch cultures of cmyc-cho and neo-cho

Cell line	Maximum Viable Cell Density (cells/ml)	Growth Rate (day^{-1})	Glucose Utilisation rate (mmol/10^5 cells/day)
Cmyc-cho	$1.7 \times 10^6 \pm 2.5 \times 10^4$	0.38	0.06
Neo-cho	$1.1 \times 10^6 \pm 3.4 \times 10^4$	0.27	0.1

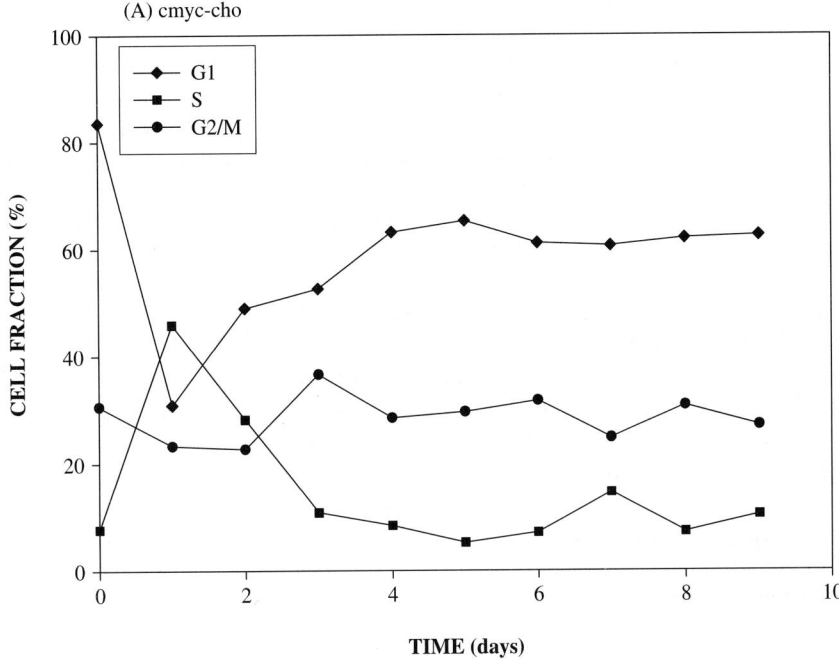

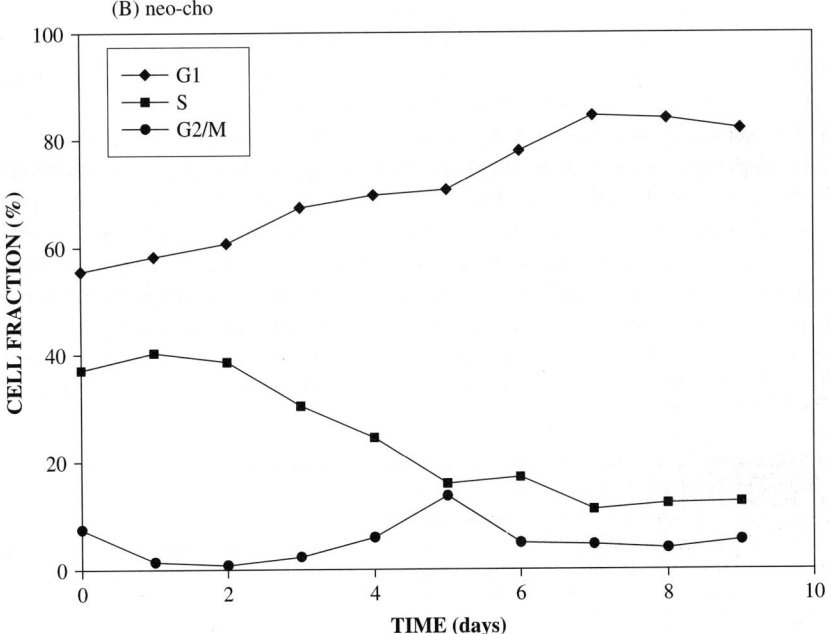

Figure 4. Cell cycle profile of cmyc-cho and neo-cho in static batch cultures

the culture the percentage of S phase cells was about 40% and it started to decline after 2 days in culture although confluence was achieved by the cells on day 4. The S phase reduction was followed by a steady increase in G1 cells although at much higher levels than in the cmyc-cho cells, while the G2/M remained at low levels, during the culture. Interestingly, the level of the dividing fraction (G2/M) is significantly higher in cmyc-cho cells than in control cells throughout the duration of culture.

In the absence of serum the effect of c-Myc on cell proliferation was more pronounced with cmyc-cho cultures reaching maximum cell number of 5.1×10^5/ml whilst the control cultures reached only 2.2×10^5/ml, which corresponds to a 130% increase in cell number. Moreover, the difference in viability between cmyc-cho and neo-cho cultures was higher in the absence than in the presence of serum, with the neo-cho cell viability declining, approximately, 20% faster than the cmyc-cho cells. Table 2, shows a summary of the results in static cultures in the absence of serum; cmyc-cho cells reached higher cell numbers and exhibited higher growth rates with less glucose utilisation when compared to the control cell line.

The percentage of G1 cells in cultures under serum deprivation increased with neo-cho cells accumulating in G1 throughout the duration of the culture at higher levels than in the cmyc-cho cultures (Figure 5). This increase was followed by a rapid decrease in S phase cells, and a small increase in G2/M cells which corresponded to the beginning of cell death for the neo-cho culture (day 3). The neo-cho culture did not grow well in the absence of serum, which was evident in the cell cycle distribution, with the accumulation of cells in G1 and G2/M and the presence of very small percentages of cells in S phase.

In the cmyc-cho cultures, there was a sharp drop in G1 cells at the beginning of the culture followed by an increase in G2/M phase cells. The percentage of S cells decreased gradually but at much lower rates than in the control cell line. Eventually, as the culture was reaching the death phase, there was an accumulation of cells in G1.

Figure 6 shows the effect of growth factors on the proliferation of cmyc-cho and neo-cho cultures in relation to serum. The results show that c-Myc over-expression leads to a 73% increase in cell number and this increase is very high when growth factors are added. In fact it is apparent that c-Myc over-expression works synergistically with the growth factors leading to 3-fold increase in cell numbers in the cmyc-cho cultures when compared with the neo-cho cultures. Additionally, in the cultures with IGF-1 there is an additional 13% increase in relation to the

Table 2. Maximum viable cell density, Growth and Glucose utilisation rates in static batch cultures of cmyc-cho and neo-cho under serum deprivation conditions

Cell line	Maximum Viable Cell Density (cells/ml)	Growth Rate (day^{-1})	Glucose Utilisation rate (mmol/10^5 cells/day)
Cmyc-cho	$5.1 \times 10^5 \pm 1.2 \times 10^4$	0.8	0.14
Neo-cho	$2.1 \times 10^5 \pm 1.4 \times 10^4$	0.08	0.5

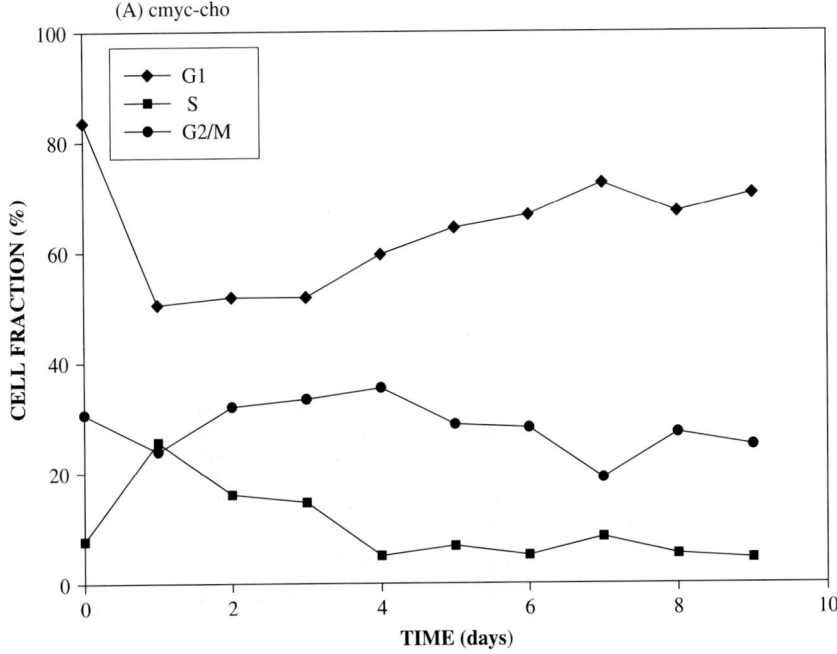

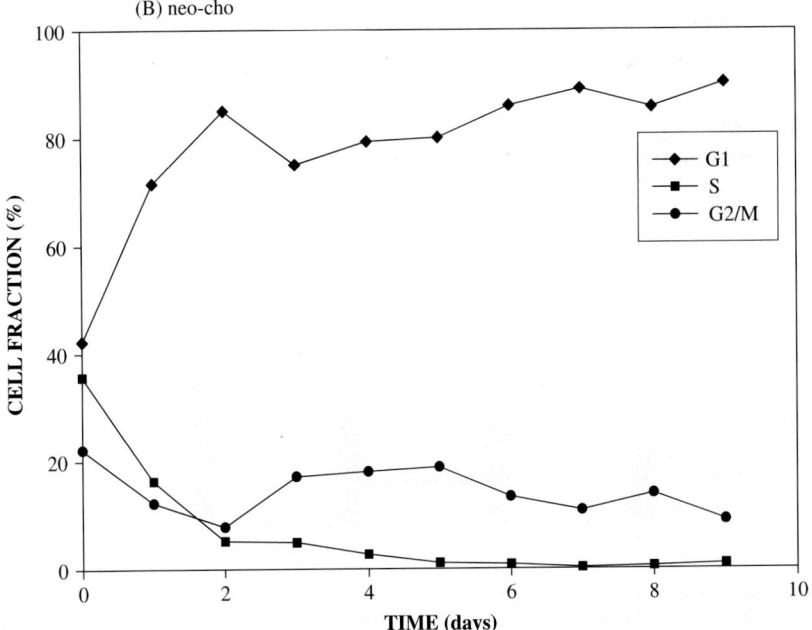

Figure 5. Cell cycle profile of cmyc-cho and neo-cho cultures in the absence of serum

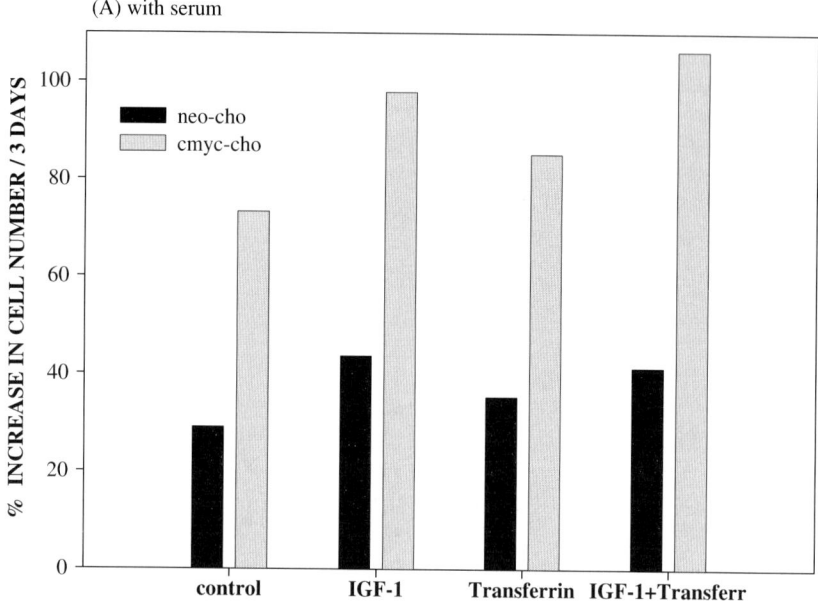

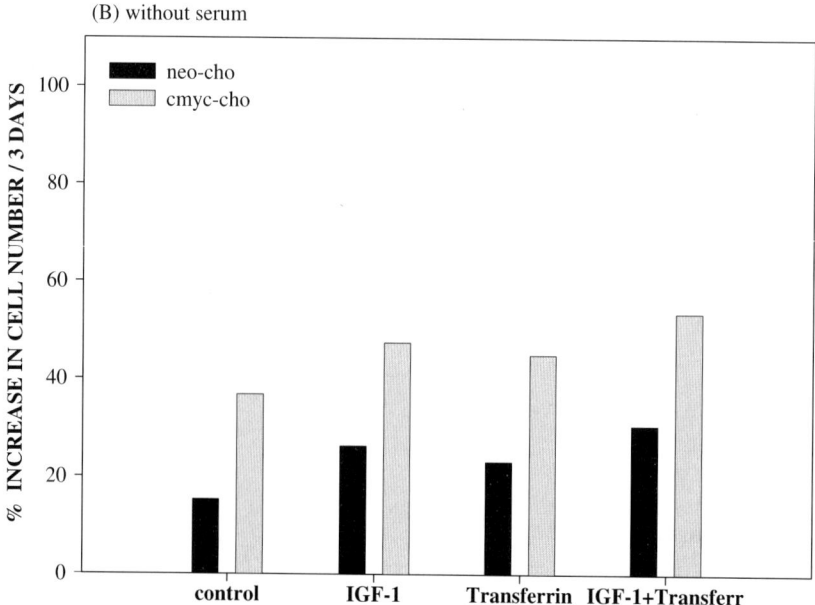

Figure 6. The effect of growth factors on cmyc-cho and neo-cho proliferation in relation to serum (control in the x-axis indicates culture conditions for both cell lines without the addition of growth factors

cultures where only transferrin was added, while both growth factors lead to an impressive 106% increase in cell number in a 3-day culture.

The same pattern is observed in the cultures grown in the absence of serum, although the percentage of increase in all conditions is lower than in the presence of serum. Nevertheless, the cmyc-cho cultures exhibit a 2-fold increase in cell number in all observed conditions in relation to the neo-cho cultures.

A more apparent difference in cell number between c-Myc transfectants and control cells is shown in fed-batch cultures (fed with Ham's F12, 5% FCS) with cmyc-cho reaching maximum cell number of 2.8 fold higher the control culture. The difference in viability was most apparent on day 13 when viability of cmyc-cho and neo-cho was approximately 90% and 60% respectively.

3.2.2 Suspension batch cultures

Prior to adaptation to suspension, it was observed that in the absence of serum, neither the control nor cmyc-cho cells could survive for prolonged period in contrast to the results obtained in static cultures. While in the process of adapting the cultures to suspension it was obvious that cmyc-cho adapted to grow in suspension cultures faster than the control cells. In fact the time required to obtain a neo-cho single cell suspension culture was double the time required by cmyc-cho.

In batch suspension cultures with serum supplementation maximum cell number has been shown in Table 3 to increase by almost 30% in the cmyc-cho cultures as compared with neo-cho cultures; whilst viability remained above 95% for up to 10 days, with a slight decline to 90% at the end of the culture for the cmy-cho cultures. The neo-cho cultures followed a similar pattern, but with at least a 5% less viability in comparison to the cmyc-cho values, for the duration of the cultures. The glucose utilization rates in suspension cultures were also lower for the cmyc-cho cells than the control, a pattern that has been observed in all the types of culture of the two cell lines.

In all the various conditions of cultures it was shown that the difference in maximum cell number observed between cmyc-cho and the control cell line range from 25-35%, with up to 60% difference in the serum deprivation experiments, with of course, the highest numbers exhibited by cmyc-cho. Interestingly, although the cell numbers for the cmyc-cho cultures were significantly higher in the presence of serum, the positive effect of c-Myc on cell proliferation was more profound in the cultures grown in the absence of serum. As mentioned above, cmyc-cho

Table 3. Maximum viable cell density, Growth and Glucose utilisation rates in suspension cultures of cmyc-cho and neo-cho after adaptation

Cell line	Maximum Viable Cell Density (cells/ml)	Growth Rate (day^{-1})	Glucose Utilisation rate (mmol/10^5 cells/day)
Cmyc-cho	$6.27 \times 10^5 \pm 1.7 \times 10^4$	0.19	0.010
Neo-cho	$4.9 \times 10^5 \pm 1.1 \times 10^4$	0.09	0.028

cell density increased by 60% compared to the control cell line. Those results were indeed surprising, because it had been widely accepted that in the absence of growth and survival factors c-Myc over-expression leads to apoptosis and inhibits cell proliferation[39]. Indeed, in the most well known study, Evan et al.[86] found that fibroblasts that express c-Myc grow well in serum-supplemented medium but undergo apoptosis upon withdrawal of serum. Moreover, they also demonstrated that apoptosis induced by c-Myc in low serum conditions was inhibited by IGF-1 and PDGF[27], both factors that are present in foetal bovine serum. However, several lines of evidence[40] indicate that c-Myc is capable of promoting survival of a variety of cells, such as B cells and keratinocytes, under growth limiting conditions. One theory is that although c-Myc promotes proliferation and induces apoptosis under growth limiting conditions, this dual ability is cell type specific. Indeed that seemed to be the case with the cmyc-cho cell line when cultured in conditions where IGF-1 and transferrin were added to the media in both the presence and absence of serum. IGF-1 has been characterised as a growth factor for a number of cell lines, and it has been reported that it can replace insulin as the main mitogenic factor in serum-free cultures of CHO cells[87]. Transferrin is another factor present in serum and has been described as essential for the growth requirements of cells since it acts as an iron vehicle[87]. The over-expression of c-Myc was enough to drive cell proliferation in the absence of serum without the addition of growth factors. In fact, there was a 2-fold increase in cell number when compared to the control cell line. Therefore, it can be suggested that, although the cmyc-cho proliferation rate is governed and affected by the same factors as the control cell line, over-expression of the c-Myc protein has provided the cell line with a specific advantage under growth-limiting conditions. However, growth of the cell line was not totally independent of cell-cell contact and growth factors present in serum as observed in static batch and fed batch cultures. In fact, the static fed batch results show that feeding affects the cell number and the duration of the culture as would have been expected; but again, c-Myc over-expression had a positive effect on proliferation rate, which was further amplified by the addition of nutrients and growth factors. Interestingly, feeding had no significant effect on neo-cho cultures.

4.   IMPACT OF METABOLIC ENGINEERING
ON BIOTECHNOLOGY

Large-scale animal cell cultures have been used extensively in the biotechnology and biopharmaceutical industry for the production of therapeutic biologics. Understanding the physiological processes that govern cell proliferation, growth and metabolism of individual cell line used in those environments, has been essential for over-coming the problem associated with the culture systems. In the case of CHO-K1 cells, the anchorage and growth factor dependency that characterises this cell line should be taken into consideration when it is used for manufacturing processes[80,81]. Further adaptation should be carried out to obtain a cell line that can easily grow in suspension and serum-free conditions. While adapting a cell line to grow in suspension can be a time consuming process, the most important factor to be considered is the elimination of serum from culture when the product is for therapeutic purposes.

There is one very good reason why the cultures are supplemented with serum and that is because serum is a source of nutrients, growth factors, carrier proteins and attachment factors; and while this is universally accepted, the replacement of serum in large-scale culture has been a field of much interest. Several factors such as risk of contamination from viruses and mycoplasma; its undefined composition, batch to batch variation, and high cost; and finally it interference with the purification process of the end product have been only but a few reasons for wanting to replace it. However, removal of serum also results in removal of essential factors that regulate proliferation, growth and survival of the cell in culture. Hence the alternative is to use metabolic engineering to create a robust cell line, able to grow in cultures without exhibiting requirements of growth factors and adhesion substrata.

The results of this present study have shown that over-expression of c-Myc in CHO-K1 cells results in a cell line that combines the robust nature of CHO cells with the high proliferative and transforming ability of c-Myc.

5. REFERENCES

1. Majno, G., and Joris, I. Apoptosis, oncosis, and necrosis-an overview of cell death. *Amer.J.Path.*, **146** 3–15 (1995).
2. Kerr, J.F.R., Wyllie, A.H., and Currie, A.R. Apoptosis: a basic biological phenomenon with wide ranging implications in tissue kinetics. *British Journal of Cancer*, **26** 239–260 (1972).
3. Al-Rubeai, M., Mills, D., and Emery, A.N. Electron microscopy of hybridoma cells with special regard to monoclonal antibody production. *Cytotechnology*, **4** 13–28 (1990).
4. Alberts, B., Johnson A., Lewis, J., Raff, M., Roberts K., and Walter P. The Cell Cycle and Programmed Cell Death, in *Molecular Biology of the Cell, Fourth Edition*. Garland Science, Taylor & Francis Group, New York, pp 983–1025 (2002).
5. Evan, G.I., Harrington, E., McCarthy, N., Gilbert, C., Benedict, M.A., and Nuñez, G. Integrated control of cell proliferation and apoptosis by oncogenes. In N.S.B., Thomas ed. *Apoptosis and cell cycle control in cancer*. BIOS Scientific Publishers, Oxford, pp 109–129 (1996).
6. Hayward, W.S., Neel, B.G., and Astrin, S.M. Activation of a cellular oncogene by promoter insertion in ALV-induced lymphoid leucosis. *Nature*, **290** 475–480 (1981).
7. Cole, M.D. The myc oncogenes: its role in transformation and differentiation. *Ann. Rev. of Genetics*, **20** 361–384 (1986).
8. Small, M.B., Hay, N., Schwab, M., and Bishop, M.J. Neoplastic transformation by the human gene N-myc. *Mol. Cell. Biol.*, **7** 1638–1645 (1987).
9. Spencer, C.A., and Groudine, M. Control of c-myc regulation in normal and neoplastic cells. *Adv. Cancer Res.*, **56** 1–48 (1991).
10. Bouchard, C., Staller, P., and Eilers, M. Control of proliferation by Myc. *Trends Cell Biol.*, **8** 202–206 (1998).
11. Furhrmann, G., Rosenberg, G., Grusch, M., Klein, N., Hofmann, J., and Krupitza, G. The MYC dualism in growth and death. *Mutation Research*, **437** 205–217 (1999).
12. Cochran, B.H., Reffel, A., and Stiles, C. Molecular cloning of gene sequences regulated by platelet-derived growth factor. *Cell*, **33** 939–947 (1983).
13. Blackwood, E.M., Kretzner, L., and Eisenman, R.N. Myc and Max function as a nucleoprotein complex. *Curr. Opin. Genet. Dev.*, **2** 227–235 (1992).
14. Kato, G.J., and Dang, C.V. Function of the c-Myc oncoprotein. *FASEB J.*, **6** 3065–3072 (1992).

15. Amati, B., Dalton, S., Brooks, M.W., Littlewood, T.D., Evan, G.I., and Land. Transcriptional activation by the human c-Myc oncoprotein in yeast requires interaction with Max. *Nature,* **359** 423–426 (1992).

16. Meichle, A., Philipp, A., and Eilers, M. The functions of Myc proteins. *Biochim. Biophys. Acta,* **1114** 129–146 (1992).

17. Lüscher, B., and Larsson, L.G. The basic region/helix-loop-helix/leucine zipper domain of Myc proto-oncoproteins: function and regulation. *Oncogene,* **18** 2955–2966 (1999).

18. Grandori, C., Cowley, S.M., James, L.P., and Eisenman, R.N. The Myc/Max/Mad network and the transcriptional control of cell behavior. *Ann. Rev. Cell Dev. Biol.,* **16** 653–699 (2000).

19. Cole, M.D., and McMahon, S.B. The Myc oncoprotein: a critical evaluation of transactivation and target gene regulation. *Oncogene,* **18** 2916–2924 (1999).

20. Baudino, T.A., and Cleveland, J.L. The Max network gone Mad. *Mol. Cell. Biol.,* **21** 691–702 (2001).

21. Marcu, K.B., Bossone, S.A., and Patel, A.J. *myc* function and regulation. *Ann. Rev. of Biochem.,* **61** 809–860 (1992).

22. Nasi, S., Ciarapica, R., Jucker, R., Rosati, J., and Soucek, L. Making decisions through Myc. *FEBS Lett.,* **490** 153–162 (2001).

23. Rabbits, P.H., Watson, J.V., Lamond, A., Forster, A., Stinson, M.A., Evan, G., Fischer, W., Atherton, E., Sheppard, R., and Rabbits, T.H. Metabolism of *c-myc* gene products: *c-myc* mRNA and protein expression in the cell cycle. *EMBO J.* **4** 2009–2015 (1985).

24. Mateyak, M.K., Obaya, A.J., Adachi, S., and Sedivy, J.M. Phenotypes of c-myc deficient rat fibroblasts isolated by targeted homologous recombination. *Cell Growth Differ.,* **8** 1039–1048b (1997).

25. Seth, A., Gupta, S., and Davis, R.J. Cell cycle regulation of the c-Myc transcriptional activation domain. *Mol. Cell. Biol.,* **13** 4125–4136 (1993).

26. Born, T.L., Frost, J., Schonthal, A., Prendergast, G., and Feramisco, J.R. C-Myc cooperates with activated Ras to induce the cdc2 promoter. *Mol. Cell. Biol.,* **14** 5710–5718 (1994).

27. Harrington, E.A., Bennet, M.R., Fanidi, A., and Evan, G.I. C-Myc induced apoptosis in fibroblasts is inhibited by specific cytokines. *EMBO J.* **13** 3286–3295 (1994).

28. Lemaitre, J.M., Buckel, R.S., and Mechali, M. C-Myc in the control of cell proliferation and embryonic development. *Adv. Cancer Res.,* **70** 96–144 (1996).

29. Amati, B., Konstantinos, A., and Vlach, J. Myc and the cell cycle. *Front. Biochem. Sci.,* **3** d250–d268 (1998).

30. Obaya, A.J., Mateyak, M.K., and Sedivy, J.M. Mysterious liaisons: the relationship between c-Myc and the cell cycle. *Oncogene,* **18** 2934–2941 (1999).

31. Dang, C.V. C-Myc target genes involved in cell growth, apoptosis, and metabolism. *Mol. Cell. Biol.,* **19** 1–11 (1999).

32. Johnston, L.A., Prober, D.A., Cheng, P.F., Edgar, B.A., Eisenman, R.N., and Gallant, P *Drosophila myc* regulates growth during development. *Cell,* **98** 779–790 (1999).

33. Schuhmacher, M., Staege, M.S., Pajic, A., Polack, A., Weidle, U.H., Bornkamm, G.W., Eick, D., and Khlhuber, F. Control of cell growth by c-Myc in the absence of cell division. *Curr. Biol.,* **9** 1255–1258 (1999).

34. Rosenwald, I.N., Rhoads, D.B., Callanan, L.D., Isselbacher, K.J., and Schmidt, E.V. Increased expression of eukaryotic translation initiation factors eIF–4E and eIF-2 alpha in response to growth induction by c-myc. *Proc. Natl. Acad. Sci. USA,* **90** 6175–6178 (1993).

35. Jones, R.M., Branda, J., Johnston, K.A., Polymenis M., Gadd, M., Rustgi, A., Callanan, L., and Schmidt, E.V. An essential E box in the promoter of the gene encoding the mRNA cap-binding protein (eukaryotic initiation factor 4E) is a target for activation by c-myc. *Mol. Cell Biol.,* **16** 4754–4764 (1996).

36. Coller, H.A., Grandori, C., Tamayo, P., Colbert, T., Lander, E.S., Eisenman, R.N., and Golub, T.R. Expression analysis with oligonucleotide microarrays reveals that MYC regulates genes involved in growth, cell cycle, signaling and adhesion. *Proc. Natl. Acad. Sci. USA,* **97** 3260–3265 (2000).

37. Schmidt, E.V. The role of c-myc in cellular growth control. *Oncogene*, **18** 2988–2996 (1999).
38. Gomez-Roman, N., Grandori, C., Eisenman, R.N., and White, R.J. Direct activation of RNA polymerase III transcription by c-Myc. *Nature*, **421** 290–294 (2003).
39. Hoffman, B., and Liebermann, D.A. The proto-oncogene c-myc and apoptosis. *Oncogene*, **17**: 3351–3357 (1998).
40. Prendergast, G.C. Mechanisms of apoptosis by c-Myc. *Oncogene*, **18** 2967–2987 (1999).
41. Evan, G., and Littlewood, T. A matter of life and cell death. *Science*, **281** 1317–1326 (1998).
42. Evan, G.I., and Vousden, K.H. Proliferation, cell cycle and apoptosis in cancer. *Nature*, **411** 342–348 (2001).
43. Amati, B., Littlewood, T.D., Evan, G.I., and Land, H. The c-Myc protein induces cell cycle progression and apoptosis through dimerization with Max. *EMBO J.*, **12** 5083–5087 (1993).
44. Hueber, A.O., Zornig, M., Lynon, D., Suda, T., Nagata, S., and Evan, G.I. Requirement for the CDC95 receptor-ligand pathway in c-Myc induced apoptosis. *Science*, **278** 1305–1309 (1997).
45. Juin, P., Hueber, A.O., Littlewood, T., and Evan, G. C-Myc sensitization to apoptosis is mediated through cytochrome c release. *Genes and Dev.*, **13** 1367–1381 (1999).
46. Soucie, E.L., Annis, M.G., Sedivy, J., Filmus, J., Leber, B., Andrews, D.W., and Penn, L.Z. Myc potentiates apoptosis by stimulating Bax activity at the mitochondria. *Mol. Cell. Biol.*, **21** 4725–4736 (2001).
47. Pelengaris, S., and Khan, M. The many faces of c-MYC. *Arch. Biochem. Bioph.*, **416** 129–136 (2003).
48. Zhu, L., and Skoultchi, A.I. Co-ordinating cell proliferation and differentiation. *Curr. Opin. Genet. Dev.* **10** 91–97 (2001).
49. Mougneau, E., Lemieux, L., Rassoulzadegan, M., and Cuzin, F. Biological activities of v-myc and rearranged c-myc oncogenes in rat fibroblast cell sin culture. *Proc. Natl.Acad. Sci. USA*, **81** 5758–5762 (1984).
50. Sorrentino, V., Drozdoff, V., McKinney, M.D., Zeitz, L., and Fleissner, E. Potentiation of growth factor activity by exogenous c-myc expression. *Proc. Natl.Acad. Sci. USA*, **83** 8167–8171 (1986).
51. Karn, J., Watson, J.V., Lowe, A.D., Green, S.M., and Vedeckis, W. Regulation of cell cycle duration by c-myc levels. *Oncogene*, **4** 773–787 (1989).
52. Land, H., Parada, L.F., and Weinberg, R.A. Tumourigenic conversion of primary embryo fibroblasts requires at least two cooperating oncogenes. *Nature*, **304** 596–602 (1983).
53. Facchini, L.M., and Penn, L.Z. The molecular role of Myc in growth and transformation: recent discoveries lead to new insights. *FASEB J.*, **12** 633–651 (1998).
54. Cerni, C. Telomeres, telomerase, and *myc*. An update. *Mutat.Res.*, **462**, 31–47 (2000)
55. Claasen G.F., and Hann, S.R. Myc mediated transformation: the repression connection. *Oncogene*, **18** 2925–2933 (1999).
56. Pelengaris, S., Littlewood, T., Khan, M., Elia, G., and Evan, G. Reversible activation of c-Myc in skin: induction of a complex neoplastic phenotype by a single oncogenic lesion. *Mol. Cell*, **3** 565–577 (1999).
57. Pelengaris, S., Rudolph, B., and Littlewood, T. Action of Myc in vivo-proliferation and apoptosis. *Curr. Opin. Genet. Dev.*, **10** 100–105 (2000).
58. Fotsis, T., Breit, S., Lutz, W., Rossler, J., Hatzi, E., Schwab, M., and Schweigerer, L. Down-regulation of endothelial cell growth inhibitors by enhanced MYCN oncogene expression in human neuroblastomas cells. *Eur. J. Biochem.* **263** 757–764 (1999).
59. Ngo, C.V., Gee, M., Akhtar, N., Yu, D., Volpert, O., Auerbach, R., and Thomas-Tikhonenko, A. An in vivo function for the transforming Myc protein: elicitation of the angiogenic phenotype. *Cell Growth Differ.*, **11** 201–210 (2000).
60. Janz, A., Sevignani, C., Kenyon, K., Ngo, C.V., and Thomas-Tikhonenko, A. Activation of the myc oncoprotein leads to increased turnover of thrombospondin-1 mRNA. *Nucleic Acids Res.*, **28** 2268–2275 (2000)
61. Semenza, G.L. Involvement of hypoxia-inducible factor 1 in human cancer. *Intern. Med.*, **41** 79–83 (2002).
62. Lengauer, C., Kinzler, K.W., and Vogelstein, B. Genetic instabilities in human cancers. *Nature*, **396** 643–649 (1998).

63. Mai, S., Hanley-Hyde, J., and Fluri, M. C-Myc over-expression associated DHFR gene amplification in hamster, rat, mouse and human cell lines. *Oncogene*, **12** 277–288 (1996).

64. Felsher, D.W., and Bishop, J.M. Transient excess of MYC activity can elicit genomic instability and tumourigenesis. *Proc. Natl. Acad. Sci., USA*, **96** 3940–3944 (1999b).

65. Kuschak, T.I., McMillan-Ward, E., Taylor, C., Israels, S., and Henderson, D.W. The ribonucleotide reductase r2 gene is a non-transcribed target of c-myc-induced genomic instability. *Gene*, **238** 351–365 (1999).

66. Mushinski, J.F., Hanley-Hyde, J., Rainey, G.J., Kuschak, T.I., and Taylor, C. Myc-induced cyclin d2 genomic instability in murine B cell neoplasms. *Curr. Top. Microbiol. Immunol.*, **246** 183–189 (1999).

67. Li, Q., and Dang, C.V. C-Myc over-expression uncouples DNA replication from mitosis. *Mol. Cell. Biol.*, **19** 5339–5351 (1999).

68. Paulovich, A.G., Toczyski, D.P., and Hartwell, L.H. When checkpoints fail. *Cell*, **88** 315–321 (1997).

69. Iguchi-Ariga, S.M.M., Itani, T., Kiji, Y., and Ariga, H. Possible function of the *c-myc* product: promotion of cellular DNA replication. *EMBO J.* **6** 2365–2371 (1987).

70. Packham, G., and Cleveland, J.L. C-Myc and apoptosis. *Biochim. Biophys. Acta*, **1242** 11–28 (1995).

71. Grandori, C., and Eisenman, R.N. Myc target genes. *Trends Biochem.*, **22** 177–181 (1997).

72. Al-Rubeai, M. Apoptosis and cell culture technology. *Adv. Biochem.Eng./Biotech.*, **59** 226–249 (1998).

73. Birch, J.R., and Froud, S.J. Mammalian cell culture systems for recombinant protein production. *Biologicals*, **22** 127–133 (1994).

74. Ifandi, V., and Al-Rubeai, M. Regulation of cell proliferation and apoptosis in CHO-K1 cells by the coexpression of c-Myc and Bcl-2. *Biotechnol Prog*, **21** 6671–677 (2006).

75. Darnbrough, C., Watts, P., and MacDonald, C. Cloning of mouse hybridoma cells infected with myc and ras containing retroviruses yields cell lines with improved growth and antibody production. In Spier, R.E., Griffiths, J.B., and Macdonald, C., (eds). *Animal Cell Technology: Developments, Processes and Products.* Butterworth-Heinemann, Ltd, Oxford, pp. 20–22 (1992).

76. Singh, R.P., Al-Rubeai, M., and Emery, A.N. Apoptosis: exploiting novel pathways to the improvement of cell culture processes. *Genet. Eng, Biotech*, **16** 227–251 (1996).

77. Lubiniecki, A.S. Historical reflections on the cell culture engineering. *Cytotechnology*, **28** 139–145 (1998).

78. Pak, S.C.O., Hunt, S.M.N., Bridges, M.W., Sleigh, M.J., and Gray, P.P. Super-CHO: a cell line capable of autocrine growth under fully defined protein-free conditions. *Cytotechnology*, **22** 139–146 (1996).

79. Rasmussen, B., Davis, R., Thomas J., and Reddy, P. Isolation, characterization and recombinant protein expression in veggie-CHO: a serum-free CHO host cell line. *Cytotechnology*, **28** 31–42 (1998).

80. Fussenegger, M., and Bailey, J.E. Molecular regulation of cell cycle progression and apoptosis in mammalian cells: implications for biotechnology. *Biotech. Prog.*, **14** 807–833 (1998).

81. Fussenegger, M., and Bailey, J.E. Control of mammalian cell proliferation as an important strategy in cell culture technology, cancer therapy and tissue engineering. In: Cell Engineering, vol. 1. Ed: M. Al-Rubeai, Kluwer Academic Publishers (1999).

82. Fussenegger, M., Bailey, J.E., Hauser, H., and Mueller, P. Genetic optimization of recombinant glycoprotein production by mammalian cells. *TIBTECH*, **17**:35–43 (1999).

83. Bi Jing-Xiu, Shuttleworth J., Al-Rubeai M. Uncoupling of cell growth and proliferation results in enhancement of productivity in p21-arrested CHO cells. *Biotechn. Bioeng.*, **85** 741–749 (2004).

84. Crea F., Sarti D., Falciani F., Al-Rubeai M. Over-expression of hTERT in CHO K1 results in decreased apoptosis and reduced serum dependency. *J. Biotechn.*, **121** 109–123 (2006).

85. Ifandi, V., and Al-Rubeai, M. Stable transfection of CHO cells with the *c-myc* gene results in increased proliferation rates, reduces serum dependency, and induces anchorage independence. *Cytotechnology*, **41** 1–10 (2003).

86. Evan, G.I., Wyllie, A.H., Gilbert, C.S., Littlewood, T.D., Land, H., Brooks, M., Waters, C.M., Penn, L.Z., and Hancock, D.C. Induction of apoptosis in fibroblasts by c-myc protein. *Cell*, **69** 119–128 (1992).
87. Sunstrom, N.-A. S., Gay, R.D., Wong, D.C., Kitchen, N.A., DeBoer, L., and Gray, P.P. Insulin like growth factor I and transferring mediate growth and survival of Chinese hamster ovary cells. *Biotech. Progr.*, **16** 698–702 (2000).

CHAPTER 6

THE MOLECULAR RESPONSE(S) DURING CELLULAR ADAPTATION TO, AND RECOVERY FROM, SUB-PHYSIOLOGICAL TEMPERATURES

SARAH J SCOTT, ROSALYN J MARCHANT, MOHAMED B AL-FAGEEH, MICHÈLE F UNDERHILL, AND C MARK SMALES

Protein Science Group, Department of Biosciences, University of Kent, Canterbury, Kent CT2 7NJ, UK

Abstract: The cold-shock response has now been studied in various organisms for several decades, but unlike the heat-shock response, the mechanisms involved in the response to sub-physiological temperatures are poorly understood. Despite this, in recent years we have begun to broadly understand those molecular mechanisms that appear to be key players in the response to cold-shock. Most notably, all organisms studied to date respond by attenuating global transcription and translation whilst selectively inducing the elite and rapid over-expression of a specific set of proteins called the cold-shock proteins (CSPs). The response involves careful coordination of specific cellular processes in a time and temperature dependent manner. When mammalian cells cultured *in vitro* are subjected to mild hypothermic temperatures (approximately 32 °C), cells response by modulating transcription, translation, the cell cycle, metabolism and the cell cytoskeleton as part of the adaptive process. Here we describe the global cold-shock response, and then focus upon the responses in yeast and mammalian cells, highlighting particular areas of interest to the cell engineering community. The key responses in *Escherichia coli* upon temperature downshift are also briefly described

Keywords: Cold-shock response, mammalian cells, cellular responses, control of gene expression, cold-shock proteins, sub-physiological temperature

1.   INTRODUCTION

Whilst much is known about the conserved cellular mechanisms activated in both prokaryotic and eukaryotic systems in response to, and recovery from, heat stress[1], comparatively little is known about the response(s) to cold-stress. This is partially a reflection of the small number of studies, in relatively few organisms, that have

185

M. Al-Rubeai and M. Fussenegger (eds.), Systems Biology, 185–212.
© 2007 *Springer.*

been reported to date[2], and the variable responses observed over a wide range of sub-physiological temperatures. This lack of study and knowledge is somewhat surprising, as a change in environmental temperature is perhaps the most common form of stress encountered by a large number of organisms. One might also argue that exposure to cold-shock temperatures (either mild or severe) is likely to be more commonly encountered by a larger variety of organisms as opposed to heat stress conditions. Further, reduced temperature changes are utilized in the biotechnology industry for the culturing and preservation of cells, and in the medical field for the preservation of tissue and treatment of brain damage[3], without necessarily understanding the cold-shock responses and mechanisms being harnessed during such uses.

Over the last decade or so there has been an increased drive to readdress our lack of knowledge in this area. Several groups have recently focused on furthering our understanding of the detailed mechanistic effects by which gene expression is modulated in response to low temperature in a number of model systems. For example, a variety of experimental approaches have been applied to identify a number of plant cold-shock inducible genes and proteins[1,4] and approximately 26 cold-shock genes have now been identified in *E. coli*[5].

Despite these reports, our current understandings of the mechanisms that control the cold-shock response remain rudimentary at best. However, reports showing that reduced temperature cultivation of mammalian cells can enhance heterologous protein production have recently fueled efforts in this area to further our understanding of the mechanisms at play during cold-shock culturing of these expression systems[3]. This research has been largely driven in order to elucidate the mechanism(s) that result in increased heterologous protein production at reduced culture temperatures in order to identify new targets for cell engineering that would exploit these responses for the enhanced production of complex recombinant proteins.

The cold-shock response is therefore now being subjected to a raft of 'omic' type approaches in bacterial, yeast and mammalian systems as never before in order to (i) further our understanding of the molecular response(s) governing cellular adaptation to cold-shock across a variety of organisms, (ii) elucidate the effects on, and mechanisms that determine, heterologous protein production in various expression systems at sub-physiological temperatures, and (iii) to identify new and rational targets for cell engineering and novel gene expression technologies. These approaches are helping develop a detailed mechanistic understanding of the effects of low temperature on various organisms, and this will undoubtedly provide information useful in the development of new cell engineering strategies, and for the further optimization of the phenotypic performance of cells in culture at reduced temperatures.

2.   THE GENERAL RESPONSES TO COLD-SHOCK

Despite our limited understanding of the mechanisms controlling the cold-shock response, there are a number of general responses initiated upon temperature reduction in both prokaryotic and eukaryotic cells. The general cold-shock response in both prokaryotes and eukaryotes is to suppress global transcription and translation, while at the same time inducing the synthesis of a small and select group of specific proteins, the so termed cold-shock proteins (CSPs)[6]. However, unlike their heat-shock counterparts, the cold-shock proteins do not appear to be as rigorously conserved across eukaryotic and prokaryotic systems. Further, the sensors and mechanisms by which the majority of cold-shock proteins are induced are not yet elucidated.

By far the most well characterized cold-shock response is that of *Escherichia coli* to temperatures between 10–15 °C, whereby approximately 27 cold-shock proteins have been identified to date[5]. The cold-shock proteins in *E. coli* are regulated in a sequential fashion upon temperature downshift and control a number of fundamental functions that determine cell fate such as DNA replication, transcription, translation, RNA stabilization and ribosome assembly[5]. A number of plant and yeast cold-shock responsive genes have now also been identified[1]. On-the-other-hand, there are few cold-shock proteins characterized in mammalian cells, however the response appears to involve a coordinated series of events that control and regulate transcription, translation, the cell cytoskeleton and metabolic processes upon temperature downshift.

The general suppression of protein synthesis upon temperature downshift is therefore now well documented[1,4,5,7–13]. Concurrent with this, cells go through a period of growth arrest during an acclimatization period, whilst at the same time inducing the expression of the cold-shock proteins[14]. This sequence of events has been particularly well demonstrated in *E. coli*, whereby upon temperature downshift to 10 °C, growth is actively arrested for a period of 1–4 h duration[15]. Following this acclimatization period during which CSPs are expressed while bulk protein synthesis is arrested, growth is again resumed, albeit at a reduced rate compared to that observed at 37 °C[15]. Upon growth resumption, bulk protein synthesis is also resumed.

The 5'untranslated regions of mRNAs are known to form stable secondary structures upon being exposed to reduced temperatures, interfering with ribosome scanning and translation elongation steps[6]. Cold-shock also results in the suppression of translation initiation via the normal cap-dependent mechanism, presumably by reducing the active pool of key initiation factors. Further, although the exact role of the cold-shock proteins has not been determined, the general consensus is that these play a role in mRNA stability and in maintaining and promoting the accurate and enhanced translation of specific mRNAs at reduced temperatures[2]. The exact mechanism(s) by which the cold-shock proteins achieve this are not yet fully described.

In addition to those general responses described above, the integrity of the cell membrane is changed and becomes compromised when *in vitro* cultured cells are subjected to reduced temperatures[16]. In particular the rigidity of the membrane is altered, which in turn results in a drop in the efficiency of various membrane-associated cellular functions[16]. In microorganisms membrane integrity is maintained upon temperature downshift by initiating a coordinated response involving the induction of fatty acid desaturases and dehydrases to ensure a more-or-less constant degree of membrane fluidity[17–20].

In the last decade genomic type approaches utilizing DNA microarray or proteomic technologies have been applied to the cold-shock response in a number of prokaryotic systems. For example, both proteomic and transcriptomic approaches have been applied to the cold-shock response and acclimatization phase in *E. coli*[21–23]. Whilst these studies have identified a number of key genes and proteins that are implicated in the cold-shock response, further studies at the transcriptional, and both the global proteomic and individual protein level, are required to help understand the mechanism(s) by which changes to such proteins/genes occur and the significance (or not) of changes in expression levels. A proteomic approach has also been employed to show that proteins involved in the signal transduction pathway of bacterial sporulation are induced upon temperature downshift in the thermophilic bacterium *Bacillus stearothermophilus*[24]. DNA microarray and proteomic approaches have also now identified cold-inducible genes and proteins in both *Bordetella bronchiseptica*[25] and *Yersinia pestis*[26]. Therefore, a large amount of interest and work is currently on going in a range of prokaryotic systems to identify further control mechanisms and additional key molecular players in the cold-shock response in prokaryotic systems.

Finally, it should be noted that a reduction in temperature actually submits cells to two stresses. In addition to stresses resulting from reduced temperature, cells are also concomitantly exposed to changes in dissolved oxygen concentrations[27]. One should therefore consider these potential stresses and the responses to these, as part of a coordinated response. Even in a bioreactor, whereby dissolved oxygen concentrations may be kept constant over a range of temperatures, this does not mean that there are not changes to intracellular oxygen levels upon temperature downshift. At 30 °C both oxygen and ATP consumption are reduced in *in vitro* cultured Chinese hamster ovary cells, but this is probably a reflection of reduced metabolic rates under these conditions and not a direct response to oxygen levels[28]. Further, it has now been suggested that the mechanisms of cold-stress induction share common components with pathways involved in the hypoxic response[29], but not with those of the hyperoxic response. This is further substantiated by reports that two mammalian cold-shock proteins, RNA-binding motif protein 3 (Rbm3) and cold inducible RNA-binding protein (Cirp), are induced upon exposure to hypoxic conditions[30].

3.   THE COLD-SHOCK RESPONSE IN *E. COLI*

There are a growing number of investigations into the effect of cold-shock upon prokaryotic organisms, although by far the most well characterized organisms are *Escherichia coli* and *Bacillus subtilis*[4]. The cold-shock responses in prokaryotes, particularly that in *E. coli*, has previously been comprehensively reviewed[4,5,11,12]. At temperatures of around 10–15 °C, *E. coli* selectively expresses around 27 cold-shock proteins over 1-4 hours[5]. At the same time, global transcription and translation are down-regulated[5]. Further, recent reports have shown that at sub-optimal temperatures the *E. coli* translational apparatus undergoes modifications allowing the selective translation of the transcripts of cold shock-induced genes, while bulk protein synthesis is drastically reduced[7]. However, once cells have adapted to the reduced temperature, cold-shock protein synthesis is once again down regulated whilst global protein synthesis is restarted, albeit at a rate reduced in comparison to that observed at 37 °C (Figure 1). At lower temperatures *E. coli* also contain increased numbers of functioning ribosomes, presumably to compensate for the observed decrease in global translation rates[31]. Interestingly, this is reversed at higher temperatures where the number of functioning ribosomes actually decreases while 'non-functioning' ribosome content remains the same[31].

The first identified cold-shock protein was cold-shock protein A (CspA), which is the major cold-shock protein expressed in *E. coli* upon temperature shift to 10 °C[32]. Cold-shock protein A is an RNA binding protein and although the exact role of the various bacterial cold-shock proteins are not yet elucidated, it is widely acknowledged that CSPs do function as RNA chaperones in bacteria[4,5,11,12,33–38]. Further, bacterial CSPs are highly similar and all have shown the ability to bind single-stranded nucleic acids[39]. Finally, although a number of the bacterial CSPs are essential for growth at 10 °C, none have been found to have an exclusive role in cold-adaptation[5]. Although a number of studies into various other bacterial have now been published, including photosynthetic, non-photosynthetic, cold loving and cyanobacteria, for the purposes of his chapter we focus here on the induction of

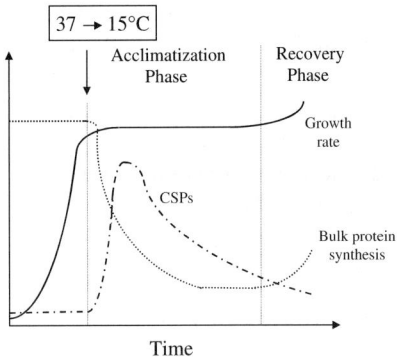

Time

Figure 1. **The global response in prokaryotic systems to cold-shock at 10–15°C.** Upon temperature shift from 37°C to 10-15°C, the cold-shock response is initiated in prokaryotic systems. The response appears to be largely conserved across prokaryotic systems and is characterized by two phases, an acclimatization phase and a recovery phase. The acclimatization phase is characterized by a halt in cell growth and rapid expression of a select set of cold-shock proteins (CSPs), whilst global transcription and protein synthesis are dramatically reduced. After 1–4 h at reduced temperature, cells enter the recovery phase whereby CSP synthesis is reduced and cell growth and bulk protein synthesis resume, albeit at a reduced rate compared to that observed at 37°C

cold-shock protein A (CspA) in *E. coli* as an example of the molecular mechanisms involved in the responses to reduced temperature in prokaryotic cell systems.

3.1   The Major Cold-shock Protein in *E. coli*, CspA

Cold-shock protein A (CspA) is a small 7.4 kDa protein which has a β-barrel structure[40], and is only synthesized under cold-shock conditions in *E. coli*[8,15]. CspA is homologous to the major *Bacillus subtilis* cold-shock protein, CspB[40], and is highly similar to other members of the bacterial cold-shock protein family (CspB-CspI). Of this family, only CspB, CspG and CspI are induced upon cold-shock[35,41]. On-the-other-hand, upon temperature downshift to 10 °C, CspA synthesis in *E. coli* is up-regulated such that it constitutes approximately 13% of total protein synthesis[41].

The molecular mechanisms involved in *CspA* induction have been thoroughly investigated and have recently been reviewed by Gualerzi et al[5], however the exact mechanisms remain an issue of debate. The promoter region was initially thought to contain a sequence −35 to −92 upstream of the *CspA* promoter that *trans*-acting factors bound to upon temperature downshift, inducing expression via up-regulation of *CspA* mRNA levels[42]. However, since this proposal further investigations have lead to the general consensus that this mechanism plays little part in *CspA* induction upon cold-shock[5], and further, have shown that the *CspA* promoter is active at 37 °C[43]. A schematic of the proposed key elements in the *CspA* promoter and 5′-untranslated region is shown in Figure 2.

The *CspA* promoter does not therefore appear to play a major role in the induction of *CspA*, suggesting that *CspA* regulation is controlled post-transcriptionally. A number of other reports also confirm that transcriptional regulation plays little role in the cold-shock induction of *CspA*[5,37]. Further, the 5′-untranslated region (5′-UTR)

Figure 2. **Schematic of the *CspA* promoter and untranslated regions.** Early reports suggested that at low temperatures a *trans*-acting factor bound preferentially to a sequence -35 to -73 upstream of the *CspA* promoter and was involved in the exclusive transcriptional regulation of *CspA*. This has since been shown to be incorrect. Further reports suggested that the promoter region of the *CspA* gene also contained an AT-rich sequence (-47 to -38) immediately upstream of the −35 region, which acted as an UP-element and facilitated and enhanced efficient transcription initiation of *CspA* at variable temperatures, however the effect this enhancer has upon transcription enhancement is modest at best. The induction and control of *CspA* has now been shown to be primarily due to post-transcriptional events. The downstream box was at one stage thought to play an important role in such regulation, however this has since been disproved. The lengthy 5′-UTR of the *CspA* transcript is now thought to play a major role in preferential translation at low temperatures due to the presence of *cis* and structural elements and due to targets for *trans*-acting factors within the translational apparatus. The presence of a 'cold-box' has also been implicated in such control mechanisms.

of the *CspA* transcript negatively affects CspA protein expression at 37 °C, but has a positive effect upon cold-shock[44], whilst *CspA* mRNA half-life is short at 37 °C, but increases upon cold-shock[34]. It is therefore thought that the 5'-UTR of *CspA* stabilizes *CspA* mRNA at low temperature[43], harbouring sequences that are absolutely required for mRNA stability at reduced temperatures[5,7,45]. Interestingly, truncation or deletion of the 5'-UTR results in an increase in *CspA* mRNA stability at 37 °C[5]. An additional feature, the cold box region (Figure 2), is also thought to be involved in the formation of secondary structure essential for *CspA* transcript stability at reduced temperatures[46,47].

With regard to cell engineering, Qing et al have recently developed a set of cold-shock vectors for the expression of target proteins in *E. coli* at reduced temperatures[48]. These vectors are based upon the unique features of *CspA* promoter and untranslated region control components, and protein expression is under the control of the *CspA* promoter. The vectors also contain the *CspA* 5'-untranslated region and the *CspA* 3' end transcription terminator site. An additional feature in several of the vectors is the inclusion of a translation-enhancing element (TEE) after the *CspA* 5'-UTR. Protein expression is induced by cold-shock, reducing the temperature to 15 °C, concomitant with the addition of isopropyl-β-thiogalactopyranoside (IPTG)[48]. The authors show that this system can be used to tightly control the induced synthesis of recombinant material and results in the expression of significant amounts of material[48]. In view of the fact that the *CspA* promoter does not appear to play a major role in cold-shock induction (see discussion above), in our view it is likely that post-transcriptional control mechanisms tightly control recombinant protein expression at reduced temperature when using these expression vectors, as opposed to transcriptional regulation.

3.2   Additional Prokaryotic Cold-shock Mechanisms

There are now an increasing number of partially characterized prokaryotic cold-shock responses in addition to the CspA response[4,11]. For example, the over-expression of the cold-shock inducible proteins CspL, CspP, and CspC in *Lactobacillus plantarum* has recently been investigated[49]. These studies suggest that each CSP has a distinct role in the different phases of the cold-shock response. The over-expression of CspP resulted in improved freezing survival, whilst the over-expression of CspL transiently prevented the typical growth rate reduction observed upon a temperature downshift to 8 °C[49]. Further, in *E. coli*, the stress protein, protein Y, inhibits translation initiation upon cold-shock but not at physiological temperature by competing with translation initiation factors required for ribosomal subunit disassembly[50]. Studies in *Bacillus subtilis* also suggest that cold-shock proteins are a direct link between transcription and translation, coupling transcription to translation initiation[51]. Finally, with the increasing number of studies now being reported in the literature on the cold-shock response, and mechanisms involved in the control of the cold-shock response in prokaryotes, a plethora of data is now emerging that is extending our understanding of the cold-shock response in prokaryotic systems.

However, we are still a long way from a comprehensive understanding of the mechanisms that control the cold-shock response in the same way that we understand the corresponding heat-shock response in prokaryotic organisms.

4.   THE YEAST COLD-SHOCK RESPONSE

In recent years the response of yeast cells to reduced temperature cultivation has received much attention, mainly through microarray based investigations. Further, low temperature can also be utilized to drive the chemo- and stereo-selective synthesis of small organic compounds in yeast. For example, Florey and colleagues have previously demonstrated that the reduction of 3-oxo-3-phenylpropanenitrile exclusively yields (S)-3-hydroxy-3-phenylpropanenitrile when the reaction is mediated by *Saccharomyces cerevisiae* at 4°C, but generates a mixture of reduction and alkylation products when the reaction is allowed to proceed at room temperature[52]. Reduced temperature cultivation of yeast cells has also recently been used to improve the yield of heterologous membrane bound proteins[53].

As stated above, the majority of investigations into the cold-shock response in yeast have utilized microarray-based technology to identify cold-responsive genes in *Saccharomyces cerevisiae*. These studies have shown that yeast respond to cold-shock in a much more moderate manner than bacterial systems due to the fact that yeast cells are generally adapted to lower temperatures[33]. Further, several yeast cold-shock responsive genes have now been identified[9]. These studies have also shown that yeast cells initiate two distinct responses to temperature downshift, dependent upon the severity of the temperature change. At severe low temperature (<10°C), yeast initiates the near freezing response whilst at milder reduced temperatures (10–18°C) yeast initiates the cold-shock response[54]. The major traits characterizing each response are outlined in Figure 3. Others have also investigated the response of yeast cells to much lower temperatures, cryopreservation, and have followed the responses upon recovery from such storage[55].

4.1   The Near Freezing Response in Yeast Cells

The near freezing response is induced in yeast cells at temperatures below 10°C and is characterized by a rapid increase in the levels of the chemical chaperone trehalose[10,33,54]. Trehalose is essential for maintaining cell viability at zero degrees or less, and the levels of trehalose accumulated correspond to the ability of the cell to resist freezing damage and death[54]. The levels of trehalose are increased in response to near freezing temperatures via the induction of the enzymes responsible for the synthesis of this chemical chaperone, *Tps1* and *Tps2*. When the environmental temperature is returned to 30°C, trehalose levels are rapidly decreased and returned to pre-near freezing response levels[54]. Interestingly, trehalose induction plays an important role in cell survival in *E. coli* upon cold-shock[56]. In *E. coli* trehalose synthesis is increased upon temperature downshift to 16°C via the induction of the *OtsA* and *OtsB* genes[56].

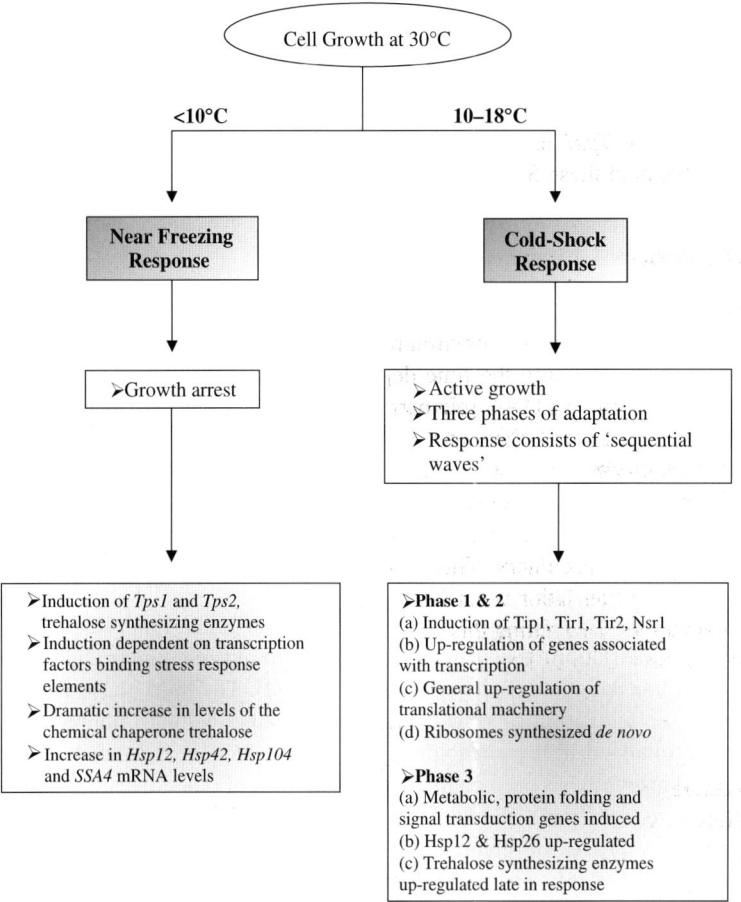

Figure 3. **The response of yeast cells to sub-optimal temperatures (<30 deg C) is temperature dependent.** Extreme low temperature (<10°C) initiates the near freezing response. The near freezing response is characterized by the synthesis of relatively large amounts of the chemical chaperone trehalose. Less extreme sub-physiological temperatures (10–18°C) initiate a vastly different response (the cold-shock response), characterized by the stepwise regulation of key genes and proteins.

Trehalose synthesis in yeast cells is therefore up-regulated in response to an increase in *Tps1* and *Tps2* mRNA levels upon temperature downshift to less than 10 °C. At 0 °C the levels of these two genes is increased more than 20-fold around 20 hours after temperature downshift[54]. Although an increase in the transcription of these genes plays a role in this dramatic rise in mRNA levels, mRNA stabilization also appears to play a crucial role[54]. *Tps1* and *Tps2* mRNA is extremely stable at 0 °C but extremely unstable at 30 °C. Thus, upon a return of the temperature to 30 °C, *Tps1* and *Tps2* mRNA levels are returned to basal levels in less than five minutes.

The induction of *Tps1* and *Tps2* transcriptional activity at near freezing response temperatures is regulated via stress response elements (STREs) in the promoter regions of these genes[54]. When the environmental temperature drops below 10 °C, transcription factors from the Msn2,4 pathway bind to these STREs, up-regulating the transcription of the *Tps1* and *Tps2* genes. In particular the Msn2 and Msn4 transcription factors bind these STREs at near freezing response temperatures.

4.2   The Cold-shock Response in Yeast Cells

The cold-shock response in yeast cells may be defined as the response observed to temperature downshift whereby the environmental temperature shifted to is more than 10 °C. An investigation into the time dependent response of yeast cells to a temperature of 10 °C utilizing DNA microarray technology showed that *S. cerevisiae* responds in three distinct phases to such environmental change[2]. The major responses that characterize each phase of the cold-shock response in yeast are described in Figure 3. Each of these phases appears to be carefully coordinated in a cooperative manner. The initiate response is to up-regulate several genes associated the transcriptional machinery. This is then followed by an up-regulation of key components of the translational machinery. In particular, yeast cells appear to synthesize ribosomes *de novo* during this phase of the cold-shock response, presumably to allow the translation of key mRNAs at the required rate under conditions whereby general translational efficiency is compromised[2]. The third phase is characterized by the up-regulation of genes involved in distinct cellular bioprocesses, including genes involved in protein folding, signal transduction and metabolism[2]. This coordinated response has been confirmed by a number of other investigators[9,57].

Investigations to date have shown that yeast cells actively respond to cold-shock temperatures (10 °C or greater) by expressing a number of cold-shock genes including *Tip1, Tir1, Tir2,* and *Nsr1*. These genes are involved in key processes during the yeast cold-shock response, including cell wall maintenance, rRNA processing and ribosome biogenesis. In addition to these cold-shock genes, microarray based approaches have identified many additional genes whose expression levels are also modulated upon temperature downshift to 10 °C[2,9,10,57,58]. Despite the large number of genes these studies have highlighted, many are probably changed upon temperature downshift simply as a result of the global reduction in transcription and translation that accompanies such changes in temperature. Therefore, further studies need to be undertaken before embarking upon cell engineering of yeast cells based upon the data generated to date, in order to identify those genes which are cold-shock inducible (or repressed) as opposed to those whose expression changes as part of a more global down-regulation response.

It is worth noting that during the cold-shock response almost all of the heat-shock protein genes are down regulated[2]. The exceptions to this are the *Hsp12* and *Hsp26* genes. During the near freezing response a number of additional heat shock genes are up-regulated. However, despite the transcriptomic data now available on the response of yeast cells to temperature downshift, the response at the protein level has been largely ignored to date. Several studies have shown that the correlation between mRNA and protein levels in yeast is poor[59,60], therefore the specific alterations in protein expression that allow yeast cells to adapt to, and survive, low temperature stress remain to be elucidated.

5.   THE RESPONSE OF *IN VITRO* CULTIVATED MAMMALIAN CELLS TO SUB-PHYSIOLOGICAL TEMPERATURES

5.1   The General Response of *in Vitro* Cultured Mammalian Cells to Reduced Environmental Temperature

Perhaps somewhat surprisingly, whilst the heat-shock response is well charac-terized in mammalian systems[61,62], until relatively recently very few studies had been undertaken into the molecular responses of mammalian cell systems to sub-physiological temperature. Further, although cold-stress not only subjects cells to a temperature stress, but also potentially to an oxygen stress (at lower temperatures oxygen dissolves at higher concentrations)[27], this facet of the cold-shock response has been almost entirely ignored to date. Thus, despite the potential importance of the cold-shock response to organ/tissue storage and transplant, hibernation, the treatment of brain damage, and for enhanced recombinant protein production from mammalian expression systems, little is currently known about the cellular mech-anisms involved in the mammalian cold-shock response. The general responses of *in vitro* cultured mammalian cells to sub-physiological temperatures ($<37\,°C$) are highlighted in Table 1 below. As a rule, cold-shock results in the attenuation of tran-scription and translation, reduced metabolism, disruption of the cell cytoskeleton, cell cycle arrest (see Figure 4) and the induction of specific cold-shock proteins[63], as in bacterial and yeast systems. It should be noted that the severity of the response is temperature dependent, and that the majority of comments in the remainder of this chapter will refer to mild hypothermic temperature responses ($25–35\,°C$) as opposed to more severe cold-shock temperatures ($<10\,°C$).

Table 1. The global responses of *in vitro* cultured mammalian cells to sub-physiological temperatures; the cold-shock response

Characteristic	Effect of Cold-Shock
Viability	Increased and prolonged
Cell cycle	Arrest in G1/G0
Metabolism	Reduced, inhibition of ATP expenditure
Membranes	Altered
Transcription	Attenuated
Translation	Globally reduced at initiation and elongation steps
Apoptosis	Delayed
Cytoskeleton	Disassembled at low temperature
Cold-shock proteins	Up-regulated expression

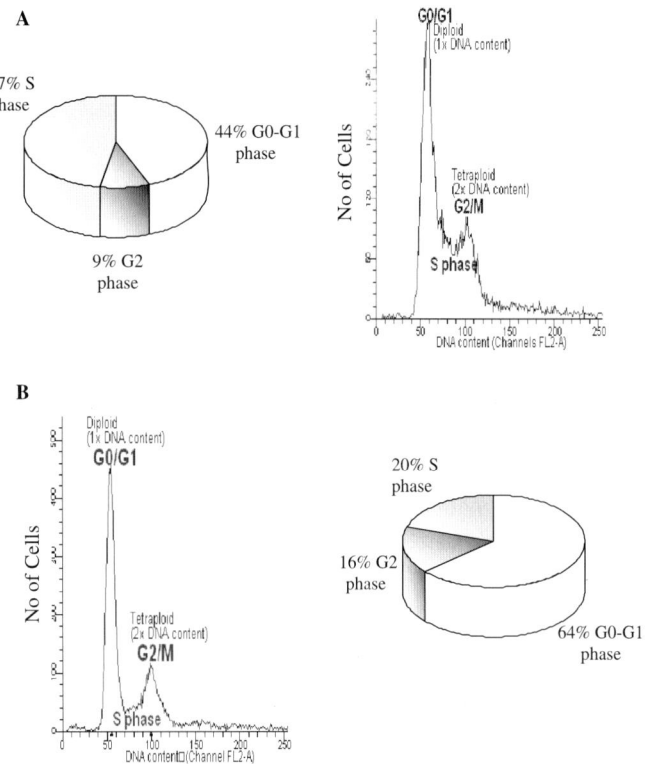

Figure 4. **Subjecting in vitro cultured mammalian cells to mild-hypothermic conditions results in cell cycle arrest.** A typical cell cycle profile of Chinese hamster ovary cells cultured at 37°C is shown above (Figure A). After temperature downshift to 32°C for several days, an increase in cells residing in the G0/G1 and G2 phases is observed with a concurrent decrease in the percentage of cells found in S phase (Figure B)

As in prokaryotes, it would therefore appear that the global responses upon cold-shock are similar to those observed upon heat-shock. Mammalian cells also appear to initiate different responses at extreme low temperatures (< 10 °C) as opposed to those initiated at more mild hypothermic temperatures (25–35 °C)[3], as described in yeast cells. The similarity between the yeast and mammalian cold-shock responses appears to end there. Interestingly, as outlined above, a heat shock protein response does not appear to be induced upon subjecting mammalian cells to cold-shock. However, the recovery of HeLa cells from transient cold-shock at 4 °C actives a heat shock response[64]. The onset, magnitude and duration of the observed heat-shock response upon rewarming were directly related to the severity of the cold-shock in terms of time and temperature[64]. The authors of this study suggest that mild heat-shock treatment before freeze storage of mammalian cells may be beneficial in terms of improving the recovery of mammalian cells from storage[64].

There are five proposed 'global' or 'general' mechanisms by which mammalian cells respond to sub-physiological temperatures[62]. These are (1) increased transcription of specific cold-shock genes via specific elements within the promoter regions of these genes, (2) the general attenuation of transcription and translation, (3) alternative splicing of heteronuclear-RNA, (4) a decrease in RNA degradation and an increase in mRNA stability, and (5) the preferential translation of specific mRNA's[62]. However, despite the recent efforts of investigators to date, the mechanisms controlling these responses are poorly understood and there are currently only two well-characterized mammalian cold-shock proteins described to date.

5.2 Mammalian Cold-shock Proteins

To date it is somewhat surprising that only two mammalian cold-shock proteins have been characterized, Cold Inducible RNA binding protein (Cirp) and RNA Binding Motif Protein 3 (Rbm3)[65]. Although other cold-shock responsive genes have been reported[1,62], the majority of these have diverse functions and therefore only Cirp and Rbm3 have been identified as cold-shock proteins (as opposed to cold-shock responsive). Cirp and Rbm3 belong to the glycine rich RNA-binding protein family and the two proteins are highly similar. Members of this family contain two distinct domains, a glycine rich C-terminal domain and a consensus sequence RNA-binding domain at the N-terminal[66]. Although the exact function of these two highly similar proteins has not yet been fully elucidated, it is thought that Cirp and Rbm3 are modulators of gene expression (transcription and translation) at mild hypothermic temperatures[1,65]. Like their bacterial counterparts, these cold-shock proteins bind RNA and are thought to act as RNA chaperones.

5.2.1 The mammalian cold-shock protein Cirp

Cirp was the first cold-shock protein identified in mammalian cells and has recently been investigated as a potential target for cell engineering strategies to improve heterologous protein expression from Chinese hamster ovary (CHO) cells. Although Cirp was identified before Rbm3, the control of Rbm3 expression has been studied in far more detail that of Cirp. What is known is that Cirp is a nuclear protein that appears to be conserved across mammalian species and is ubiquitously expressed in various tissues at 37 °C[67]. Furthermore, it has been reported that *Cirp* mRNA stability is not enhanced at mild hypothermic temperatures[68,69], however transcription and translation of *Cirp* is dramatically induced upon temperature down-shift[1]. *Cirp* mRNA and protein levels are increased in response to temperature reduction in the range of 32–35 °C, but are not induced in response to more extreme temperature reduction (<15 °C).

Several investigators have now proposed that Cirp functions as an RNA chaperone upon mild temperature downshift, acting to prevent the formation of mRNA secondary structures[1]. As stated above, *Cirp* mRNA stability is not changed upon temperature reduction and *Cirp* gene expression appears to be controlled transcriptionally by cold-responsive elements within its promoter[1] and possibly translationally by elements within the transcript. Initial investigations into *Cirp* expression suggested that it plays a role in the cell cycle arrest associated with the mild-hypothermic response in mammalian cells. Early studies showed that upon a temperature downshift from 37 °C to 32 °C, the growth of BALB/3T3 cells was impaired[70]. This growth impairment could be, at least partially, reversed by using antisense oligos to *Cirp* mRNA to prevent Cirp synthesis[70]. The over-expression of Cirp at 37 °C also resulted in G1 phase arrest of BALB/3T3 cells, further evidence that Cirp acts as a cell cycle arrest agent[70]. However, recent reports by Tan and colleagues have shown that the over-expression or knockdown of Cirp did not affect the doubling time of CHO cells, but did affect that of BALB/3T3 cells, suggesting that Cirp may not act as a cell cycle arrest mechanism in all cell types[71]. Further, over-expression of Cirp in CHO cells corresponded with an increase in the amount of recombinant interferon-γ mRNA and an associated increase in interferon-γ protein production at 32 °C[71].

5.2.2   *The mammalian cold-shock protein Rbm3*

Much more is known about the control and expression of the *Rbm3* gene than the *Cirp* gene. The levels of the *Rbm3* transcript have been shown to increase in a range of cell types including HeLa, NC65 and HepG2, upon temperature downshift to 32 °C[72]. Although much is known about the translational regulatory elements of *Cirp* expression, the precise function or roles of Rbm3 are less well understood. Like Cirp, Rbm3 binds mRNAs and has been implicated in the control of cytokine-dependent proliferation[30]. Further, transcription of *Rbm3* has recently been reported to be induced in response to hypoxia, as has *Cirp* transcription[30].

Chappell and colleagues have extensively characterized the regulatory elements within the 5′-untranslated region (5′-UTR) of the *Rbm3* transcript[73]. The 5-UTR of *Rbm3* is extremely long, and detailed analysis showed that within the 5′-UTR were 12 upstream open reading frames[73]. Classical deletion analysis of a putative internal ribosome entry segment (IRES) in the 720 base pair 5′-UTR of *Rbm3* mRNA identified a 22-nucleotide IRES module, 9 *cis*-acting sequences, 2 inhibitory sequences and a 10-nucleotide enhancer[74]. When the full 5′-UTR was placed directly upstream of a reporter gene, the *Rbm3* 5′-UTR was shown to facilitate enhanced translation at 32 °C[73]. The 5′-UTR of *Rbm3* mRNA is therefore thought to contain various elements that facilitate translation of the transcript even when global translation is compromised under cold-shock conditions[73,74]. Recent evidence also suggests that Rbm3 itself plays a pivot role in facilitating the translation of target mRNAs upon cold-shock by altering the levels of microRNA levels under such conditions[65]. MicroRNAs compromise the translation of target mRNAs and it is suggested that Rbm3 may remove this obstacle under cold-shock conditions by binding microRNAs, preventing the disruption to the translation of key target mRNAs[65].

5.3   The Cell Cytoskeleton and Cold-shock

Cold-shock results in the down-regulation or loss of protein synthetic machinery in mammalian cells[75]. The global down-regulation in mRNA translation upon cold-shock is likely to be partially due to increased mRNA stability, and the presence of stable mRNA secondary structures, under these conditions which slow or hinder ribosome scanning and progression. However, at mild-hypothermic temperatures mRNA translation is most probably controlled and regulated via the well-characterized phosphorylation and dephosphorylation of key initiation and elongation factors. This mechanism of translational control is known to be the key factor limiting translation initiation when conditions of environmental stress are prevalent[76,77].

Although classical translational control via initiation and elongation factors is undoubtedly of importance in the cold-shock response, the role of the cytoskeleton has until recently been largely neglected. Components of the translation machinery such as mRNAs, initiation factors and polyribosomes all co-localize with the cell cytoskeleton[78], forming a permanently and spatially orchestrated structure in eukaryotic cells[79]. These observations have lead to the suggestion that the cell cytoskeleton actually functions to organize the protein synthesis machinery and at the same time actively facilitates translation[78].

There is now a growing body of evidence to support the idea that the cell cytoskeleton plays a crucial role in mRNA translation and the cold-shock response. For example, an intact F-actin system is required for efficient protein synthesis in mammalian cells and more severe cold-shock is known to disrupt the F-actin system[78]. Translation elongation factor 1α is also known to act as a linker between the cell cytoskeleton and the mRNA translation machinery[80]. Cold-shock can also result in microtubule disassembly, leading to changes in the organization of the translation machinery and a reduction of F-actin[78,81], all of which influence translational efficiency.

Such a close link between the cell cytoskeleton and translational efficiency in mammalian cells is not a unique observation. Investigators working with yeast systems have now shown that yeast cells regulate translation initiation and the actin cell cytoskeleton simultaneously, yet independently, upon encountering glucose limiting conditions[82]. The maintenance of the cell cytoskeleton and translation apparatus would therefore appear to play an important and linked role in modulating the cold-shock response and maintaining translation efficiency upon temperature down-regulation.

5.4   'Omic' Approaches Towards Understanding the Cold-shock Response in Mammalian Cells

To date there have only been a handful of 'omic' investigations into the molecular responses of mammalian cells to cold-shock. Undoubtedly this will change in the near future, with a number of groups currently instigating and undertaking such investigations. One of the few transcriptomic studies reported to date on the response of mammalian cells to temperature downshift investigated changes in the transcriptome of 3T3 cells to a temperature drop from 37 to 32 °C[83]. The results showed that the expression levels of around 10% of all genes investigated changed upon such a temperature shift. A number of proteomic studies have also recently been reported on the response of CHO cells to mild hypothermic conditions and show that these cells do actively respond to such environmental stress by up-regulating the synthesis of a number of proteins [84,85]. Despite these reports, our current understanding of the biological significance of such changes in protein and mRNA expression is extremely limited. However, as further and more advanced 'omic' studies are reported, the data derived from such experiments is likely to be key in furthering our understanding of the changes in functional gene expression and protein modification(s) implicated in the molecular response to, and recovery from, cold-shock in cultured mammalian cells. Further, such data will be pertinent to developing our understanding of how these cold-shock responses are coordinated and controlled, and will provide novel targets for cell engineering.

5.5   The Use of Mild-Hypothermic Temperatures for the Enhanced Production of Heterologous Proteins from *in Vitro* Cultured Mammalian Cell Systems

5.5.1   *The effect of reduced temperature on cell specific and total product yields from mammalian expression systems*

Ever since mammalian cell systems have been utilized for the production of target products, investigators have manipulated culture variables in an effort to increase product yields. Culture temperature as a variable has been investigated for many years. The earliest experiments investigating the effect of sub-physiological temperatures on mammalian cell systems were reported more than forty years ago[86]. Initial investigations into the effect of *in vitro* sub-physiological temperature culturing on target product yield were rather negative and showed at best no change in product yield, and at worst a decrease in cell specific productivity[87]. However, in the last twenty years there have been repeated claims and reports that mild *in vitro* hypothermic temperature culturing (25–35 °C) of mammalian cells has a positive effect on cell specific heterologous protein production and/or protein/product yield. Table 2 below summarizes the findings from a number of these reports.

What is strikingly obviously from the data presented in Table 2 are the variable effects on cell specific and product yield reported. Indeed, it is now widely acknowledged, particularly in CHO cells where most work has been carried out to date, that the effect of mild-hypothermic temperatures on cell specific and total product yield is cell line specific[87]. For example, Yoon and colleagues showed that in parental CHO clones expressing a humanized monoclonal antibody, culturing at 32 °C resulted in a 4- to 25-fold increase in the cell specific productivity observed[88]. However, upon amplification of the parental clones, amplified clones showed a variable cell specific productivity response to culturing at 32 °C, with some exhibiting an increase over that observed at 37 °C whilst several showed no enhancement at all[88].

Table 2. The effect of mild cold-shock upon cell specific productivity and product yield compared to that observed at 37 °C when using mammalian expression systems

Expression System	Temp (°C)	Product	Effect on qP*	Total Yield Effect	Comments	Source
Hybridoma	33	MAb	21% ↓			Barnabe and Butler[90]
Hybridoma	34	Anti-interleukin-2	None			Bloemkolk et al.[91]
Hybridoma	Various	MAb	↓			Reuveny et al.[92]
Hybridoma	29	MAb	-	95% ↓		Sureshkumar & Mutharasan[93]
	33		42% ↓	20% ↓		
	35		No X	10% ↑		
NIH3T3	32	Amphotropic MLV-A provirus		10 fold ↓		Beer et al.[83]
	32	Amphotropic MLV-A provirus		5–10 fold ↑		Kaptein et al.[94]
BHK-21	Various	Antithrombin III		No X		Weidemann et al.[95]
CHO	33	Endogenous retrovirus	1.5 log$_{10}$ particles/cell/day			Brorson et al.[96]
CHO	34	Not disclosed	2–3 fold ↑		Cell line specific. Decrease for others	Chuppa et al.[97]
CHO	32	Not disclosed	6 fold ↑			Ducommun et al.[98]
CHO	33	Granulocyte macrophage colony stimulating factor	2.1 fold ↑	2.3 fold ↑		Fogolin et al.[99]
CHO	32	Interferon-γ	2 fold ↑	40–90% ↑		Fox et al.[87]
CHO	32	Interferon-γ		4.9–7.7 fold ↑	Under conditions of active growth	Fox et al.[100,101]
CHO	35	C-terminal amidating enzyme		1.8 fold ↑		Furukawa and Ohsuye[102]

(*continued*)

Table 2. (Continued)

Expression System	Temp (°C)	Product	Effect on qP*	Total Yield Effect	Comments	Source
CHO	32	Tissular plasminogen activator		1.7 fold ↑		Hendrick et al.[103]
CHO	30	SEAP	1.7 fold ↑	3.4 fold ↑		Kaufmann[85]
CHO	28	Chimeric Fab		14 fold ↑	14-fold in serum, 38-fold in serum free medium	Schatz et al.[104]
				38 fold ↑		
CHO	30	Erythropoietin	5.6 fold ↑			Yon et al.[105]
	33		4 fold ↑	2.5 fold ↑		
CHO	30	Anti-4-1BB	No X	3.9 fold ↓		Yoon et al.[89]
	33		1.2 fold ↑	No X		
CHO	32	Humanized antibody	4–25 fold ↑		Parental clones 4–25 fold ↑, amplified clones variable	Yoon et al.[88]
CHO	32.5	Erythropoietin		1.4-fold ↑	pH 7	Yoon et al.[106]

*Cell specific productivity; ↓, decrease; ↑, increase; Mab, monoclonal antibody; No X, no change

Further, although some of the amplified clones did exhibit enhanced cell specific productivity at 32 °C, the enhancement was reduced in all those amplified clones relative to that observed in the parental clones[88]. This data suggests that clonal selection plays a crucial role in deriving CHO cell lines that exhibit enhanced cell specific productivity at 32 °C relative to that observed at 37 °C.

Further evidence of the CHO cell line specific effect is detailed in Table 2. Perhaps even more striking than the CHO cell line specific effect is the difference in the data presented for the CHO and hybridoma cell lines. Although not always the case, in general mild hypothermic culturing of CHO cells appears to result in enhanced cell specific productivity, if not enhanced product yield, whilst the effect on hybridoma cells would appear to always be detrimental with regard to cell specific and volumetric productivity (Table 2). The reasons for this discrepancy are currently unclear, but may be related to the nature of the product. Antibodies are one of the most complex products to be expressed in mammalian cell systems. Successful expression requires the coordinated synthesis of heavy and light chain transcripts, followed by translation, folding and assembly of the chains to yield the final, biologically active product. The multi-chain and multi-step process may be compromised at reduced temperatures. The fact that production of an antibody in CHO cells at mild hypothermic temperatures was also shown to have little, or a detrimental, effect on productivity[89], further suggests that multi-chain antibody production is compromised at reduced temperatures.

5.5.2   How does sub-physiological temperature in vitro culturing of mammalian cells lead to enhanced cell specific productivity?

Sub-physiological temperature culturing does appear to offer the potential for enhancing cell specific and volumetric productivity of a number of target proteins. However, the exact mechanism(s) by which recombinant protein production may be increased, whilst global transcription and translation are reduced, are currently unknown. One of the earliest suggestions was that at lower temperatures extra- and intra-cellular proteolytic attack of recombinant proteins might be reduced, and as such protein degradation would also be reduced[97]. However, recent studies suggest that such an effect does not play a significant role in the enhanced productivity observed at 32 °C relative to that at 37 °C in CHO cells producing interferon-γ[100].

A further suggested mechanism by which mild hypothermic conditions may enhance cell specific and/or volumetric productivity was via cell cycle arrest. As discussed above, although mammalian cells can continue to grow at mild hypothermic temperatures, they tend to arrest in the G0/G1 phase of the cell cycle[85]. Interestingly, over-expression of the mammalian cold-shock protein Cirp results in cell cycle arrest at 37 °C[1]. Previous reports have correlated cell cycle arrest with enhanced productivity in mammalian cell systems[85,107], and a number of investigators have suggested modulating the cell cycle as a method for improving recombinant protein production from mammalian cell systems[108]. The theory is that upon arrest, cellular proliferation processes that use large amounts of energy are ceased; 'freeing up' energy and cellular machinery that can be utilized for, and redirected towards, recombinant protein production. However, it now seems unlikely that cell cycle arrest is a major contributor to the increase in recombinant protein production observed at lower temperatures.

There are several reasons why cell cycle arrest now seems to be a minor player in enhanced recombinant protein production at reduced culture temperature. Firstly, if cell cycle arrest was the predominant mechanism leading to enhanced productivity one would expect to see a much more reproducible effect across cell lines. Secondly, it is reported that hybridoma cell lines actually exhibit highest cell specific productivities in the G0/G1 cell cycle phase[109]. Despite such reports, Table 2 shows that reduced temperature cultivation of hybridoma cells adversely effects productivity, despite the fact that such conditions should induce cell cycle arrest in the G0/G1, most productive phase, of the cell cycle. Finally, a recent study has shown that active hypothermic growth results in enhanced recombinant protein productivity in CHO cells[100,101]. Fox and colleagues showed that specific productivity was optimized at 32 °C when cells were actively growing, using the percentage of S-phase cells as an indicator of active growth[100,101]. Productivity

was highest when cells were both actively growing, with a relatively high proportion of cells in S-phase of the cell cycle, and when also exposed to reduced
temperature $(32\,°C)$[100,101].

The use of conditions that promote active growth at $32\,°C$ could therefore be
used to circumvent the adverse effect of the reduced number of viable cells at such
temperatures. Although mild-hypothermic temperatures may lead to an increase
in cell specific productivity, the resulting cell cycle arrest that such conditions
induce ultimately results in lower cell concentrations. Thus, any improvement in
cell specific productivity may be more than offset by a decrease in the total number
of viable cells, leading to a drop in volumetric or total product yield. For example,
despite the fact that Table 2 shows that Furukawa and Ohsuye reported an increase
in the product yield of C-terminal amidating enzyme of 1.8-fold at $35\,°C$ relative
to that observed at $37\,°C$, the highest cell specific productivity was observed at
$32\,°C$. At $32\,°C$ CHO cells expressing the C-terminal amidating enzyme had a cell
specific productivity more than 50% of that observed at $35\,°C$[102]. On-the-other-
hand, mammalian cells cultured at lower temperatures exhibit prolonged viability
(compared to $37\,°C$), and this may partially offset the reduced maximum viable cell
count effect.

Fox and colleagues suggest that the use of low temperature with conditions
or systems that promote active growth at such temperatures may be a method of
optimizing and enhancing product yield[100]. Such approaches could harness the beneficial enhancement in cell specific productivity that mild-hypothermic temperatures
impart, whilst improving overall or total yield by maximizing the number of viable
cells with enhanced productivity characteristics. An alternative strategy suggested
by the same group is one of temperature shift optimization. Such a strategy involves
allowing active cell growth at $37\,°C$ until a predetermined optimized point at which
time the temperature is shifted to $32\,°C$[87]. This biphasic approach initially utilizes
an active growth phase at $37\,°C$ whereby cell biomass is allowed to accumulate,
followed by a second reduced temperature phase to achieve maximum cell specific
productivity of the accumulated biomass.

Despite the development of such approaches, the detailed molecular mechanisms
at play that result in enhanced recombinant protein production at reduced temperatures remain unknown. Studies to date in mammalian cell systems have not utilized
cold inducible promoters or elements that enhance translation upon cold-shock.
The reason why recombinant protein production is enhanced when global protein
synthesis is reduced therefore currently remains somewhat of a mystery. However,
recently the first light has been shed upon possible mechanisms that may, at least
partially, be responsible for the observed increase in cell specific productivity at
mild hypothermic temperatures.

The first hints as to the mechanisms at play leading to enhanced heterologous protein production at reduced temperature came from a study by Yoon and
colleagues. In this study it was reported that erythropoietin production from CHO
cells was increased by more than 2.5-fold at $33\,°C$ relative to that observed at
$37\,°C$, however, more importantly, relative erythropoietin mRNA content was also

increased[105]. This result suggests that either transcription of the target gene, or stability of the mRNA transcript, at 33 °C, or both, were increased at the lower temperature and could, at least partially, account for the improved cell specific productivity observed. In view of the importance of mRNA transcription and stability in the cold-shock response as discussed previously, it is perhaps not surprising that such factors may play a role in enhanced recombinant protein production at such temperatures.

More recently Fox and colleagues have reported that enhanced production of interferon-γ from CHO cells at reduced temperature (32 °C) is the result of elevated mRNA levels compared to those observed at 37 °C[101]. The mechanism by which these mRNA levels are increased has not yet been elucidated; however enhanced transcript stability is likely to be involved. However, the authors of this study showed that enhanced transcript stability appears to be a global cold-shock response[101], and this mechanism would not therefore explain why recombinant proteins were expressed at higher levels at 32 °C relative to endogenous proteins. The authors suggest that the viral promoter used in this study to control recombinant gene expression may play a crucial role by circumventing cold-shock control systems of endogenous promoters that reduce general transcription at such temperatures[101]. Thus, the recombinant gene may be transcribed at an enhanced rate at 32 °C relative to that of endogenous genes.

These observations therefore provide a working hypothesis that may be tested, and could provide a partial mechanistic explanation as to why recombinant proteins may be expressed at enhanced levels under mild hypothermic temperatures in the face of globally reduced protein synthesis. Further, there are few reports to date on mRNA levels upon cold-shock, and it remains to be seen if this relationship holds true for other recombinant genes in other cell lines. Finally, there appear to be no reports to date on mRNA levels of heavy and light chain genes in antibody producing cell lines or hybridoma cell lines. It therefore remains to be seen if the lack of effect observed to date in hybridoma and other cell lines engineered to produce antibodies under reduced temperature cultivation is a reflection of the mRNA transcript level, or whether other mechanisms prevent an increase in productivity at reduced temperatures.

We also note that sub-physiological temperature can be utilized to prolong the length of culture, and prevent or prolong the period of time before cell death. Such an approach has been investigated in an attempt to extend the production phase of *in vitro* cultured mammalian cells[89]. Usually, mammalian cells will rapidly die as a result of apoptosis late in culture, and this limits the length of the production phase in such expression systems[110]. For example, a shift in the culture temperature of CHO cells from 37 °C to 30 °C has been shown to delay the onset of apoptosis[28]. Further, expression of the anti-apoptotic protein Bcl-2 is reported to be enhanced in hippocampal neurons in response to mild hypothermia[111]. However, like many of the effects upon cold-shock, the effect of reduced temperature on apoptosis varies from one cell line to another and others have reported that reduced apoptosis upon temperature downshift is largely due to the global reduction in cellular metabolism[28].

The role (if any) of an apoptotic response, or the mechanism by which this is activated upon cold-shock, therefore remains to be elucidated.

6.   FUTURE ENGINEERING OPPORTUNITIES
IN THE COLD-SHOCK FIELD

Our understanding of the molecular mechanisms involved in the cold-shock response across a large range of organisms has come a long way in the last few years. However, in terms of cell engineering, the delicate balance between the mechanisms at play during the cold-shock response, the definitive role of those cold-shock proteins that are selectively expressed at sub-physiological temperatures, and the relationship between the *in vitro* reduced temperature culturing of bacterial, yeast and mammalian cells and the rate of recombinant protein production remain unclear. Thus, although we now have a broad understanding of the molecular responses to cold-shock in a number of organisms and expression systems, if these cold-shock responses are to be actively harnessed for the further enhancement of recombinant protein production it will be necessary to further develop this understanding. That more investigators than ever before are now addressing the effect of sub-physiological temperatures at the cellular level bodes well for the continued development of our understanding of the molecular responses governing cellular adaptation to cold-shock.

We predict that future cell engineering strategies in this area are likely to involve harnessing elements of the promoter regions and untranslated regions of specific cold-shock proteins to drive the expression of target genes and proteins at reduced temperatures, particularly in *in vitro* cultured mammalian cell systems. Further, such elements could be utilized for the tightly controlled inducible expression of target proteins and/or toxic targets. Additional strategies may include optimization of bi-phasic production approaches whereby cells are originally grown at 37 °C until a critical biomass is achieved, and then lowering the temperature to further enhance and induce target expression. The development of cell lines where active growth is promoted at lower temperatures may also result in bioprocessing benefits at reduced temperatures. Finally, as mRNA stability and translation have been shown to play an important role in the cold-shock response, we suggest that future developments in this area are likely to offer opportunities for cell engineering. This could, for example, involve the use of cold-specific IRESes to enhance translational efficiency of specific mRNAs at reduced temperatures, or manipulation of the 5′-untranslated region of specific transcripts to enhance translation and alter stability at reduced temperatures.

7.   REFERENCES

1. J. Fujita, Cold shock response in mammalian cells, *J Mol Microbiol Biotechnol* 1, 243–255 (1999).
2. T. Sahara, T. Goda and S. Ohgiya, Comprehensive expression analysis of time-dependent genetic responses in yeast cells to low temperature, *J Biol Chem* 277, 50015–50021 (2002).

3. M. B. Al-Fageeh, R. J. Marchant, M. J. Carden and C. M. Smales, The cold-shock response in cultured mammalian cells: Harnessing the response for the improvement of recombinant protein production, *Biotechnol Bioeng* 22 (2006).

4. S. Phadtare, Recent developments in bacterial cold-shock response, *Curr Issues Mol Biol* 6, 125–136 (2004).

5. C. O. Gualerzi, A. M. Giuliodori and C. L. Pon, Transcriptional and post-transcriptional control of cold-shock genes, *J Mol Biol* 331, 527–539 (2003).

6. D. N. Ermolenko and G. I. Makhatadze, Bacterial cold-shock proteins, *Cell Mol Life Sci* 59, 1902–1913 (2002).

7. A. M. Giuliodori, A. Brandi, C. O. Gualerzi and C. L. Pon, Preferential translation of cold-shock mRNAs during cold adaptation, *RNA* 10, 265–276 (2004).

8. E. L. Golovlev, Bacterial cold shock response at the level of DNA transcription, translation and chromosome dynamics, *Mikrobiologiia* 72, 5–13 (2003).

9. T. Homma, H. Iwahashi and Y. Komatsu, Yeast gene expression during growth at low temperature, *Cryobiology* 46, 230–237 (2003).

10. Y. Murata, T. Homma, E. Kitagawa, Y. Momose, M. S. Sato, M. Odani, H. Shimizu, M. Hasegawa-Mizusawa, R. Matsumoto, S. Mizukami, K. Fujita, M. Parveen, Y. Komatsu and H. Iwahashi, Genome-wide expression analysis of yeast response during exposure to 4 degrees C, *Extremophiles*, (2006) published online, *in press*.

11. S. Phadtare, J. Alsina and M. Inouye, Cold-shock response and cold-shock proteins, *Curr Opin Microbiol* 2, 175–180 (1999).

12. M. H. Weber and M. A. Marahiel, Bacterial cold shock responses, *Sci Prog* 86, 9–75 (2003).

13. N. N. Ulusu and E. F. Tezcan, Cold Shock Proteins, *Turk J Med Sci* 31, 283–290 (2001).

14. C. L. Rieder and R. W. Cole, Cold-shock and the Mammalian cell cycle, *Cell Cycle* 1, 169–175 (2002).

15. P. G. Jones, R. A. VanBogelen and F. C. Neidhardt, Induction of proteins in response to low temperature in Escherichia coli, *J Bacteriol* 169, 2092–2095 (1987).

16. J. R. Hazel, Thermal adaptation in biological membranes: is homeoviscous adaptation the explanation?, *Annu Rev Physiol* 57, 19–42 (1995).

17. S. V. Avery, D. Lloyd and J. L. Harwood, Temperature-dependent changes in plasma-membrane lipid order and the phagocytotic activity of the amoeba Acanthamoeba castellanii are closely correlated, *Biochem J* 312, 811–816 (1995).

18. S. M. Carty, K. R. Sreekumar and C. R. Raetz, Effect of cold shock on lipid A biosynthesis in Escherichia coli. Induction at 12 degrees C of an acyltransferase specific for palmitoleoyl-acyl carrier protein, *J Biol Chem* 274, 9677–9685 (1999).

19. A. R. Cossins and A. G. Macdonald, The adaptation of biological membranes to temperature and pressure: fish from the deep and cold, *J Bioenerg Biomembr* 21, 115–135 (1989).

20. B. F. Dickens and G. A. Thompson, Jr., Rapid membrane response during low-temperature acclimation. Correlation of early changes in the physical properties and lipid composition of Tetrahymena microsomal membranes, *Biochim Biophys Acta* 644, 211–218 (1981).

21. S. Phadtare and M. Inouye, Genome-wide transcriptional analysis of the cold shock response in wild-type and cold-sensitive, quadruple-csp-deletion strains of Escherichia coli, *J Bacteriol.* 186, 7007–7014 (2004).

22. A. Polissi, W. De Laurentis, S. Zangrossi, F. Briani, V. Longhi, G. Pesole and G. Deho, Changes in Escherichia coli transcriptome during acclimatization at low temperature, *Res Microbiol* 154, 573–580 (2003).

23. Y. H. Kim, K. Y. Han, K. Lee and J. Lee, Proteome response of Escherichia coli fed-batch culture to temperature downshift, *Appl Microbiol Biotechnol* 68, 786–793 (2005).

24. S. Topanurak, S. Sinchaikul, B. Sookkheo, S. Phutrakul and S. T. Chen, Functional proteomics and correlated signaling pathway of the thermophilic bacterium Bacillus stearothermophilus TLS33 under cold-shock stress, *Proteomics* 5, 4456–4471 (2005).

25. D. Stubs, T. M. Fuchs, B. Schneider, A. Bosserhoff and R. Gross, Identification and regulation of cold-inducible factors of Bordetella bronchiseptica, *Microbiology.* 151, 1895–1909 (2005).

26. Y. Han, D. Zhou, X. Pang, L. Zhang, Y. Song, Z. Tong, J. Bao, E. Dai, J. Wang, Z. Guo, J. Zhai, Z. Du, X. Wang, P. Huang and R. Yang, DNA microarray analysis of the heat- and cold-shock stimulons in Yersinia pestis, *Microbes Infect* 7, 335–348 (2005).

27. Y. Ohsaka, S. Ohgiya, T. Hoshino and K. Ishizaki, Phosphorylation of c-Jun N-terminal kinase in human hepatoblastoma cells is transiently increased by cold exposure and further enhanced by subsequent warm incubation of the cells, *Cell Physiol Biochem* 12, 111–118 (2002).

28. A. Moore, J. Mercer, G. Dutina, C. J. Donahue, K. D. Bauer, J. P. Mather, T. Etcheverry and T. Ryll, Effects of temperature shift on cell cycle, apoptosis and nucleotide pools in CHO cell batch cultues, *Cytotechnology* 23, 47–54 (1997).

29. Y. Gon, S. Hashimoto, K. Matsumoto, T. Nakayama, I. Takeshita and T. Horie, Cooling and rewarming-induced IL-8 expression in human bronchial epithelial cells through p38 MAP kinase-dependent pathway, *Biochem Biophys Res Commun* 249, 156–160 (1998).

30. S. Wellmann, C. Buhrer, E. Moderegger, A. Zelmer, R. Kirschner, P. Koehne, J. Fujita and K. Seeger, Oxygen-regulated expression of the RNA-binding proteins RBM3 and CIRP by a HIF-1-independent mechanism, *J Cell Sci* 117, 1785–1794 (2004).

31. H. S. Yun, J. Hong and H. C. Lim, Regulation of Ribosome Synthesis in *Escherichia coli*: Effects of Temperature and Dilution Rate Changes, *Biotechnol Bioeng* 52, 615–624 (1996).

32. W. Jiang, Y. Hou and M. Inouye, CspA, the major cold-shock protein of Escherichia coli, is an RNA chaperone, *J Biol Chem* 272, 196–202 (1997).

33. M. Inouye and S. Phadtare, Cold shock response and adaptation at near-freezing temperature in microorganisms, *Sci STKE* 2004(237), pe26 (2004).

34. A. Brandi, P. Pietroni, C. O. Gualerzi and C. L. Pon, Post-transcriptional regulation of CspA expression in Escherichia coli, *Mol Microbiol* 19, 231–240 (1996).

35. J. P. Etchegaray and M. Inouye, CspA, CspB, and CspG, major cold shock proteins of Escherichia coli, are induced at low temperature under conditions that completely block protein synthesis, *J Bacteriol* 181, 1827–1830 (1999).

36. K. Nakashima, K. Kanamaru, T. Mizuno and K. Horikoshi, A novel member of the cspA family of genes that is induced by cold shock in Escherichia coli, *J Bacteriol* 178, 2994–2997 (1996).

37. S. Phadtare and K. Severinov, Extended -10 motif is critical for activity of the cspA promoter but does not contribute to low-temperature transcription, *J Bacteriol.* 187, 6584–6589. (2005).

38. C. Yang and F. Carrier, The UV-inducible RNA-binding protein A18 (A18 hnRNP) plays a protective role in the genotoxic stress response, *J Biol Chem* 276, 47277–47284 (2001).

39. R. Hofweber, G. Horn, T. Langmann, J. Balbach, W. Kremer, G. Schmitz and H. R. Kalbitzer, The influence of cold shock proteins on transcription and translation studied in cell-free model systems, *FEBS J* 272, 4691–4702 (2005).

40. H. Schindelin, W. Jiang, M. Inouye and U. Heinemann, Crystal structure of CspA, the major cold shock protein of Escherichia coli, *Proc Natl Acad Sci U S A* 91, 5119–5123 (1994).

41. J. Goldstein, N. S. Pollitt and M. Inouye, Major cold shock protein of Escherichia coli, *Proc Natl Acad Sci U S A* 87, 283–287 (1990).

42. H. Tanabe, J. Goldstein, M. Yang and M. Inouye, Identification of the promoter region of the Escherichia coli major cold shock gene, cspA, *J Bacteriol* 174, 3867–3873 (1992).

43. M. Mitta, L. Fang and M. Inouye, Deletion analysis of cspA of Escherichia coli: requirement of the AT-rich UP element for cspA transcription and the downstream box in the coding region for its cold shock induction, *Mol Microbiol* 26, 321–335 (1997).

44. K. Yamanaka, M. Mitta and M. Inouye, Mutation analysis of the 5' untranslated region of the cold shock cspA mRNA of Escherichia coli, *J Bacteriol* 181, 6284–6291 (1999).

45. L. Fang, W. Jiang, W. Bae and M. Inouye, Promoter-independent cold-shock induction of cspA and its derepression at 37 degrees C by mRNA stabilization, *Mol Microbiol* 23, 355–364 (1997).

46. W. Jiang, L. Fang and M. Inouye, The role of the 5'-end untranslated region of the mRNA for CspA, the major cold-shock protein of Escherichia coli, in cold-shock adaptation, *J Bacteriol* 178, 4919–4925 (1996).

47. B. Xia, H. Ke, W. Jiang and M. Inouye, The Cold Box stem-loop proximal to the 5'-end of the Escherichia coli cspA gene stabilizes its mRNA at low temperature, *J Biol Chem* 277, 6005–6011 (2002).

48. G. Qing, L. C. Ma, A. Khorchid, G. V. Swapna, T. K. Mal, M. M. Takayama, B. Xia, S. Phadtare, H. Ke, T. Acton, G. T. Montelione, M. Ikura and M. Inouye, Cold-shock induced high-yield protein production in Escherichia coli, *Nat Biotechnol* 22, 877–882 (2004).

49. S. Derzelle, B. Hallet, T. Ferain, J. Delcour and P. Hols, Improved adaptation to cold-shock, stationary-phase, and freezing stresses in Lactobacillus plantarum overproducing cold-shock proteins, *Appl Environ Microbiol* 69, 4285–4290 (2003).

50. A. Vila-Sanjurjo, B. S. Schuwirth, C. W. Hau and J. H. Cate, Structural basis for the control of translation initiation during stress, *Nat Struct Mol Biol* 11, 1054–1059 (2004).

51. M. H. Weber, A. V. Volkov, I. Fricke, M. A. Marahiel and P. L. Graumann, Localization of cold shock proteins to cytosolic spaces surrounding nucleoids in Bacillus subtilis depends on active transcription, *J Bacteriol* 183, 6435–6443 (2001).

52. P. Florey, A. J. Smallridge, A. Ten and M. A. Trewhella, Chemo- and stereoselective reduction of an alpha-cyanoketone by bakers' yeast at low temperature, *Org Lett* 1, 1879–1880 (1999).

53. N. Bonander, K. Hedfalk, C. Larsson, P. Mostad, C. Chang, L. Gustafsson and R. M. Bill, Design of improved membrane protein production experiments: quantitation of the host response, *Protein Sci* 14, 1729–1740 (2005).

54. O. Kandror, N. Bretschneider, E. Kreydin, D. Cavalieri and A. L. Goldberg, Yeast adapt to near-freezing temperatures by STRE/Msn2,4-dependent induction of trehalose synthesis and certain molecular chaperones, *Mol Cell* 13, 771–781 (2004).

55. M. Odani, Y. Komatsu, S. Oka and H. Iwahashi, Screening of genes that respond to cryopreservation stress using yeast DNA microarray, *Cryobiology* 47, 155–164 (2003).

56. O. Kandror, A. DeLeon and A. L. Goldberg, Trehalose synthesis is induced upon exposure of Escherichia coli to cold and is essential for viability at low temperatures, *Proc Natl Acad Sci U S A.* 99, 9727–9732 (2002).

57. S. Rodriguez-Vargas, F. Estruch and F. Randez-Gil, Gene expression analysis of cold and freeze stress in Baker's yeast, *Appl Environ Microbiol* 68, 3024–3030 (2002).

58. L. Zhang, A. Ohta, H. Horiuchi, M. Takagi and R. Imai, Multiple mechanisms regulate expression of low temperature responsive (LOT) genes in Saccharomyces cerevisiae, *Biochem Biophys Res Commun* 283, 531–535 (2001).

59. T. J. Griffin, S. P. Gygi, T. Ideker, B. Rist, J. Eng, L. Hood and R. Aebersold, Complementary profiling of gene expression at the transcriptome and proteome levels in Saccharomyces cerevisiae, *Mol Cell Proteomics* 1, 323–333 (2002).

60. S. P. Gygi, Y. Rochon, B. R. Franza and R. Aebersold, Correlation between protein and mRNA abundance in yeast, *Mol Cell Biol* 19, 1720–1730 (1999).

61. G. C. Sieck, Molecular biology of thermoregulation, *J Appl Physiol* 92, 1365–1366 (2002).

62. L. A. Sonna, J. Fujita, S. L. Gaffin and C. M. Lilly, Effects of heat and cold stress on mammalian gene expression, *J Appl Physiol* 92, 1725–1742 (2002).

63. F. Van Breukelen and S. L. Martin, Invited review: molecular adaptations in mammalian hibernators: unique adaptations or generalized responses?, *J Appl Physiol* 92, 2640–2647 (2002).

64. A. Y. Liu, H. Bian, L. E. Huang and Y. K. Lee, Transient cold shock induces the heat shock response upon recovery at 37 degrees C in human cells, *J Biol Chem* 269, 14768–14775 (1994).

65. J. Dresios, A. Aschrafi, G. C. Owens, P. W. Vanderklish, G. M. Edelman and V. P. Mauro, Cold stress-induced protein Rbm3 binds 60S ribosomal subunits, alters microRNA levels, and enhances global protein synthesis, *Proc Natl Acad Sci U S A* 102, 1865–1870 (2005).

66. S. Danno, H. Nishiyama, H. Higashitsuji, H. Yokoi, J. H. Xue, K. Itoh, T. Matsuda and J. Fujita, Increased transcript level of RBM3, a member of the glycine-rich RNA-binding protein family, in human cells in response to cold stress, *Biochem Biophys Res Commun* 236, 804–807 (1997).

67. J. H. Xue, K. Nonoguchi, M. Fukumoto, T. Sato, H. Nishiyama, H. Higashitsuji, K. Itoh and J. Fujita, Effects of ischemia and H2O2 on the cold stress protein CIRP expression in rat neuronal cells, *Free Radic Biol Med* 27, 1238–1244 (1999).

68. H. Nishiyama, S. Danno, Y. Kaneko, K. Itoh, H. Yokoi, M. Fukumoto, H. Okuno, J. L. Millan, T. Matsuda, O. Yoshida and J. Fujita, Decreased expression of cold-inducible RNA-binding protein (CIRP) in male germ cells at elevated temperature, *Am J Pathol* 152, 289–296 (1998).

69. H. Nishiyama, J. H. Xue, T. Sato, H. Fukuyama, N. Mizuno, T. Houtani, T. Sugimoto and J. Fujita, Diurnal change of the cold-inducible RNA-binding protein (Cirp) expression in mouse brain, *Biochem Biophys Res Commun* 245, 534–538 (1998).

70. H. Nishiyama, K. Itoh, Y. Kaneko, M. Kishishita, O. Yoshida and J. Fujita, A glycine-rich RNA-binding protein mediating cold-inducible suppression of mammalian cell growth, *J Cell Biol* 137, 899–908 (1997).

71. H. K. Tan, M. G. S. Yap and D. I. C. Wang, CIRP expression on growth and productivity of CHO cells, *MIT on line Library* (2005).

72. S. Danno, K. Itoh, T. Matsuda and J. Fujita, Decreased expression of mouse Rbm3, a cold-shock protein, in Sertoli cells of cryptorchid testis, *Am J Pathol* 156, 1685–1692 (2000).

73. S. A. Chappell, G. C. Owens and V. P. Mauro, A 5' Leader of Rbm3, a Cold Stress-induced mRNA, Mediates Internal Initiation of Translation with Increased Efficiency under Conditions of Mild Hypothermia, *J Biol Chem* 276, 36917–36922 (2001).

74. S. A. Chappell and V. P. Mauro, The internal ribosome entry site (IRES) contained within the RNA-binding motif protein 3 (Rbm3) mRNA is composed of functionally distinct elements, *J Biol Chem* 278, 33793–33800 (2003).

75. R. H. Burdon, Temperature and animal cell protein synthesis, *Symp Soc Exp Biol* 41, 113–133 (1987).

76. R. C. Wek, H. Y. Jiang and T. G. Anthony, Coping with stress: eIF2 kinases and translational control, *Biochem Soc Trans* 34, 7–11 (2006).

77. M. F. Underhill, J. R. Birch, C. M. Smales and L. H. Naylor, eIF2alpha phosphorylation, stress perception, and the shutdown of global protein synthesis in cultured CHO cells, *Biotechnol Bioeng* 89, 805–814 (2005).

78. R. Stapulionis, S. Kolli and M. P. Deutscher, Efficient mammalian protein synthesis requires an intact F-actin system, *J Biol Chem* 272, 24980–24986 (1997).

79. E. Barbarese, D. E. Koppel, M. P. Deutscher, C. L. Smith, K. Ainger, F. Morgan and J. H. Carson, Protein translation components are colocalized in granules in oligodendrocytes, *J Cell Sci* 108, 2781–2790 (1995).

80. G. Liu, W. M. Grant, D. Persky, V. M. Latham, Jr., R. H. Singer and J. Condeelis, Interactions of elongation factor 1alpha with F-actin and beta-actin mRNA: implications for anchoring mRNA in cell protrusions, *Mol Biol Cell* 13, 579–592 (2002).

81. R. C. Weisenberg, Microtubule formation in vitro in solutions containing low calcium concentrations, *Science* 177, 1104–1105 (1972).

82. Y. Uesono, M. P. Ashe and E. A. Toh, Simultaneous yet independent regulation of actin cytoskeletal organization and translation initiation by glucose in Saccharomyces cerevisiae, *Mol Biol Cell* 15, 1544–1556 (2004).

83. C. Beer, P. Buhr, H. Hahn, D. Laubner and M. Wirth, Gene expression analysis of murine cells producing amphotropic mouse leukaemia virus at a cultivation temperature of 32 and 37 degrees C, *J Gen Virol* 84, 1677–1686 (2003).

84. J. Y. Baik, M. S. Lee, S. R. An, S. K. Yoon, E. J. Joo, Y. H. Kim, H. W. Park and G. M. Lee, Initial transcriptome and proteome analyses of low culture temperature-induced expression in CHO cells producing erythropoietin, *Biotechnol Bioeng* 93, 361–371 (2006).

85. H. Kaufmann, X. Mazur, M. Fussenegger and J. E. Bailey, Influence of low temperature on productivity, proteome and protein phosphorylation of CHO cells, *Biotechnol Bioeng* 63, 573–582 (1999).

86. P. N. Rao and J. Engelberg, HeLa cells: Effects of temperature on the life cycle, *Science* 148, 1092–1094 (1965).

87. S. R. Fox, U. A. Patel, M. G. Yap and D. I. Wang, Maximizing interferon-gamma production by Chinese hamster ovary cells through temperature shift optimization: experimental and modeling, *Biotechnol Bioeng* 85, 177–184 (2004).

88. S. K. Yoon, S. O. Hwang and G. M. Lee, Enhancing effect of low culture temperature on specific antibody productivity of recombinant Chinese hamster ovary cells: clonal variation, *Biotechnol. Prog.* 20, 1683–1688 (2004).

89. S. K. Yoon, S. H. Kim and G. M. Lee, Effect of low culture temperature on specific productivity and transcription level of anti-4-1BB antibody in recombinant Chinese hamster ovary cells, *Biotechnol Prog* 19, 1383–1386 (2003).

90. N. Barnabe and M. Butler, Effect of Temperature on Nucleotide Pools and Monoclonal-Antibody Production in a Mouse Hybridoma, *Biotechnol Bioeng* 44, 1235–1245 (1994).

91. J. W. Bloemkolk, M. R. Gray, F. Merchant and T. R. Mosmann, Effect of Temperature on Hybridoma Cell-Cycle and Mab Production, *Biotechnol Bioeng* 40, 427–431 (1992).

92. S. Reuveny, D. Velez, J. D. Macmillan and L. Miller, Factors affecting cell growth and monoclonal antibody production in stirred reactors, *J Immunol Methods* 86, 53–59 (1986).

93. G. K. Sureshkumar and R. Mutharasan, The Influence of Temperature on a Mouse Mouse Hybridoma Growth and Monoclonal-Antibody Production, *Biotechnol Bioeng* 37, 292–295 (1991).

94. L. C. Kaptein, A. E. Greijer, D. Valerio and V. W. van Beusechem, Optimized conditions for the production of recombinant amphotropic retroviral vector preparations, *Gene Ther* 4, 172–176 (1997).

95. R. Weidemann, A. Ludwig and G. Kretzmer, Low temperature cultivation–a step towards process optimisation, *Cytotechnology* 15, 111–116 (1994).

96. K. Brorson, C. De Wit, E. Hamilton, M. Mustafa, P. G. Swann, R. Kiss, R. Taticek, G. Polastri, K. E. Stein and Y. Xu, Impact of cell culture process changes on endogenous retrovirus expression, *Biotechnol Bioeng* 80, 257–267 (2002).

97. S. Chuppa, Y. S. Tsai, S. Yoon, S. Shackleford, C. Rozales, R. Bhat, G. Tsay, C. Matanguihan, K. Konstantinov and D. Naveh, Fermenter temperature as a pool for control of high-density perfusion cultures of mammalian cells, *Biotechnol Bioeng* 55, 328–338 (1997).

98. P. Ducommun, P. A. Ruffieux, A. Kadouri, U. von Stockar and I. W. Marison, Monitoring of temperature effects on animal cell metabolism in a packed bed process, *Biotechnol Bioeng* 77, 838–842 (2002).

99. M. B. Fogolin, R. Wagner, M. Etcheverrigaray and R. Kratje, Impact of temperature reduction and expression of yeast pyruvate carboxylase on hGM-CSF-producing CHO cells, *J. Biotechnol.* 109, 179–191 (2004).

100. S. R. Fox, M. X. Yap, M. G. Yap and D. I. Wang, Active hypothermic growth: a novel means for increasing total interferon-gamma production by Chinese-hamster ovary cells, *Biotechnol Appl Biochem* 41, 265–272 (2005).

101. S. R. Fox, H. K. Tan, M. C. Tan, S. Wong, M. G. S. Yap and D. I. C. Wang, A detailed understanding of the enhanced hypothermic productivity of interferon-gamma by Chinese-hamster ovary cells, *Biotechnol Appl Biochem* 41, 255–264 (2005).

102. K. Furukawa and K. Ohsuye, Effect of culture temperature on a recombinant CHO cell line producing a C-terminal alpha-amidating enzyme, *Cytotechnology* 26, 153–164 (1998).

103. V. Hendrick, P. Winnepenninckx, C. Abdelkafi, O. Vandeputte, M. Cherlet, T. Marique, G. Renemann, A. Loa, G. Kretzmer and J. Werenne, Increased productivity of recombinant tissular plasminogen activator (t-PA) by butyrate and shift of temperature: a cell cycle phases analysis, *Cytotechnology* 36, 71–83 (2001).

104. S. M. Schatz, R. J. Kerschbaumer, G. Gerstenbauer, M. Kral, F. Dorner and F. Scheiflinger, Higher expression of Fab antibody fragments in a CHO cell line at reduced temperature, *Biotechnol Bioeng* 84, 433–438 (2003).

105. S. K. Yoon, J. Y. Song and G. M. Lee, Effect of low culture temperature on specific productivity, transcription level, and heterogeneity of erythropoietin in Chinese hamster ovary cells, *Biotechnol Bioeng* 82, 289–298 (2003).

106. S. K. Yoon, S. L. Choi, J. Y. Song and G. M. Lee, Effect of culture pH on erythropoietin production by Chinese hamster ovary cells grown in suspension at 32.5 and 37.0 degrees C, *Biotechnol Bioeng* 89, 345–356 (2004).

107. M. Fussenegger, The impact of mammalian gene regulation concepts on functional genomic research, metabolic engineering, and advanced gene therapies, *Biotechnol Prog* 17, 1–51 (2001).

108. N. Ibarra, S. Watanabe, J. X. Bi, J. Shuttleworth and M. Al-Rubeai, Modulation of cell cycle for enhancement of antibody productivity in perfusion culture of NS0 cells, *Biotechnol Prog* 19, 224–228 (2003).

109. F. W. F. Lee, C. B. Elias, P. Todd and D. S. Kompala, Engineering Chinese hamster ovary (CHO) cells to achieve an inverse growth â<002>" associated production of a foreign protein, $\hat{I}^2$-galactosidase, *Cytotechnology* 28, 73–80 (1998).

110. J. Goswami, A. J. Sinskey, H. Steller, G. N. Stephanopoulos and D. I. Wang, Apoptosis in batch cultures of Chinese hamster ovary cells, *Biotechnol Bioeng* 62, 632–640 (1999).

111. Z. Zhang, R. A. Sobel, D. Cheng, G. K. Steinberg and M. A. Yenari, Mild hypothermia increases Bcl-2 protein expression following global cerebral ischemia, *Brain Res Mol Brain Res* 95, 75–85 (2001).

CHAPTER 7

MOLECULAR RESPONSE TO OSMOTIC SHOCK

SUSAN T. SHARFSTEIN, DUAN SHEN, THOMAS R. KIEHL
AND RUI ZHOU

Rensselaer Polytechnic Institute, Department of Chemical and Biological Engineering, 110 8th Street, Troy, NY 12180-3590

Abstract: The cellular responses of cultured mammalian cells and non-mammalian organisms to changes in osmolarity are discussed. A number of common themes including activation of protein kinase cascades can be observed in a diverse group of organisms. A combination of physiological and transcriptional studies has been performed to identify regulatory factors and proteins that play a causal role in the cellular responses to osmotic changes. These factors may serve as targets for cellular engineering strategies to improve the productivity of cultured mammalian cells, particularly in response to osmotic shock

Keywords: osmolarity, protein kinase cascades, microarray

1.   INTRODUCTION

All organisms respond and adapt to their external environments. Prokaryotes and lower eukaryotes may experience dramatic changes in their external conditions, necessitating a wide range of cellular responses in order to survive. Higher, multicellular eukaryotes have evolved a more sophisticated strategy for maintaining control of their extracellular environment, however they are still subject to perturbations, and some cell types may also experience extreme deviations. When mammalian cells are removed from their organismal context and placed into cell cultures, they may be subjected to a wide range of environmental changes, including variations in temperature, pH and osmolarity. The observation that perturbations in the cellular environment of cultured mammalian cells used as hosts to produce therapeutic or diagnostic proteins may increase specific productivities has created an impetus to understand the physiological responses to these environmental changes.

213

M. Al-Rubeai and M. Fussenegger (eds.), Systems Biology, 213–236.
© 2007 *Springer.*

While the details of cellular responses vary from organism to organism, there are several conserved motifs including extracellular sensors and protein kinase cascades for signal transduction. Our laboratory and others have attempted to decipher these signaling cascades with an ultimate objective of performing cellular engineering to optimize cellular responses to external perturbations, particularly hyperosmotic stress.

In this chapter, we review responses of both non-mammalian organisms and cultured mammalian cells to changes in environmental osmolarity. Both physiological changes and transcriptional changes whose physiological response may or may not be known are discussed. In particular, we highlight a microarray study performed in our laboratory and discuss the physiological phenomena observed both in our laboratory and others which support the transcriptional changes we observed.

2.   OSMOTIC RESPONSES IN NON-MAMMALIAN ORGANISMS

2.1   *Saccharomyces cerevisiae* and Other Yeast

Yeasts grow on numerous substrates and in many, rapidly changing environments, making effective osmoadaptation a necessary trait[1]. Budding yeast produce glycerol for osmotic stabilization via the high osmolarity glycerol signaling system (HOG)[2,3]. This system produces glycerol as a solute to decrease intracellular water potential. This mitogen activated protein (MAP) kinase pathway is controlled by two upstream osmosensing branches containing the two membrane bound proteins, Sln1p and Sho1p. These branches meet at the Pbs2 MAP kinase kinase and lead to the activation of the HOG1 MAPK as shown below in Figure 1.

Under normal conditions, the osmosensing histidine kinase Sln1 acts as an inhibitory signal to the HOG MAPK cascade. Sln1 responds with cytokinin receptor Cre1 to reduced turgor pressure[4]. Some find it curious that this branch utilizes the Sln1p-Ypd1p-Ssk1p phosphorelay system for osmosensing rather than a two-component system as in bacteria[1].

Reiser et al. state that activation of the HOG1 MAPK cascade under these conditions (reduced turgor pressure) is strictly through the SLN1 branch and not the SHO1 branch. Hohman draws some similarities between Sln1 and two bacterial histidine kinases, KdpD and EnvZ which are discussed briefly in section 2.3[1].

The function of the SHO1 branch is not as clear. It is apparent that Sho1p is not itself an osmosenser[5]. Raitt et al. describe a process whereby Sho1p binds Pbs2p, forming a membrane bound complex which may activate the GTPase Cdc42p as well as several other molecules which are involved in maintaining cell wall integrity including the PAK kinase(s) Ste20p and Cla4p, as well as the MAPKKK Ste11p[6]. More recent work by Uhlik et al. concludes that this branch, and its utilization of a CDC42-STE50-STE11-Pbs2 scaffold, bears similarity to the activation of mammalian p38 via the Rac-OSM-MEKK3-MKK3 scaffold[7] (described in Section 3.1).

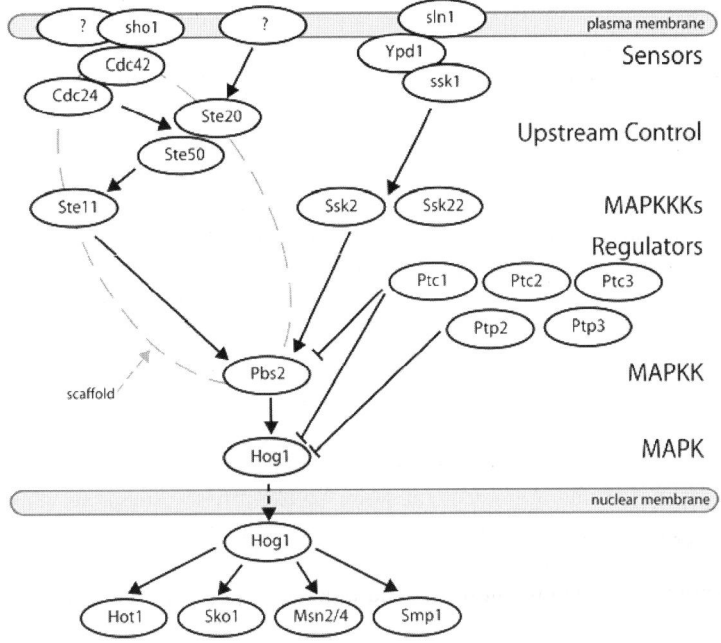

Figure 1. Outline of the HOG pathway

It is interesting to note the relative sensitivities of the two branches. The Sln1 branch is much more sensitive, with the SHO branch reacting only to very high osmolarity[1].

The resulting MAP kinase cascade activates Hog1p1. After Hog1p is phosphorylated it has been observed to rapidly, within one minute, translocate to the nucleus[8] which induces transcription of two isoforms of glycerol-3-phosphate dehydrogenase, GPD1 and GPD2, as well as two isoforms of glycerol-3-phosphotase, GPP1 and GPP2[9]. The net effect is the increased metabolism of glucose to glycerol, with the resulting increase in glycerol having an osmoprotective effect.

In order for this production of glycerol to have the desired affect, the glycerol channel aquaglyceroporin Fps1 is also regulated by osmolarity. Fps1 is closed by hyperosmotic stress and is highly active in hypoosmotic conditions. Under normal conditions Fps1 knockouts behave as though they are experiencing hypoosmotic conditions. Another osmotolerant yeast, *Z. rouxii*, employs an Fps1 homologue which, when expressed in *S. cerevisiae fps1Δ* mutant behaves similarly to wild-type *S. cerevisiae*.

The HOG pathway is modulated by phosphotyrosine phosphatases Ptp2p and Ptp3p and phosphoserine/threonine phosphatases Ptc1p and Ptc3p. These species act upon phosphorylated Hog1p to limit inappropriate activation of the HOG pathway[1].

The transcription factor, Smp1, has been identified as a downstream target for Hog1. Nadel et al. demonstrated that overexpression of *SMP1* induced a

Hog-1 dependant expression of osmoresponsive genes such as the sugar transporter homolog encoded by *STL1*[10]. Rep et al. have identified a number of other transcription factors that are expected to be controlled by Hog1p. These include Hot1p, Msn2p and Msn4p. They also demonstrated reduced induction of GPD1 and GPP2 in a *hot1* mutant[9].

In addition to these specific phenomena, Causten et al. identified 179 genes which are differentially expressed under sorbitol or salt stress[11]. Rep et al. found 186 genes which show at least a three-fold increase and 100 genes which decrease at least 1.5 fold in response to hyperosmolarity[9].

2.2   *Arabidopsis thaliana* and Other Plant Species

Plants encounter osmotic stresses in various forms such as under drought conditions and in cases of high soil salt concentrations. In protecting themselves from osmotic stress, plants employ various strategies for accumulating osmoprotectants while maintaining turgor and a gradient for water uptake. These osmoprotectants serve multiple purposes including the stablization of cellular structures and maintaining solubility of proteins during dehydrated conditions.

While osmotic responses in plants have been recently reviewed[12,13], the current science on *A. thaliana*, and plants in general, leaves many open questions as to specific signaling pathways related to stressors such as drought, cold and high salinity. The current understanding of signaling pathways and signal transduction is summarized in Figure 2 below.

Little has been determined in the way of specific osmosensors for plants. While *A. thaliana* harbors AtHK1, a homolog to the yeast histidine kinase SLN1, no direct evidence supports an osmosensing role for that protein[14]. It has been proposed that the tobacco *Nicotiana tabacum* NtC7 receptor-like protein may form dimers allowing it to sense changes in membrane architecture[15].

Calcium has been shown to be an important second messenger during stress. The Ca^{2+} signature varies across cell types and stress types[16]. There is a growing body of work describing the various players involved in transduction of salt stress signals via Ca^{2+}. Knight et al. demonstrated that a Ca^{2+} signal is initiated upon salt stress[17]. Zhu and Chinnusamy et al. have recently reviewed the salt-overly-sensitive (SOS) pathway in *A. thaliana*[12,18]. Within this pathway the components SOS1, SOS2, and SOS3 have been identified. SOS3 encodes a Ca^{2+}-binding protein which contains an N-myristoylation motif and three Ca^{2+}-binding EF hands. SOS2 encodes a serine/threonine protein kinase. SOS2 is activated by SOS3 and subsequently activates SOS1[19]. SOS1 encodes a plasma membrane Na^+/H^+ antiporter that is used to achieve sodium-ion homeostasis and to maintain salt tolerance. Additionally, the *A. thaliana* genome encodes at least eight SOS3-like Ca^{2+}-binding proteins and 23 proteins with significant homology to SOS2[12,20].

Many physiological processes in plants are regulated by the phytohormone abscisic acid (ABA)[21]. This hormone inhibits normal growth and increases stress tolerance and adaptation. Drought conditions activate many ABA-dependent and

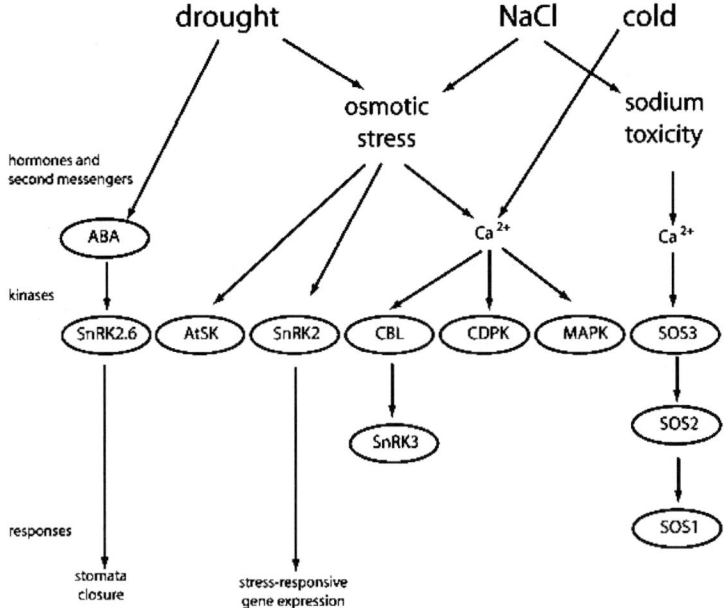

Figure 2. Stress regulation network in Arabidopsis

ABA-independent events[22]. Responses induced by ABA indicate that this hormone may play a role in osmoregulation. Stomatal closure is regulated by increased ABA levels induced by osmotic stress. Stomatal closure can even be induced by exogenous application of ABA[21]. One downstream family of ABA responsive genes contains an EF-hand Ca^{2+} binding domain[23].

As in other systems, MAP kinases make a good target for investigation in *A. thaliana*. One such kinase, ATMEKK1 has been shown to be rapidly induced within five minutes of stress treatment[24]. Covic et al. explored the effects of this kinase when induced in yeast and hypothesized that plants may harbor a pathway similar to the yeast HOG pathway.

2.3   *Escherichia coli* and Other Bacteria

Bacteria are subjected to a wide variety of environmental conditions requiring effective management of abiotic stress. As in plant and yeast species, bacterial responses to osmolarity are largely dependant on the organism's ability to accumulate osmoprotectants. A large number of transporters are believed to be involved in transporting osmoprotectants. The EnvZ/OmpR two-component system is one of the primary osmosensing systems of *E. coli*. EnvZ is a transmembrane histidine kinase which phosphorylates OmpR. Activated OmpR then functions as a transcription factor regulating the expression of the major porins OmpF and OmpC[25]. This regulatory connection has been well established going back to the work of Forst et al.[26]. These porins are also present in other bacteria such as *S. typhimurium*. Most recently these porins were further explored in *S. marcescens*[27]. It has also been reported that high osmotic strength and high temperature can increase production of OmpC while repressing OmpF in *E. coli*[28].

Trehalose accumulation is correlated with Rpos and glycine betaine uptake is mediated by transporters ProP and PropU[29]. BetU and TrkG contribute to osmoregulatory capacity of *E. coli*.[30]. MscL and MscS channels release pressure during hypoosmotic shock[31].

Some recent studies have sought to quantify the ability of the *E. coli*. chemotaxis system to sense changes in osmolarity. Building on prior work, Vaknin and Berg[32] used FRET studies to show that osmotic agents act as repellants, enhancing kinase activity. They proposed that this response is primarily mechanical as the chemoreceptors and the cytoplasmic membrane interact under osmotic conditions. Several other mechanosensitive channels are also implicated in osmotic regulation. In particular, MscS has been shown to be gated by tension in the lipid bilayer[33]. Earlier it was shown that MscL also responds to mechanical stress[34].

Osmoregulatory K^+ uptake can be mediated by activated Trk and Kdp systems[35]. These systems involve numerous subunits, homologues and operons allowing for complex regulation mechanisms. The Kdp system includes the KdpD sensor kinase and the KdpFABC ATPase which is encoded by the *kdpFABCDE* operon. Meanwhile the TrkA subunit corresponds to the SapG protein in *S. typhimurium* in sequence similarities and in function[35].

2.4   Various Fish Species

Fish and other aquatic species offer interesting systems for studies of osmoregulation. More specifically, gill cells are an important system for the study of osmotic-stress adaptation. Evans et al. present an important review of gill cells in different species. Outlined in this work are species that allow for studies in freshwater, saltwater as well as euryhalines which may adapt to a wide range of salt water concentrations[36]. Below we provide small snapshots of a few species and recent work in identifying molecular players indicated in osmotic responses.

Recent work by Fiol, in *Oreochromis mossambicus* (Tilapia), identified two transcription factors that are rapidly and transiently induced by hyperosmotic stress. They observed six-fold and four-fold increases in mRNA levels for osmotic stress transcription factor 1 (OSTF1) and a homolog of TFIIB, respectively[37].

The small euryhaline killifish, *Fundulus heteroclitus*, has a unique reputation of being able to adapt to highly variable salinity and temperatures. Kultz and Avila exposed gill epithelium cells to different conditions over varying time periods and investigated the prevalence of three subgroups of MAPKs (described more fully in Section 3.1): extracellular regulated kinase 1 (ERK1), stress-activated protein kinase 1/Jun N-terminal kinase (SAPK1), and stress activated protein kinase 2/p38 (SAPK2)[38]. Kultz and Avila found SAPK2 increased under hyperosmotic conditions, but not hypoosmotic stress while ERK1 and SAPK1 are stable under both hyper- and hypoosmotic conditions. At the same time, the overall activity of MAPKs decreased significantly under hyperosmotic, but increased significantly during hypoosmotic stress.

In skate, *Raja erinacea*, it has been shown that the red blood cell anion exchanger (AE), skAE1, appears on the surface of erythrocytes in hypoosmotic medium. This appearance is strongly associated with volume expansion under these conditions[39]. AEs are shown to participate in the efflux of solutes during regulatory volume decreases. In particular, AEs appear to contribute to the loss of aurine from red blood cells of a number of species. Recently, the expression of skate AE1 was utilized in *Xenopu laevis* oocytes to better quantify the osmolarity of these oocytes[40].

3. STRESS RESPONSES IN MAMMALIAN CELLS TO HYPEROSMOTIC STRESS

3.1 Stress Activated Protein Kinases

Signal transduction in mammalian cells is frequently accomplished by the use of protein kinase cascades that convey an extracellular signal into the nucleus, activating transcription of the appropriate gene products. The first kinase cascade to be identified was the mitogen activated protein kinase (MAPK) cascade. These proteins are also known as the extracellular signal-regulated kinases (ERK)[41]. More recently, a group of protein kinases has been identified which respond specifically to environmental stresses; these kinases are termed the stress-activated protein kinases (SAPK, also known as JNK)[42–44]. These kinases are activated in response to UV-light[42,43], and tumor necrosis factor-α (TNF-α), an inflammatory cytokine produced in the whole-body response to stressful stimuli[44]. The SAPKs bind to and phosphorylate the c-Jun activation domain[43], presumably inducing transcription of genes with an AP-1 binding site in their promoter. Unlike the mitogen-activated protein kinases (MAPK)[41,45,46], the SAPKs are not strongly activated by mitogens[44]. While the MAP kinases are activated by some stress inducers (albeit less strongly than the SAP kinases)[44], they do not activate c-Jun[47]. The SAP and MAP kinase families show moderate sequence homology ($\sim$40% identity, $\sim$65% similarity) and are both proline-directed Ser/Thr kinases[43,46]. Because of their similarities, the SAP kinases are often referred to as members of either the MAPK family or the MAPK superfamily.

An additional member of the MAPK superfamily, p38, has recently been identified by its phosphorylation in response to osmotic stress[48]. p38 has a high degree of homology to HOG1, a yeast gene that is essential for growth after osmotic shock. p38 complements a mutation in the yeast HOG1 gene, although not as effectively as the native gene[48]. All three groups of kinases (MAPKs, SAPKs, and p38) require phosphorylation on Thr and Tyr for activation; however the MAPKs have the consensus sequence TEY, the SAPKs (JNK), TPY, and p38 and HOG1, TGY[48].

The upstream activators of the MAPKs have been reasonably well elucidated; the activators for JNK and p38 are less well established. The upstream activators

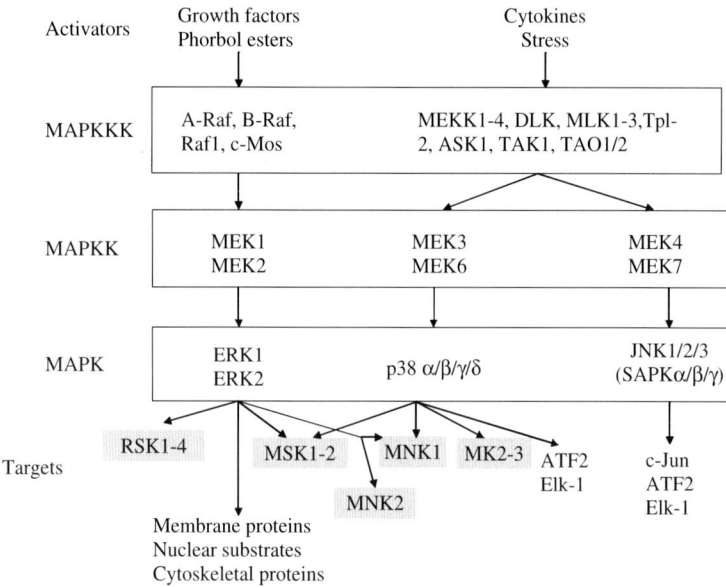

Figure 3. MAP kinase superfamily signal transduction pathways. Members of the superfamily are activated either by mitogens or stress. Each family of MAPKs is composed of a set of three evolutionarily conserved, sequentially acting kinases: a MAPK, a MAPK kinase (MAPKK), and a MAPKK kinase (MAPKKK). The MAPKKKs are serine/threonine kinases which phosphorylate and activate the MAPKKs. The MAPKKs are dual acting threonine/tyrosine kinases that, in turn, phosphorylate and activate the MAPKs. The MAPKs have a variety of targets including transcription factors (e.g. c-Jun, Elk-1), membrane proteins, and cytoskeletal proteins. In addition, the ERK1/2 and p38 families also phosphorylate a variety of kinases, termed MAPK-activated protein kinases (MKs), shown in grey. MAPK-mitogen activated protein kinase, ERK-extracellular signal-regulated kinase. SAPK-stress associated protein kinase, JNK-Jun amino terminal kinase

for all three groups are members of the MEK (MAP or ERK kinase) family also known as MKK or MAPKK family, shown in Figure 3. MEK1 and MEK2 appear to activate the ERK or MAPK subfamily[49]. MKK3 and MKK6 activate p38[50,51], and MKK4 (also known as SEK1 or SAPK1) appears to activate the SAPKs and possibly p38[51–53]. The upstream activator of MEK1 is Raf-1. MEKK[54], SPRK[55], and MLK-3[56] appear to serve as activators for MKK4. The putative signal transduction pathways for the various MAPK families are shown in Figure 3. Part of the difficulty in establishing the upstream activators is that many experiments have either been performed in vitro or with high levels of overexpression in vivo, leading to nonphysiological effects[57].

The role of the different pathways in osmotic stress responses is unclear. Exposure of Chinese hamster ovary cells to hyper-osmolar medium caused a strong increase in JNK activity[58]. JNK also complements a HOG1 mutation in yeast, although a kinase-negative form of JNK does not. As previously mentioned, p38 was cloned in response to its phosphorylation upon osmotic shock, and it too complements a

HOG1 mutation in yeast[48]. Matsuda and coworkers have found that in response to osmotic shock, fibroblast cells show strong activation of both the MAPK and the SAPK families[54].

Rosette and Karin[59] have found that exposure to UV light or osmotic shock led to clustering and internalization of epidermal growth factor (EGF), interleukin-1 (IL-1) and tumor necrosis factor (TNF) receptors. This clustering was both necessary and sufficient for activation of JNK in response to UV light or osmotic stress. In a chicken B-cell line (DT40), Qin et al. found that osmotic stress strongly induced JNK activity whereas oxidative stress induced both JNK and ERK/MAPK activity[60].

3.2   Stress Response Proteins

In addition to the stress-activated protein kinases, a number of other factors are involved in cellular stress responses, most notably, the stress response proteins. The stress response proteins were first identified in Drosophila in response to heat shock; hence, they are generally described as heat shock proteins (hsp)[61]. However, these proteins are induced by a variety of stress factors including heavy metals, ethanol, and anoxia[62]. Subsequent work has shown that these proteins (or closely related proteins) are also expressed under normal physiological conditions, and that their levels are elevated in response to stress[63]. Recently, a new member of the heat shock family of proteins was cloned in response to osmotic stress[64]. This protein, Osp94, is a member of the hsp110 subfamily of the hsp70 family of heat shock proteins. Members of the hsp70 family of proteins are believed to act as molecular chaperones, assisting in the proper folding of newly synthesized proteins[65,66].

Within the endoplasmic reticulum (ER), there are several stress response proteins believed to play a role in post-translational processing of secreted proteins. One of the first to be identified was BiP, heavy chain binding protein (also known as GRP78). BiP was associated with free heavy chains in pre-B derived cell lines and incompletely assembled Ig molecules in secreting cell lines[67]. BiP appears to bind incompletely assembled antibodies and prevent their secretion; upon release from BiP, immunoglobulins are free to move from the endoplasmic reticulum to the Golgi apparatus. BiP was subsequently identified as part of a larger set of ER proteins known as the glucose-regulated proteins (GRPs). These proteins are induced by a variety of cell stresses which lead to the accumulation of misfolded or underglycosylated proteins in the ER. They are also induced by over-expression of recombinant proteins that require glycosylation[68]. Another ER protein, protein disulfide isomerase (PDI) which catalyzes the formation of disulfide bonds is induced in CHO cells in response to cellular stresses[69]. Some effort has been made to understand the relationships between the levels of ER proteins and protein secretion for recombinant proteins[70–73] and for hybridomas[74,75], but the results remain unclear, suggesting that it may be protein and/or cell type specific. A recent study of nonsecreted immunoglobulin light chains demonstrated that BiP was bound to the variable region of the light chain and that the higher the binding affinity for BiP, the longer the half-life of the protein[76].

4.   OSMOTIC RESPONSES IN INDUSTRIALLY RELEVANT CELL LINES

A variety of process strategies have been explored to improve specific productivity from mammalian cell cultures. One of the most widely studied approaches is the application of hyperosmotic stress, which can be easily induced by addition of salts or sugars to the culture medium. The effect of hyperosmotic stress on increasing the specific antibody has been observed in many hybridoma cell lines[77–79]. Although the extent that antibody productivity is increased is cell-line specific[80], the mechanisms by which osmotic stress increases specific antibody production are likely to be similar in many different hybridoma cell lines. Moreover, other mammalian cells such as Chinese hamster ovary cells also exhibit increased specific productivity in response to hyperosmotic stress[81,82] and they are likely to share a common mechanism of action. Generally, sodium chloride is used to increase the osmolarity, but similar results have been achieved by using other ionic or nonionic compounds such as KCl and sucrose[83,84]. Both our laboratory and others have made significant progress in elucidating the mechanisms for the increase in antibody production, which we believe will provide useful guidance for engineering cells and culture strategies for higher production of monoclonal antibodies and other recombinant proteins produced in mammalian cells.

4.1   Global Transcriptional and Proteomic Effects of Osmolarity

Global gene expression analysis provides an unprecedented approach to understand the global physiological changes that occur in response to environmental stimuli such as hyperosmolarity. There are two complementary techniques that contribute to the understanding of global gene expression, genomic mRNA expression studies and proteomic protein expression studies. Both genomic and proteomic approaches have been reported in the study of mammalian responses to osmotic stress.

Although the concentration of proteins and their interactions are the true causative forces in the cell, the information extracted by proteomic studies is still limited. Lee et al. reported a proteomic analysis of CHO cells in response to hyperosmotic pressure[82]. In this study, they established a CHO cellular protein map comprising 23 identified proteins. From this map, three proteins were differentially expressed under hyperosmotic stress; pyruvate kinase and glyceraldehyde-3-phosphate dehydrogenase were up-regulated, and tubulin was down-regulated. The limited number of proteins identified indicates that even with the recent significant developments in the technologies used to quantify protein abundance, protein identification and quantification still lags behind the high-throughput experimental techniques used to determine mRNA expression levels.

Currently, mRNA expression analysis is still the approach that provides the best coverage and highest reliability. We reported a genome-wide analysis of the transcriptional response of mouse hybridoma OKT3 to hyperosmotic stress using GeneChip® MOE430A from Affymetrix[85]. The 22,690 probe sets on the array cover almost all of the well substantiated mouse genes. In our study, 215 genes showed a two-fold or greater change with a p-value < 0.05. These genes were considered significantly differentially expressed; 137 genes were primarily up-regulated and 78 genes were primarily down-regulated. As compared with the hyperosmotic response of *Saccharomyces cerevisiae*[9] (reviewed in Section 2.1), the transcriptome of murine hybridoma OKT3 was significantly less affected by osmotic shock than that of yeast. Considering that the yeast genome is substantially smaller that mouse genome, both the proportion of genes that were regulated and the degree of regulation was substantially lower in the hybridoma cells than in the yeast cells. This may be due to global differences in transcriptional responses between mammalian cells and yeast. Alternatively, the decreased transcriptional response may be due to mammalian cells not requiring as much osmoadaptation as yeast cells as mammalian cells have evolved to be maintained in an environment with much less osmotic variation.

Although the 215 significantly differentially expressed genes can be roughly clustered into two groups, up-regulated and down-regulated, the temporal profiles for the genes in each cluster showed a large variation[85]. This reflects the complexity of the transcriptional response of hybridoma cell to hyperosmotic stress. As shown in Table 1, the majority of these differentially expressed genes can be classified into several basic functional groups. In some groups, correlations were not evident, perhaps, in part, because of the complexity of the categories. However, there were several functional groups in which the majority of the genes behaved in

Table 1. Functional distribution of hyperosmolarity regulated genes

Category	Number of genes	
	Up-regulated	Down-regulated
Antibody-related genes	3	1
Metabolic and catabolic regulators	19	12
Carbohydrate	7	1
Protein and amino acid related	6	1
Nucleic acid and nucleotide	1	7
Lipid	3	2
Others	2	1
Cell cycle regulators	10	5
Apoptotic regulators	8	2
Transcriptional regulators	18	13
Translational regulators	2	2
Transporters and signal pathways	24	12
Electron transporter	2	1
Protein transporter	4	1
Metal ion transporter	6	1
Kinase related	6	3
Phosphotase	1	4
GTP related	4	1
Others	1	2
Cellular response	6	4
Structural constituents	18	7
Protein processing genes	6	2

similar manner, indicating the existence of biological processes globally induced or repressed by hyperosmotic stress.

In the following sections, detailed effects of osmolarity on mammalian cell physiology will be reviewed and the related transcriptional response will be discussed.

4.2   Effects of Osmolarity on Cell Growth and Viability

The first thorough investigation of the behavior of mammalian cells in response to osmotic stress was made by Ozturk and Palsson. As shown in Figure 4, they observed a notable decrease in the specific growth rate and maximum viable cell density when cells were subjected to hyperosmolarity[77]. Similar results have been found in many follow-up studies; notably a decrease in cell growth rate was repeatedly reported no matter what cell line was used[83,84,86].

Meanwhile, the response of cell viability varied between different cell lines. Although Ozturk and Palsson observed a significant increase in cell death rate (as shown in Figure 4), in our studies on two murine hybridoma cell lines, we observed slight decrease in cell viability, which was not statistically significant on repeated experiments[85,87].

Moreover, we performed a flow cytometric analysis and found no increase in apoptosis in the osmotically stressed cells from which we inferred that the decrease in viable cell concentrations is not caused by apoptotic death[87]. However, the apoptotic response to hyperosmolarity also differs dramatically between cell lines. In a study of mouse hybridoma AB143.2, apoptosis was indicated as the main mechanism of hybridoma cell death under elevated osmolarity; a dose-dependent increase in the occurrence of apoptosis was observed as osmolarity was increased[88].

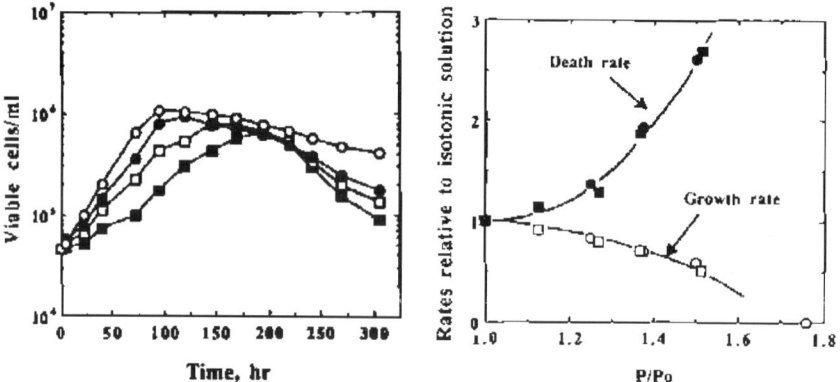

Figure 4. Left panel: The time profiles of viable cells for the batches of different osmolarities. Legend shows 290 mOsm (○), 326 mOsm (●), 380 mOsm (□), and 435 mOsm (■). Right panel: Effect of osmolarity on cell, growth rates and death rates were normalized using the values at Po = 290 mOsm and plotted as a function of normalized osmolarity, P/Po. Legend shows PBS addition (circle) and sucrose addition (square). Adapted with permission from Palsson et al., 1991[77]

Our transcriptome analysis identified 10 apoptotic genes that were significantly regulated by hyperosmotic stress[85]. Among those genes, eight were induced; only two caspase genes were significantly repressed. Since the apoptosis process is mediated by the activation of a series of caspases, the down-regulation of the caspases might be a cellular defense response to elevated apoptosis. However, the majority of the significantly induced genes act as apoptosis inducers, so the overall effect of the osmotic-induction of these genes upon apoptosis regulation is still not clear.

Oyaas et al. used glycine betaine as an osmoprotectant to overcome the drawbacks of hyperosmotic stress on cell growth. Their experiments demonstrated that glycine betaine was able to alleviate the growth repression in osmotically stressed cultures[83]. Several other compounds such as sarcosine, proline and glycine can also act as osmoprotective reagents. Other culture strategies such as adaptation of hybridoma cells to hyperosmolarity have also shown some positive results on cell viability and density[86,89].

4.3   Effects of Osmolarity on Protein Production

Although the majority of hyperosmotic studies observed a significant increase in the specific antibody productivity, the overall antibody production was not dramatically improved since the maximum viable cell density was significantly decreased[77,83,84,86]. A typical result is shown in Figure 5 in which Ozturk and Palsson observed an approximately two-fold increase in specific antibody production by increasing the medium osmolarity from approximately 300 mOsm to 450 mOsm by the addition of sodium chloride; however, this increase in specific productivity was not accompanied by an increase in overall yield[77].

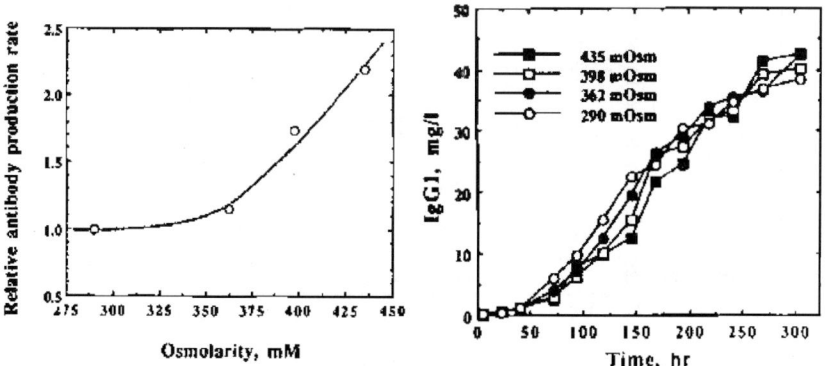

Figure 5. Left panel: The influence of media osmolarity on specific antibody production rate. Right panel: The time profiles of monoclonal antibody concentrations for the batches of different osmolarities. Legend shows 290 mOsm (o), 326 mOsm (●), 380 mOsm (□), and 435 mOsm (■). Adapted with permission from Palsson et al., 1991[77]

In a study of the function of glycine betaine as an osmoprotectant, Oyaas et al. demonstrated that glycine betaine was able to alleviate the growth repression in osmotically stressed cultures, and consequently, improved overall antibody production[83]. Their results further suggest that medium osmolarity, rather than growth rate, determines the specific antibody production rate. Hence, the combination of osmotic stress and glycine betaine can result in an overall increase in monoclonal antibody yield.

4.4   Metabolic Effects of Hyperosmotic Stress

As shown in Figure 6, Ozturk and Palsson observed a substantial increase in glucose uptake, lactate production, glutamine uptake and ammonia production after a increase in osmolarity from 290 to 435 mOsm[77]. Many other groups reported similar increases in metabolic rates with higher osmolarities, even with the repressed growth rate[84,90,91]. More specific investigations suggested osmolarity plays a predominant role in the increases in the specific glutamine consumption and ammonia production rates of hybridoma cells in hyperosmotic culture[88].

Oyaas et al. demonstrated that the increase in metabolic rates could be reversed by the addition of glycine betaine as an osmoprotective reagent when the osmolarity was mildly increased[83]. Moreover, the return of metabolic rates to normal levels showed no negative effects on antibody production, rather, the antibody productivity was further increased. Thus, the increased antibody productivity is not likely caused by the increased metabolic rate alone.

As increased cellular metabolic rates in response to hyperosmotic culture conditions were observed in many studies[84,90,91], it is possible that the increase in

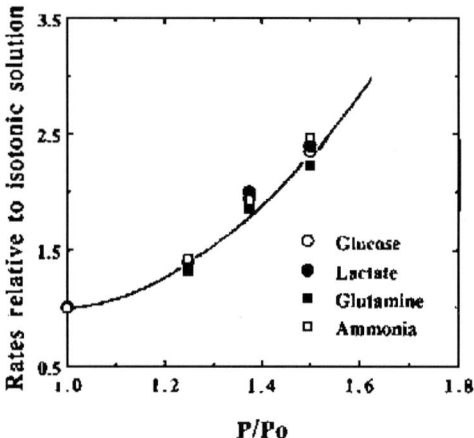

Figure 6. Effect of osmolarity on cell metabolism. The rates were normalized using the values at Po = 290 mOsm and plotted as a function of normalized osmolarity, P/Po. Legend shows glucose uptake (o), lactate production (•), glutamine uptake (■), and ammonia production (□). Adapted with permission from Palsson et al., 1991[77].

metabolic rates is regulated at the transcriptional level. Our microarray analysis identified 31 metabolic genes differentially expressed in response to hyperosmolarity[85]. These genes are related to carbohydrate, fatty acid, nucleotide, amino acid and protein metabolism; 19 of them were up-regulated, and the other 12 were down-regulated. Eight genes involved in carbohydrate metabolism were regulated, only one of them was repressed. Among the seven genes related to protein or amino acid and amino acid-derivative metabolism/catabolism, only Oazin was down-regulated. The induction of Gstt3, which is involved in coenzyme metabolism, was possibly due to an increase cofactor demand.

There were seven nucleic-acid and nucleotide-metabolism genes differentially expressed; all were repressed except Hnrpdl. This may be correlated with the repressed cell growth and proliferation. Five genes for lipid metabolism were also regulated with two down-regulated. Studies of salt-tolerant yeast *Candida membranefaciens* found hypersalinity signals affect the lipid composition which in turn affects the membrane fluidity of *C. membranefaciens*, while higher membrane fluidity favors osmotic adaptation against NaCl stress[92]. Hence, the differential expression of these lipid-metabolism-related genes might result in the alteration of the lipid composition of cellular membranes upon osmotic stress, which, in turn, would change the fluidity or the water permeability of the membrane.

Glyceraldehyde-3-phosphate dehydrogenase (GAPDH) is often assumed to be a nonregulated gene and is frequently considered as a housekeeping gene in gene expression analysis. We observed no significant change in the expression of GAPDH at the transcriptional level by either microarray or quantitative PCR assays[85]. However, considering the possible regulation at translational and posttranslational levels, this does not necessarily mean the expression of GAPDH was unregulated. In fact, in the proteomic analysis of CHO cells in response to hyperosmotic pressure, Lee and coworkers found GAPDH was significantly up-regulated under osmotic shock[82]. Many other groups have also observed that the expression level of GAPDH varied in different circumstances in various mammalian cells[93–95]. Recently, Klipp et al. reported that both the mRNA and enzyme activity levels of GAPDH increased in yeast in response to hyperosmotic shock[3]. Increased GAPDH activity leads to the increased metabolism of glucose to glycerol; the accumulation of glycerol, in turn, helps protect the cell from hyperosmotic shock.

4.5   Effects of Hyperosmotic Stress on the Cell Cycle

Hyperosmotic stress affects cell cycle control and alters the relative fractions of the cell population in G0/G1, S, and G2/M phases. Ryu et al. studied two CHO cell lines and found the cells were gradually arrested in G1 phase and the cell population in S and G2/M phases was decreased with increasing medium osmolarity, suggesting that the cell proliferation was suppressed at hyperosmolarities[96]. In continuous cultures of mouse hybridoma OKT3, Cherlet and Marc reported that hyperosmotic pressure induced a G1-phase arrest. When the osmolarity of the medium was gradually increased from 335 mOsm to 425 mOsm, the distribution of cells in the cell cycle adjusted from 33% in G1 phase and 54% in S phase to 51% in G1 phase and 38% in S phase[97]. In contrast, we measured the cell cycle distribution of hybridoma cell 167.4G5.3 in batch culture for control and osmotically stressed cultures at various times after osmotic shock[87]. At 18, 24 and 32 hours after osmotic stress, the fraction of cells in S phase increased with a corresponding decrease in the G1/G0 population; the cell cycle distributions were very similar in the two populations at the earlier and later stages of the culture. This pattern suggests that, in hybridoma 167.4G5.3, the decrease in growth rate is caused not by G1 arrest but by the additional time required for DNA replication under hyperosmotic conditions.

We observed fifteen cell-cycle-related genes significantly regulated by hyperosmolarity in our microarray analysis[85]. These genes regulate the cell cycle by several different mechanisms including kinase activity, ATP binding and DNA binding. All five repressed genes were most altered at 4 hours after osmotic shock, while the ten up-regulated genes showed three different expression patterns. Lmyc1, which also acts as transcriptional regulator, was continuously induced; its expression levels under hyperosmotic shock were more than two-fold higher than the expression levels at normal osmolarity at all four time points studied. Three of the up-regulated genes showed the greatest fold change at four hours after the introduction of osmotic shock, suggesting that they are among the primary response genes of the cell toward hyperosmotic stress. The other six up-regulated genes were most altered at 12 hours after the osmolarity was elevated; hence, they were more likely to be regulated by some primary response genes. Further pathway analysis based on these genes and their functions will provide us a better understanding of cell cycle regulation caused by osmotic shock.

4.6   Effects of Hyperosmotic Stress on Gene Expression and Protein Processing

Our microarray study identified 31 transcriptional regulatory genes differentially expressed in response to hyperosmolarity[85]. The expression patterns of these 31 genes varied widely. This is partly because eukaryotic transcription is a highly evolved process, involving thousands of interrelated genes and pathways. The majority of these 31 genes regulate transcription in a DNA-dependent manner or have DNA-binding activity. There are seven genes that regulate transcription via RNA polymerase II transcription-factor activity, among them, only transcription repressor Cri1 was down-regulated, the other six were up-regulated; Cited2 and Pparbp function as positive regulators of transcription from Pol II promoters, while Idb1, Idb2, Dr1 and Sin3a act as negative regulators. This may be a compromise that the cell makes for better adjustment of RNA transcription under hyperosmotic stress. Additionally, two genes for DNA methylation and one for RNA processing were down-regulated; another two genes for DNA repair were regulated, one induced and the other repressed.

In comparison, far fewer translational regulation genes were affected by hyperosmotic stress; only two translation-initiation regulators (Bzw1 and Ppp1r15b) were induced and two mRNA-splicing factors were repressed (Sfrs3 and Sfrs7). It is surprising that so few translational regulation genes were differentially expressed. We expected the expression of many translational regulation genes should be altered by hyperosmotic stress since we observed an increase of the total protein synthesis rate in osmotically stressed cultures compared to the control culture in our previous study of another hybridoma cell line[87]. This is partly because imposing a requirement for a two-fold change in expression for analysis may have neglected some genes whose transcription was altered to a lesser degree. The study of yeast cells' responses to increased external osmolarity showed similar behavior; the salt-dependent regulation of protein expression was found to be in good agreement with transcript levels, indicating that saline control of expression for these proteins is mainly at the transcriptional level during growth[98].

The effects of osmotic stress on protein translation, posttranslational processing and protein secretion have been examined as well. Oh et al reported that stressed cultures contained an enhanced level of total RNA, of which $\sim 80\%$ is ribosomal RNA[84]. Higher rRNA content could in turn increase the translation rates of proteins. We observed the total protein synthesis rate was increased in hyperosmotic cultures and sustained at least 48 hours after the osmotic shock, while there was no further increase in the immunoglobulin biosynthesis rate[87]. Lee and Lee did not quantify the total protein synthesis rate, however, they found heavy chain and light chain mRNA specific translation rates per cell at higher osmolarity increased by 172% and 240%, respectively, compared with those at control osmolarity[91]. These results suggest that hyperosmotic pressure enhanced translation rates. In the transcriptional analysis, we found protein-folding genes were globally induced; six out of seven differentially expressed protein folding genes were up-regulated by hyperosmotic stress[85]. This may be due to the increased demand caused by higher protein synthesis rates.

We also found osmotic stress did not decrease the degradation rate in either the heavy chain or the light chain, indicating that this is not a significant mechanism of increased antibody production in response to hyperosmotic stress. The antibody assembly and secretion rates were commensurate with the increase in protein translation observed in response to osmotic shock. Hence, posttranslational processing appears not to be a rate-limiting step in the increase of antibody production in response to hyperosmotic stress[87].

4.7   Effects of Hyperosmotic Stress on Transport and Signaling Pathways

In our microarray work, NaCl was added to the medium to raise the osmolarity. Consequently, the ion density was increased as well; thus, it is not surprising that some ion transporter genes were regulated in our experiment. Moreover, since Na^+ is a common signaling ion, the signaling pathways were also involved. In total, we identified 36 differentially expressed genes involved in transport and signaling pathways. Ten transporter genes were significantly regulated; eight of them were induced. Hence, the transport systems might be generally activated in OKT3 in response to elevated osmolarity and ionic strength. In addition to these transporter genes, the expression of five ion-binding genes was altered by elevated osmolarity as well; one iron ion-binding gene was down-regulated, while four zinc ion-binding genes were up-regulated with three of them most altered at 12 hours after osmotic shock. The general up-regulation of transcription of these transporter genes is in consensus with an intracellular response study, in which Oh et al. observed that the increase in osmotic pressure enhanced the transport of amino acids into cells. This enhancement occurred via Na^+-dependent transport systems, while no enhancement was observed in Na^+-independent transport systems[84]. Sucrose and KCl showed a similar effect, but to a much smaller extent.

Eight genes with kinase activity were differentially expressed; six were induced and two were repressed. Complementing the general induction of kinase activity, four out of five differentially expressed phosphatase genes were down-regulated. One kinase-regulation gene was down-regulated as well. This suggests that regulation of enzyme activity occurs widely in OKT3 in response to osmotic shock. One specific gene that is noteworthy is Dusp16, an up-regulated phosphatase which can lead to inactivation of MAPK[99]. As shown in Figure 7, we observed the activation of several kinases from MAPK superfamily, beginning at 5 minutes after osmotic shock, peaking around 15 to 30 minutes with a subsequent decline. It is possible that the induction of Dusp16 was the result of activated MAPK kinases, which would then be inactivated by the increased expression of Dusp16. Four GTP-related genes were also induced by hyperosmolarity; their functions include GTP binding, GTPase activity and GTPase regulation.

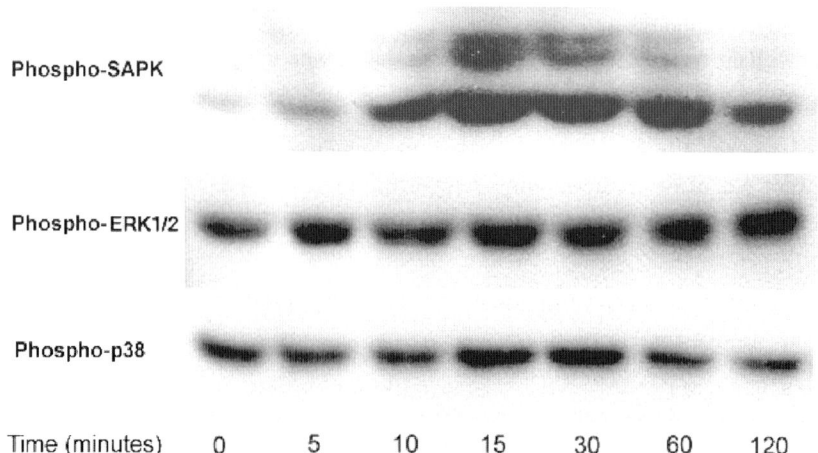

Figure 7. Western blot of phosphorylated MAPK, SAPK and p38. Cells were harvested at indicated time after the onset of osmotic stress. Total cellular protein from 2×10^5 cells was loaded onto each lane and separated by SDS-PAGE. The blot was probed by antibodies against phosphorylated kinases. The first lane is without osmotic stress

As is apparent from Figure 7, in addition to the transcriptional responses to hyperosmotic stress, there are likely a wide range of post-translational regulatory processes, particularly utilizing kinase-activation cascades, as seen in virtually all organisms in response to hyperosmolarity.

5. CONCLUSIONS

The physiological and transcriptional studies of mammalian cells and non-mammalian organisms suggest a wide range of targets for potential cellular engineering strategies. What remains to be determined is whether or not these targets play a causal role in either the increase in protein production or the reduction of cell growth in response to hyperosmolarity. Current studies in our laboratory using silencing RNA constructs, pathway analysis, and computational tools are attempting to identify targets where a causal relationship can be determined. Using a combination of cellular engineering approaches along with process engineering strategies, we believe it will be possible to take advantage of the molecular responses to hyperosmolarity to increase protein production from cultured mammalian cells.

ACKNOWLEDGEMENTS

This work was partially supported by the National Science Foundation (BES-0409969) and the Office of Research at Rensselaer Polytechnic Institute. TRK is the recipient of a National Science Foundation graduate fellowship.

6. REFERENCES

1. Hohmann S., Osmotic stress signaling and osmoadaptation in yeasts. *Microbiol. Mol. Biol. Rev.* **66**(2), 300–372 (2002).
2. Westfall P.J., Ballon D.R. & Thorner J., When the stress of your environment makes you go HOG Wild. *Science.* **306**(5701), 1511–1512 (2004).
3. Klipp E., Nordlander B., Kruger R., Gennemark P. & Hohmann S., Integrative model of the response of yeast to osmotic shock. **23**(8), 975–982 (2005).
4. Reiser V., Raitt D.C. & Saito H., Yeast osmosensor Sln1 and plant cytokinin receptor Cre1 respond to changes in turgor pressure. *J. Cell Biol.* **161**(6), 1035–1040 (2003).
5. Raitt D.C., Posas F. & Saito H., Yeast Cdc42 GTPase and Ste20 PAK-like kinase regulate Sho1-dependent activation of the Hog1 MAPK pathway. *the EMBO Journal.* **19**(17), 4623–4631 (2000).
6. Reiser V., Salah S.M. & Ammerer G., Polarized localization of yeast Pbs2 depends on osmostress, the membrane protein Sho1 and Cdc42. *Nature Cell Biology.* **2**(9), 620–627 (2000).
7. Uhlik M.T., Abell A.N., Johnson N.L., Sun W., Cuevas B.D., Lobel-Rice K.E., Horne E.A., Dell'Acqua M.L. & Johnson G.L., Rac-MEKK3-MKK3 scaffolding for p38 MAPK activation during hyperosmotic shock. *Nature Cell Biology.* **5**(12), 1104–1110 (2003).
8. Reiser V., Ruis H. & Ammerer G., Kinase activity-dependent nuclear export opposes stress-induced nuclear accumulation and retention of Hog1 mitogen-activated protein kinase in the budding yeast Saccharomyces cerevisiae. *Mol. Biol. Cell.* **10**(4), 1147–1161 (1999).
9. Rep M., Krantz M., Thevelein J.M. & Hohmann S., The transcriptional response of Saccharomyces cerevisiae to osmotic shock. Hot1p and Msn2p/Msn4p are required for the induction of subsets of high osmolarity glycerol pathway-dependent genes. *J. Biol. Chem.* **275**(12), 8290–8300 (2000).

10. Nadal E.d., Casadome L. & Posas F., Targeting the MEF2-like transcription factor smp1 by the stress-activated Hog1 mitogen-activated protein kinase. *Mol. Cell. Biol.* **23**(1), 229–237 (2003).

11. Causton H.C., Ren B., Koh S.S., Harbison C.T., Kanin E., Jennings E.G., Lee T.I., True H.L., Lander E.S. & Young R.A., Remodeling of yeast genome expression in response to environmental changes. *Mol. Biol. Cell.* **12**(2), 323–337 (2001).

12. Chinnusamy V., Schumaker K. & Zhu J.-K., Molecular genetic perspectives on cross-talk and specificity in abiotic stress signalling in plants. *J. Exp. Bot.* **55**(395), 225–236 (2004).

13. Boudsocq M. and Lauriere C., Osmotic signaling in plants. multiple pathways mediated by emerging kinase families. *Plant Physiol.* **138**(3), 1185–1194 (2005).

14. Urao T., Yakubov B., Satoh R., Yamaguchi-Shinozaki K., Seki M., Hirayama T. & Shinozaki K., A Transmembrane hybrid-type histidine kinase in Arabidopsis functions as an osmosensor. *Plant Cell.* **11**(9), 1743–1754 (1999).

15. Tamura T., Hara K., Yamaguchi Y., Koizumi N. & Sano H., Osmotic stress tolerance of transgenic tobacco expressing a gene encoding a membrane-located receptor-like protein from tobacco plants. *Plant Physiol.* **131**(2), 454–462 (2003).

16. Kiegle E., Moore C.A., Haseloff J., Tester M.A. & Knight M.R., Cell-type-specific calcium responses to drought, salt and cold in the Arabidopsis root. *The Plant Journal.* **23**(2), 267–278 (2000).

17. Knight H., Trewavas A.J. & Knight M.R., Calcium signalling in Arabidopsis thaliana responding to drought and salinity. *The Plant Journal.* **12**(5), 1067–1078 (1997).

18. Kreps J.A., Wu Y., Chang H.-S., Zhu T., Wang X. & Harper J.F., Transcriptome changes for Arabidopsis in response to salt, osmotic, and cold stress. *Plant Physiol.* **130**(4), 2129–2141 (2002).

19. Guo Y., Qiu Q.-S., Quintero F.J., Pardo J.M., Ohta M., Zhang C., Schumaker K.S. & Zhu J.-K., Transgenic evaluation of activated mutant alleles of SOS2 reveals a critical requirement for its kinase activity and c-terminal regulatory domain for salt tolerance in Arabidopsis thaliana. *Plant Cell.* **16**(2), 435–449 (2004).

20. Guo Y., Halfter U., Ishitani M. & Zhu J.-K., Molecular characterization of functional domains in the protein kinase SOS2 that is required for plant salt tolerance. *Plant Cell.* **13**(6), 1383–1400 (2001).

21. Leung J. and Giraudat J., Abscisic acid signal transduction. *Annual Review of Plant Physiology and Plant Molecular Biology.* **49**(1), 199–222 (1998).

22. Riera M., Valon C., Fenzi F., Giraudat J. & Leung J., The genetics of adaptive responses to drought stress: abscisic acid-dependent and abscisic acid-independent signalling components. *Physiol. Plant.* **123**(2), 111–119 (2005).

23. Frandsen G., Müller-Uri F., Nielsen M., Mundy J. & Skriver K., Novel plant Ca^{2+}-binding protein expressed in response to abscisic acid and osmotic stress. *J. Biol. Chem.* **271**(1), 343–348 (1996).

24. Covic L., Silva N.F. & Lew R.R., Functional characterization of ARAKIN (ATMEKK1): a possible mediator in an osmotic stress response pathway in higher plants. *Biochimica et Biophysica Acta (BBA) – Molecular Cell Research.* **1451**(2–3), 242–254 (1999).

25. Hasegawa P.M., Bressan R.A., Zhu J.-K. & Bohnert H.J., Plant cellular and molecular responses to high salinity. *Annual Review of Plant Physiology and Plant Molecular Biology.* **51**(1), 463–499 (2000).

26. Forst S., Delgado J. & Inouye M., Phosphorylation of OmpR by the osmosensor EnvZ modulates expression of the ompF and ompC genes in Escherichia coli. *PNAS.* **86**(16), 6052–6056 (1989).

27. Begic S. and Worobec E.A., Regulation of Serratia marcescens ompF and ompC porin genes in response to osmotic stress, salicylate, temperature and pH. *Microbiology.* **152**(2), 485–491 (2006).

28. Nikaido H., Molecular basis of bacterial outer membrane permeability revisited. *Microbiol. Mol. Biol. Rev.* **67**(4), 593–656 (2003).

29. Culham D.E., Lu A., Jishage M., Krogfelt K.A., Ishihama A. & Wood J.M., The osmotic stress response and virulence in pyelonephritis isolates of Escherichia coli: contributions of RpoS, ProP, ProU and other systems. *Microbiology.* **147**(6), 1657–1670 (2001).

30. Ly A., Henderson J., Lu A., Culham D.E. & Wood J.M., Osmoregulatory systems of Escherichia coli: Identification of betaine-carnitine-choline transporter family member BetU and Distributions of betU and trkG among pathogenic and nonpathogenic isolates. *J. Bacteriol.* **186**(2), 296–306 (2004).

31. Booth I.R.L., P, Managing hypoosmotic stress: aquaporins and mechanosensitive channels in Escherichia coli. *Current Opinion in Microbiology*. **2**(2), 166–169 (1999).

32. Vaknin A. and Berg H.C., Osmotic stress mechanically perturbs chemoreceptors in Escherichia coli. *Proc. Natl. Acad. Sci.* **103**(3), 592–596 (2006).

33. Sukharev S., Purification of the small mechanosensitive channel of Escherichia coli (MscS): the subunit structure, conduction, and gating characteristics in liposomes. *Biophys. J.* **83**(1), 290–298 (2002).

34. Blount P., Schroeder M.J. & Kung C., Mutations in a bacterial mechanosensitive channel change the cellular response to osmotic stress. *J. Biol. Chem.* **272**(51), 32150–32157 (1997).

35. Wood J.M., Osmosensing by bacteria: signals and membrane-based sensors. *Microbiol. Mol. Biol. Rev.* **63**(1), 230–262 (1999).

36. Evans D.H., Piermarini P.M. & Choe K.P., The multifunctional fish gill: dominant site of gas exchange, osmoregulation, acid-base regulation, and excretion of nitrogenous waste. *Physiol. Rev.* **85**(1), 97–177 (2005).

37. Fiol D.F. and Kultz D., Rapid hyperosmotic coinduction of two tilapia (Oreochromis mossambicus) transcription factors in gill cells. *Proc. Natl. Acad. Sci.* **102**(3), 927–932 (2005).

38. Kultz D. and Avila K., Mitogen-activated protein kinases are in vivo transducers of osmosensory signals in fish gill cells. *Comparative Biochemistry and Physiology Part B: Biochemistry and Molecular Biology*. **129**(4), 821–829 (2001).

39. Musch M.W., Koomoa D.-L.T. & Goldstein L., Hypotonicity-induced exocytosis of the skate anion exchanger skAE1: Role of lipid raft regions. *J. Biol. Chem.* **279**(38), 39447–39453 (2004).

40. Dana-Lynn T. Koomoa M.W.M., & Goldstein, L., Osmotic stress stimulates the organic osmolyte channel in *Xenopus laevis* oocytes expressing skate *Raja erinacea* AE1. *Journal of Experimental Zoology Part A: Comparative Experimental Biology*. **303A**(4), 319–322 (2005).

41. Pelech S.L. and Sanghera J.S., Mitogen-activated protein kinases: versatile transducers for cell signaling. *Trends Biochem. Sci.* **17**, 233–238 (1992).

42. Hibi M., Lin A., Smeal T., Minden A. & Karin M., Identification of an oncoprotein- and UV-responsive protein kinase that binds and potentiates the c-Jun activation domain. *Genes & Devel.* **7**, 2135–2148 (1993).

43. Derijard B., Hibi M., Wu I.-H., Barrett T., Su B., Deng T., Karin M. & Davis R.J., JNK1: A protein kinase stimulated by UV Light and Ha-Ras That binds and phosphorylates the c-Jun activation domain. *Cell.* **76**, 1025–1037 (1994).

44. Kyriakis J.M., Banerjee P., Nikolakaki E., Dai T., Rubie E.A., Ahmad M.F., Avruch J. & Woodgett J.R., The stress-activated protein kinase subfamily of c-Jun kinases. *Nature.* **369**(6476), 156–60 (1994).

45. Nishida E. and Gotoah Y., The MAP kinase cascade is essential for diverse signal transduction pathways. *Trends Biochem. Sci.* **18**, 128–131 (1993).

46. Davis R.J., The mitogen-activated protein kinase signal transduction pathway. *J. Biol. Chem.* **268**, 14553–14556 (1993).

47. Minden A., Lin A., Smeal T., Derijard B., Cobb M., Davis R. & Karin M., c-Jun N-terminal phosphorylation correlates with activation of the JNK Subgroup but not the ERK subgroup of mitogen-activated protein kinases. *Mol. Cell. Biol.* **14**, 6683–6688. (1994).

48. Han J., Lee J.-D., Bibbs L. & Ulevitch J., A MAP kinase targeted by endotoxin and hyperosmolarity in mammalian cells. *Science.* **265**, 808–811 (1994).

49. Crews C.M., Alessandrini A. & Erikson R.L., The primary structure of MEK, a protein kinase that phosphorylates the ERK product. *Science.* **258**, 478–480 (1992).

50. Raingeaud J., Whitmarsh A.J., Barrett T., Derijard B. & Davis R.J., MKK3- and MKK6-regulated gene expression is mediated by the p38 mitogen-activated protein kinase signal transduction pathway. *Mol. Cell. Biol.* **16**(3), 1247–55 (1996).

51. Derijard B., Raingeaud J., Barrett T., Wu I.-H., Han J., Ulevitch R.J. & Davis R.J., Independent human MAP kinase signal transduction pathways defined by MEK and MKK isoforms. *Science.* **267**, 682–684 (1995).

52. Lin A., Minden A., Martinetto H., Claret F.-X., Lange-Carter C., Mercurio F., Johnson G.L. & Karin M., identification of a dual specificity kinase that activates the Jun kinases and p38-Mpk2. *Science*. **268**, 286–290 (1995).

53. Sanchez I., Hughes R.T., Mayer B.J., Yee K., Woodgett J.R., Avruch J., Kyriakis J.M. & Zon L.I., Role of SAPK/ERK kinase-1 in the stress-activated pathway regulating transcription factor c-Jun. *Nature*. **372**, 794–800 (1994).

54. Matsuda S., Kawasaki H., Moriguchi T., Gotoh Y. & Nishida E., Activation of protein kinase cascades by osmotic shock. *J. Biol. Chem.* **270**, 12781–12786 (1995).

55. Rana A., Gallo K., Godowski P., Hirai S., Ohno S., Zon L., Kyriakis J.M. & Avruch J., The mixed lineage kinase SPRK phosphorylates and activates the stress-activated protein kinase activator, SEK-1. *J. Biol. Chem.* **271**, 19025–19028 (1996).

56. Tibbles L.A., Ing Y.L., Kiefer F., Chan J., Iscove N., Woodgett J.R. & Lassam N.J., MLK-3 activates the SAPK/JNK and p38/RK pathways via SEK1 and MKK3/6. *EMBO J.* **15**(24), 7026–35 (1996).

57. Zanke B.W., Rubie E.A., Winnett E., Chan J., Randall S., Parsons M., Boudreau K., McInnis M., Yan M., Templeton D.J. & Woodgett J.R., Mammalian mitogen-activated protein kinase pathways are regulated through formation of specific kinase-activator complexes. *J. Biol. Chem.* **271**(47), 29876–81 (1996).

58. Galcheva-Gargova Z., Derijard B., Wu I.-H. & Davis R., An osmosensing signal transduction pathway in mammalian cells. *Science*. **265**, 806–808 (1994).

59. Rosette C. and Karin M., Ultraviolet light and osmotic stress: activation of the JNK cascade through multiple growth factor and cytokine receptors. *Science*. **274**, 1194–1197 (1996).

60. Qin S., Minami Y., Hibi M., Kurosaki T. & Yamamura H., Syk-dependent and -independent signaling cascades in B cells elicited by osmotic and oxidative stress. *J. Biol. Chem.* **272**(4), 2098–103 (1997).

61. Tissiáeres A., Mitchell H.K. & Tracy U.M., Protein synthesis in salivary glands of Drosophila melanogaster: relation to chromosome puffs. *Journal of molecular biology*. **84**(3), 389–98 (1974).

62. Nover L. and Scharf K.D., in *Heat Shock Response*, L. Nover, Editor. 1991, CRC Press: Boca Raton, Fl. p. 41–128.

63. Lindquist S. and Craig E.A., The heat-shock proteins. *Ann. Rev. Genet.* **22**, 631–677 (1988).

64. Kojima R., Randall J., Brenner B.M. & Gullans S.R., Osmotic stress protein 94 (Osp94): A new member of the Hsp110/SSE gene subfamily. *J. Biol. Chem.* **271**, 12327–12332. (1996).

65. Gething M.J. and Sambrook J., Protein folding in the cell. *Nature*. **355**, 33–44. (1992).

66. Becker J. and Craig E.A., Heat-shock proteins as molecular chaperones. *Eur. J. Biochem.* **219**, 11–23 (1994).

67. Bole D.G., Hendershot L.M. & Kearny J.F., Posttranslational association of immunoglobulin heavy chain binding protein with nascent heavy chains in nonsecreting and secreting hybridomas. *J. Cell Biol.* **102**, 1558–1566 (1986).

68. Dorner A.J., Wasley L.C. & Kaufman R.J., Increased synthesis of secreted proteins induces expression of glucose-regulated proteins in butyrate-treated Chinese hamster ovary cells. *J. Biol. Chem.* **264**(34), 20602–7 (1989).

69. Dorner A.J., Wasley L.C., Raney P., Haugejorden S., Green M. & Kaufman R.J., The stress response in Chinese hamster ovary cells. Regulation of ERp72 and protein disulfide isomerase expression and secretion. *J. Biol. Chem.* **265**(35), 22029–34 (1990).

70. Morris J.A., Dorner A.J., Edwards C.A., Hendershot L.M. & Kaufman R.J., Immunoglobulin binding protein (BiP) function is required to protect cells from endoplasmic reticulum stress but is not required for the secretion of selective proteins. *J. Biol. Chem.* **272**(7), 4327–34 (1997).

71. Li L.-J., Li X., Ferrario A., Rucker N., Liu E.S., Wong S., Gomer C.J. & Lee A.S., Establishment of a Chinese hamster ovary cell line that expresses grp78 antisense transcripts and suppresses A23187 induction of both GRP78 and GRP94. *J. Cell. Physiol.* **153**, 575–582 (1992).

72. Dorner A.J., Krane M.G. & Kaufman R.J., Reduction of endogenous GRP78 levels improves secretion of a heterologous protein in CHO Cells. *Mol. Cell. Biol.* **8**(10), 4063–4070 (1988).

73. Dorner A.J., Wasley L.C. & Kaufman R.J., Protein dissociation from GRP78 and secretion are blocked by depletion of cellular ATP levels. *PNAS*. **87**(19), 7429–32 (1990).

74. Kitchin K. and Flickinger M.C., Alteration of hybridoma viability and antibody secretion in transfectomas with inducible overexpression of protein disulfide isomerase. *Biotechnol Prog.* **11**, 565–574 (1995).

75. Lambert N. and Merten O.-W., Effect of serum-free and serum-containing medium on cellular levels of ER-based proteins in various mouse hybridoma cell lines. *Biotechnol. Bioeng.* **54**(2), 165–180 (1997).

76. Skowronek M.H., Hendershot L.M. & Hass I.G., The variable domain of nonassembled Ig light chains determines both their half-life and binding to the chaperone BiP. *Proceedings of the National Academy of Science, USA.* **95**, 1574–1578 (1998).

77. Ozturk S.S. and Palsson B.O., Effect of medium osmolarity on hybridoma growth, metabolism, and antibody production. *Biotech. Bioeng.* **37**, 989–993 (1991).

78. Kim N.S. and Lee G.M., Response of recombinant Chinese hamster ovary cells to hyperosmotic pressure: effect of Bcl-2 overexpression. *J. Biotechnol.* **95**, 237–248 (2002).

79. Wu M.-H., Dimopoulos G., Mantalaris A. & Varley J., The effect of hyperosmotic pressure on antibody production and gene expression in the GS-NS0 cell line. *Biotechnology and Applied Biochemistry.* **40**, 41–46 (2004).

80. Lee G.M. and Park S.Y., Enhanced specific antibody productivity of hybridomas resulting from hyperosmotic stress is cell line-specific. *Biotechnol. Lett.* **17**(2), 145–150 (1995).

81. Ryu J.S., Kim T.K., Chung J.Y. & Lee G.M., Osmoprotective effect of glycine betaine on foreign protein production in hyperosmotic recombinant Chinese hamster ovary cell cultures differs among cell lines. *Biotechnol Bioeng.* **70**(2), 167–75 (2000).

82. Lee M.S., Kim K.W., Kim Y.H. & Lee G.M., Proteome analysis of antibody-expressing CHO cells in response to hyperosmotic pressure. *Biotechnol Prog.* **19**(6), 1734–1741 (2003).

83. Oyaas K., Ellingsen T.E., Dyrset N. & Levine D.W., Hyperosmotic hybridoma cell cultures: Increased monoclonal antibody production with addition of glycine betaine. *Biotech. Bioeng.* **44**, 991–998 (1994).

84. Oh S.K.W., Chua F.K.F. & Choo A.B.H., Intracellular responses of productive hybridomas subjected to high osmotic pressure. *Biotech. Bioeng.* **46**, 525–535 (1995).

85. Shen D. and Sharfstein S.T., Genome-wide analysis of the transcriptional response of murine hybridomas to osmotic shock. *Biotechnol. Bioeng.* **93**(1), 132–145 (2006).

86. Reddy S. and Miller W.M., Effects of abrupt and gradual osmotic-stress on antibody-production and content in hybridoma cells that differ in production kinetics. *Biotechnol Prog.* **10**(2), 165–173 (1994).

87. Sun Z., Zhou R., Liang S., McNeeley K.M. & Sharfstein S.T., Hyperosmotic stress in murine hybridoma cells: effects on antibody transcription, translation, posttranslational processing, and the cell cycle. *Biotechnol Prog.* **20**(2), 576–89 (2004).

88. deZengotita V.M., Schmelzer A.E. & Miller W.M., Characterization of hybridoma cell responses to elevated pCO(2) and osmolality: intracellular pH, cell size, apoptosis, and metabolism. *Biotechnol Bioeng.* **77**(4), 369–80 (2002).

89. Oh S.K.W., Vig P., Chua F., Teo W.K. & Yap M.G.S., Substantial overproduction of antibodies by applying osmotic pressure and sodium butyrate. *Biotechnol. Bioeng.* **42**, 601–610 (1993).

90. Wurm F.M., Production of recombinant protein therapeutics in cultivated mammalian cells. *Nat Biotechnol.* **22**(11), 1393–8 (2004).

91. Lee M.S. and Lee G.M., Hyperosmotic pressure enhances immunoglobulin transcription rates and secretion rates of KR12H-2 transfectoma. *Biotechnol Bioeng.* **68**(3), 260–8 (2000).

92. Khaware R.K., Koul A. & Prasad R., High membrane fluidity is related to NaCl stress in Candida membranefaciens. *Biochemistry and Molecular Biology International.* **35**(4), 875–880 (1995).

93. Graven K.K., Troxler R.F., Kornfeld H., Panchenko M.V. & Farber H.W., Regulation of endothelial cell glyceraldehyde-3-phosphate dehydrogenase expression by hypoxia. *J. Biol. Chem.* **269**, 24446–24453 (1994).

94. Mansur N.R., Meyer-Siegler K., Wurzer J.C. & Sirover M.A., Cell cycle regulation of the glyceraldehyde-3-phophate dehydrogenase/uracil DNA glycosylase gene in normal human cells. *Nucleic Acids Res.* **21**, 993–998 (1993).

95. McNulty S.E. and Toscano W.A., Transcriptional regulation of glyceraldehyde-3-phosphate dehydrogenase by 2,3,7,8-tetrachlorodibenzo-p-dioxin. *Biochem. Biophys. Res. Commun.* **212**, 165–171 (1995).

96. Ryu J.S., Lee M.S. & Lee G.M., Effects of cloned gene dosage on the response of recombinant CHO cells to hyperosmotic pressure in regard to cell growth and antibody production. *Biotechnol Prog.* **17**(6), 993–999 (2001).

97. Cherlet M. and Marc A., Hybridoma cell behaviour in continuous culture under hyperosmotic stress. *Cytotechnology.* **29**, 71–84 (1999).

98. Blomberg A., Osmoresponsive proteins and functional assessment strategies in Saccharomyces cerevisiae. *Electrophoresis.* **18**(8), 1429–1440 (1997).

99. Masuda K., Shima H., Watanabe M. & Kikuchi K., MKP-7, a novel mitogen-activated protein kinase phosphatase, functions as a shuttle protein. *J. Biol. Chem.* **276**(42), 39002–39011 (2001).

CHAPTER 8

METABOLOMICS

An Emerging Tool for Understanding Metabolic Systems

SOO HEAN GARY KHOO[1] AND MOHAMED AL-RUBEAI[2]

[1] *Department of Chemical Engineering, University of Birmingham, Edgbaston Birmingham B15 2TT, United Kingdom*
[2] *School of Chemical and Bioprocess Engineering, University College Dublin, Belfield Dublin 4, Ireland*

Abstract: Metabolomics, the 'global' study of metabolite changes in a biological system, has drawn a significant amount of interest over the last few years. It can be said to be an amalgam of traditional areas like metabolite analysis, bioanalytical development and chemometrics. Thus, piecing these areas together into the cohesive science of metabolome analysis has proven to be difficult. Most work to date has been focused on plant, microbial as well as tissue and biofluid samples. However, the diverse potential of metabolomics in many fields, including cell engineering, has made it a universal tool for industrial, medical and research purposes. It is also a vital component of a 'systems biology' approach, as it is believed to be a good reflection of the phenotype of any cell or tissue. Unfortunately, with present methods and technologies, metabolomics still lacks significant comprehensiveness and development. This chapter gives a closer look at the arguments encompassing its definitions, its potential uses as well as the complex challenges of metabolome analysis. The diverse methodologies involved in metabolome analysis; sample preparation, instrument analysis, data processing and bioinformatics tools, represent part of its complexity. Thus, as the nascent field of metabolomics moves forward, significant integration and uniformity issues will need to be addressed

Keywords: systems biology, metabolomics, metabolite analysis, metabolism

1.   INTRODUCTION

With the systematic genome sequencing of various organisms, an unprecedented amount of information has been revealed. Deciphering such a blueprint via the

237

M. Al-Rubeai and M. Fussenegger (eds.), Systems Biology, 237–273.
© *2007 Springer.*

understanding of functions and interactions within a complex biological system has been the focus of the post-genomic era. That has fuelled the growth of other 'omic' sciences like transciptomics and proteomics which, together with the rapid development of bioinformatics and statistical tools, have lead to the technology platforms that can now be used in various applications. One of those relatively new 'omic' sciences is the field of metabolomics. Metabolomics has been viewed as the vital piece of the 'Rossetta stone'[1] needed for deciphering the puzzle of complex systems. It is proving to be a diverse field, with numerous applications in industry and medicine. However, just like the other 'omic' sciences, building knowledge rather than information gathering has been the greatest challenge of them all.

> "Science is built up with facts, as a house is with stones. But a collection of facts is no more a science than a heap of stones is a house"
> Jules Henri Poincaré (1854–1912)
> *La Science et l'hypothése.*

The field of metabolomics could be said to fuse metabolite analysis, bioanalytical science and chemometrics. Thus the following sections give an overview to different elements of this field, highlighting methods and its current status as an 'omic' science. It is essential to emphasize that this rapidly changing field is still in its infancy but its potential application in industrial cell engineering, as seen in yeast and plant cultures, makes it a vital tool for the future.

2.   DEFINITIONS

The metabolome was first described by Oliver et al[2] as being the set of all low molecular mass compounds synthesized by an organism. Metabolome analysis was considered "the measurement of the change in the relative concentrations of metabolites as the result of the deletion or over expression of a gene...(that) should allow the target of a novel gene product to be located on the metabolic map". Fiehn[3] subsequently made distinctions between different metabolite analyses and referred to metabolomics as the comprehensive and quantitative analysis of all metabolites of an organism. These distinctions in analysis techniques are explained in detail in Section 4. Another definition of the metabolome states that it consists of 'only of those native small molecules (definable non-polymeric compounds) that are participants in general metabolic reactions and that are required for the maintenance, growth and normal function of a cell[4]. This would exclude peptides and even some larger lipids. Realistically, the 'global' analysis of metabolites seems to be a long way off and thus, in the pure sense, metabolomics now only consists of fragments of biochemical and metabolic analysis. Even as the debate continues, the field of metabolomics grows towards the objective of extracting useful knowledge from metabolite pools. Therefore, for the purpose of utility, metabolomics is best described as an area of science, rather than an analytical approach, that characterizes a metabolic phenotype under a specific set of conditions which links these phenotypes to their correspondent genotypes[5].

There is also some question as to whether targeted analysis of metabolites constitutes metabolomics. Targeted analysis refers to the analysis of predetermined metabolites for the purpose of unambiguous quantification. One such analysis is metabolic flux analysis. In flux analysis, a set of metabolites in a defined metabolic network are studied for the purpose of determination of flux. As it is, flux analysis has been recently been termed 'fluxomics'. Thus, the distinction between targeted analysis and untargeted analysis (like metabolomics) has been made.[6]. Whatever the case, narrowing the gap of understanding of gene functions (functional genomics) is essential in an "omic"-related research. Such aims can only be achieved by using high-throughput techniques which will be able to give a 'globally' normalized set of data from which significant parameters can be determined.

3. THE IMPORTANCE OF METABOLOMICS

In order to fully appreciate the rapidly growing interest in metabolomics, it is necessary to align it with the other 'omic' sciences. While the genome gives rise to a blueprint to any living organism, the transcriptome is result of the transcription factors responding to stimulus from upstream receptors and kinases. It represents how the cell transfers information from the genome to the rest of the cell. Similarly, the concentration of proteins, termed the proteome, is determined by the corresponding levels of mRNA transcripts. It represents the activity of the translational apparatus, protein kinases, phosphatases and proteases. Such a transfer of information cannot be considered to be linear[7] and furthermore, not all the proteins are enzymatically active. Metabolite concentrations are determined by enzymes, which form a part of the proteome. Thus, metabolomics is 'complementary' to transcriptomics and proteomics (see Figure 1) as it is downstream of the other two 'omics', placing it 'close' to cell's physiological state[8]. The interactions from gene to phenotype can therefore be better understood by including the analysis of the metabolome. Consequently, it maybe also be used to 'phenotype' mutant organisms into useful classifications. Fiehn[3] states that in order to study the pleiotropic effects of certain genetic or environmental perturbations, a complete analysis of the biological system in question needs to be done. Hence, metabolomics will be able to give a snapshot of the events happening at any moment in time.

As the metabolic network in a living cell is viewed as being a complex network of reactions that are tightly connected[9], any perturbations in the proteome can cause significant changes in the concentration of metabolites. This makes the metabolome data a form of integrated data which gives it strength as an 'omic' science. Metabolite concentration and fluxes have been shown to not be regulated by gene expression alone (e.g. glycolysis in trypanosomes)[10]. This may again disprove the idea of a simple 'hierarchical' regulation of function by gene expression as is the case with proteins and mRNA. Another reason for the great interest in metabolomics lies in the fact that metabolites have the same chemical structures irrespective of the organism from which they are extracted and therefore learning how to identify and quantify them gives rise to a platform that spans all species barriers[11].

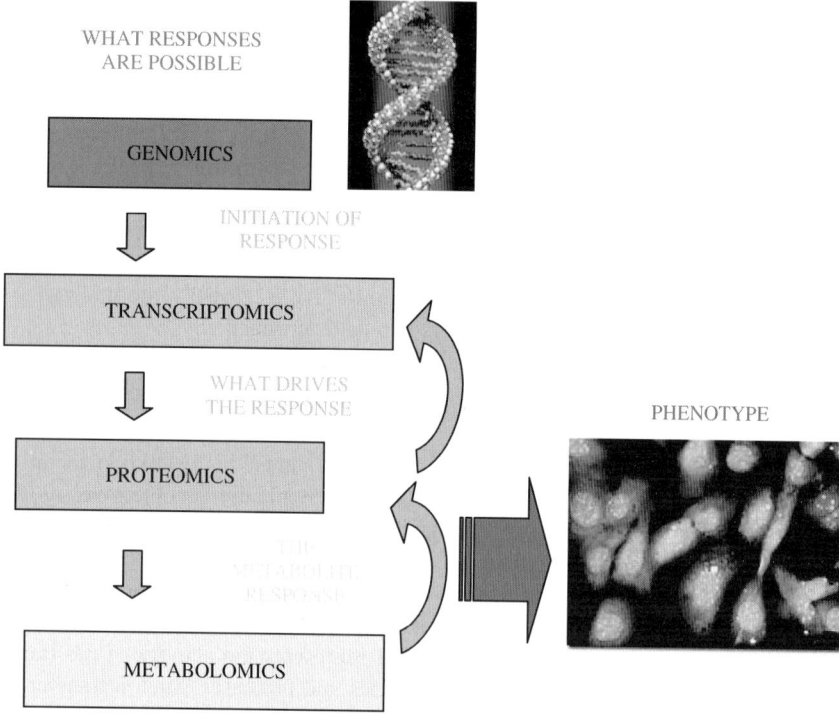

Figure 1. Overview of interactions between different 'omic's wtihin a cell

Profiling of the transcriptome and proteome has received some criticism due to its inability to always predict gene function. It appears that metabolomics can fill the gap in information and become a vital functional genomics tool[12]. Functional genomics refers to the elucidating of the causal relationships between genes and their eventual function as determined by proteins. Gene knockouts and genetic strategies using RNAi can be used with phenotypic screens like metabolomics to determine the physiological functions of enzymes. It has been used to map silent mutations[13,14] as well as to observe pleiotropic effects of single genetic alterations[13]. In the past, gene functions were often inferred from homology to previously investigated sequences. Metabolome analysis might go some way in positively identifying the enzymatic activity of the protein it encodes[15]. Therefore an integrative approach with transcriptomics and proteomics is often preferred when identifying enzymes, product and substrates[16]. This was demonstrated in plant biology by the use of expression sequence tag (EST) databases together with metabolic profiles[17,18].

A number of other diverse applications of metabolomic analysis exist. These include the commercial applications in agriculture, industrial biotechnology and xenobiochemistry[19], medical applications like biomarker discovery and nutritional health as well as environmental applications like environmental toxicology, development of organisms and pathogen-host interactions. Comparisons between mutant or transgenic plants with wild-type plants prove insightful as in the case of invertase in potato tubers[20] and in the transformation of ubiquitous naringenin to isoflavone genistein as a means of introduction of isoflaviods into legumes[21,22]. Similar mutant comparisons have been carried out on yeast mutants[23] and strains of mice[24]. These transformed organisms can then be used in commercial applications where metabolic engineering becomes useful. Even cell transfection can be monitored using similar techniques[25]. Other potential cell engineering applications could involve the monitoring of bioreactors or the development of protein media. Drug metabolism studies in rat livers[26] or in endometrial cells[27] allow us to create models for therapeutic drug development for the future. Medical applications for diagnosis of diseases by noninvasive methods seem a long way from reality, but metabolomics offers one type of solution by the analysis of urine or plasma. There has been significant progress in the advancement of biomarker detection for cancers[8,28,29] as well as coronary heart disease[30]. Examples of these biomarkers can be found in Figure 2. Other clinical uses include the monitoring of nutrition and health[31-33] in patients in a fairly rapid fashion. Symbiotic associations[34,35], fruit ripening[36] and embryo growth[37] are just examples of developmental biology studies done where metabolomics is used as a novel tool. Response of organisms to environmental pathogens or stressors can be easily monitored as is the case of the Withering syndrome in abalones[38] or the exposure of *Silene cucubalus* cell

Metabolite*	Metabolic function	Associated tumours/characteristics
Alanine	In conjunction with lactate, increases in tissues during hypoxia; made by transamination of pyruvate to prevent further increases in lactate[81]	Hepatoma and brain tumours, including astrocytomas, metastases, gliomas, meningiomas and dysembryoplastic neuroepithelal tumours
Saturated lipids	An important constituent of cell membranes, although membrane lipids are poorly resolved by NMR; lipid peaks detected by NMR are believed to either be present in cell-membrane microdomains or in cytoplasmic vesicles	Alterations in levels associated with proliferation, inflammation, malignancy, necrosis and apoptosis[20,28,33,38,82]
CCMs	Include choline, phosphocholine, phosphatidylcholine and glycerophosphocholine; these are key constituents of cell membranes	Levels change during apoptosis and necrosis; the tumour types that these changes have been found in include brain, sarcomas, prostate and hepatoma[31-34]
Glycine	An amino acid and an essential precursor for de novo purine formation	Decreased following disruption of the HIF-1 signalling pathway[16]
Lactate	An end product of glycolysis	Increases rapidly during hypoxia and ischaemia; poorly vascularized tumours have a low intracellular pH as a result of increased lactate production; increased rates of lactate production have been associated with a range of tumours and, in particular, certain types of neoplasms[43]
Myo-inositol	Involved in osmoregulation and volume regulation	Increased in colon adenocarcinoma, glioma, schwannomas, ovarian carcinoma, astrocytoma and endometrial cells[43,35]; decreased in breast tumours[84]
Nucleotides	Used to manufacture DNA and RNA; also key metabolic intermediates in fatty-acid and glycogen metabolism; changes in ATP concentration also indicate the energetic status of the tumour	Increased in glioma during apoptosis[41]; CDP-choline is also increased during apoptosis[14,85]
PUFAs	Constituents of cell membranes, especially mitochondrial	Increased in glioma cells undergoing apoptosis[20,21], and in dedifferentiated and pleomorphic liposarcomas[86]
Taurine	Important in osmoregulation and volume regulation; hypotaurine is also an antioxidant and might protect cells from free-radical damage	Increased in squamous-cell carcinoma[17], prostate cancer and liver metastasis[89]

*Metabolites identified by NMR spectroscopy of tissue extracts or magnetic resonance spectroscopy in vivo. CCM, choline-containing metabolites; CDP-choline, cytidine diphosphocholine; HIF-1, hypoxia-inducible factor 1; NMR, nuclear magnetic resonance; PUFAs, polyunsaturated fatty acids.

Figure 2. Metabolic biomarkers of tumours. Reprinted by permission from: Nature Reviews Cancer 'Metabolic profiles of cancer cells', Griffin J.L and Shockcor J.P., Vol.4 Issue 7, 2004

cultures to cadmium[39]. These examples represent only a small fraction of existing and potential applications of metabolomic technologies.

4.   CATEGORIES OF METABOLITE ANALYSIS

Metabolomics, the unbiased identification and quantification of all the metabolites present in a specific biological sample (from an organism).[40], is difficult to be carried out with the present analytical technologies. It is necessary to clearly identify what constitutes an ideal metabolome analysis. Metabolomics should:
- give an instantaneous snapshot of all metabolites, in any given system,
- use analytical methods that have high recovery, experimental robustness, reproducibility, high resolving power and high sensitivity[3] while being able to be applied universally,
- provide the unambiguous quantification and identification of metabolites, and
- allow distinguishing factors to be highlighted while easily being incorporated into biochemical network models.

Since metabolomics cannot be carried out in its totality, the use of different analytical approaches (categories) can help answer specific types of questions. These categories represent non-targeted approaches that are generally accepted as being constituent parts of the metabolomics field. They represent the approaches that contribute towards the goal of metabolomics and fuel growth in technological and methodological improvements. These categories include metabolite or metabolic profiling, metabolic fingerprinting and metabonomics. Metabolite Footprinting, which refers to analysis of extracellular media in which cells are grown in, is somewhat ambiguous as it falls at the border of the definitions given. For cell culture, this would be the analysis of cell-free supernatants or fermentation broths.

Metabolite fingerprinting can be used to screen a large number of clones in genomic or plant breeding programs, as well as to enable diagnostic usage in industry or clinical routines[41]. In such cases, it might not be necessary to determine the levels of all metabolites individually. Rather, such an approach looks at the evidence of major metabolic effects as a result of perturbations. Sometimes, this is termed pattern recognition. Rapid classification of samples according to their origin or their biological relevance allows the maintenance of high throughput analysis. Typically, fingerprinting analysis uses non-calibrated detector readings obtained from complex metabolite mixtures for sample classification and biomarker screening.[6]. Readings are then referenced against controls so that relative changes in signals may be compared. As a result of the emphasis on classification rather than on identity, the underlying physiological and metabolic differences are mostly not elucidated. In addition, the technologies used in this analysis have multiple signals for a single metabolite; therefore the masking effect of abundant metabolites may be evident. Artifacts affect data significantly and cannot be ruled out. In addition, metabolite extraction and recovery is usually not controlled.

Metabolic profiling is used to elucidate the function of a whole pathway or intersecting pathways and does not require the characterization of the entire metabolome.

Thus, analysis focuses on a chosen class of compounds (like amino acids or carbo-hydrates). Unlike targeted analysis where only specific metabolites are analyzed, profiling looks at general classes of compounds. For example, extraction of cells with perchloric acid ensures that all the metabolites are generally polar and soluble. Sample preparation and cleanup can be focused on the chemical properties of these compounds so as to reduce matrix effects. However, unlike metabolic fingerprinting, there is greater emphasis on identifying as many metabolites as possible. This is possible with hyphenated technologies. Profiling is normally used in the context of drug research in the description of catabolic degradation of an applied chemical[41]. In essence, doing several forms of profiling for different classes of compounds can add up to a comprehensive set of metabolites.

A similar definition to metabolite profiling is termed Metabonomics. Metabo-nomics[42] was first coined by Nicholson and his colleagues in 1996 to describe the studies of metabolite profiles in biofluids from whole organisms such as plasma or urine. As further elaborated, metabonomics is 'the quantitative measurement of the dynamic mutliparametric metabolic responses of living systems to patho-physiological stimuli or genetic modification' and the key feature in this type of analysis is pattern recognition.[43]. The metabolon refers to coordinated channeling of substrates through tightly connected enzyme complexes[44]. This has been higher useful in medical applications as pioneered by Nicholson and colleagues. Another more recent category has been termed discovery metabolite profiling (DMP). This has the objective of evaluating the global metabolic effects of enzyme inactivation in vivo[45], which determines enzyme substrate and products directly.

5.   COMPLEXITY OF METABOLOME ANALYSIS

In a more practical sense, 'global' analysis of the metabolome implies the ability to distinguish, somewhat structural and quantitatively, a comprehensive coverage of biochemical and molecular characterizations in a relatively high-throughput fashion. Coverage of a large number of metabolites is difficult when the number of metabolites in a given system is not known. Present textbook biochemical path-ways are insufficient to provide such information, and we can only estimate these from emerging data. It is believed that the number of metabolites for a partic-ular cell type should be lower than the number of genes and proteins in a cell[46]. However, if an analysis platform is to be used universally for various organisms, this number increases very significantly. Of the estimated 5000 different primary and secondary metabolites anticipated in a typical *Arbidopsis* leaf, Bino et al.[12] estimate that only *circa* 10% of the metabolome has been annotated using current technologies. It is further speculated that there are an estimated 200000 different metabolites in the plant kingdom[41], with the numbers in the mammalian systems being smaller. To put this into perspective, present microarray technologies for transcripts have an upper limit of about 15-20000 ESTs per array while a two dimensional polyacrylamide gel electrophoresis (2-DE) can readily differentiate a few thousand proteins with 10000 proteins as an upper limit.[47]. Hence, one could easily say that metabolomic technologies should be able to handle the analysis of at least 10000 metabolites, with the limit rising up to two or three times that number. The next issue about metabolite analysis is their diverse chemical compositions. Genes compose of a linear 4-letter code while proteins have a 20 letter code of primary amino acids as a foundation. Metabolites do not have any fixed codes and thus they can only be identified by their basic chemical atomic compositions like carbon, hydrogen and nitrogen. This makes a universal method of characterization difficult while most methods use the chemical properties of these entities to separate, identify and decipher their structures. These molecules can range from ionic inorganic species to hydrophilic carbohydrates, hydrophobic lipids and other amphoteric natural products. Furthermore, the elemental composition, the order of atoms and the stereochemical orientation may have to be elucidated *de novo* for metabolites[11]. Thus, as there is no single technology platform that has the ability to profile the metabolome simultaneously, a combinatorial approach is used. This would mean that selective extraction and analysis is done on a single sample, thus leading to several pitfalls and obstacles. Selective extraction could lead to biasness while the use of different instruments would mean that data integration and normalization throughout all data becomes difficult. As the metabolome has high connectivity this integration is complex[48] but necessary, especially since substrate metabolites may differ from properties of the product metabolite after an enzymatic conversion.

Analytical variance is a major factor when using different technology platforms and refers to the coefficient of variance associated with a particular experimental approach. This could also result when using the sample protocol with instruments in different laboratories. Therefore, since a combinatorial analytical approach is preferred, the knowledge and correction of analytical variance between experimental approaches is necessary. Another significant factor in these analyses is the biological variance. Typically, biological variance is larger than analytical variations. At times, biological variance within replicates may be larger than the discriminating differences between a control and treated sample, making it impossible to identify functional differences. One way of reducing this variance is to pool biological replicates, thus minimizing random variations through statistical averaging[15]. However, this could dilute any specific up or down regulated metabolites amongst the replicates. Homogeneous samples like cell cultures can help reduce this variation but will fail to factor in the interactions that a complete biological system could have.

Quantification and resolution of metabolites is also technological challenging. The dynamic range of a technology platform defines the boundaries of concentration that can be quantified linearly and unequivocally. This range is limited by the sample matrix (matrix effect) or the presences of interfering and competing compounds. The matrix effect is an ubiquitous observation in chemical and enzymatic analyses of complex biological samples. This effect can lead to false quantification due to either the stabilization of labile compounds (matrix stabilization causes over prediction of concentration) or the suppression of compounds (matrix suppression leads to under prediction of concentration). These can happen at any stage of the analytical methodology from sample extraction to the analysis itself. It is also known that recovery of chemical compounds during extraction is never complete, resulting in biasness. Thus, the recovery of metabolites becomes a factor in sample preparation or separation steps[49]. Moreover, to become an 'omic' science, metabolomic technological platforms need to cater to a wide dynamic range where a single sample may have an abundant concentration of a metabolite while having a minute amount of another. The presence of these excessive metabolites causes interferences to the identification of other metabolites. The presence of contaminating compounds can create significant background noise such that the presence of less abundant compounds is overlooked. Co-elution or unresolved peaks can further complicate problems. Thus, the resolving power of the analytical techniques need to be carefully assessed. In essence, the importance of these issues may make a difference in the identification of biomarkers associated with a phenotypic state, especially at an early stage of detection.

Interfering or competing metabolites may be only present in a small amount but can significantly lower the performance of the instruments used. These include the presence of salts in proton nuclear magnetic resonance (NMR) as well as in electrospray ionization type mass spectrometers (ESI-MS). The pH of samples in proton NMR can affect the identification of the metabolites as peaks shift[50]. Proteins not removed or chemicals used in extraction may also lead to desensitization of the instruments. Thus, biasness towards certain metabolites may result due to the presence of certain contaminants or chemicals. These have to be reduced to a minimum and experimental protocols thoroughly validated. Indeed, very often, a compromise between sensitivity and throughput needs to be addressed. The greater the resolution and sensitivity of instruments, the lower the number of metabolites it will be able to analyze in a given period of time. This can be said to be a universal problem in all 'omic' sciences and an optimal balance between accuracy and coverage needs to be found[51].

6.   METHODOLOGY

The different analytical objectives and diverse types of metabolites give rise to multiple analytical methodologies. Sources of metabolite samples also differ and may originate from tissues, cell cultures and biofluids. Therefore, some amount of sample preparation is necessary. Since this is a rapidly growing field, new methodologies, particularly in the area of instrumentation and data processing are constantly being developed. In an authoritative commentary about metabolomics, Brown et al.[52] have suggested a 'pipeline' (see Figure 3) for how experiments can lead to successful knowledge building. This includes the proper design of experiment, instrument optimization, data storage and manipulation, data processing and validation. In general, all metabolite approaches take the similar methodology of sampling, sample preparation, analysis using instrumentation, data processing and finally deriving knowledge using modeling or bioinformatics tools. This section aims to give an overview of these processes.

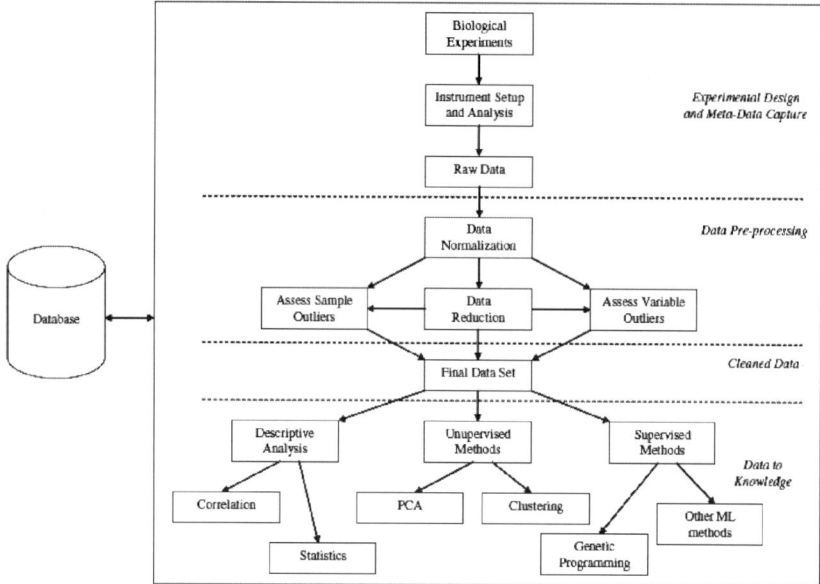

Figure 3. An overview of a pipeline for the design, performance, storage and analysis of metabolomics experiments and their attendant data. With permission from: Metabolomics, 2005, Vol. 1, 39-51, 'A metabolome pipeline: from concept to data to knowledge', Brown M et al

6.1   Sample Preparation and Extraction

The first step to ensuring the simultaneous detection of a large number of metabolites is to have adequate methods for sample preparation and extraction. Erroneous sample preparation can lead to misleading or inaccurate data, even with the most sensitive instruments. Sampling techniques, disruption of tissues, extraction, storage and pre-analysis preparation are just some of the necessary steps taken before instrument analysis. The amount of sample needed for the detection of unidentified peaks of very low abundance and the amount of sample required for structural elucidation of unknown compounds may differ.

There are two types of sampling, invasive (intra-cellular metabolites) and noninvasive sampling (urine, metabolite footprint). Footprinting is used if metabolites are naturally excreted into the supernatant. Invasive sampling can provide an instantaneous snapshot of the metabolome if cellular processes are stopped immediately. As metabolic processes are rapid, the first necessary step is to rapidly stop any inherent enzymatic activity or any changes in the metabolite levels. This is sometimes termed quenching. In addition, sample methods should not be bias towards any group of molecules but this challenge is presently unresolved. The time and method of sampling are important issues to be considered to ensure reproducibility in the analytical sample, especially since large number of biological replicates is commonly used.

To quench enzymatic activity, several methods are used, including freeze clamping (with lower temperature receptacles), immediate freezing in liquid nitrogen or by acidic treatments like perchloric or nitric acid[53]. Freezing in liquid nitrogen is generally considered to be the easiest way of stopping enzyme activity provided that cells or tissues are not allowed to thaw partially before extracting metabolites. In order to prevent this from happening, enzyme activity is inhibited by lyophilization or by immediate addition of organic solvents while applying heat. Metabolites from cellular compartments can even be distinguished if non-aqueous fractions of lyophilized samples are separately analyzed.[54]. However, some loss of metabolites may still possible due to cold shock[55]. In addition, lyophilization of samples before extraction may result in the irreversible adsorption of metabolites on cell walls and membranes[41]. Freeze clamping is a faster process of freezing cells, thus avoiding potential artifacts caused by wound response[41]. This may be an issue if tissue samples are used and have to be weighted before freezing. However, for a large number of samples, freeze clamping is not easily enforced. Acidic treatments can severely reduce the number of metabolites detected as a result of degradation due to the low pH. Acidic treatments also pose severe problems for many analytical methods that follow, thus the acids have to be removed. Cold organic solvents may be directly added to tissue samples and kept below temperatures of $-20\,^{\circ}\mathrm{C}$ during the entire sample preparation.

Cells are subsequently disrupted, releasing the metabolites. For plant cells with relatively hardy cell walls, this may be a difficult task. Frozen samples may be grounded down by sonication, homogenization by mechanical means (like mortars or ball mills) in pre-chilled holders[56] or directly in an extraction solvent[57]. Most frequently, polar organic solvents like alcohols are added directly to frozen samples to extract polar compounds while non-polar solvents such as chloroform or dichloromethane allow the extraction of lipids and other hydrophobic compounds. Sometimes adding a mixture of polar and non-polar solvents allows for extraction of both classes of metabolites. Hot alcoholic extractions are also performed routinely. A procedure for the extraction and separation of metabolites, proteins and mRNA from a single sample has also been reported[58].

As mentioned previously, a compromise is often required between complete recovery of some metabolites and the prevention of degradation of more labile metabolites. Gentle and cold conditions may cause the chemical degradation of some compounds while aromatic compounds may require significant amounts of energy (heat) to be released from membranes or protein complexes. Some vitamins are prone to oxidation, while other metabolites only remain stable in acidic conditions. Once the metabolites are extracted, the mixture of cell debris, protein and the desired metabolites need to be separated. This is done by centrifugation. Removal of debris and proteins eliminates contaminants which affect analyses. Samples are subsequently stored or ran on instruments immediately. Care must be taken to avoid further sample contamination from the environment as well as the degradation of the sample before analysis. Samples may also be lyophilized and frozen till ready for analysis. Freeze/thawing of samples should be avoided. Because some high-throughput instruments have autosamplers, it has to be noted that samples awaiting analysis should not be allowed to degrade. Samples may require pre-instrument preparation steps like being placing in a suitable buffer for proton NMR analyses or chemical derivatization for chromatographic analysis. All these factors influence the precision, stability and reproducibility of results and thus need to be carefully controlled and validated[46]. Organism specific validation of extraction methods have been carried out on yeast[59] and *E. coli*[60] where hot ethanol, hot and cold methanol, perchloric acid, alkaline and methanol-chloroform extraction methods were compared and recoveries of classes of metabolites studied.

6.2   Analytical Instrument Platforms

Without a single acceptable technology to visualize the metabolome, it is first important to highlight the characterisitcs of an ideal analytical process. Firstly, the instruments need to have excellent sensitivity (see Figure 4), that is, to be able to analyze of multiple metabolite classes without lose of resolution. Peaks should therefore not represent the merger of several components. It should also allow easy identification and quantification of metabolites and allows the comparison of relative changes in metabolite abundances in comparative experiments. Lastly, it should have a rather short analysis time to increase the number of samples analyzed. High throughput can be understood in two ways, rapid analysis and/or able to have a wide coverage of components. One way of increasing resolution is by reducing the number of metabolites that are simultaneously analyzed by the instrument and reducing analysis time. This can be achieved by using simple fractionation steps (as seen in sample preparation steps where lipophilic and hydrophilic metabolites are separated) and by separating metabolites by means of suitable chromatographic

NMR	10^{-6} mol
LC/UV	10^{-9} mol
GC/MS	10^{-12} mol
LC/MS	10^{-15} mol
LC/L.I.F	10^{-19} mol
CE/L.I.F	10^{-23} mol

Figure 4. A comparison of the relative sensitivities of various metabolomic tools. The limit of detection indicated on the right side of the figure corresponds to the instrument on the left. Reprinted from: Phytochemistry, Vol. 62, Sumner L.W. et. al, Plant metabolomics: large-scale phytochemistry, 825

methods. Therefore, by fractionation and applying the increased resolution technique stated above, the criteria for metabolomic analyses is met[41]. Instruments used can be stand-alone systems like the mass spectrometry (MS) or nuclear magnetic resonance (NMR) spectroscopy. A short summary of spectroscopic methods can be found in Figure 7. However, to determine unknown structures of metabolites, information from the intact molecule needs to be obtained, such as size and elemental composition. Therefore, hyphenated instruments have shown to be extremely useful, particularly established technologies like gas chromatography-mass spectrometry (GC-MS). Having hyphenated instruments allow for an extra dimensionally of analysis. Dimensionally does two things; increase the separation of compounds by different characteristics like retention time and different physical properties like mass and provide additional structural information like in the case of 2-D J-resolved types methods that minimize ambiguity in spin multiplicity and coupling constants in NMR spectra. Therefore, may emerging instruments now boast multi-dimensionality (LC-LC-MS, GC-MS-MS *etc.*). It is also possible to combine data from separate instruments, to form a comprehensive study on metabolism[61–63]

6.2.1   *Nuclear magnetic resonance (NMR) spectroscopy*

Nuclear magnetic resonance spectroscopy (NMR) is a non-invasive, highly discriminatory, high-throughput method that can analyze rather crude samples. Strong magnetic fields and radio frequency pulses are applied to the nucleus of atoms. This causes the nucleus to possess a spin (nuclear spin) and cause them to be excited, that is, to go from a low energy level to a high energy level. The nucleus will align itself in the magnetic field and there are only a few permissible orientations as it is a quantum mechanical system. If the correct energy quanta are applied, the nucleus will be able to resonate, constantly flipping between energy states with an induced emission of quantal energy occurring.[43]. When relaxation of the pulse begins, emission of radiation from the atoms is detected by the instrument to give rise to spectra. This is termed 'free induction decay' and is detected in the time domain. Fourier transformation is subsequently carried out to convert it to a frequency spectrum. NMR has a good chemical specificity for compounds containing elements with non-zero magnetic moments such as ^1H, ^{13}C, ^{15}N, and ^{32}P. Hydrogen, carbon, nitrogen and phosphorous atoms commonly form the backbone of biological metabolites[64]. A single metabolite will result in several signals in the NMR, causing the problem of resolving individual metabolites. This is disadvantageous as it makes identification of metabolites difficult in convoluted complex spectra. Acquisition time for NMR can be between 10–15minutes per sample and is favorable with transcriptomic and proteomic approaches. However, NMR falls short of resolution and sensitivity compared to MS methods. Sensitivity depends on the natural abundance of the atoms studied (^1H, ^{31}P, ^{13}C *etc.*) or the artificial introduction of the isotopes into the sample. Significant amounts of culture or tissue are required for metabolite extraction. By increasing the strength of magnetic fields, increased specificity with greater resolution and separation of signature chemical shifts is possible. In addition, longer analysis times or the use of cryogenic probes

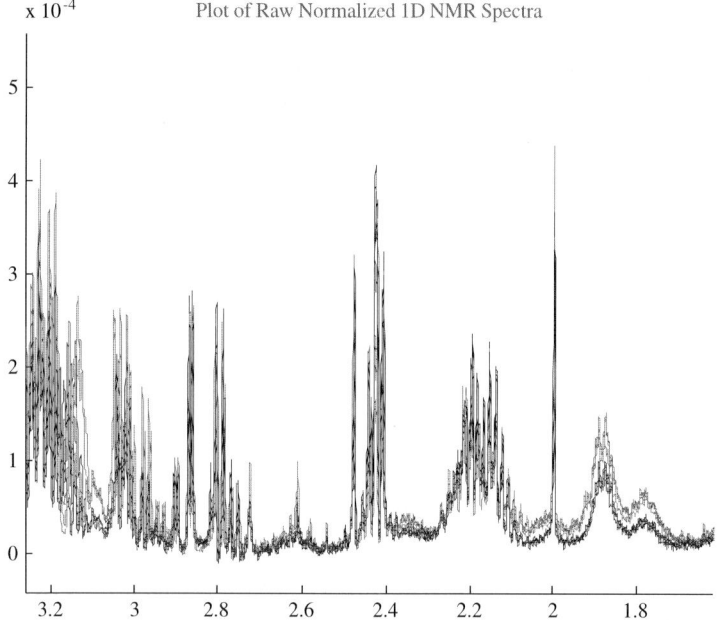

Figure 5. Overlaid spectra of replicate extractions of metabolites extracted from Mouse myeloma NS0 cells. Horizontal axis represents parts per million while the vertical axis represent the intensity in arbitrary units

can help[65]. Another disadvantage of NMR based analysis is the fact that the probes are generally large in volume and hence the sample volumes required are comparatively large. The NMR spectrum (see Figure 5) typically appears as a chemical shift (in parts per million) and represents the difference between the resonance frequency of the observed proton and that of a reference proton present in the reference compound. How the individual group of atoms shift depends on the effect of shielding by electrons orbiting the nucleus, thus similar chemical groups of different metabolites could exhibit varying chemical shifts. The most common reference for proton NMR is tetramethylsilane (TMS) which is taken to be 0 ppm. The presence of different compounds in the crude sample does not affect the chemical shift and quantification is based on the signal intensity of the peaks.

The resulting spectra are complex, with overlapping peaks and thus require significant data processing. However, to remove ambiguity to the assignment of metabolites, an addition dimension maybe added. These include the J-resolved, homonuclear shift-correlated spectroscopy (COSY), total correlation spectroscopy (TOCSY) and nuclear overhauser effect spectroscopy (NOESY)[50]. Intact cells or tissues are often studied with the application of magic angle spinning[43] where the sample is rapidly spun at an angle in the spectrometer and short relaxation times produces resonance broadening and reduces spectral information. Most NMR based programs use ^{1}H proton and appear to be focused on biomarker analyses rather than on the comprehensive analyses of a large number of metabolic pathways. Experiments are typically done on metabolite fingerprints of mammalian samples, and also include the study of *in vivo* metabolites[66]. It has also been shown to have reasonably good inter-laboratory reproducibility as seen in the Consortium for Metabonomic Toxicology (COMET) project[67]. In a ground breaking use of the NMR, metabolite concentrations in yeast mutant were followed and correlated to the function of genes. This was termed Functional Analysis by Co-responses in Yeast (FANCY)[14].

6.2.2   Mass spectrometry

Mass spectrometric methods are by far the most widely used instrument in the field of metabolomics (including hyphenated technologies). It operates on the principle of ion formation and then the separation of ions according to their mass-to-charge ratio. Direct injection mass spectrometry (DIMS) is the direct injection of samples into low-resolution electrospray ionization (ESI) MS instruments, resulting in a quick (less than 1 minute per sample) and useful way of getting high throughput (hundreds of samples) with sufficient information. This approach was taken in characterizing olive oils and yeast strains[68–70] and can only be used as a screening method. Ion suppression is, however, a major problem in mass spectrometry. Recent technological advancements have made time-of-flight (TOF) MS acquisition time faster and mass determination very accurate. TOF instruments can provide mass resolutions greater than 4000 at mass 200, which allow the resolution and the detection of metabolites of the same nominal mass but different monoisotopic masses[46]. Matrix-assisted laser desorption ionization (MALDI) MS methods are advantageous as it is a time-of-flight (TOF) instrument that gives direct mass to charge ratios but has substantial interference from the matrix used. However, to circumvent the problem, a matrix-free system was developed called the 'desorption ionization on silicon' (DIOS) method[71] which when coupled with a TOF instrument, becomes a powerful tool for metabolite quantification[72,73]. When multiple ionization techniques coupled to Fourier transform mass spectrometry (FTMS) were used together, it was able to identify metabolites specifically associated with the development and ripening of strawberry fruit[36]. FTMS has high resolution which allows for the separation and differentiation of very complex mixtures and has high mass accuracies that allow for the calculation of elemental compositions for compound structural elucidation and characterization. Unfortunately, the major problem with MS methods is that they cannot differentiate chemical isomers with identical mass to charge ratios such as those of common hexoses. Furthermore, due to the disruption of chemical bonds during ionization, structural identification from intact masses of the molecule is lost. The use of mass spectra libraries goes some way toward reducing this problem. Routine spectral deconvolution algorithms[74,75] have been developed to find peaks without prior knowledge of their abundance, mass spectral characteristics or retention times.[40]. To improve the quality of mass spectra from complex mixtures, genetic search algorithms have been shown to reduce matrix effects by systematically varying mass spectral conditions[76].

Even with TOF instruments, the mass accuracy is typically only 5-ppm accuracy. This means that without separation of metabolites, overlapping peaks could result in mass differences lower than that of the threshold. Fourier Transform Ion Cyclotron Resonance (FTICR) MS can overcome this problem. FTICR has a huge resolving power ($100,000+$) and high mass accuracy (< 1 ppm external calibration, < 0.1 ppm internal calibration). It has a better limit of detection than a high performance quadrapole TOF (Q-TOF) instrument. It would therefore be possible for all metabolite peaks to be resolved and their molecular formula calculated very accurately. Masses up to 500Da can be measured and FTICR allows for accurate measurements in MS^N which simplifies structural elucidation and identification[77]. The disadvantages of FTICR are that it has smaller dynamic ranges and still fails when structural isomers having the same monoisotopic mass are employed. The smaller dynamic range is still considered relatively acceptable and can be increased by increasing the time for analysis. The typical issues with sample ionization will still exist with the FTICR. Choosing one of the two most universal sources (ESI or MALDI) and optimizing the analysis to yield the most information seems to be the only choice until a universal ionization mode for all metabolites is found. This new instrument will become increasingly popular in metabolomic analyses and when coupled to software that can exploit the information in isotope patterns, it can produce the empirical formula for metabolites directly[36].

6.2.3   Chromatographic and other column separations

Thin Layer Chromatography (TLC) has been employed to detect changes in 70 most abundant ^{13}C-labeled compounds found in *E. coli* under various culture conditions[78]. However, TLC is a simple and low resolution tool that cannot accommodate complex mixtures. The most common forms of chromatography are gas chromatography (GC) and liquid chromatography (LC). In liquid chromatography, there is a shift from the standard High-Performance Liquid chromatography (HPLC) to the Ultra-Performance liquid chromatography (UPLC) which can significantly increase resolution sensitivity and peak capacity[79] while decreasing sample volumes and mobile phases[80]. UPLC systems operated at high operating pressures and use sub-$2\mu m$ porous packing. The move towards smaller bore capillary size columns is also advantageous as complex samples require high sensitivities. Unlike pressured systems like the LC, capillary electrophoresis (CE) makes use of an electric field to move molecules towards the detector, much like gel electrophoresis. Capillary electrophoresis coupled with ultra violet (UV) or laser induced fluorescence (LIF) detectors are highly sensitive but lack selectivity[15]. With different detection modes available, it is possible to add two detectors to a single chromatography step as in the example of liquid chromatographic instruments (LC/UV/PAD) where PAD is an electrochemical method termed photodiode array detection[81]. However, LC/UV/PAD methods require compounds to contain chromophores.

Electrochemical detectors together with LC instruments are used commonly as an alternative detection step. There are several modes of electrochemical detection such as amperometric and coulometric[82]. Electrochemical detection is based on the monitoring of changes in an electrical signal as the result of the equilibrium concentrations of the reduced and oxidized forms of a redox couple. Coulometric array electrochemical detection identify only redox active compounds, a property that not many detectors can utilize[83]. Thus, many optically inactive compounds like carbohydrates can be detected using this method. Flow from a HPLC may also be split between the electrochemical array detector and an MS instrument[84]. Recent validation of the variation and analytical precision of a HPLC-coupled coulometric array detector show low peak shift coefficient of variance (CV) but the quantification of peaks show a larger variance[85], due mostly to biological variation. Pulsed electrochemical detection is also another mode used commonly[86]. Such detection can also be used for carotenoids, polyphenols and flavonoids, allowing one to circumvent the limitations of MS[87,88].

6.2.4   Other vibrational spectroscopies: FT-IR, NIR, Raman

Vibrational spectroscopies are relatively insensitive but FT-IR allows for high-throughput screening of biological samples in an unbiased fashion. In FT-IR, a sample is permeated with light (or electromagnetic radiation). Chemical bonds at specific wavelengths absorb this light and vibrate in one of a number of ways, such as stretching or bending vibrations. These absorptions/vibrations can then be correlated to single bonds or functional groups of a molecule for the identification of unknown compounds, mostly polar bonds. Spectra are examined for a number of clearly defined peaks in the major region of interest, mostly in the mid-IR, which is usually defined as 4000–$6000 \, \text{cm}^{-1}$. Samples require little or no preparation, with as little at $20 \, \mu\text{l}$ of sample required. However, IR has some drawbacks. Similar to NMR, water signals pose a problem and must be subtracted electronically or attenuated total reflectance may be used. Compared to the other instruments, it is one of the least sensitive and selective but its ability to analyze large numbers of samples in a day (up to a 1000 spectra) make it a plausible method. FT-IR has been used in the preliminary analysis of the yeast metabolome[2] (Figure 6 gives an example of a near IR spectra from yeast fermentations) as well as in a non-invasive manner to study the over production of metabolites from E. coli and S. aureus in vivo[89]. Metabolite fingerprinting techniques for diagnosis of disease by the analysis of tissues and biofluids using FTIR are also possible[90–92].

FT-IR studies outnumber the studies done with Raman or Near infrared (NIR) instruments. However, Raman spectroscopy seems complementary to FT-IR as it measures predominantly non-polar bonds but has not often been used for metabolite analysis. Near infrared (NIR) has shown greater potential in this area. Studies conducted on lactate in human blood[93] show that NIR could possibly be used for metabolite measurements as it measures the overtones and interaction of vibrations between metabolites.

1st derivative supernatant spectra (5600-6200cm^{-1}), 1mm, 16cm^{-1}, 128 co-added scans

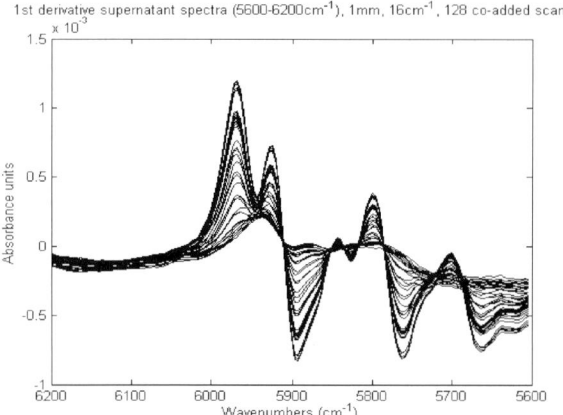

Figure 6. First derivative spectra of NIR spectra from supernatants of yeast fermentations. Courtesy of McLeod G, University of Birmingham

6.2.5   *Hyphenated mass spectrometric instruments GC-MS, LC-MS, CE-MS*

Liquid chromatography-mass spectrometry (LC-MS) and gas chromatography-mass spectrometry (GC-MS) methods are the most common hyphenated technologies. Furthermore, GC-MS is considered the 'gold standard' in metabolite detection and quantification[44]. They offer good sensitivity (limits of detection being pmol or nmol concentrations) and selectivity but have relatively longer analysis times[15]. LC-MS methods typically have a somewhat lower chromatographic resolution than do GC-MS methods, but as GC-MS methods require samples to be volatilized; LC-MS methods can analyze higher mass ranges. GC-MS is a relatively low cost alternative that provides high separation efficiencies that can resolve complex biological mixtures. Most samples require chemical derivatization to allow for easy volatization under flame ionization, but this comes at the cost of additional time, processing, and variance. Two step derivatization is employed to cater to the wide range of chemical functionalities[20]. Ideally, an automated system employing a 'derivatization-when-required' approach is used to ensure maximum stability[46]. Therefore, there is biasness against non-volatile and high molecular weight compounds. Typically, GC-MS is performed with affordable single quadrupole MS and takes about 30 minutes. Not all compounds can be quantified by GC-MS methods, particularly thermolabile or large molecules such as bis- and triphosphates, CoA adducts or lipids[41]. Each sample injection can be as low as 1 μl and separation occurs on capillary columns with different polarities and stationary phases. To search for physical properties unique to each metabolite while distinguishing it from neighboring peaks, a deconvoluting software is often used on the mass spectra[94,95] (AMDIS: http://chemdata.nist.gov/mass-spc/amdis). MSFACTS[96] is another program that can elucidate a list of metabolites from a database such as KEGG based on the MS profiles and specific GC retention times. Recently, a database for GC-MS data on metabolites was proposed[97]. This database

Technique	Description	Advantages	Disadvantages
Fourier-transform infrared (FT-IR) spectrometry	Uses vibrational frequencies of metabolites to produce a fingerprint of metabolism	Cheap and good for high-throughput first screening; used to differentiate yeast respiratory-chain mutants from wild-type strains[77]	Very difficult to identify which metabolites are responsible for causing changes; very poor at distinguishing metabolites within a class of compounds
Gas chromatography–mass spectrometry (GC–MS)	The method of choice for plant metabolomics; uses gas chromatography to separate metabolite mixtures prior to mass spectrometry to identify the different metabolites	A relatively cheap and reproducible method that also has a high degree of sensitivity	Sample preparation can be time consuming; not all compounds are suitable for gas chromatography
Liquid chromatography–mass spectrometry (LC–MS)	A similar approach to GC–MS, except separation occurs during liquid chromatography	This method is increasingly being used in place of GC–MS as sample preparation is not as time consuming; similar in sensitivity to GC–MS	More costly than GC–MS and depends on the reproducibility of the liquid chromatography (potentially more difficult to control than gas chromatography); also can suffer from ion suppression, where metabolites are poorly ionized when in the presence of cations and anions
Metabolite arrays	These devices use a 96-well plate assay system for phenotyping; such arrays have been used to phenotype *Escherichia coli* by 700 different assay mixtures ('assay-on-a-chip')[78]	Good as a screening tool when produced for a given situation	The number of metabolites that can be measured is limited by the number placed on the chip; difficult to screen for unknowns and follow metabolism of xenobiotics
Nuclear magnetic resonance (NMR) spectroscopy	This approach has been widely used by the pharmaceutical industry and in the screening of patient urine and blood plasma samples	A non-invasive technique – the use of NMR spectroscopy demonstrates that metabolomic analysis of tissues in humans is possible; it can be fully automated and has a high degree of reproducibility; relatively easy to identify metabolites from simple one-dimensional spectra	Lower sensitivity than mass spectrometry; co-resonant metabolites can be difficult to quantify; drug metabolites can be co-resonant with metabolites of interest
Raman spectroscopy[79]	An extension of FT-IR and ultraviolet/visible-light spectroscopy; relies on light scattering following irradiation with a laser	Has the advantage over FT-IR in that water has only a weak Raman spectrum and, therefore, many functional groups can be observed (for example, better distinction of carbon–carbon bonds)	Very difficult to determine which metabolites are responsible for causing changes; very poor at distinguishing classes of compounds
Thin-layer chromatography (TLC)	Used to follow the metabolic fate of ^{14}C-glucose in *E. coli* under different culture conditions[80]	Inexpensive	Inter-assay variation, limited in terms of the metabolites that can be quantified

Figure 7. Different spectroscopic methods used in metabolomic analysis. Reprinted by permission from: Nature Reviews Cancer 'Metabolic profiles of cancer cells', Griffin J.L and Shockcor J.P., Vol.4 Issue 7, 2004

would make use of mass spectral and retention time index (MSRI) which would also contain mass spectral metabolite tags (MSTs) profiles from various sources. Quantification is carried out by either external calibration or response ratio (peak area of metabolite/peak area of internal standard). External calibration is time consuming and not all metabolites are available for the creation of standards. GS-MS methods have been shown to have relatively low variance as seen in the example of quantification of 149 polar metabolites with mean deviations of ~8%[98]. By matching the retention time or retention index and mass spectrum of the sample peak with those of a pure compound previously analysed on the same or different instrument under identical instrumental conditions, compounds can be identified. Further structural information is determined if an electron impact mass spectrometer is used. This is possible via the interpretation of fragment ions and fragmentation patterns after collision induced dissociation (CID).

LC-MS can overcome the shortcomings of GC-MS methods. LC/UV detectors can detect double bonded or aromatic substances while different separation modes can be chosen. Most ionization methods are electrospray (ESI) in nature while a few rely on atmospheric pressure chemical ionization (APCI). Electrospray instrumentation operates in positive and negative ion modes (either as separate experiments or by polarity switching during analyses) and only detects those metabolites that can be ionized by addition or removal of a proton or by addition of another ionic species. Metabolites are generally detected in one ion mode, so wider metabolome coverage can be obtained by analysis in both modes[46]. ESI MS cannot tolerate high concentrations of salt, acid or base, so LC mobile phase compositions can be rather limited. Liquid phase methods suffer from significant matrix effects, a major one being the presence of non-volatiles, which may reduce the evaporation of volatile ions during the electrospray process[99]. This effect is termed ion suppression and can only be circumvented by reducing the size of liquid droplets[100]. Sample derivatization is not necessary but may be advantageous as it could be used to increase resolution and sensitivity[101] Typically a 10 μl sample can be injected into a 2.1 mm × 10 cm column and a gradient elution carried out. Capillary columns, which have a smaller diameter are also used to increase sensitivity[102]. Reverse phase liquid chromatographic separation allows the separation of most compounds but more polar compounds require normal phase or ion-exchange separations. Ion exchange modes are not preferred due to their high salt content which interfere with the MS, thus hydrophilic interaction liquid chromatography (HILIC) is used instead[103]. Unlike GC-MS systems where deconvolution software has been successfully introduced, LC-MS instruments do not possess these algorithms. Sample quantification requires external calibration with peak areas being used. Since ESI does not result in fragmentation of molecular ions, direct metabolite identification by comparison of mass spectra is not possible. Two dimensional maps have been created using the retention time and mass-to-charge ratios as dimensions. ESI mass spectral libraries are not commonly available as is the case for GC-MS, therefore accurate mass measurements by tandem MS (MS/MS or MS^n) provide collisional induced dissociation (CID) data that could lead to the identification of a specific metabolite[63]. CID resultant mass spectra from different instruments and different manufacturers have been shown to be similar but variations could arise from the ESI sources.

Capillary electrophoresis-mass spectrometry (CE-MS) is another method that can be used to separate a variety of cationic, anionic, nucleotides and coenzyme metabolites while identifying and quantifying them on the MS. In this instance, samples from *Bacillus subtilis* extracts were run in CE-MS instruments operating in different modes, proving that this can be advantageous.[104] CE was also used to determine the metabolites in rice leaves using MS and PAD detectors.[105] A further development of the CE setup for the study of carbohydrates[106] was used, where a microfluidic chip was used with a PAD detector.

6.2.6   *Other multidimensional instruments*

The use of hyphenated instruments has led to the development of other multidimensional instruments. Since LC-MS and NMR analyses are complementary[79], on-line LC-NMR and LC-NMR-MS approaches have been developed[107,108]. LC-NMR techniques were first developed together with other online separation techniques[109,110] as they were able to overcome the shortcomings of NMR spectroscopy. Special probes for the NMR instruments are required and flow cells are larger than those in normal UV detectors. Solvents used must also be carefully chosen and liquid flows have to be temporarily diverted or 'held' in loops to allow for the NMR acquisition to occur. These coupled on-line setups have been built to cope with mixed solvents and a separate sample compartment containing a deuterated liquid to provide a signal for stabilizing the magnetic field[107]. One way of overcoming the solvent issues was to use an online solid phase extraction column (SPE)[108]. For the double coupling of the MS and NMR, a compromise is required for the choice of solvents used with both instruments. This allowed for the nondestructive NMR analysis to occur before the destructive MS analysis. LC-NMRs have to operate at lower duty cycles of the NMR and have elevated expenses due to the need for deuterated mobile phases. However, once these methods were optimized, pharmacokinetic studies where sample like urine[111] have commonly been examined. *De novo* identification of secondary plant metabolites has also been carried out on an LC-NMR system[112] and a direct biochemical characterization of broths of *Streptomyces citricolor* has been studied using a similar method[62].

LC-MS-MS methods have been used successfully to elucidate the biochemical pathways of propanolol degradation in rats[113]. ^{13}C isotopes have been used together with LC-ESI-MS-MS instruments to follow the metabolism of yeast cells thereby eliminating the drawbacks such as nonlinear responses or matrix effects. This isotopic experiment was termed 'Mass isotopomer ratio analysis of U-^{13}C labeled extracts' or MIRACLE for short[114]. In addition, the use of multi-column separations seems logical as this adds an addition level of separation in an automated online fashion[79]. GC $\times$ GC time of flight (TOF) MS is another innovative multidimensional system that has been developed[115].

6.3 Data Processing

The overall objective of metabolomics is to associate the relative changes in quantitative or qualitative metabolite level with functional assignments. These functional assignments are used to distinguish groups according to the design of the comparative experiment. Therefore, data processing is an important process for making sense of the 'soup'[116] and the steps taken depends on what question was initially asked (hypothesis approach). For example, chemical identification for novel metabolites may not need peak quantification. If the reconstruction of metabolic networks or gene-phenotype correlations is required, both may be necessary. In a discovery approach, general steps for processing may be carried out but care has to be taken to ensure vital information is not filtered out. Statistical analysis must be performed to ensure analytical rigor[15]. How this is done remains a debate, as there is no agreement as to what constitutes the best statistical procedure, since different methods could give varying results. Certainly, for useful knowledge to be derived, chemometric, comparative and visualization tools will help. Table 1 shows some of the common methods used for preprocessing, reduction and processing of data. Many of the statistical analyses performed are pattern recognition and multivariate discrimination analysis. As with all 'omic' sciences, large datasets will be required to be

Table 1. Common methods for data preprocessing, reduction and processing

Data preprocessing

Normalization of data-data transforms
Normalization of data-using internal standard(s)
Baseline correction, peak shifting and noise removal
Missing value correction
Deconvolution of peaks
Data reduction
Limiting data analysis to specified representative region of data
Excluding variables or samples that have not consistent in replicates or lie outside the analytical limits
Exclude sample outliers
Data processing methods
Univariate and multivariate statistics
Coefficient of variation
ANOVA or MANOVA
Correlation or regression
Unsupervised methods
PCA, ICA and subtypes
Clustering-HCA, k-means
Self organizing maps
Supervised methods
Fisher discriminant analysis
Partial Least Squares
Neural Networks (ANN, PNN)
Genetic programming and algorithms

stored, retrieved and analyzed in an open consistent format. In addition, with the use of multiple instruments, data standardization is necessary for integrated data analysis. Likewise, with transcriptomic data, universal data standardization is pertinent if data sharing and databases are to be established[117]. The Standard Metabolic Reporting Structures working group has outlined a format as to how metabolic data is reported, which includes the experimental and data processing stages.

The diverse data from instruments is typically made up of spectra or plots. While some instruments possess built in software for deconvolution or analysis, most others do not. To begin, raw data will require normalization, alignment, corrections, noise removal and transformation. These procedures are of great importance[118] and are typically termed preprocessing steps. For HPLC chromatograms peaks can be normalized using a stretching algorithm[119], which accounts for the retention time drifts. A high-speed peak alignment algorithm for GC data has also been developed[120]. In GC and LC techniques, peaks are integrated to give an area for use in a transformed matrix. In NMR spectra, new methods for peak alignment include segment-wise peak alignment (SWA) or an improvised version by beam search algorithm and peak alignment by reduced set mapping (PARS)[121]. Subsequently, the NMR data can be subjected to data segmentation (at 0.005ppm bin width) and a logarithmic transformation[122]. To deal with a large-scale data set, it is very important to carry out the correct preprocessing steps as batch to batch differences in data become a clear problem. False positives due to statistical overfitting may also result. Thus, a validation strategy was introduced by Bijlsma et al[123] for a large data set of 600 plasma samples. This process involved data clean-up and alignment of the LC-Ms data, normalization with an internal standard (or with total spectrum intensity), combining duplicates, determination of missing values (peaks not present or below threshold or missed by integration function), centering and scaling. In a separate validation study, it was determined that careful sample preparation and post-acquisition peak alignment were important factors as chemometric methods were bound to highlight the small differences which could lead to inaccurate results[49].

The lines between pattern recognition and multivariate analysis are not distinct as there is a great deal of similarities between the two. Both aim to reduce the dimensionality of the measurement vector. This can be done by selecting a subset of variables directly as a representation of the total process, combining (transforming) the original measurements to from a new set with fewer features (feature extraction) or by a combination of both methods[124]. The probability distribution is estimated and this probability density function (or its estimates) would help in the process of validation and testing the system against the available data. A classification system is determined during this process and now derives 'rules' that allocate new observations to pre-defined categories. Error estimation gives a good prediction for the future classification performance of the classifier. This will provide a confidence index as to how good the classifier rules are. Pattern recognition methods include non-linear mapping (NLM), hierarchical cluster analysis (HCA), principle component analysis (PCA)[43], k-means clustering and self-organization maps (SOM)[125]. These are nonsupervised methods that do not make use of any prior knowledge for classification. PCA has become the norm in visualization of data (see Figure 8). It makes use of a statistical method for reducing multidimensional data (such as multiple spectra) down to a few dimensions that can be readily comprehended. Each dimension, called Principle Components (PCs), are variables created from linear combinations of the starting variables with appropriate weighting coefficients[43] or scores. PCA is an additive model where each PC accounts for a portion of the total variance of the data set. Plotting the PCs provides a rapid way of visualizing differences. Once these PCs are identified, the loadings can be analyzed for the specific differences on the origin spectral axes. However, PCA is only powerful if the biological objective represents the highest variance in the dataset. Several techniques were developed to overcome this shortcoming, namely non-linear PCA[126] or locally linear embedding[127]. Still, linear methods were more reliable due to the limited number of samples that most experiments have. Therefore, another more recently used method is independent component analysis (ICA)[128]. Each independent component or dimension does not overlap and is ideal for datasets having a large number of samples and only a small number of variables. Since the derived components do not have to be orthogonal to each other, it can be optimized for statistical independence[118]. However, applying ICA directly usually gives results that have no practical relevance. Thus, the dimensionality of data should be reduced

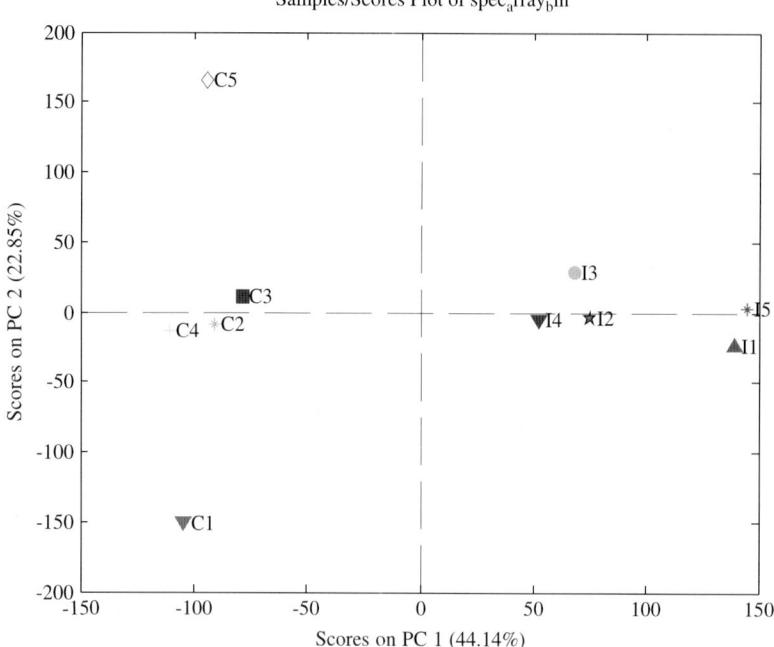

Figure 8. PCA plot of mouse myeloma NS0 samples. C represents control cultures while I represents the induced cultures which have been activated to overexpress p21 cytostatic gene

before apply ICA. Still ICA has been successfully implemented in the study of brain tumors[129]. Work done on yeast using the FANCY approach[14] made use of PCA and discriminant function analysis (DFA) to discriminate data from several sources, including FTIR, proton NMR and ESI-MS. To reduce the effects of physical and biological variation, orthogonal signal correction can be applied to data after PCA and/or Partial least squares-discriminant analysis (PLS-DA)[130]. Multilevel component analysis (MCA) has been recently developed[131] to deal with datasets that have variations in different levels, like variations between organism and variations in time. A subset of MCA, multilevel simultaneous component analysis (MSCA) allows the time-resolved variation and variations in animals to be analyzed simultaneously as subsets.

In hierarchical cluster analysis (HCA), similar samples are grouped together by the measurement of distances. Initially, two samples which are deemed the closest based on the distance measurement are grouped together as an initial pair. This initial pair is then grouped with another data point, forming a higher level pair. This process carries on until all pairs end up being linked at a common point. The pair-wise grouping can be done by several distance algorithms such as the Euclidean distance, correlations or the Manhattan distance. Results differ with each type of distance used and the results are visualized as a dendogram or a tree. In these visualizations, the length of branches is proportional to the distances between groups. K-means clustering is another method that uses grouping, this time with a fixed number (K) of groups. The principle is the same as HCA but the method for forming pairs differs. Similarly, there are several algorithms for forming pairs, each giving varying results. SOMs are similar to k-means clusters as data is grouped based on a predetermine number of groups (2^n groups). SOMs are gaining popularity due to their enhanced ability to differentiate and visualize data compared to PCA.

Supervised methods or machine-learning methods exist, but require a training dataset for it to function well. These methods can only be used if prior knowledge about the samples is provided. Despite this drawback, supervised methods are generally more powerful than unsupervised methods. Examples of these include soft independent modeling of class analogy (SIMCA) and neural networks (NNs), partial least square (or partial least squares-discriminant analysis PLS-DA), support vector machines, genetic programming and genetic algorithms. Information on the presupposed classes allow the maximum separation of data based on factors identified from the data. Probablistic neural networks (PNNs) are useful for nonlinear classification problems as well as Bayesian probability distributions.[43]. PNN makes no mathematical assumptions and uses all data provided without over fitting. Work by McGovern et al.[132] used genetic programming and genetic algorithms to decipher spectral data from different spectroscopic sources. In metabolomics, the supervised method most used extensively is artificial neural networks (ANN). Partial least square regression together with artificial neural networks were used together in the discrimination of spectral data of metabolites[89,133]. ANNs were used to discriminate ESI-MS data of plant extracts[134] and classify toxins in urine sample[135]. A new approach of identification of biomarkers using Statistical Total Correlation Spectroscopy (STOCSY) was introduced by Cloarec et al[136]. In this method, a supervised method, orthogonal projection on latent structure-discriminant analysis (O-PLS-DA) was first applied to the NMR data. Subsequently, the data was combined with STOCSY to help identify the structural information of biomarkers. From the various studies carried out, it appears that by using a combination of multivariate techniques appears to give the best results. Beckonert et al.[137] used a combination of PCA, HCA and k-means nearest neighbour (kNN) techniques to study the toxicity of rat unrine samples. PCA was used as a visual overview, HCA allowed similarities to be visualized while classification was done with kNN.

Once discriminate analysis is done, it has been suggested that statistical significance can be determined by applying classical statistics like Student's t test, ANOVA or multiple analysis of variance (MANOVA)[41]. Given that large statistical variation exists in biological samples, this may not be fruitful. In addition, as instrumental and procedural variances are not taken into consideration, the results may also become questionable. Another way of dealing with data after the preprocessing step is to detect for significant correlations of components. This is based on the covariance and/or correlations within a data matrix.[118]. Correlation analysis is one approach being used to discover novel networks[138]. Integration and comparison of data from different analytical platforms is a key concern in meteabolomics. Several methodologies have been suggested as to how data can be standardized and integrated. The use of the data matrix at the end of the preprocessing step may be one likely mode of integration. However, standardizing raw data has been suggested by Birkemeyer et al.[6] where stable isotopomers are added for each metabolite. This allows for quantitative metabolite profiling by mass isotopomer ratios and thus allow for the comparison of quantitative results from diverse analytical sources and platforms. Statistical heterospectroscopy (SHY) could be used to integrate analysis of data from different analysis platforms. This method analyzes the intrinsic covariance between signal intensities in the same and related molecules measured by different techniques. It was successfully applied to NMR and LC-MS data sets from Rat urine samples[139]. For a larger number of samples with sets of identified and quantified metabolites, the correlations between them can be found using Person's correlation coefficient[40]. This correlation allows the distance of biological connectivity of all the measured metabolites to be found and enables such 'distance maps' to be visualized.

Future developments in data processing need to address 3 issues[116]. Firstly, improved, open domain algorithms need to be used. This includes easy to use interfaces as well as controllable degrees of automation are criteria for such 'programs'. Secondly, 'spectrum-to-structure' recognition software is required. This requires the development of databases and libraries for cross-species, cross-instrument metabolites. Lastly, methods for metabolic network reconstruction are sorely lacking and need urgent attention. These methods for metabolic network reconstruction will need to address the presence of stochastic noise[40] as it leads to false correlations being witnessed. Some of these developments are discussed in the next section.

6.4   Interpretation of Metabolomics Data: Modeling and Bioinfomatics

For the proper assignment of metabolites, whether in terms of identity or models of networks, bioinfomatic tools and databases are required. Databases allow for the referencing of metabolites and the elucidation of chemical structures. However, for databases to be widely used by different communities, curation should be user-friendly and biology-orientated, thus avoiding different computational standards which will limit access to data. One way of introducing standards is the use of Systems Biology Markup Language (SBML)[140] which will allow interoperability between different models. Two main types of databases are used in metabolomics, the reference biochemical databases and the metabolite profile databases. There are many online reference databases of both types as show in Table 2. ArMET, architecture for metabolomics, is one way where a framework for reporting metabolomic data and experiments for data storage is introduced[141]. Databases should be able to have a variety of features[11,142]:

- able to store detailed metabolite profiles, including raw data and metadata,
- able to store simple single species profiles, complex profiles from many species and different physiological states, and
- able to list metabolites and compile established biochemical facts.

Table 2. Online databases of biochemical references and metabolite profile references

Name	URL
Biochemical references	
KEGG	www.genome.ad.jp/keg
BRENDA	www.brenda.uni-koeln.de
The EMP project	www.empproject.com
IUBMB Enzyme Nomenclature	www.chem.qmul.ac.uk/iubmb/enzyme
EcoCyc	biocyc.org/ECOLI/class-subs-instances?object=Pathways
MetaCyc	metacyc.org
Metabolite profile references	
ArMet	www.Armet.org
DOME	medicago.vbi.vt.edu/dome.html
AMDIS	chemdata.nist.gov/mass-spc/amdis
MetAlign	www.metalign.nl
MetaGeneAlyse	metagenealyse.mpimp-golm.mpg.de
MeT-RO	www.metabolomics.bbsrc.ac.uk/MeT-RO.htm
Metabolic tools	
AraCyc	www.Arabidopsis.org/tools/aracyc/
MapMan	gabi.rzpd.de/projects/MapMan
MetNet	metnet.vrac.iastate.edu/
Biosilico	biosilico.kaist.ac.kr

In particular, metabolite profiles need to be validated and the metadata complete so that comparisons can be done when using the same protocol. Precision is vital as no simple standards exist for metabolome wide measurements and each sample is dependent on the source and sample preparation. It has also been suggested that freshly made up cocktails of standards, together with the use of advanced transformation[143], could solve this problem. Yet it is difficult to simulate identical conditions of sample pH or extraction chemical signatures which inevitably influence the data acquisition process.

Data visualization tools allow the direct visual identification of properties of the data sets. Visualizations may be instrumental in identifying the functions of mutated genes directly by visual inspection. Diagrams representing the network of reactions linking metabolites give a visual understanding of relationships between metabolites[142]. These diagrams can also be linked with gene expression data to give a functional genomic understanding[144]. However, in order that accurate interpretations are made, visualization diagrams should be complete, that is, take into account all the reactions surrounding the metabolites in question are shown. The traditional metabolic pathway diagrams, such as the ones in the popular database KEGG[145] may not show all 'side" reactions. The question is whether we know which reactions to include or omit. Graphing tools allow the visualization of biochemical pathways by edges and nodes. Some tools allow automatic structuring of data into forms of metabolic cycles or hierarchies[146]. However, as the amount of data points (edges or nodes) increase, these techniques become unproductive as a visualization tool.

Therefore visualization tools should inevitably include more complex modeling of the biological system.

The inverse problem of metabolic network construction from metabolome data is generally hard but a growing number of approaches and tools are now available for modeling metabolic systems[147]. Shifting from metabolic pathways to metabolic networks and neighborhoods[148] is important in understanding the relationships between metabolites and the structural properties of networks. Figure 9 shows how an ideal model should incorporate both complexity and accuracy and how present interaction based modeling fails in their accuracy. Topology, the pattern of inter-actions between components, acts as a basic feature of any network. Stochastic models have been use for interpreting metabolic networks[149,150]. One such analo-gous model system interprets metabolite correlations after multivariate data mining and connect these correlations to the underlying enzymatic pathway structure have been developed[40,118]. The observation of correlated metabolites offers the chance to investigate whole metabolite network topologies. This is done by the creation of a correlation matrix that acts as a fingerprint of the enzymatic and regulatory network. In a study of whether correlation analysis of metabolomics data can recover features

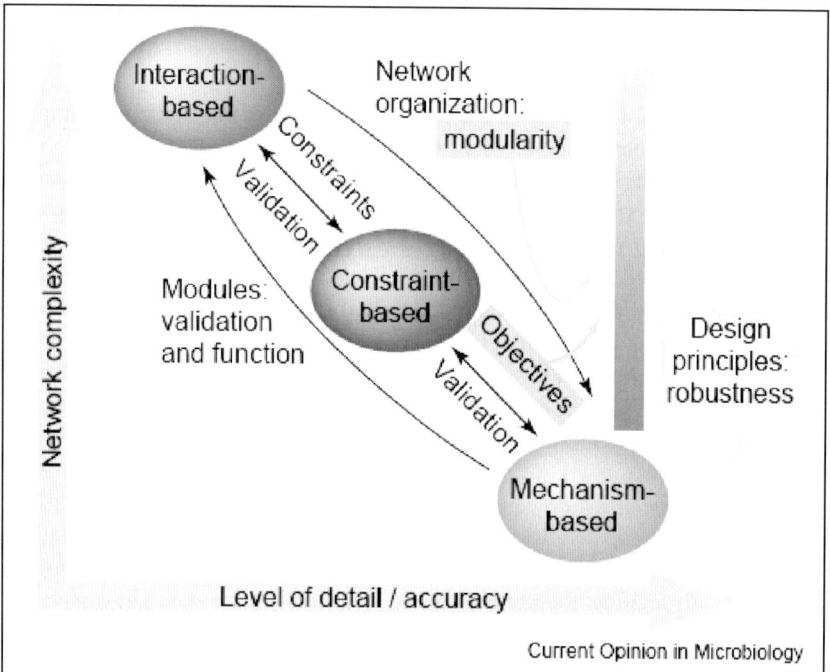

Figure 9. Mathematical modeling: scope and interactions. The black arrow represents possible inter-actions; the green arrow represents the desirable progress towards a genotype. Reprinted from Current Opinion in Microbiology, Vol. 7, Stelling J. 'Mathematical modeling in microbial systems biology',516, 2004

of the biochemical system, Mendes et al. showed that four major metabolic regulation configurations gave strong metabolic correlations[151]. Different strategies using metabolic control analysis (MCA) or variations of this method have also been introduced, including the examination of correlations between metabolites[152]. However, there are many limitations of MCA based models as detailed by Fiehn[41].

Graph theory[148] provides quantitative methods for analyzing these static representations of components (nodes) and interactions (links). Graph theory uses a 'scale-free' global cross organism topology of cellular networks that parallels the structure of complex engineered systems[153]. Their topology is dominated by a few highly connected nodes which link the rest of the less connected nodes to the system but can be rather 'static'. This was further developed into hierarchical modules which would form a larger less cohesive unit[154]. Alternatively, restricting the metabolic networks and their regulations by the application of stoichiometric constraints and 'elementary modes'[155–157], analyses can be made manageable. Genome-scale models may represent one of these constrained models. Genome-scale metabolic reconstructions give rise to metabolic network topology which can then be used to reveal patterns following transcriptional responses[158]. 'Network motifs'[159] have also been introduced to look at possible feedback arrangements of a restricted subset of a network. Network motifs are arrangements of reactions, including feedback structures, which are functionally important in biology. Another graphing technique which draws in the use of graph theory is clique correlation analysis. Clique-metabolite matrices instead of edges and nodes[160] are used, allowing the computation of subgraphs, isolated groups and missing edges. This area of research is constantly improving itself but the real challenge of the future is to integrate these approaches and obtain complete integrated functional genomic systems to better understand and visualize systems biology[161].

7.   TOWARDS SYSTEMS BIOLOGY

Systems biology is similar to functional genomics in its approach but is slightly different in its objectives. Systems biology encompasses a holistic approach to the study of biology and the objective is to simultaneously monitor all biological processes operating as an integrated system. It involves the iterative interplay between linked activities which, allow for the study of signal processing elements that lie 'downstream' of the signal initiator. These help one understand how cross-talk may be occurring between pathways[162]. Hence, in order for metabolomics to be used in systems biology, holistic approaches will need to be reinforced in methodologies. This includes the integration of metabolomic measurements with measurements of time-dependent concentrations of other types of components (transcriptomic and/or proteomic) within a system of interest[12,163]. The exciting prospect of this happening is limited by the present state of technological development and the lack of integrative approaches in dealing with information. Yet, in due time, the widespread use of metabolomics in understanding biological processes will see boundless benefits.

8.   REFERENCES

1. Trethewey R.N., Krotzky A.J. & Willmitzer L. Metabolic profiling:a Rossetta Stone for genomics? *Current Opinion in Plant Biology* **2**, 83–85 (1999).
2. Oliver S.G., Winson M.K., Kell D.B. & Bagnanz F. Systematic functional analysis of the yeast genome. *Trends in Biotechnology* **16**, 373–377 (1998).
3. Fiehn O. Combining genomics, metabolome analysis, and biochemical modelling to understand metabolic networks. *Comparative and Functional Genomics* **2**, 155–168 (2001).
4. Beecher C.W.W. Metabolic Profiling: its role in biomarker discovery and gene function analysis. Harrigan G.G & Goodacre R. (eds.), pp. 311–319 (Kluwer Academic Publishers,2003).
5. Villas-Bôas S.G., Rasmussen S & Lane G.A. Metabolomics or metabolite profiles? *Trends in Biotechnology* **23**, 385–386 (2005).
6. Birkemeyer C, Luedemann A, Wagner C, Erban A & Kopka J. metabolome analysis: the potential of *in vivo* labeling with stable isotopes for metabolite profiling. *Trends in Biotechnology* **23**, 28–33 (2005).
7. Gygi S.P., Rochon Y, Franza B.R. & Aebersold R. Correlation between protein and mRNA abundance in Yeast. *Molecular and Cell Biology* **19**, 1720–1730 (1999).
8. Schmidt C.W. Metabolomics: What's happening dowstream of DNA. *Environmental Health Perspectives* **112**, A410–A415 (2004).
9. Nielsen J. It is all about metabolic fluxes. *Journal of Bacteriology* **185**, 7031–7035 (2003).
10. ter Kuile B.H. & Westerhoff H.V. Transcriptome meets metabolome: hierarchical and metabolic regulation of the glycolytic pathway. *FEBS letters* **500**, 169–171 (2001).
11. Goodacre R, Vaidyanathan S., Dunn W.B., Harrigan G.G & Kell D.B. Metabolomics by number: acquiring and understnading global metabolite data. *Trends in* Biotechnology **22**, 245–252 (2004).
12. Bino r.J. et al. Potential of metabolomics as a functional genomics tool. *Trends in Plant Science* **9**, 418–425 (2004).
13. Weckwerth W, Loureiro M.E., Wenzel K & Fiehn O. Differential metabolic networks unravel the effects of silent plant phenotypes. *Proceedings of the National Academy of Science* **101**, 7809–7814 (2004).
14. Raamsdonk L.M. et al. A functional genomics strategy that uses metabolome data to reveal the phenotype of silent mutations. *Nature Biotechnology* **19**, 45–50 (2001).
15. Sumner L.W, Mendes P & Dixon R.A. Plant metabolomics: large-scale phytochemistry in the functional genomics era. *Phytochemistry* **62**, 817–836 (2003).
16. Fridman E. & Pichersky E. Metabolomics, genomics, proteomics, and the identification of enzymes and their substrates and products. *Current Opinion in Plant Biology* **8**, 242–248 (2005).
17. Martin D.M., Faldt J & Bohlmann J. Functional characterization of nine Norway Spruce TPS genes and evolution of gymnosperm terpene synthases of the TPS-d subfamily. *Plant Physiology* **135**, 2024 (2004).
18. Richman A.S. et al. Functional genomics uncovers three glucosyltransferases involved in the synthesis of the major sweet glucosides of *Stevia rebaudiana*. *Plant Journal* **41**, 56–67 (2005).
19. Bugrim A, Nikolskaya T & Nikolsky Y. Early prediction of drug metabolism and toxicity: systems biology approach and modeling. *Drug Discovery Today* **9**, 127–135 (2004).
20. Roessner U, Wagner C, Kopka J, Trethewey R.N. & Willmitzer L. Simultaneous analysis of metabolites in potato tuber by gas chromatography mas spectrometry. *Plant Journal* **23**, 131–142 (2000).
21. Jung W et al. Identification and expression of isoflavone synthase, the key enzyme for biosynthesis of isoflavones in legumes. *Nature Biotechnology* **18**, 208–212 (2000).
22. Liu C-J, Blount J.W, Steele C.L. & Dixon R.A. Bottlenecks for metabolic engineering of isoflavone glycoconjugates in *Arabidopsis*. *Proceedings of the National Academy of Science* **99**, 14578–14583 (2002).
23. Allen J et al. High-throughput classification of yeast mutants for functional genomics using metabolic footprinting. *Nature Biotechnology* **21**, 692–696 (2003).

24. Gavaghan C.L., Holmes E, Lenz E, Wilson I.D & Nicholson J.K. An NMR-based metabonomic approach to investigate the biochemical consequences of genetic strain differences:application to the C57BL10J and Alpk:ApfCD mouse. *FEBS letters* **484**, 169–174 (2000).
25. Griffin J.L., Mann C.J., Scott J, Shoulders C.C & Nicholson J.K. Choline containing metabolites during cell transfection: an insight into magnetic resonance spectroscopy detectable changes. *FEBS letters* **509**, 263–266 (2001).
26. Schnackenberg L, Beger R.D & Dragan Y. NMR-based metabonomic evaluation of livers from rats chronically treated with tamoxifen, mestranol, and phenobarbital. *Metabolomics* **1**, 87–94 (2005).
27. Griffin J.L, Pole J.C.M, Nicholson J.K & Carmichael P.L. Cellular environment of metabolites and a metabonomic study of tamoxifen in endometrial cells using gradient high resolution magic angle spinning ¹H NMR spectroscopy. *Biochimica et Biophysica Acta* **1619**, 151–158 (2003).
28. Griffin J.L & Shockcor J.P. Metabolic profiles of cancer cells. *Nature Reviews Cancer* **4**, 551–561 (2004).
29. Yang J et al. Diagnosis of liver cancer using HPLC-based metabolomics avoiding false-positive result from hepatitis and hepatocirrhosis diseases. *Journal of Chromatography B* **813**, 59–65 (2004).
30. Brindle J.T et al. Rapid and noninvasive diagnosis of the presence and severity of coronary heart disease using ¹H-NMR-based metabonomics. *Nature Medicine* **8**, 1439–1444 (2002).
31. German J.B, Watkins S.M & Fay L-B. Metabolomics in practice: emerging knowledge to guide futrue dietetic advice toward individualized health. *Journal of the American Dietetic Association* **105**, 1425–1432 (2005).
32. Watkins S.M & German J.B. Toward the implementation of metabolomic assessments of human health and nutrition. *Current Opinion in Biotechnology* **13**, 512–516 (2002).
33. Davis C.D & Milner J. Frontiers in nutrigenomics, proteomics, metabolomics and cancer prevention. *Mutation Research* **551**, 51–64 (2004).
34. Harrison M.J. & Dixon R.A. Isoflavoniod accumulation and expression of defense gene transcripts during the establishment of vesicular arbuscular mycorrhizal associations in roots of *Medicago truncatula*. *Molecular Plant-Microbe Interactions* **6**, 654 (1993).
35. Maier W, Schmidt J., Wray V, Walter M.H & Strack D. The arbuscular mycorrhizal fungus, *Glomus intraradices*, induces the accumulation of cyclohexenone derivatives in tobacco roots. *Planta* **207**, 620–623 (1999).
36. Aharoni A et al. Non-targeted metabolomic profiling using Fourier transform ion cyclotron mass spectrometry (FTMS). *OMICS: A Journal of Integrative Biology* **6**, 217–234 (2002).
37. Pincetich C.A., Viant M.R., Hinton D.E. & Tjeerdema R.S. Metabolic changes in Japanese medaka (*Oryzias latipes*) during embryogenesis and hypoxia as determined by *in vivo* ³¹P NMR. *Comparative Biochemistry and Physiology, Part C* **140**, 103–113 (2005).
38. Viant M.R., Rosenblum E.S & Tjeerdema R.S. NMR-based metabolomics: A powerful approach for characterizing the effects of environmental stressors on organism health. *Environmental Science and Technology* **37**, 4982–4989 (2003).
39. Bailey N.J.C, Oven M, Holmes E, Nicholson J.K & Zenk M.H. Metabolomic analysis of the consequences of cadmium exposure in *Silene cucubalus* cell cultures via ¹H NMR spectroscopy and chemometrics. *Phytochemistry* **62**, 851–858 (2003).
40. Weckwerth W. Metabolomics in Systems Biology. *Annual Review of Plant Biology* **54**, 669–689 (2003).
41. Fiehn O. Metabolomics-the link between genotypes and phenotypes. *Plant Molecular Biology* **48**, 155–171 (2002).
42. Nicholson J.K, Lindon J.C. & Holmes E. 'Metabonomics': understanding the metabolic responses of living systems to pathophysiological stimuli via multivariate statistical analysis of biological NMR spectroscopic data. *Xenobiotica* **29**, 1181–1189 (1999).
43. Mitchell S, Holmes E & Carmichael P. Metabonomics and medicine: the biochemical oracle. *Biologist* **49**, 217–221 (2002).
44. Fiehn O. & Spranger J. Metabolite profiling: Its role in biomarker discovery and gene function analysis. Harrigan G.G & Goodacre R (eds.), pp. 199–216 (Kluwer Academic Publishers, London, UK, 2003).

45. Saghatelian A & Cravatt B.F. Discovery metabolite profiling-forging functional connections between the proteome and metabolome. *Life Sciences* **77**, 1759–1766 (2005).

46. Dunn W.B. & Ellis D.I. Metabolomics: current analytical platforms and methodologies. *Trends in Analytical Chemistry* **24**, 285–294 (2005).

47. Klose J & Kobalz U. Two-dimenisonal electrophoresis of proteins : an updated protocol and implications for a functional analysis of the genome. *Electrophoresis* **16**, 1034–1059 (1995).

48. Nielsen J & Oliver S.G. The next wave in metabolome analysis. *Trends in Biotechnology* **23**, 544–546 (2005).

49. Defernez M & Colquhoun I.J. Factors affecting the robustness of metabolite fingerprinting using ^{1}H NMR spectra. *Phytochemistry* **62**, 1009–1017 (2003).

50. Fan T.W-M. Metabolite profiling by one- and two-dimensional NMR anlaysis of complex mixtures. *Progress in Nuclear Magnetic Resonance Spectroscopy* **28**, 161–219 (1996).

51. Oksman-Caldentey K-M & Saito K. Integrated genomics and metabolomics for engineering plant metabolic pathways. *Current Opinion in Biotechnology* **16**, 174–179 (2005).

52. Brown M et al. A metabolome pipline: from concept to data to knowledge. *Metabolomics* **1**, 39–51 (2005).

53. ap Rees T & Hill S.A. Metabolic control analysis of plant metabolism. *Plant Cell Environment* **17**, 587–599 (1994).

54. Gerhardt R & Heldt H.W. Measurement of subcellular metabolite levels in leaves by fractionation of freeze-stopped material in nonaqueous media. *Plant Physiology* **75**, 542–547 (1984).

55. Wittman C, Krömer J.O, Kiefer P, Binz T & Heinzle E. Impact of the cold shock phenomenon on quantification of intracellular metabolites in bacteria. *Analytical biochemistry* **327**, 135–139 (2004).

56. Fiehn O., Kopka J, Trethewey R.N. & Willmitzer L. Identification of uncommon plant metabolties based on calculation of elemental compositions using gas chromatography and quadrupole mass spectrometry. *Analytical Chemistry* **72**, 3573–3580 (2000).

57. Orth H.C.J, Rentel C & Schmidt P.C. Isolation, purity analysis and stability of hyperforin as a standard material from *Hypericum perforatum L. Journal of Pharmacy and Pharmacology* **51**, 193–200 (1999).

58. Rossi D.T & Sinz M.W. Mass spectrometry in drug discovery. Marcel Dekker, New York, USA (2002).

59. Villas-Bôas S.G., Højer-Pedersen J, kesson M, Smedsgaard J & Nielsen J. Global metabolite analysis of yeast: evaluation of sampel preparation methods. *Yeast* **22**, 1155–1169 (2005).

60. Maharjan R.P & Ferenci T. Global metabolite analysis: the influence of extraction methodology on metabolome profiles of *Escherichia coli. Analytical biochemistry* **313**, 145–154 (2003).

61. Kaderbhai N.N, Broadhurst D.I, Ellis D.I, Goodacre R & Kell D.B. Functional genomics via metabolic footprinting: monitoring metabolite secretion by *Escherichia coli* tryptophan metabolism mutants using FT-IR and direct injection electrospray mass spectrometry. *Comparative and Functional Genomics* 376–391 (2003).

62. Abel C.B.L et al. Characterization of metabolites in intact *Streptomyces citricolor* culture supernatants using high-resolution nuclear magnetic resonance and directly coupled high-pressure liquid chromatography-nuclear magnetic resonance spectroscopy. *Analytical biochemistry* **270**, 220–230 (1999).

63. Lenz E.M, Bright J, Knight R, Wilson I.D & Major H. Cyclosporin A induced changes in endogeneous metasbolites in rat urine: a metabonomic investigation using high field ^{1}H NMR spectroscopy, HPLC-TOF/MS and chemometrics. *Journal of Pharmaceutical and Biomedical Analysis* **35**, 599–608 (2004).

64. Bligny R & Douce R. NMR and plant metabolism. *Current Opinion in Plant Biology* **4**, 191–196 (2001).

65. Keun H.C et al. Cryogenic probe ^{13}C NMR spectroscopy of urine for metabonomic studies. *Analytical Chemistry* **74**, 4588–4593 (2002).

66. Wang Y et al. Spectral editing and pattern recognition methods applied to high-resolution magic-angle spinning ^{1}H nuclear magnetic resonance spectroscopy of liver tissues. *Analytical biochemistry* **323**, 26–32 (2003).

67. Lindon J.C et al. Contemporary issues in toxicology: the role of metabonomcis in toxicology and its evaluation by the COMET project. *Toxicology and Applied Pharmacology* **187**, 137–146 (2003).

68. Goodacre R, Vaidyanathan S., Bianchi G & Kell D.B. Metabolic profiling using direct infusion electrospray mass spectrometry. *Analyst* **127**, 1457–1462 (2002).

69. Castrillo J.I, Hayes A, Mohammed S, Gaskell S.J. & Oliver S.G. An optimized protocol for metabolome analysis in yeast using direct infusion electrospray mass spectrometry. *Phytochemistry* **62**, 929–937 (2003).

70. Allen J.K et al. High-throughput characterization of yeast mutants for functional genomics using metabolic footprinting. *Nature Biotechnology* **21**, 692–696 (2003).

71. Wei J, Burlak J.M & Siuzdak G. Desorption-ionization mass spectrometry on porous silicon. *Nature* **399**, 243–246 (1999).

72. Go E.P et al. Desorption/ionization on silicon time-of-flight /time-of-flight mass spectrometry. *Analytical Chemistry* **75**, 2504–2506 (2003).

73. Go E.P, Shen Z, Harris K & Siuzdak G. Quantitative analysis with desorption/ionization on silicon mass spectrometry using eletrospray deposition. *Analytical Chemistry* **75**, 5475–5479 (2003).

74. Tong C.S & Cheng K.C. Mass spectral search method using the neural network approach. *Chemometrics and Intelligent Laboratory Systems* **49**, 135–150 (1999).

75. Stein S.E & Scott D.R. Optimization and testing of mass-spectral library search algorithms for compound identification. *Journal of the American Society for Mass Spectrometry* **5**, 859–866 (1994).

76. Vaidyanathan S., Broadhurst D.I, Kell D.B & Goodacre R. Explanatory optimization of protein mass spectrometry via genetic search. *Analytical Chemistry* **75**, 6679–6686 (2003).

77. Brown S.C, Kruppa G & Dasseux J-L. Metabolomics applications of FT-ICR mass spectrometry. *Mass Spectrometry Reviews* **24**, 223–231 (2005).

78. Tweeddale H, Notley-McRobb L & Ferenci T. Effect of slow growth on metabolism of *Escherichia coli*, as revealed by global metabolite pool ("Metabolome") analysis. *Journal of Bacteriology* **180**, 5109–5116 (1998).

79. Wilson I.D et al. HPLC-MS-based methods for the study of metabonomics. *Journal of Chromatography B* **817**, 67–76 (2005).

80. Nováková L, Matysová L & Solich P. Advantages of application of UPLC in Pharmaceutical analysis. *Talanta* **68**, 908–918 (2006).

81. Fraser P.D, Pinto M.E, Holloway D.E & Bramley P.M. Application of high-performance liquid chromatography with photodiode array detection to the metabolic profiling of plant isoprenoids. *Plant Journal* **24**, 551–558 (2000).

82. González de la Huebra M-J, Bordin G & Rodriguez A.R. Comparative study of coulometric and amperometric detection for the determination of macrolides in human urine using high-performance liquid chromatography. *Analytical and Bioanalytical Chemistry* **375**, 1031–1037 (2003).

83. Harrington D.J et al. Determination of the urinary aglycone metabolites of vitamin K by HPLC with redox-mode electrochemical detection. *Journal of Lipid Research* **46**, 1053–1060 (2005).

84. Kaddurah-Daouk R et al. Bioanalytical advances for metabolomics and metabolic profiling. *Pharmagenomics* **January**, 46–52 (2004).

85. Shurubor Y.I, Paolucci U, Krasnikov B.F, Matson W.R & Kristal B.S. Analytical precision, biological variation, and mathematical normalization in high data density metabolomics. *Metabolomics* **1**, 75–85 (2005).

86. Kaushik R, Lacourse W.R & Levine B. Determination of ethyl glucuronide in urine using reverse-phase HPLC and pulsed electrochemical detection (Part II). *Analytica Chimica Acta* **556**, 267–274 (2006).

87. Brenes M, Garcia A, Garcia P & Garrido A. Rapid and complete extracton of henols from olive oil and determination by means of a coulometric electrode array system. *Journal of Agriculture and Food Chemistry* **48**, 5178–5183 (2000).

88. Ferruzzi M.G, Sander L.C, Rock L.C & Schwartz S.J. Carotenoid dtermination in biological microsamples using liquid chromatography with a coulometric electrochemical array detection. *Journal of Chromatography B* **760**, 289–299 (2001).

89. Winson M.K. et al. Diffuse reflectance absorbance spectroscopy taking in chemometrics (DRASTIC). A hyperspectral FT-IR-based approach to rapid screening for metabolite overproduction. *Analytica Chimica Acta* **348**, 273–282 (1997).

90. Heise H.M et al. Multivariate calibration for the determination of analytes in urine mid-infrared attenuated total reflection spectroscopy. *Applied Spectroscopy* **55**, 434–443 (2001).

91. Chiriboga L et al. Infrared spectroscopy of human tissue. I. Differentiatio and muturation of epithelial cells in the human cervix. *Biospectroscopy* **4**, 47–53 (1998).

92. Schmitt J et al. Identification of scrapie infection from blood serum by Fourier Transform Infrared Spectroscopy. *Analytical Chemistry* **74**, 3865–3868 (2002).

93. Lafrance D, Lands L.C & Burns D.H. Measurement of lactate in whole human blood with near-infrared transmission spectroscopy. *Talanta* **60**, 635–641 (2003).

94. Stein S.E. An intergrated method for spectrum extraction and compound identification from gas chromatography/mass spectrometry data. *Journal of American Society for Mass Spectrometry* **10**, 109–110 (1999).

95. Halket J.M et al. Deconvolution gas chromatography/mass spectrometry of urinary organic acids – potential for pattern recognition and automated identification of metabolic disorders. *Rapid Communications in Mass Spectrometry* **13**, 279–284 (1999).

96. Duran A.L, Yang J, Wang L-J & Sumner L.W. Metabolomics spectral formatting, alignment and conversion tools (MSFACTs). *Bioinformatics* **19**, 2283–2293 (2003).

97. Schuaer N et al. GC-MS libraries for the rapid identification of metabolites in complex biological samples. *FEBS letters* **579**, 1332–1337 (2005).

98. Fiehn O. et al. Metabolite profiling for plant functional genomics. *Nature Biotechnology* **18**, 1157–1161 (2000).

99. Niessen W.M. Progress in liquid chromatography-mass spectrometry instrumentation and its impact on high-throughput screening. *Journal of Chromatography A* **1000**, 413–436 (2003).

100. Bahr U, Pfenninger A, Kara M & Stahl B. High sensitivity analysis of meutral underivatized oligosaccharides by nanoelectrospray mass spectrometry. *Analytical Chemistry* **69**, 4530–4535 (1997).

101. Leavens W.J, Lane S.J, Carr R.M, Lockie A.M & Waterhouse I. Derivatization for liquid chromatography/electrospray mass spectrometry: synthesis of tris(trimehtylethoxyphenyl)phosphonium compounds and their derivatives of amine and carboxylic acids. *Rapid Communications in Mass Spectrometry* **16**, 433–441 (2002).

102. Tolstikov V.V, Lommen A, Nakanishi K, Tanaka N & Fiehn O. Monolithic silica-based capillary reversed phase liquid chromatography/electrospray mass spectrometry for plant metabolomics. *Analytical Chemistry* **75**, 6737–6740 (2003).

103. Tolstikov V.V & Fiehn O. Analysis of highly polar compounds of plant origin: combination of hydrophilic interaction chromatography and electrospray ion trap mass spectrometry. *Analytical biochemistry* **301**, 298–307 (2002).

104. Soga T et al. Quantitative metabolome analysis using capillary electrophoresis mass spectrometry. *Journal of Proteome Research* **2**, 488–494 (2003).

105. Sato S, Soga T, Nishioka T & Tomita M. Simultaneous determination of the main metabolites in rice leaves using capillary electrophoresis mass spectrometry an capillary electrophoresis diode array detection. *Plant Journal* **40**, 151–163 (2004).

106. Garcia C.D & Henry C.S. Comparison of pulsed electrochemical detection modes coupled with microchip capillary electrophoresis. *Electroanalysis* **17**, 223–230 (2005).

107. Lindon J.C, Nicholson J.K & Wilson I.D. Directly coupled HPLC-NMR and HPLC-NMR-MS in pharmaceutical research and development. *Journal of Chromatography B* **748**, 233–258 (2000).

108. Corcoran O & Spraul M. LC-NMR-MS in drug discovery. *Drug Discovery Today* **8**, 624–631 (2003).

109. Albert K. On-line use of NMR detection in separation chemistry. *Journal of Chromatography A* **703**, 123–147 (1995).

110. Lindon J.C, Nicholson J.K & Wilson I.D. Direct coupling of chromatographic separations to NMR spectroscopy. *Progress in Nuclear Magnetic Resonance Spectroscopy* **29**, 1–49 (1996).

111. Dear G.J et al. Mass directed peak selection, an efficient method of drug metabolite identification using directly coupled liquid chromatography-mass spectrometry-nuclear magnetic resonance spectroscopy. *Journal of Chromatography B: Biomedical Sciences and Applications* **748**, 281–293 (2000).

112. Wolfender J.L, Rodriguez S & Hostettmann K. Liquid chromatography coupled to mass spectrometry and nuclear magnetic resonances for the screening of plant constituents. *Journal of Chrmatography A* **794**, 299–316 (1998).

113. Beaudry F et al. Metabolite profiling study of propranolol in rat using LC/MS/MS analysis. *Biomedical Chromatography* **13**, 363–369 (1999).

114. Mashego M.R et al. MIRACLE: Mass isotopomer ratio analysis of U-^{13}C-labeled extracts. A new method for accurate quantification of changes in concentrations of intracellular metabolites. *Biotechnology and Bioengineering* **85**, 620–628 (2004).

115. Shellie R.A et al. Statistical methods for comparing conprehensive two-dimensional gas chromatography-time-of-flight mass spectrometry results: Metabolomic analysis of mouse tissue extracts. *Journal of Chromatography A* **1086**, 83–90 (2005).

116. Kell D.B. Metabolomics and systems biology: making sense of the soup. *Current Opinion in Microbiology* **7**, 296–307 (2004).

117. Lindon J.C et al. Summary of recommendations for standardization and reporting of metabolic analyses. *Nature Biotechnology* **23**, 833–838 (2005).

118. Weckwerth W & Morgenthal K. Metabolomics: from pattern recognition to biological interpretation. *Drug Discover Today: Targets* **10**, 1551–1558 (2005).

119. Vigneau-Callahan K.E, Shestopalov A.I, Milbury P.E, Matson W.R & Kristal B.S. Characterization of diet-depedent metabolic serotypes.: Proof of principle in female and male rats. *Journal of Nutrition* **131**, 924S–932S (2001).

120. Johnson K.J, Wright B.W, Jarman K.H & Synovec R.E. High-speed peak matching algorithm for retention time alignment of gas chrmatographic data for chemometric analysis. *Journal of Chromatography A* **996**, 141–155 (2003).

121. Forshed J et al. A comparison of methods for alignment of NMR peaks in the context of cluster analysis. *Journal of Pharmaceutical and Biomedical Analysis* **38**, 824–832 (2005).

122. Viant M.R. & Viant. Improved methods for the aquisition and interpretation of NMR metabolomic data. *Biochemical and Biophysical Research Communications* **310**, 943–948 (2003).

123. Bijlsma S et al. Large-scale human metabolomics studies: a strategy for data (pre-)processing and validation. *Analytical Chemistry* **78**, 567–574 (2006).

124. El-Deredy W. Pattern recognition approachers in biomedical and clinical magnetic resonance spectroscopy: a review. *NMR in Biomedicine* **10**, 99–124 (1997).

125. Kohonen T. Self-organizing maps. Springer-Verlag, New York (1995).

126. Scholz M & Vigario R. Nonlinear PCA: a new hierarchical approach. *Proceedings of ESANN* 439–444 (2002).

127. Roweis S & Saul L. Nonlinear dimensionality reduction by locally linear embedding. *Science* **290**, 2323–2326 (2000).

128. Scholz M, Gatzek S, Sterling A, Fiehn O. & Selbig J. Metabolite fingerprinting: detecting biological features by indepedent component analysis. *Bioinformatics* **20**, 2247–2454 (2004).

129. Szabo de Edelenyi F, Simonetti A.W, Postma G, Huo R & Buydens L.M.C. Application of independent component analysis to ^{1}H MR spectroscopic imaging exams of brain tumours. *Analytica Chimica Acta* **544**, 36–46 (2005).

130. Beckwith-Hall B.M et al. Applications of orthogonal singal correction to minimise the effects of physical and biological variation in high resolution ^{1}H NMR spectra of biofluids. *Analyst* **127**, 1283–1288 (2002).

131. Jansen J.J, Hoefsloot H.C.J, van der Greef J, Timmerman M.E. & Smilde A.K. Multilevel component analysis of time-resolved metabolic fingerprinting data. *Analytica Chimica Acta* **530**, 173–183 (2005).

132. McGovern A.C et al. Monitoring of complex industrial bioprocesses for metabolite concentrations using modern spectroscopies and machine learning: Application to Gibberellic acid production. *Biotechnology and Bioengineering* **78**, 527–538 (2002).

133. Goodacre R & Kell D.B. Pyrolysis mass spectrometry and its application in biotechnology. *Current Opinion in Biotechnology* **7**, 20–28 (1996).

134. Goodacre R, York E.V, Heald J.K & Scott I.M. Chemometric discrimination of unfractionated plant extracts analyzed by electrospray mass spectrometry. *Phytochemistry* **62**, 859–863 (2003).

135. Anthony M.L, Rose V.S, Nicholson J.K & Lindon J.C. Classification of toxin-induced changes in ^{1}H NMR spectra of urine using an artificial neural network. *Journal of Pharmaceutical and Biomedical Analysis* **13**, 205–211 (1995).

136. Cloarec O et al. Statistical total correlation spectroscopy: An exploratory approach for latent biomarker identification from metabolic ^{1}H NMR data sets. *Analytical Chemistry* **77**, 1282–1289 (2005).

137. Beckonert O et al. NMR-based metabonomic toxicity classification: hierarchical cluster analysis and k-nearest-neighbour approaches. *Analytica Chimica Acta* **490**, 3–15 (2003).

138. Weckwerth W & Fiehn O. Can we discover novel pathways using metabolomics analysis? *Current Opinion in Biotechnology* **13**, 156–160 (2002).

139. Crockford D.J et al. Statistical heterospectroscopy, an approach to the integrated analysis of NMR and UPLC-MS data sets: Applications in metabonomic toxicology stidues. *Analytical Chemistry* **78**, 363–371 (2006).

140. Hucka M et al. The systems biology markup language (SBML): a medium for representation and exchange of biochemical network models. *Bioinformatics* **19**, 524–531 (2003).

141. Jenkins H et al. A proposed framework for the description of plant metabolomics experiments and their results. *Nature Biotechnology* **22**, 1601–1606 (2004).

142. Mendes P. Emerging bioinformatics for the metabolome. *Briefings in Bioinformatics* **3**, 134–145 (2002).

143. Goodacre R & Kell D.B. Correction of mass spectral drift using artificial neural networks. *Analytical Chemistry* **68**, 271–280 (1996).

144. Wolf D, Gray C.P & de Saizieu A. Visualising gene expression in its metabolic context. *Briefings in Bioinformatics* **1**, 297–304 (2000).

145. Kanehisa M, Goto S, Kawashima S & Nakaya A. The KEGG databases at GenomeNet. *Nucleic Acid Research* **30**, 42–46 (2002).

146. Becker M.Y & Rojas I. A graph layout algorithm for drawing metabolic pathways. *Bioinformatics* **17**, 461–467 (2001).

147. Mendes P & Kell D.B. Non-linear optimization of biochemical pathways: applications to metabolic engineering and parameter estimation. *Bioinformatics* **14**, 869–883 (1998).

148. Barabási A-L & Oltvai Z.N. Network biology: understanding the cell's functional organization. *Nature Revews Genetics* **5**, 101–113 (2004).

149. Vance W, Arkin A.P & Ross J. Determination of causal connectivities of species in reaction networks. *Proceedings of the National Academy of Science* **99**, 5816–5821 (2002).

150. Arkin A, Shen P & Ross J. A test case of correlation metric construction of a reaction pathway from measurements. *Science* **277**, 1275–1279 (1997).

151. Mendes P, Camacho D & de la Fuente A. Modelling and simulation for metabolomics data analysis. *Biochemical Society Transactions* **33**, 1427–1429 (2005).

152. Camacho D, de la Fuente A & Mendes P. The orgins of correlations in metabolomics data. *Metabolomics* **1**, 53–63 (2005).

153. Jeong H, Tombor B, Albert R, Oltvai Z.N & Barabási A-L. the large-scale organization of metabolic networks. *Nature* **407**, 651–654 (2000).

154. Ravasz E, Somera A.L, Mongru D.A, Oltvai Z.N & Barabási A-L. Hierarchical organization of modularity in metabolic networks. *Science* **297**, 1551–1555 (2002).

155. Klamt S & Stelling J. Two approaches for metabolic pathway analysis? *Trends in Biotechnology* **21**, 64–69 (2003).

156. Stelling J, Klamt S, Bettenbrock K, Schuster S & Gilles E.D. Metabolic network structure determines key aspects of functionality and regulation. *Nature* **420**, 190 (2002).

157. Papin J.A, Price N.D, Wiback S.J, Fell D.A & Palsson B.O. Metabolic pathways in the post-genome era. *Trends in Biochemical Science* **28**, 250–258 (2003).

158. Patil K.R & Nielsen J. Incovering transcriptional regulation of metabolism by using metabolic network topology. *Proceedings of the National Academy of Science* **102**, 2685–2689 (2005).
159. Shen-Orr S, Milo R, Mangan S & Alon U. Network motifs in transcriptional regulation network of *Escherichia coli*. *Nature Genetics* **31**, 64–68 (2002).
160. Kose F, Weckwerth W, Linke T & Fiehn O. Visualizing plant metabolomic correlation networks using clique-metabolite matrices. *Bioinformatics* **17**, 1198–1208 (2001).
161. Voit E.O & Radiovoyevithc T. Biocehmical system analysis of genome-wide expression data. *Bioinformatics* **16**, 1023–1037 (2000).
162. Kell D.B. Metabolomics, machine learning and modelling: towards an understanding of the language of cells. *Biochemical Society Transactions* **33**, 520–524 (2005).
163. Saghatelian A & Cravatt B.F. Global stretagies to integrate the proteome and metabolome. *Current Opinion in Plant Biology* **9**, 62–68 (2005).

CHAPTER 9

METABOLIC FLUX ANALYSIS OF MAMMALIAN CELLS

D.E. MARTENS

Food and Bioprocess Engineering Group, Wageningen University, Bomenweg 2, 6703 HD Wageningen, The Netherlands

Abstract: Metabolic flux analysis has become a standard tool for analyzing metabolism and optimizing bioprocesses. Metabolic flux analysis makes use of a metabolic reaction network in combination with extra-cellular measurements and mass balancing to calculate flux distributions in metabolism. It is a useful tool to analyze metabolism of cells and can be used to optimize the bioprocess in terms of medium design and metabolic engineering of the cells. In this chapter first the fundamental aspects of metabolic networks and the mathematical methods are described. Next a metabolic model for mammalian cells is discussed, Finally, applications of metabolic flux balancing are reviewed. Further extension of the metabolic network models, possibly towards genome-scale models, will further increase the value of these models

Keywords: Metabolic flux analysis, mammalian cells, Biopharmaceutical proteins, metabolism

1. INTRODUCTION

Mammalian cells are used increasingly for the production of vaccines and biopharmaceutical proteins that require specific post-translational modifications, like glycosilation. Production of these proteins cannot be done by other micro-organisms, because they cannot perform these post-translational modifications in a proper way. The market for vaccines and biopharmaceutical proteins is rapidly growing. In 2000 84 biopharmaceuticals were approved in the US and Europe, whereas in the last three following years alone, 60 products got market approval and another 500 products were in clinical evaluation[1]. Besides this mammalian cells are used to study the effect of toxic or beneficial compounds *in vitro*. Finally, cells are used in tissue engineering to repair damaged tissue. Certainly for the first two applications detailed knowledge of cell physiology including cell metabolism is essential either to optimize the production process or to understand the influence of a certain compound

M. Al-Rubeai and M. Fussenegger (eds.), Systems Biology, 275–299.
© 2007 *Springer.*

on cell metabolism. However, the metabolism of animal cells is complex. First of all metabolic processes take place in different compartments, like for example the mitochondria and cytosol. Secondly, coming from a multi-cellular environment, the cells require a complex medium that contains with respect to primary metabolism the essential and non-essential amino acids and glucose. Finally, animal cells have a substantial waste metabolism excreting lactate, alanine and sometimes other amino acids like proline and glutamate. Metabolic flux analysis can be a useful tool to reduce this complexity and obtain a better understanding of animal cells.

To be able to do metabolic flux analysis first a metabolic network must be constructed. Such a network contains the stoichiometry of all relevant biochemical reactions taking place in the cell. This requires detailed knowledge on which enzymes may be expressed in the cell, the reactions they catalyze and sometimes also their sub-cellular localization. Information on whether a reaction may be present or not can be obtained from literature and biochemistry hand-books. Furthermore, when the genome of an organism from which a cell line has been derived has been sequenced also the genome information can be used to construct a metabolic network. This will result in large networks (if possible genome-scale). To simplify this network reactions can be lumped, for which two basic forms can be discerned:

1. Serial reactions in linear pathways can be lumped into a single overall reaction. Reactions forming a branch point in metabolism should preferably not be lumped, since this would result in loss of information.
2. Biomass compounds having the same biosynthetic pathway are lumped into a single compound with an average composition. For example, all different proteins in the cell are lumped into one protein molecule with an average amino acid composition.

By writing mass balances over all compounds involved in the network and assuming that no accumulation of intracellular compounds occurs, fluxes can be calculated based on the measurement of uptake and production rates using mass balancing.

This chapter gives an overview of metabolic flux analysis in animal cells. In paragraph 2 the mathematical aspects of flux balancing are described in relation to animal cells. In paragraph 3 the metabolic network for animal cells is presented and applications of flux balancing are reviewed.

2.   MATHEMATICAL ASPECTS

2.1   Mathematical Translation of Network

The final network contains m metabolites (network nodes) and n reactions. Next the reactions are translated into a stoichiometry matrix S. The m columns of S contain the compounds, while the n rows contain the reactions. By doing an elemental check the reactions can be checked for elemental consistency.

$$S \cdot E = 0 \qquad\qquad (1)$$

Where E is the elemental composition matrix containing the elemental composition (columns) of all compounds (m rows).

Next mass balances are written for all m compounds present in the network. For a single compound this becomes:

$$\frac{dc_i}{dt} = -r_i + \sum_{j=1}^{n} s_{ij} \cdot x_j \tag{2}$$

Where c_i is the amount of compound i per cell (mol.cell^{-1}), r_i is the production rate of compound i (mol.cell^{-1}.s^{-1}), x_j is the flux through reaction j (mol.cell^{-1}.s^{-1}) and s_{ij} is the amount of compound i taking part in reaction j.

Assuming no accumulation of compounds occurs in the biomass, the left side of Eq. (2) becomes zero and Eq. (2) can be written in matrix notation as follows

$$\left[S^T - I_r - I_I \right] \cdot \begin{pmatrix} x \\ r \\ r_I \end{pmatrix} = \bar{0} \Rightarrow A' \cdot \begin{pmatrix} x \\ r \\ r_I \end{pmatrix} = \bar{0} \tag{3}$$

Where A' is a matrix with dimension $[(n+m) \times m]$ composed of the transpose of the stoichiometry matrix, S^T and the negative identity matrix, -I, which in turn is subdivided into a sub-matrix belonging to r, I_r, and a sub-matrix belonging to r_I, I_I, x is a vector containing the fluxes, r is a vector containing all production rates of compounds that are at the end of a metabolic pathway and can be net formed or consumed and r_I are the production rates for the intermediate compounds. For intermediates the net production rate is equal to the growth rate multiplied with the intracellular concentration of the compound. Since intracellular concentrations are always very low, the production rates of the intermediates can be neglected compared to the flux values and are therefore set to zero. Thus r_I and the corresponding columns in A' can be removed giving:

$$\left[S^T - I_r \right] \cdot \begin{pmatrix} x \\ r \end{pmatrix} = A \cdot \begin{pmatrix} x \\ r \end{pmatrix} = 0 \tag{4}$$

For compounds that are excreted or taken up theoretically a separate transport flux should be included. However, for simplicity this transport flux is often replaced directly by the net production rate of the compound. This includes the rates for formation of the macromolecules that make up biomass. Equation (4) is the basic equation describing the mass balances in the network. The number of reactions or unknowns $q = (x+r)$ is usually larger than the rank of matrix A. Furthermore, the number or rows in A is usually larger than the rank of A due to conserved moieties in the cell like NAD and NADH. Linear dependencies in A can be, but do not have to be, removed without loss of information. A complete set of linearly independent solutions for $(x\ r)^T$ forms the null space, K, of A with dimension $[q \times (q\text{-Rank}(A))]$. Several interesting features of the metabolic network can be derived from analyzing K. For example, if a row in K contains only zeros this means that the corresponding flux or rate must always be zero[2]. This is called a dead end and can either be removed or points to an error in constructing the metabolic network. Furthermore, elementary flux modes can be calculated from K, which are a minimal set of enzymes that fulfill the steady state conditions[3-5]. Elementary flux modes are useful to detect structural features of the network.

Since not all compounds that are excreted or taken up are and can be measured (for instance water), Eq. (4) is usually written as:

$$A \cdot \begin{pmatrix} r_c \\ r_m \end{pmatrix} = [A_C \ A_M] \cdot \begin{pmatrix} r_c \\ r_m \end{pmatrix} = 0 \tag{5}$$

Where, r_c contains the fluxes and the rates that are not measured and r_m contains all measured rates. By splitting the matrix A into a part belonging to the unknown fluxes and rates and a part belonging to the measured rates the next equation is obtained:

$$A_c \cdot r_c = -A_m \cdot r_m \tag{6}$$

Equation (6) is the central equation for flux mass balancing. According to van der Heijden et al.[6] and Klamt et al.[2] networks and rates can now be classified as follows:

- Underdetermined: $\text{rank}(A_c) <$ number unknowns. There are not enough linear independent relations to calculate all the unknowns, r_c, from the measured rates r_m. Note that some of the unknowns may be calculable.
- Determined: $\text{rank}(A_c) =$ number unknowns. All unknown rates can be calculated from the measured rates.
- Redundant: At least two rates can be balanced, i.e there is at least one relation between two or more measured rates. This means the measurements can be inconsistent, or in other words there is no r_c for the given r_m that exactly matches Eq. (6).
- Not redundant: There is no relation between rates and the system is always consistent

In relation with the above also rates can be classified:
- Unknown rates can be calculable or non-calculable. If all unknown rates can be calculated the system is determined. If one of the rates cannot be calculated the system is underdetermined.
- Measured rates can be balanceable or non-balanceable. If two or more rates can be balanced the system is redundant.

An important characteristic of the matrix A_c is the condition number, which is obtained by dividing the largest eigenvalue of A_c by the smallest eigenvalue. The condition number is a measure for the sensitivity of the unknowns for the measured rates[7]. Thus in the case of large condition numbers, errors in the measured rates may result in very large errors in the calculated values for the unknowns. Whether this actually will occur depends on the measured values[7,8].

In case of a determined system enough rates are measured to fix the system and calculate the unknown fluxes and rates. This can be done using:

$$r_c = -A_c^{-1} \cdot A_m \cdot r_m \tag{7}$$

Equation (7) can only be used if the matrix A_c is square and not singular. For a determined system the matrix is always non-singular. However, it may contain more rows than columns, in which case it cannot be directly inverted. If the network is not redundant and determined, but non-square, Eq. (7) can be replaced by:

$$r_c = -\left[A_c^T \cdot A_c\right]^{-1} \cdot A_c^T \cdot A_m \cdot r_m \tag{8}$$

In case the system is also redundant, the measured rates may be inconsistent. In this case Eq. (8) should not be applied directly and another approach can be followed, which is described in paragraph 2.3.

2.2   Underdetermined Networks

If the degree of freedom of the system, which equals the number of unknowns minus the number of independent relations in $A_c (= \text{rank}(A_c))$ is larger than zero the system is underdetermined. This means that at least part of the unknown fluxes and rates cannot be calculated from the measured rates. Two main causes can be discerned:

1. Insufficient rates are measured. This can thus be solved by measuring extra rates if this is experimentally possible.
2. The set of reactions is inherently independent from the measured rates. That is, even if all measurable rates are measured the matrix A_c still contains linear dependent relations and Eqs. (7) and (8) cannot be used, because A_c will still be singular.

For underdetermined systems usually part of the fluxes and unknown rates can be calculated. By calculating the null space of matrix A_c the fluxes and rates that cannot be calculated can be found. The null space can be found using singular value decomposition as described by van der Heijden et al.[6] or a more extensive analysis of the null space can be done as described by Klamt et al.[2]

Values for the calculable fluxes and rates in an underdetermined system can be obtained using the pseudo inverse. A general solution of Eq. (6) is than given by:

$$r_c = -A_c^\# \cdot A_m \cdot r_m + K_c \cdot \lambda \tag{9}$$

Where $A_c^\#$ is the so-called pseudo inverse, λ is a row vector with arbitrary values and K_C is a matrix of which the columns span the null space of A_C. Equation (9) gives values for the calculable rates, while for the non-calculable rates an infinite range of solutions exist. For $\lambda = 0$ a special solution is obtained called the minimum norm solution. For this solution the sum of the fluxes squared (the Euclidian norm) is minimal. There are a number of other options to calculate values for the fluxes in the underdetermined part:

- Measurement of the fluxes using isotopic labeling. This is by far the best option, since the actual fluxes are measured. However, it is also time consuming and difficult. Methods for measuring fluxes using isotopic tracers have been extensively described[9-14]. Vallino et al.[7,8] and van de Winden et al.[11] developed a method to determine, which fluxes should be measured in order to minimize the errors in the calculated fluxes.

- Using biochemical information like thermodynamic constraints[15], expression data[16,17] and enzyme activities[18,19] it can be found that some enzymes are not present and the corresponding reaction does not take place or that reactions can only proceed in one direction. These extra constraints either result in a determined system or at least narrow down the solution space, which can then be used in the method described next.

- Linear optimization. In addition to the extra constraints based on biochemical and thermodynamic information an objective function is formulated resulting in a flux distribution that optimizes the objective function[20-23]. Examples of these are maximizing ATP or NADPH and NADH production. The minimum norm solution obtained with Eq. (9) is also an example of this approach. The sum of fluxes squared is minimized in this case. The biological meaning given to this solution is that the organism strives to distribute the fluxes evenly over the network and minimize the total flux activity. However, the minimum norm solution depends on the extent of lumping and has been found not to agree well with experimentally found flux distributions.[22]

2.3   Consistency

In flux balancing one should always strive to have a redundant system, since this results in the possibility to check the consistency of the model and measurements. Redundancy relations contain a sum of rates that always must be zero. Redundancy relations can be found from[24-26]:

$$R = A_m - A_c \cdot A_c^{\#} \cdot A_m \qquad (10)$$

The biological meaning of the relations in R obtained from Eq. (10) is usually not easy to see. Usually R contains elemental balances like for example the nitrogen, carbon and redox balance. To obtain these relations from R, one has to bring it in the row echelon format with compounds on the first position that only contribute in one of the balances. Sometimes extra relations are found. For example, in the case of animal cells, if the only fate of an amino acid is to be build into protein, this will result in a redundancy relation containing the rates of the amino acid and of biomass protein and product. If such a redundancy relation is to be used one should accurately measure the protein composition of both product and biomass protein and not settle for an average literature composition, since otherwise the consistency test (see further on) will not be passed. When the redundancy relations are multiplied with the measured rates the result should ideally be zero. However, this is in principle never the case, because the measurements will have a measurement error. If the vector of measurement errors is given by $\delta(\text{mol.cell}^{-1}.\text{s}^{-1})$, the true measurement vector $r_t(\text{mol.cell}^{-1}.\text{s}^{-1})$ is related to the measured rate vector by:

$$r_m = r_t + \delta \qquad (11)$$

It is assumed that the measurement errors are normally distributed with an expected value of zero and a variance-covariance matrix, ψ_δ. The presence of the random errors in the measurements will cause a vector of residuals, $\varepsilon(\text{mol.cell}^{-1}.\text{s}^{-1})$, in the redundancy equations giving:

$$R \cdot r = R \cdot (r_t + \delta) = 0 + R \cdot \delta = \varepsilon \qquad (12)$$

It can be shown now that also ε is normally distributed[24] with an expected value of zero and a variance covariance matrix ψ_ε given by:

$$\Psi_e = R \cdot \Psi_d \cdot R^T \qquad (13)$$

Next a test function can be constructed:

$$h_e = e^T \cdot \Psi_e^{-1} \cdot e \qquad (14)$$

The test-function has a Chi-square distribution[24], with the degree of freedom being equal to the number of redundancy relations. Next, the zero hypotheses that the residuals are a consequence of measurement errors can be tested, where high values of h_e point to the alternative hypothesis. Critical values for the Chi-square distribution can be obtained from tables. If the test is not passed this can have different causes being:[24]

1. Errors of the first kind: At least one of the measurements has a gross error. Such an error is, for instance, due to a wrong calibration curve or a faulty pipette.
2. Errors of the second kind: The system description is incorrect, which may be due to a component that is not included or a wrong stoichiometry of a reaction.
3. Errors of the third kind: The measurement errors are larger than estimated.

If more than one balance is present in the redundancy matrix, a balance that is far off can still result in passing the test if the other balances close very well. Thus, the most stringent situation occurs if the balances are used separately for testing, where each individual balance as well as all balances together must be passed.

If the test is passed, which means that the residuals can be completely explained from measurement variances, the rates should be balanced. In other words the measurement error vector δ should be found to calculate the true rate vector, r_t, from the measured one, r_m. The best estimate is obtained by minimising the sum of errors squared. Because, some measurements will be more accurate than others, this sum of squares is weighed according to the average measurement error in each measurement or in other words:

$$Minimize \ J = \delta^T \psi_\delta^{-1} \delta \ subject \ to \ R \cdot \delta = \varepsilon \tag{15}$$

The solution of this minimisation is given by:

$$\hat{\delta} = \psi_\delta R^T \psi_\varepsilon^{-1} \varepsilon \tag{16}$$

The true rate vector can now be calculated using Eq. (11). The variances of this vector can be calculated by:

$$\hat{\psi}_\delta = \psi_\delta - \psi_\delta \cdot R^T \cdot \psi_\varepsilon^{-1} \cdot R \cdot \psi_\delta \tag{17}$$

It can be shown that the new variances are smaller than the variances of the original measurements[24]. Next the fluxes should be calculated using this balanced rate vector. The variances of the calculated fluxes can be calculated using:

$$\Psi_{rc} = A_c^\# \cdot A_m \cdot \hat{\Psi}_\delta \cdot A_m^T \cdot (A_c^\#)^T \tag{18}$$

3.   FLUX ANALYSIS IN ANIMAL CELLS

3.1   Metabolic Network

Many different metabolic networks have been proposed for animal cells. They mainly concern mammalian cells. For insect cells to the best of the author's knowledge only two publications on flux balancing have been published[27,28]. This probably relates to the fact that at the moment insect cells are hardly applied in industrial processes and the fact that insect cell metabolism is not as well studied as mammalian metabolism. Furthermore, insect cell media are usually more complex, containing next to all amino acids and glucose also a number of citric acid cycle intermediates, disaccharides and protein hydrolysates. Thus, in order to do flux analysis more compounds must be measured. Here we will focus on metabolic flux analysis for mammalian cells.

In general the metabolic networks are roughly quite similar. They differ mainly in how underdetermined parts are treated and in the extent in which the different compartments in the cell (mainly cytosol and mitochondria) have been accounted for. In Figure 1 a metabolic network is shown, which is a summary of the different models used. The network contains the following pathways:

Glycolysis (reactions 1–7): The Embden Meyerhof pathway is lumped into six reactions including the reaction from pyruvate to lactate. In this way all important nodes (G6P, F6P, F16P, GAP, PEP and PYR) are present.

Pentose Phosphate Pathway (PPP): This pathway contains an oxidative part (reaction 8) and a non-oxidative part consisting of a series of transaldolase and transketolase reactions (reactions 9–12). The non-oxidative part is often lumped

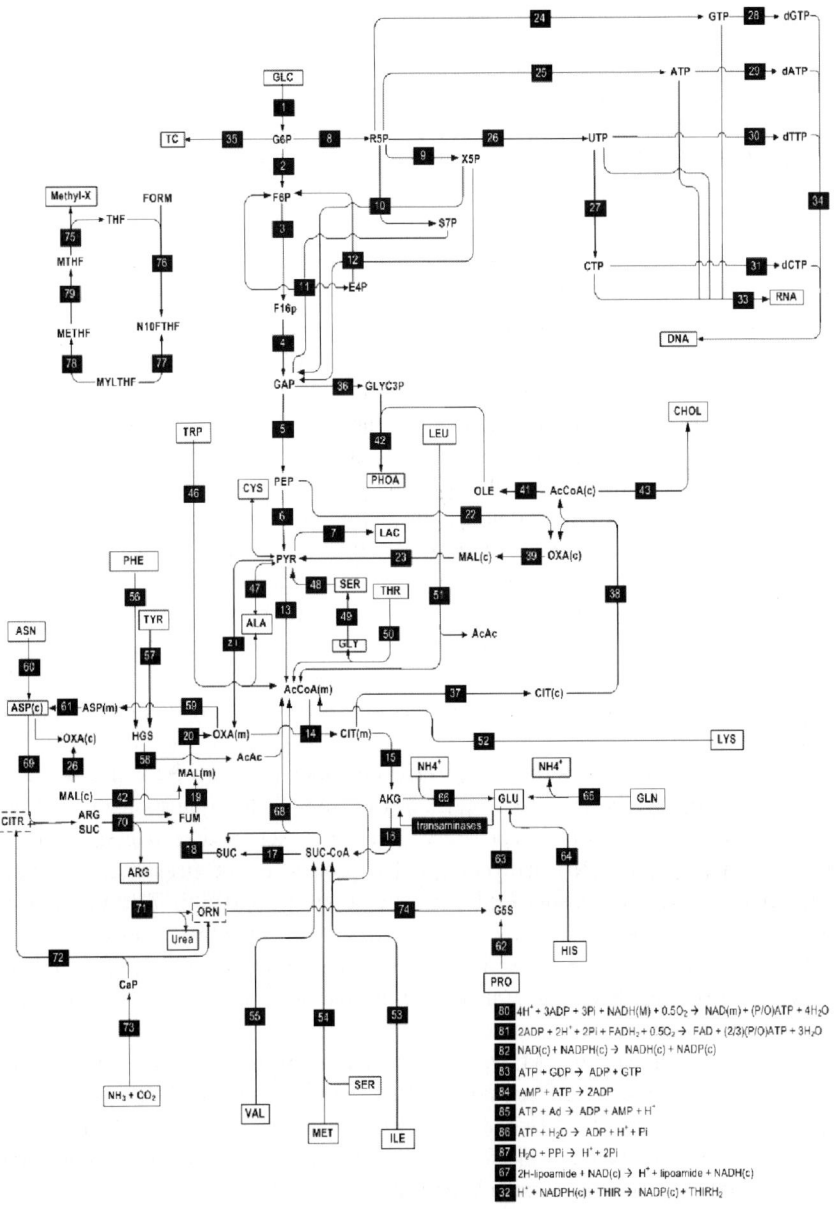

Figure 1. Metabolic network. For the corresponding reactions and meaning of the abbreviations see appendix I. Compounds surrounded by a square can be exchanged with the environment. (c) = cytosolic, (m) = mitochondrial

into one reaction with 3 R5P giving 2 F6P and 1 GAP. The pathway can operate in three different basic modes being a mode for generation of NADPH, one for generation of R5P for nucleotide synthesis and a combination of these[29].

Tricarboxylic acid (TCA) cycle (reactions 14–20): In the TCA cycle acetyl-CoA is broken down to CO_2 and water generating energy. Furthermore, it supplies intermediates for biomass synthesis. A number of reactions can convert TCA cycle intermediates malate and oxaloacetate to intermediates of glycolysis phospho-enol-pyruvate and pyruvate and vice versa (reactions 21–23).

Amino acid metabolism (reactions 46–68): All amino acids can be broken down to generate energy. In addition, non-essential amino acids can be synthesized *de novo*. In case a non-essential amino acid has a different anabolic and catabolic pathway the catabolic pathway is included in Figure 1. However, if fluxes through such a catabolic pathway become significantly smaller than zero, they should be replaced by the anabolic pathway. Note that the flux through a breakdown pathway of an essential amino acid can never become significantly smaller than zero, since they cannot be synthesized. In this case there must be an error in one or more measurements, for instance, biomass composition or medium concentration.

Urea Cycle (reactions 69–74): Industrial cell lines are immortal and have tumor-like properties. In tumor cells the urea cycle is assumed to be not active, because the enzyme carbamoyl phosphate synthase (reaction 73) is not expressed[30,31] and, consequently, urea is not found in the supernatant[31]. However, for primary cells the cycle may be active. Furthermore, the intermediates ornithine and citruline of the cycle can be excreted by cells, in which case and efflux of these compounds should be included in the model.

THF metabolism (reaction 75–79): THF functions as an acceptor and donor of C1 groups in the synthesis of nucleotides and in catabolism and anabolism of certain amino acids. A reaction is added which represents methylation of compounds in the cell, for example, DNA and proteins. The product is not measured but always calculated and serves as a sink for C1 groups. If not included this THF metabolism would contain a dead end meaning that the flux values are always zero.

Energy metabolism (reaction 80–87): This part contains all reactions related to ATP generation. It contains also the transhydrogenase reaction which makes NADH and NADPH exchangeable. The ATP balance in the cell can generally not be closed. The amount of ATP formed per NADH oxidized, the P/O ratio, is not exactly known. Miller et al.[32] assume this value to be 2. Xie and Wang[33] estimate the value to be 2 for NADH and 1 for $FADH_2$. Vriezen et al.[34] calculated the value to be 1.2 for NADH and 0.8 for $FADH_2$. However, in many papers it is assumed simply to be three for NADH and two for $FADH_2$. The amount of ATP needed for biomass formation is in principle included in the anabolic reactions in the network. However, actual ATP requirements for growth may be higher. In addition the ATP requirement for maintenance is not known. Therefore, an ATPase reaction is included which represents net formation of ATP in the model. Generally this ATPase reaction contains up to 60% of the energy turnover. However, this will depend strongly on the P/O ratio used.

Biomass: Biomass is usually divided into 6 fractions being carbohydrates, protein, lipids, Cholesterol, DNA and RNA. The average composition for the macromolecule is derived from measurements. This average composition determines the stoichiometry of the biosynthetic reaction and the molecular weight of the molecule. Based on the amount of a macromolecule present in the cell in g.cell^{-1} the number of moles of the molecule per cell can now be easily calculated. The net production rate of the compound is now given by the growth rate multiplied by the number of moles per cell. The biomass composition as published in different papers is given in Table 1. If the relative composition is more or less independent from reactor conditions, it suffices to measure it at one condition and next for other conditions only measure cell dry-weight or protein content. However, the relative composition of cells can vary depending on conditions and thus should be measured in principle for each measurement set[34]. Because also cell-size can vary between reactor conditions, expressing fluxes and rates only on a per cell basis may not provide the correct insights. In case the cell size varies, fluxes should also be expressed per g dry-weight or per g protein. Another option is to express the flux distribution relative to a particular flux, for example the glucose uptake flux. Below the different biomass fractions are next treated:

Protein (reaction 44): Here protein is assumed to consist of 10 amino acids. For the amino acid composition of cellular protein a literature average is taken here, which is given in Table 2. Note that the amino acid composition and cellular protein content together with the uptake of amino acids from the medium determine the catabolic rates of the amino acids. Therefore, wrong assumptions or measurements of cellular protein content and composition can lead to wrong flux values for a large part of the network.

Lipids and Cholesterol (reaction 36–43): The fatty acids in lipids are mainly oleic acid (C18:1, 55%)[31]. Since the amount of fatty acids with a shorter chain equals that with a longer chain[31] it is usually assumed in flux models that all fatty acid is oleic acid. This will on average not influence the demand for precursors, reducing equivalents and energy. Cholesterol is modeled as a separate compound and makes

Table 1. Biomass composition of mammalian cells found in literature. Cell dry weight is expressed in pg.cell^{-1}, biopolymer fractions are expressed weight percentage

	Savinell[8]	Zupke[35]	Xie[36]	Bonarius[31]		Vriezen[34]	Average
Dry weight		250	250	470	510	190	330
Protein	60.0	60.1	72.9	70.6	67.1	79.1	70.0
Lipids	16.7	17.0	13.5	9.7	10.0	7.7	11.6
Carbohydrate	6.7	6.4	3.5	7.1	7.0	2.6	5.3
DNA	0.9	5.6	1.4	1.4	1.5	1.7	1.4[a]
RNA	3.7	10.8	3.8	5.8	5.3	8.9	5.9
Total	88	99.9	95.1	94.6	90.9	100[b]	94.1

a. value of Zupke not included in the average b. values expressed in percentage of biopolymer. However, the sum of biopolymer equaled the dry weight.

Table 2. Amino acid composition of protein in molar percentage

	Xie[36]	Bonarius[31]	Savinell[20]	Okayasu[37a]	Average
Ala	8.5	8.1	7.5	9.1	8.3
Arg	8.3	7.7	9.0	9.0	8.5
Asn	6.0	5.4	6.9	6.5	6.2
Asp	8.1	8.3	7.5	9.0	8.2
Cys	4.2	4.4	4.6	5.7	4.7
Gln	2.2	2.5	1.7	2.3	2.2
Glu	5.3	10.1	4.6	5.5	6.4
Gly	3.2	3.9	3.5	3.7	3.6
His	1.1	0.3	1.1		0.8
Ile	6.9	5.9	7.1	6.2	6.5
Leu	5.7	4.7	6.0	5.4	5.5
Lys	4.4	4.2	4.4	4.6	4.4
Met	6.2	5.0	3.9	6.1	5.3
Phe	2.6	2.4	3.5	2.7	2.8
Pro	2.8	0.7	2.8	0.3	1.6
Ser	6.8	8.6	7.0	6.9	7.3
Thr	5.9	5.5	4.7	4.7	5.2
Trp	2.2	1.8	2.1	2.2	2.1
Tyr	4.7	5.3	5.5	5.5	5.3
Val	5.0	5.9	4.6	6.5	5.5

a. Gln, glu and Asn, ASP were measured as one fraction. Discrimination is made based on the average ratios of Gln/Glu and Asn/Asp. Trp was not measured.

up about 7% of the lipid fraction. Fatty acids and cholesterol are synthesized from acetyl-coA in the cytosol. Consequently, acetyl-CoA has to be transported from the mitochondria to the cytosol. This is done through citrate, which acts as a carrier. For this reason the fatty acid flux together with oxygen consumption determine the activity of the TCA cycle. The lipid content of cells may vary depending on conditions and should thus be measured for each condition.

Carbohydrates (reaction 35): Carbohydrates must be included to close the carbon balance and are assumed to be mainly present in glycoproteins[31]. It is assumed that they are all derived from G6P.

DNA and RNA (reaction 24–34): DNA and RNA are assumed to consist of 10 nucleotides with a G:C:A:T ratio of 2:2:3:3[31].

Product (reaction 45): the approach is the same as for biomass protein. For the product however the complete gene sequence is often known. From this sequence the amino acid composition can be derived.

Modeling different compartments (reaction 88–98): Mammalian cells contain different compartments in which different metabolic processes take place. For primary metabolism especially the mitochondria are important. Figure 2 shows a schematic representation of the metabolism around the mitochondria. The mito-chondrial membrane is impermeable for NAD(H). For transport of electrons across the membrane different shuttle systems exist being the glycerol-phosphate shuttle

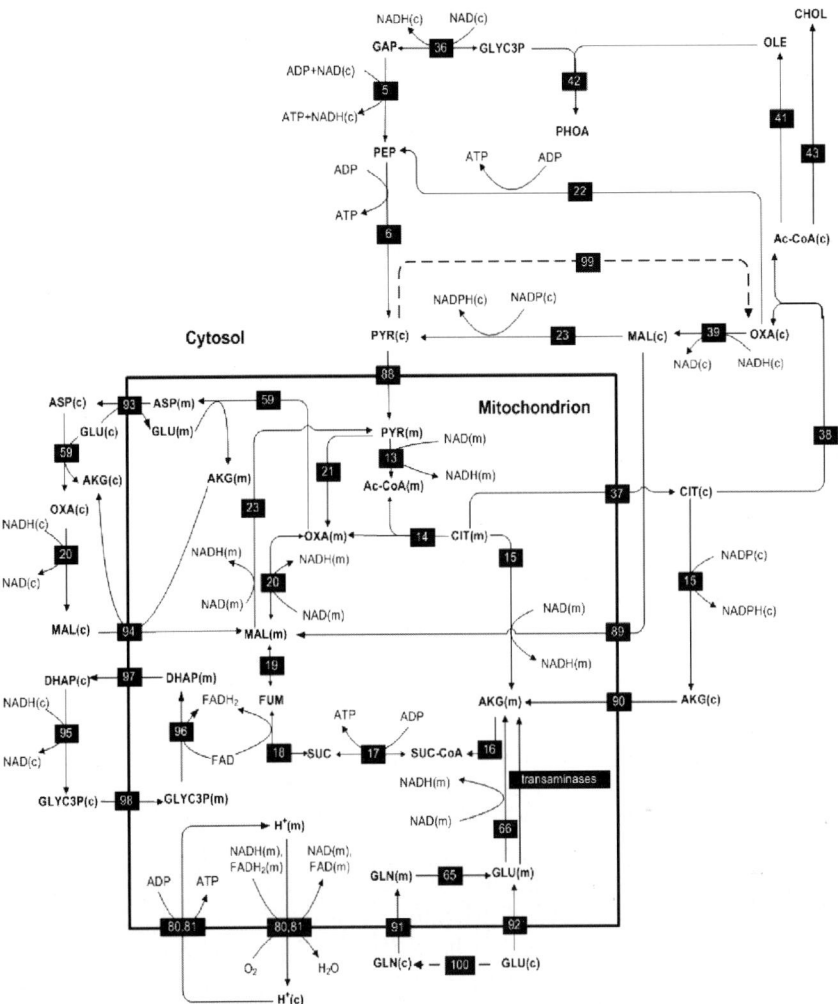

Figure 2. Detailed metabolic network showing separation into mitochondrial and cytosolic compartment. (c)=cytosolic, (m)=mitochondrial

and the aspartate-malate shuttle[29]. The difference between both systems is that the glycerol phosphate shuttle converts cytosolic NADH into mitochondrial FADH$_2$ meaning a loss of energy. The malate-aspartate shuttle turns cytosolic NADH into mitochondrial NADH without loss of energy. However, this also means this system is ready reversible and may work in both directions depending on the ratio of cytosolic to mitochondrial NADH. The extra energy loss in the case of the glycerol phosphate system makes that this system can transport electrons even if the NADH level in the mitochondria is higher than in the cytosol. Enzyme activity measurements indicate that in mammalian cells probably the aspartate-malate shuttle is the active shuttle system[38].

Another compound that cannot cross the mitochondrial membrane is acetyl-coA. For lipid synthesis, which takes place in the cytosol, acetyl-CoA is transported across the membrane in the form of citrate as described before. In the cytosol citrate is split into oxaloacetate and acetyl-CoA. Oxaloacetate can next be converted to PEP, to malate and next to pyruvate, or can be exchanged with mitochondrial citrate. Different compartments can be easily incorporated into metabolic models by assuming different pools of the same compound. However, reactions in the mitochondria sometimes have equivalent reactions in the cytosol with an identical stoichiometry or only differing in type of electron donor/acceptor used (NADH in mitochondria and NADPH in the cytosol). Since NADPH and NADH are ready exchangeable this can easily result in underdetermined sets of reactions. An example of this is the reaction of citrate to AKG (reaction 15) shown in figure 2. It is however possible to separate mitochondria experimentally from the cytosol, making direct measurements of, for example, enzyme activity in both compartments possible[38].

3.2   Underdetermined Parts

One of the first problems encountered in flux balancing is that a number of under-determined parts are commonly found in metabolic networks for animal cells.

The pentose phosphate pathway forms an underdetermined set together with glycolysis and the TCA cycle. In other words complete aerobic degradation of glucose in the PP pathway cannot be discerned from complete aerobic degradation in glycolysis and the TCA cycle. This is due to the possibility of exchange of NADPH and NADH and the fact that the ATP balance cannot be closed as mentioned before. Measurements using isotopic tracers show that at high growth rates about 23% of the glucose is shuttled through the PP Pathway[39]. In a hollow fiber bioreactor where growth rates and thus the demand for precursors for biosynthesis are likely to be much lower this was found to be 4–8%.[40,41] Zupke and Stephanopoulos[35] used another approach to calculate the flux distribution. They assumed that no transhydrogenase activity was present. Consequently, the flux through the PPP is determined by the biosynthetic demand for NADPH and PPP intermediates. For rapid growing cells in batch experiments they found that in the exponential phase about 10% of the glucose was channeled through the PPP. Isotopic tracer distributions based on the calculated flux distributions agreed with measured distributions. Finally, Bonarius et al.[22] used linear optimization to estimate the flux distribution. Although maximizing ATP and NADH formation gave overall flux distributions close to the measured distribution, the PPP-flux was not estimated well. In many papers it is assumed that the PPP is used only to supply nucleotides. In other words the transaldolase and transketolase activity is neglected and only the oxidative part is considered. Consequently, the flux through the PPP will equal the demand of ribose for nucleotides, which will be between 0.5% and 5% of the glycolytic flux depending on growth rate and extent of lactate formation. However Mazurek et al.[42]

state that it is probably the oxidative part of the PPP that is inhibited in tumor cells and that ribose-5-phosphate is formed through the transaldolases and transketolases.

Another common underdetermined part is usually present around the PEP-pyruvate-oxaloacetate-malate part of the network. Three enzymes connect the glycolytic intermediates with TCA intermediates being PEP carboxykinase (reaction 22), pyruvate carboxylase (reaction 21) and malic enzyme (reaction 23). Several publications using isotopic labeling[40,41] and enzyme activity[18] assays have measured that pyruvate carboxylase activity is negligible. Enzyme activity measurements for PEP-carboxykinase[18] suggest that the activity of this enzyme in the cytosol is sufficient to carry the flux between pyruvate/PEP and oxaloacetate/malate. The activity of malic enzyme could not be measured[18] and also no distinction based on isotopic tracer studies can be made between pep-carboxykinase and malic enzyme. The overall reaction for these two enzymes from oxaloacetate to pyruvate only differs in that malic enzyme generates NADPH and the route through PEP carboxykinase generates NADH. Since NADH and NADPH are exchangeable, this underdetermined part can be eliminated by just choosing one of the enzymes.

Other underdetermined parts are found for the non-essential amino acids. As mentioned, non essential amino acids have an anabolic and a catabolic pathway associated with them. The only difference between both pathways is usually the amount of ATP, NADH and NADPH required or formed. To solve this either the anabolic or catabolic route is left out of the model depending on whether there is net breakdown or synthesis of the amino acid.

Cellular protein and RNA are subject to constant turnover. Proteins and RNA is broken down to intermediate building blocks, which are next used to build new protein and RNA molecules through salvage pathways. These salvage pathways can differ from the *de-novo* pathways. In the end this results in an undetermined set of reactions that turn-over macromolecules at the cost of extra reducing equivalents and energy. Savinell and Palsson took the turnover of protein and RNA into account[20]. If not accounted for these extra energy requirements will end up in the ATPase reaction.

3.3   Applications of Flux Analysis

The first studies on metabolic flux analysis were especially aimed at establishing the methods for animal cells and finding the correct metabolic network, i.e. find out which reactions took place and in which compartment. A number of studies were aimed at solving the fluxes through the underdetermined parts mentioned earlier[39-41].

Flux analysis for estimation of physiological state overlaps with flux analysis for medium design. Savinell and Palsson[20,21] were the first to use metabolic flux analysis and linear optimization to analyze a hybridoma metabolic network. The objective function of maximizing growth rate was used and two situations were analyzed: One where only the main substrates were constrained to their measurement values and one where also the main waste metabolites were constrained to their measurement values. Results indicated amongst others that neither ATP demand for maintenance nor antibody production limited the growth rate but that amino acids were the limiting nutrients. Bonarius et al.[22] also applied objective functions and linear optimization to find solutions for the underdetermined part and compared the solutions to isotopic tracer studies they did. They found that flux distributions using maximization of ATP or NADH were relatively similar to experimental distributions. However, the flux through the PPP was still calculated incorrectly using the different objective functions. Notably, the minimum norm solution was very different from the experimental values.

Vriezen et al.[18] measured the activities for different enzymes in SP2/0-Ag14 cells as well as the sub-cellular localization[38] of some enzymes in combination with flux analysis. Their main conclusion was that regulation of metabolism in central metabolism occurs through modulation of enzyme activity and to a much lesser extent through enzyme level. Furthermore, they found substantial activity of PEP-carboxykinase in the cytosol indicating this may be an important enzyme connecting glycolysis and TCA. Activity of pyruvate carboxylase was found to be negligible. Finally, high activities of malate dehydrogenase and aspartate aminotransferase indicated the malate aspartate shuttle may be functional.

Xie and Wang[43] used a stoichiometric approach to design a medium that exactly balances the anabolic needs of the cells. Although in their first study they did not use and actual metabolic network and flux balancing, the approach is similar. A feed medium designed in this way resulted in a reduction of lactate and ammonia formation and increased total cell concentrations and product concentrations. In later studies they used a metabolic network and flux balancing to assess the energy metabolism of hybridoma cells[44].

Metabolic flux models have been extensively used to assess the effect of certain compounds, including many medium components, on cell metabolism. Bonarius et al.[45] studied the effect of oxygen tension on hybridoma cells. In addition, they used an artificial electron acceptor, PMS, to simulate oxidative stress. Under hypoxic conditions and when PMS was applied they found an increase in fluxes generating reducing equivalents like catabolic breakdown of amino acids Lys, Ile, Leu, Met and Arg and the glutamate dehydrogenase flux.

Vriezen et al.[34,46] used different cell lines and different concentrations of glucose and glutamine to estimate the P/O ratio and the amount of ATP required for biomass formation not accounted for in the network itself. Altamirano et al.[47] studied the effect of replacing glutamine by glutamate and the balance between glutamate and glucose metabolism on the metabolism of recombinant CHO cells producing t-PA. Amongst others, they found multiple steady states i.e. different steady states under the same conditions. Bonarius et al.[31] and Nyberg et al.[48] studied the effect of the hydrolysate Primatone on hybridoma and CHO cells, respectively, using metabolic flux analysis.

Follstad et al.[49] found multiple steady states in chemostats with hybridoma cells. At the same dilution rate two different steady states were reached displaying very different viable cell concentrations (0.74 and 1.36 million cells per ml respectively) at identical viabilities. The first low cell density state was reached by inoculating cells in rich medium. The second high cell density state was reached by first imposing nutrient limitation at low dilution rates and next return to higher dilution rates gradually. They used metabolic flux analysis to study these to steady states and found that in the high cell density state more pyruvate entered the TCA cycle leading to more efficient use of glucose. A similar situation was observed by Europa et al.[50] who induced the low cell density and high cell density steady state by starting the chemostat form a batch and fed-batch respectively. However, in terms of metabolism the situation is clearly different from that observed by Follstad et al.. While in the case of Follstad et al. cells became more efficient through a higher TCA flux, in the case of Europa et al. not only became the cells more efficient but also specific rates and fluxes became smaller. The work of Europa et al. was continued by analyzing gene expression and proteomics[51,52]. Upon the metabolic shift a decrease in expression of some metabolic genes was observed. However the number of genes and extent of decrease was moderate confirming the suggestion of Vriezen et al.[18] that changes in metabolism are regulated at the level of enzyme activity.

Nadeau et al.[53,54] used metabolic flux analysis to study metabolism of 293SF cells in different media and during growth and infection. Marked differences were observed for different media. For the infection phase flux analysis could also be done despite the highly transient nature of this phase.

Chan et al.[55,56] used metabolic flux analysis in combination with radio-labeling of substrates to study hepatocyte function in plasma supplemented with insulin, hydroxycortisone and amino acids. The used model differed in a number of aspects from the model in Figure 1. It contained reactions for β-oxidation of fatty acids, which are taken up by hepatocytes and present in the plasma. Furthermore, hepatocytes store glucose in the form of glycogen and thus the glycogen synthesis pathway was added. Gluconeogenesis is important in hepatocytes meaning that a phosphatase reaction has to be added converting F16P to F6P. Finally, since the hepatocytes are primary cells, they produce urea in contrast to industrial cell lines. Interestingly, they combined flux analysis with multivariate data analysis to find underlying structures and causal relationships within the data.

Three examples of genetic engineering of primary metabolism exist. The genetic engineering targets were not directly obtained from flux analysis, but more indirectly through an improved understanding of metabolism. Paredes et al.[58] used two approaches to reduce the amount of lactate and ammonia produced by the cell. They expressed antisense mRNA to a glucose transporter and to enolase catalyzing the conversion of 2 phosphoglycerate to PEP thus limiting the flux through glycolysis. A reduction of glucose flux to lactate was obtained, which was most clear for the inhibition of the transporter (47%). However, the mechanism was not completely clear and cell lines were unstable. Another approach taken by several groups is the introduction of glutamine synthase in hybridoma cells and next growing the cells in the absence of glutamine[58,59]. Since no glutamine is in the medium the glutaminase reaction is zero and anabolic glutamine requirements are fulfilled by the glutamine synthase (reaction 100). Expression resulted in complete reduction of ammonia excretion and excretion of other amino acids like alanine and in a 50% reduction in glycolysis flux. Yield of lactate on glucose remained the same. Notably, this modification is inherently stable since hybridoma cells without the gene cannot grow. Finally, CHO cells express the glutamine synthase at sufficient levels meaning genetic modification is not needed to grow them on glutamine-free media.

Irani et al.[19,60] introduced into BHK-21 cells a pyruvate carboxylase gene (reaction 99), which was expressed in the cytoplasm, to restore the weak link between glycolysis and the TCA cycle. Pyruvate is converted to oxaloacetate, which is subsequently converted to malate turning cytosolic NADH into NAD. Thus less pyruvate and NADH are available for lactate formation. Malate can next enter the mitochondria and is converted to oxaloacetate generating a mitochondrial NADH, which can be oxidised by oxygen generating ATP. In addition, the TCA intermediate level rises, which may lead to more pyruvate entering the cycle via acetyl-CoA. Finally, cytosolic malate may be converted back to pyruvate by malic enzyme generating NADPH. This thus results in a cycle converting NADH into NADPH. This means less NADH is available for lactate formation and more NADPH is available for biosynthesis. Cell metabolism became more efficient in the sense that glucose consumption decreased, more carbon derived from glucose entered the TCA cycle and the lactate yield on glucose decreased. In addition, more oxygen was consumed and also glutamine consumption was reduced. Cell concentrations in chemostat mode were equal for the control and the transformed cells. However, in perfusion mode cell concentrations were higher for the pyruvate carboxylase containing cells.

Provost and Bastin[61] use a metabolic network to construct a dynamic metabolic model for CHO cells. First, the consistency of the network and the measurements is tested. Next elementary flux modes are calculated and translated into relations connecting extra-cellular substrates and products. Based on these relations a dynamic model was build resulting in good predictions for the growth phase of a batch culture.

4.   CONCLUSIONS

Metabolic flux analysis has become a standard tool to analyse animal cell metabolism. Calculation and measurement methods are in principle well established. Also the reaction stoichiometry is well known. For fluxes that cannot be calculated using mass balancing, isotopic tracer methods are available to determine them. Other methods to determine these fluxes like linear optimisation or other assumptions have not been proven to be valid yet.

A first step is made for on-line flux estimation for mammalian cells in dynamic situations, which would make it possible to control cell cultures based on monitoring a critical flux. Supplementing metabolic network models with kinetic information also make them more valuable for process design. Furthermore, further extension of the networks with for example glycosylation pathways, more detailed lipid biosynthesis and compartmentation is possible. Clearly this will ask for more and more detailed measurements. This may even lead to genome-scale models like is the case for prokaryotes. Genome-scale models reflect the metabolic capabilities of the organism and are of great value in medium design and genetic engineering strategies[63-65].

ACKNOWLEDGEMENTS

The author would like to thank M.C.F. Dalm and D.M. Kisztelinski for their help in constructing Figure 1, appendix I and table 2.

5.   REFERENCES

1. G. Walsh, Biopharmaceutical benchmarks, Nature Biotechnol. **21**(11), 1396–1406 (2003).
2. S. Klamt, S. Schuster, E.D. Gilles, Calculability analysis in underdetermined metabolic networks illustrated by a model of the central metabolism in purple nonsulfur bacteria. Biotechnol. Bioeng. **77**(7), 734–751 (2002).
3. S. Schuster and C. Hilgetag, On elementary flux modes in biochemical reaction systems ate steady state. J. Biol. Syst. **2**, 165–182 (1994)
4. S. Schuster, T. Dandekar, and D.A. Fell, Detection of elementary flux modes in biochemical networks: a promising tool for pathway analysis and metabolic engineering. TibTech, **17**, 53–60 (1999).
5. S. Schuster, D. Fell, T. Dandekar, A general definition of metabolic pathways useful for systematic organization and analysis of complex metabolic networks. Nature Biotechnol. **11**, 326–332.
6. R.T.J.M. van der Heijden, J.J. Heijnen, C. Hellinga, B. Romein, K.Vh.A.M. Luyben, Linear constraint relations in biochemical reaction systems: I. Classification of the calculability and the balanceability of conversion rates. Biotechnol. Bioeng. **43**, 3–10 (1994).
7. J.M. Savinell and B.O, Palsson, Optimal selection of metabolic fluxes for in vivo measurements. I. Development of mathematical methods. J. Theor. Biol. **155**, 201–214 (1992).
8. J.M. Savinell and B.O. Palsson, Optimal selection of metabolic fluxes for in vivo measurement. II. Application to Escherichia coli and hybridoma cell metabolism. J. Theor. Biol. **155**, 215–242 (1992).
9. J. Christensen, T. Christiansen, A.K. Gombert, J. Thykaer and J. Nielsen, Simple and robust method for estimation of the split between the oxidative pentose phosphate pathway and the Embden-Meyerhof-Parnas pathway in microorganisms. Biotechnol. Bioeng. **74**, 517–523 (2001).
10. K. Schmidt, A. Marx., A.A. deGraaf, W. Wiechert, H. Sahm, J. Nielsen and J. Villadsen, C-13 tracer experiments and metabolite balancing for metabolic flux analysis: comparing two approaches. Biotechnol. Bioeng. **58**, 254–257 (1998).
11. W.A. van Winden, J.J. Heijnen, P.T. Verheijen and J. Grievink. A priori analysis of metabolic flux identifiability from C-13-labeling data. Biotechnol. Bioeng. **74**, 505–516 (2001).
12. W.A. van Winden, J.C. van Dam, C. Ras, R.J. Kleijn, J.L. Vinke, W.M. van Gulik and J.J. Heijnen, Metabolic-flux analysis of Saccharomyces cerevisiae CEN.PK113-7D based on mass isotopomer measurements of C-13-labeled primary metabolites. Fems Yeast Research **5**, 559–568 (2005).
13. S.A. Wahl, M. Dauner and W. Wiechert, New tools for mass isotopomer data evaluation in C-13 flux analysis: Mass isotope correction, data consistency checking, and precursor relationships. Biotechnol. Bioeng. **85**, 259–268 (2004).
14. T.H. Yang, E. Heinzle and C. Wittmann, Theoretical aspects of C-13 metabolic flux analysis with sole quantification of carbon dioxide labelling. Comp. Biol. Chem., **29**, 121–133 (2005).
15. D.A. Beard, E. Babson, E. Curtis and H. Qian, Thermodynamic constraints for biochemical networks. J. Theor. Biol., **228**, 327–333 (2004).

16. M. Akesson, J. Forster and J. Nielsen, Integration of gene expression data into genome-scale metabolic models. Metabolic Eng., **6**, 285–293 (2004).
17. R. Korke, M.D. Gatti, L.Y. Lau, A. J.W.E.Lim, T.K. Seow, M.C.M. Chung and W.S. Hu, Large scale gene expression profiling of metabolic shift of mammalian cells in culture., J. Biotechnol., **107**(1), 1–17 (2004).
18. N. Vriezen and J. P. Vandijken., Fluxes and enzyme activities in central metabolism of myeloma cells grown in chemostat culture. Biotechnol. Bioeng., **59** (1), 28–39 (1998).
19. N. Irani, M. Wirth, J. van den Heuvel and R. Wagner. Improvement of primary metabolism of cell cultures by introducing a new cytoplasmic pyruvate carboxylase reaction., Biotechnol. Bioeng., **66** (4), 238–246 (1999).
20. J. M. Savinell and B. O. Palsson. Network analysis of intermediary metabolism using linear optimization.I. Development of mathematical formalism. J. Theor. Biol. **254**, 421–454 (1992).
21. J. M. Savinell and B. O. Palsson. Network analysis of intermediary metabolism using linear optimization II. Interpretation of hybridoma cell metabolism. J. Theor. Biol. **154**, 455–473 (1992).
22. H. P. J. Bonarius, B. Timmerarends, C. D. de Gooijer and J. Tramper. Metabolite-balancing techniques vs. c-13 tracer experiments to determine metabolic fluxes in hybridoma cells. Biotechnol. Bioeng., **58** (2–3), 258–262 (1998).
23. R. Mahadevan and C.H. Schilling, The effects of alternate optimal solutions in constraint-based genome-scale metabolic models. Metabolic Eng., **5**, 264–276 (2003).
24. R. T. J. M. van der Heijden, B. Romein, J. J. Heijnen, C. Hellinga, and K. Vh. A. M. Luyben. Linear constraint relations in biochemical reaction systems: II. Diagnosis and estimation of gross errors. Biotechnol. Bioeng. **43**, 11–20 (1994).
25. R. T. J. M. van der Heijden, B. Romein, J. J. Heijnen, C. Hellinga, and K. Ch. A. M. Luyben. Linear constraint relations in biochemical reaction systems: III. Sequential application of data reconciliation for sensitive detection of systematic errors. Biotechnol. Bioeng. **44**, 781–791 (2001).
26. N. S. Wang and G. Stephanopoulos. Application of macroscopic balances to the identification of gross measurement errors. Biotechnol. Bioeng. **25**, 2177–2208 (1983).
27. R. Bhatia, G. Jesionowski, J. Ferrance and M. M. Ataai. Insect cell physiology. Cytotechnol., **20**, 33–41 (1996).
28. J. P. Ferrance, A. Goel and M. M. Ataai. Utilization of glucose and amino acids in insect cell cultures: Quantifying the metabolic flows within the primary pathways and medium development. Biotechnol. Bioeng. **42**, 697–707 (1993).
29. L. Sryer, Biochemistry, W.H. Freedman and company, New York (1981).
30. P.S. Coleman and B.B. Lavietes, Mewmbrane cholesterol, tumorigenisis and the biochemical phenotype of neoplasia. CRC Crit.Rev.Biochem., **11**, 341–393 (1981).
31. H. P. J. Bonarius, V. Hatzimanikatis, K. P. H. Meesters, C. D. de Gooijer, G. Schmid, and J. Tramper. Metabolic flux analysis of hybridoma cells in different culture media using mass balances. Biotechnol. Bioeng., **50**, 299–318 (1996).
32. W.M. Miller, C.R. Wilke and H.W. Blanch., Effects of dissolved oxygen concentration on hybridoma growth and metabolism in continuous culture, J. Cell. Physiol. **132**, 524–530 (1987).
33. L. Xie and D. I. C. Wang. Energy metabolism and ATP balance in animal cell cultivation using a stoichiometrically based reaction network. Biotechnol. Bioeng., **52**, 591–601 (1996).
34. N. Vriezen. Physiology of mammalian cells in suspension culture. Ph.D Thesis Technical University Delft, Delft, The Netherlands, (1998).
35. C. Zupke and G. Stephanopoulos. Intracellular flux analysis in hybridomas using mass balances and in vitro 13C NMR. Biotechnol. Bioeng., **45**, 292–303 (1995).
36. L. Xie and D. I. C. Wang. Applications of improved stoichiometric model in medium design and fed-batch cultivation of animal cells. Cytotechnol. **15**, 17–29 (1994).
37. T. Okayasu, M. Ikeda, K. Akimoto and K. Sorimach., The amino acid composition of mammalian and bacterial cells. Amino Acids **13**, 379–391 (1997).
38. N. Vriezen and J. P. Vandijken. Subcellular localization of enzyme activities in chemostat-grown murine myeloma cells. J. Biotechnol. **61**(1), 43–56 (1998).

39. HP J. Bonarius, A. Ozemre, B. Timmerarends, P. Skrabal, J. Tramper, G. Schmid, and E. Heinzle. Metabolic-flux analysis of continuously cultured hybridoma cells using (CO2)-C-13 mass spectrometry in combination with C-13-lactate nuclear magnetic resonance spectroscopy and metabolite balancing. Biotechnol. Bioeng. **74**(6), 528–538 (2001).

40. A. Mancuso, S. T. Sharfstein, S. N. Tucker, D. S. Clark, and H. W. Blanch. Examination of primary metabolic pathways in a murine hybridoma with carbon-13 nuclear magnetic resonance spectroscopy. Biotechnol. Bioeng. **44**, 563–585 (1994).

41. S. T. Sharfstein, S. N. Tucker, A. Mancuso, H. W. Blanch, and D. S. Clark. Quantitative in vivo nuclear magnetic resonance studies of hybridoma metabolism. Biotechnol. Bioeng. **43**, 1059–1074 (1994).

42. Mazurek S, Boschek C.B., Hugoc F., and Eigenbrodt E. Pyruvate kinase type M2 and its role in tumor growth and spreading. Seminars in Cancer Biol. **15**, 300–308 (2005).

43. L. Xie and D. I. C. Wang. Stoichiometric analysis of animal cell growth and its application in medium design., Biotechnol. Bioeng. **43**, 1164–1174 (1994).

44. L. Xie and D. I. C. Wang. Material balance studies on animal cell metabolism using a stoichiometrically based reaction network., Biotechnol. Bioeng., **52**, 579–590 (1996).

45. H. P. J. Bonarius, J. H. M. Houtman, G. Schmid, C. D. de Gooijer, and J. Tramper. Metabolic-flux analysis of hybridoma cells under oxidative and reductive stress using mass balances. Cytotechnol. **32**(2), 97–107 (2000).

46. N. Vriezen, B. Romein, K. C. A. M. Luyben, and J. P. Vandijken. Effects of glutamine supply on growth and metabolism of mammalian cells in chemostat culture. Biotechnol. Bioeng., **54**(3), 272–286 (1997).

47. C. Altamirano, A. Illanes, A. Casablancas, X. Gamez, J. J. Cairo, and C. Godia. Analysis of CHO cells metabolic redistribution in a glutamate-based defined medium in continuous culture. Biotechnol. Prog.. **17**(6), 1032–1041 (2001).

48. G. B. Nyberg, R. R. Balcarcel, B. D. Follstad, G. Stephanopoulos, and D. I. C. Wang. Metabolism of peptide amino acids by Chinese hamster ovary cells grown in a complex medium. Biotechnol. Bioeng. **62**(3), 324–335 (1999).

49. B. D. Follstad, R. R. Balcarcel, G. Stephanopoulos, and D. I. C. Wang. Metabolic flux analysis of hybridoma continuous culture steady state multiplicity. Biotechnol. Bioeng., **63**(6), 675–683 (1999).

50. A. F. Europa, A. Gambhir, P. C. Fu, and W. S. Hu. Multiple steady states with distinct cellular metabolism in continuous culture of mammalian cells. Biotechnol. Bioeng., **67**(1), 25–34 (2000).

51. A. Gambhir, R. Korke, J. C. Lee, P. C. Fu, A. Europa, and W. S. Hu. Analysis of cellular metabolism of hybridoma cells at distinct physiological states. J. Biosci. Bioeng., **95**(4), 317–327 (2003).

52. R. Korke, M. D. Gatti, A. L. Y. Lau, J. W. E. Lim, T. K. Seow, M. C. M. Chung, and W. S. Hu. Large scale gene expression profiling of metabolic shift of mammalian cells in culture. J. Biotechnol., **107**(1), 1–17 (2004).

53. I. Nadeau, D. Jacob, M. Perrier, and A. Kamen. 293SF metabolic flux analysis during cell growth and infection with an adenoviral vector. Biotechnol. Progr., **16** (5), 872–884 (2000).

54. I. Nadeau, P. A. Gilbert, D. Jacob, M. Perrier, and A. Kamen. Low-protein medium affects the 293SF central metabolism during growth and infection with adenovirus. Biotechnol. Bioeng., **77** (1), 91–104 (2002).

55. C. Chan, F. Berthiaume, K. Lee, and M. L. Yarmush. Metabolic flux analysis of cultured hepatocytes exposed to plasma. Biotechnol. Bioeng. **81**(1), 33–49 (2003).

56. C. Chan, F. Berthiaume, K. Lee, and M. L. Yarmush. Metabolic flux analysis of hepatocyte function in hormone- and amino acid-supplemented plasma. Metabol. Eng. **5**(1), 1–15 (2003).

57. C. Chan, D. Hwang, G. N. Stephanopoulos, M. L. Yarmush, and G. Stephanopoulos. Application of multivariate analysis to optimize function of cultured hepatocytes. Biotechnol. Prog. **19**(2), 580–598 (2003).

58. C. Paredes, E. Prats, J. J. Cairo, F. Azorin, L. Cornudella, and F. Godia. Modification of glucose and glutamine metabolism in hybridoma cells through metabolic engineering. Cytotechnol., **30**(1–3), 85–93 (1999).

59. S. L. Bell, C. Bebbington, F. Scott, J. N. Wardell, R. E. Spier, M. E. Bushell, and P. G. Sanders. Genetic engineering of hybridoma glutamine metabolism. Enzyme Microbiol. Technol. **17**, 98–106 (1995).

60. N. Irani, A. J. Beccaria, and R. Wagner. Expression of recombinant cytoplasmic yeast pyruvate carboxylase for the improvement of the production of human erythropoietin by recombinant BHK-21 cells. J. Biotechnol., **93**(3), 269–282 (2002).

61. A. Provost and G. Bastin. Dynamic metabolic modelling under the balanced growth condition. J. Proc. Control, **14**(7), 717–728 (2004).

62. I. Borodina and J. Nielsen. From genomes to *in silico* cells via metabolic networks. Curr. Opin. Biotechnol. **16**(3), 350–355, (2005).

63. E. J. Smid, D. Molenaar, J. Hugenholtz, W. M. de Vos, and B. Teusink. Functional ingredient production: application of global metabolic models. Curr. Opin. Biotechnol. **16**(2), 190–197 (2005).

64. B. Teusink, F. H. J. van Enckevort, C. Francke, A. Wiersma, A. Wegkamp, E. J. Smid, and R. J. Siezen. *In silico* reconstruction of the metabolic pathways of Lactobacillus plantarum: Comparing predictions of nutrient requirements with those from growth experiments. Appl. Environ. Microbiol., **71**(11), 7253–7262 (2005).

65. S. S. Fong, A. P. Burgard, C. D. Herring, E. M. Knight, F. R. Blattner, C. D. Maranas, and B. O. Palsson. In silico design and adaptive evolution of Escherichia coli for production of lactic acid. Biotechnol. Bioeng., **91**(5), 643–648 (2005).

APPENDIX I. MODEL REACTIONS

Glycolysis

1	$ATP + GLC \rightarrow ADP + G6P + H^+$
2	$G6P \rightarrow F6P$
3	$F6P + ATP \rightarrow F16P + ADP + H^+$
4	$F16P \rightarrow 2GAP$
5	$ADP + GAP + NAD(c) + P_i \rightarrow ATP + H_2O + H^+ + NADH(c) + PEP$
6	$ADP + H^+ + PEP \rightarrow ATP + PYR$
7	$H^+ + NADH(c) + PYR \rightarrow LAC + NAD(c)$

Pentose Phosphate Pathway

8	$2NADP + G6P + H2O \rightarrow CO2 + R5P + 2H[+] + 2NADPH$
9	$R5P \rightarrow X5P$
10	$R5P + X5P \rightarrow GAP + S7P$
11	$GAP + S7P \rightarrow E4P + F6P$
12	$E4P + X5P \rightarrow F6P + GAP$

TCA-cycle

13	$CoA + NAD(m) + PYR \rightarrow AcCoA(m) + CO_2 + NADH(m)$
14	$AcCoA(m) + H_2O + OXA(m) \rightarrow CIT(m) + CoA + H^+$
15	$CIT(m) + NAD(m) \rightarrow AKG + CO_2 + NADH(m)$
16	$AKG + CoA + NAD(m) \rightarrow CO_2 + NADH(m) + SUCCoA$
17	$GDP + P_i + SUCCoA \rightarrow GTP + CoA + SUC$
18	$SUC + FAD \rightarrow FUM + FADH_2$
19	$FUM + H_2O \rightarrow MAL(m)$
20	$MAL(m) + NAD(m) \rightarrow H^+ + NADH(m) + OXA(m)$

Connection glycolysis TCA

21	$PYR + CO_2 + H_2O + ATP \rightarrow OXA(m) + ADP + P_i + 2H^+$
22	$PEP + GDP + CO_2 \rightarrow OXA(c) + GTP$
23	$MAL(c) + NADP \rightarrow CO_2 + NADPH + PYR$

Biomass: DNA/RNA

24 $8ATP + 2GLN + 2H_2O + 2N10FTHF + ASP + CO_2 + GLY + NAD(c) + NH_4^+ + R5P \rightarrow$
    $FUM + GTP + NADH(c) + 2AMP + 2GLU + 2PPi + 2THF + 4P_i + 6ADP + 11H^+$

25 $8ATP + 2ASP + 2GLN + 2N10FTHF + CO_2 + GLY + GTP + H_2O + R5P \rightarrow ATP + GDP +$
    $PPi + 2FUM + 2GLU + 2THF + 5P_i + 8ADP + 10H^+$

26 $3ATP + ASP + GLN + R5P + 0.5O_2 \rightarrow AMP + GLU + H_2O + UTP + 2ADP + 2H^+ + 2P_i$

27 $ATP + NH_4^+ + UTP \rightarrow ADP + CTP + P_i + 2H^+$

28 $GTP + THIRH_2 \rightarrow H_2O + THIR + dGTP$

29 $ATP + THIRH_2 \rightarrow H_2O + THIR + dATP$

30 $H^+ + METHF + NADH + THIRH_2 + UTP \rightarrow H_2O + NAD + THF + THIR + dTTP$

31 $CTP + THIRH_2 \rightarrow H_2O + THIR + dCTP$

32 $H^+ + NADPH + THIR \rightarrow NADP + THIRH_2$

33 $10H_2O + 3ATP + 3UTP + 2CTP + 2GTP \rightarrow RNA + 10H^+ + 20P_i$

34 $10H_2O + 3dATP + 3dTTP + 2dCTP + 2dGTP \rightarrow DNA + 10H^+ + 20P_i$

Biomass: Total carbohydrates

35 $ATP + G6P + H_2O \rightarrow ADP + H^+ + TC + 2P_i$

Biomass: Membrane lipids + Cholesterol

36 $GAP + H^+ + NADH(c) \rightarrow GLYC3P + NAD(c)$

37 $Cit(m) \rightarrow Cit(c)$

38 $ATP + Cit(c) + CoA \rightarrow Oxa(c) + AcCoA(c) + ADP + Pi$

39 $MAL(c) + NAD(c) \rightarrow H^+ + NADH(c) + OXA(c)$

40 $MAL(c) \rightarrow MAL(m)$

41 $8H^+ + 8ATP + 16NADPH + NADH(c) + 9AcCoA(c) + O_2 \rightarrow H_2O + OLE + NAD(c) +$
    $16NADP + 8ADP + 8P_i + 9CoA$

42 $2H^+ + 2OLE + GLYC3P \rightarrow ML + 2H_2O$

43 $18ATP + 30NADPH + 7H^+ + 18AcCoA(c) + 11O_2 \rightarrow CHOL + FORM + 3H_2O + 8CO_2 +$
    $18CoA + 30NADP + 18ADP + 18P_i$

Biomass: Protein

44 $40ATP + 30H_2O + 0.85PRO + 0.82GLY + 0.82LEU + 0.80LYS + 0.79ALA + 0.62SER +$
    $0.56ARG + 0.56GLU + 0.56VAL + 0.54GLN + 0.51ASP + 0.50THR + 0.43ASN +$
    $0.43ILE + 0.37PHE + 0.24MET + 0.24TYR + 0.19HIS + 0.14CYS + 0.05TRP \rightarrow Protein +$
    $40ADP + 40H^+ + 40P_i$

Monoclonal Antibody

45 $40ATP + 30H_2O + SER + 0.85THR + 0.77VAL + 0.69LYS + 0.69PRO + 0.62GLY +$
    $0.62LEU + 0.62GLN + 0.54ALA + 0.53GLU + 0.52ASP + 0.46ASN + 0.38PHE + 0.37ILE +$
    $0.32TYR + 0.31CYS + 0.24ARG + 0.18TRP + 0.17HIS + 0.13MET \rightarrow Mab + 40ADP +$
    $40H^+ + 40P_i$

Amino acid metabolism

46 $4H_2O + 3O_2 + 2CoA + 2NAD(m) + FAD + NADPH + TRP \rightarrow ALA + FADH2 + FORM +$
    $H^+ + NADP + NH_4^+ + 2AcCoA(m) + 2NADH(m) + 3CO_2$

47 $GLU + PYR \rightarrow AKG + ALA$

48 $SER \rightarrow NH_4^+ + PYR$

49 $GLY + H_2O + METHF \rightarrow SER + THF$

50 $CoA + NAD(m) + THR \rightarrow AcCoA(m) + GLY + H^+ + NADH(m)$

51 $2H_2O + AKG + ATP + CoA + FAD + LEU + Lipoamide \rightarrow 2Hlipoamid + ADP + AcAc +$
    $AcCoA(m) + FADH_2 + GLU + H^+ + P_i$

52 $2CoA + 2NAD(m) + AKG + FAD + H_2O + LYS + O_2 \rightarrow FADH_2 + GLU + H^+ + NH_4^+ +$
    $2AcCoA(m) + 2CO_2 + 2NADH(m)$

53 $2CoA + 2H_2O + AKG + ATP + FAD + ILE + Lipoamide + NAD(m) \rightarrow 2Hlipoamid + ADP +$
    $AcCoA(m) + FADH2 + GLU + NADH(m) + P_i + SUCCoA + 2H^+$

54 $4H_2O + 2ATP + CoA + MET + NAD(m) + SER + THF \rightarrow ADP + Ad + MTHF +$
    $NADH(m) + PPi + PYR + S + SUCCoA + 2NH_4^+ + 2P_i + 5H^+$

55   $2H_2O + 2NAD(m) + AKG + CoA + FAD + Lipoamide + VAL \rightarrow 2Hlipoamid + CO_2 +$
     $FADH_2 + GLU + SUCCoA + 2H^+ + 2NADH(m)$
56   $2O_2 + AKG + H^+ + NADH(m) + PHE \rightarrow CO_2 + GLU + H_2O + HGS + NAD(m)$
57   $AKG + O_2 + TYR \rightarrow CO_2 + GLU + HGS$
58   $H_2O + HGS + O_2 \rightarrow AcAc + FUM + 2H^+$
59   $GLU + OXA(m) \rightarrow AKG + ASP(m)$
60   $ASN + H_2O \rightarrow ASP(c) + NH4[+]$
61   $ASP(m) \rightarrow ASP(c)$
62   $H_2O + NAD(m) + PRO \rightarrow G5S + H^+ + NADH(m)$
63   $2H^+ + GLU + NADPH \rightarrow G5S + H_2O + NADP$
64   $2H_2O + 2H^+ + HIS + NADH(m) + O_2 \rightarrow CO_2 + GLU + NAD(m) + 2NH_4^+$
65   $GLN + H_2O \rightarrow GLU + NH_4^+$
66   $AKG + H^+ + NADH(m) + NH_4^+ \rightarrow GLU + H_2O + NAD(m)$
67   $2Hlipoamid + NAD(m) \rightarrow H^+ + Lipoamide + NADH(m)$
68   $AcAc + CoA + SUCCoA \rightarrow SUC + 2AcCoA(m)$

Ureum cycle
69   $ASP(c) + ATP + Citr \rightarrow AMP + ArgSuc + PPi$
70   $ArgSuc \rightarrow ARG + FUM + H^+$
71   $ARG + H_2O \rightarrow ORN + Urea$
72   $CaP + ORN \rightarrow Citr + H^+ + P_i$
73   $2ATP + CO_2 + H_2O + NH_4^+ \rightarrow CaP + P_i + 2ADP + 3H^+$
74   $AKG + ORN \rightarrow G5S + GLU$

Energy, THF
75   $MTHF \rightarrow MethylX + THF$
76   $ATP + FORM + THF \rightarrow ADP + N10FTHF + P_i$
77   $H_2O + MYLTHF \rightarrow H^+ + N10FTHF$
78   $MYLTHF + NADH(c) \rightarrow METHF + NAD(c)$
79   $H^+ + METHF + NADPH \rightarrow MTHF + NADP$
80   $4H^+ + 3ADP + 3P_i + NADH(m) + 0.5O_2 \rightarrow NAD(m) + (P/O)ATP + 4H_2O$
81   $2ADP + 2H^+ + 2P_i + FADH_2 + 0.5O_2 \rightarrow FAD + (2/3)(P/O)ATP + 3H_2O$
82   $NAD(c) + NADPH \rightarrow NADH(c) + NADP$
83   $ATP + GDP \rightarrow ADP + GTP$
84   $AMP + ATP \rightarrow 2ADP$
85   $ATP + Ad \rightarrow ADP + AMP + H^+$
86   $ATP + H_2O \rightarrow ADP + H^+ + P_i$
87   $H_2O + PPi \rightarrow H^+ + 2P_i$

Compartmentation
88   $PYR(c) \rightarrow PYR(m)$
89   $MAL(c) \rightarrow MAL(m)$
90   $AKG(c) \rightarrow AKG(m)$
91   $GLN(c) \rightarrow GLN(m)$
92   $GLU(c) \rightarrow GLU(m)$
93   $ASP(m) + GLU(c) \rightarrow ASP(c) + GLU(m)$
94   $MAL(c) + AKG(m) \rightarrow MAL(m) + AKG(c)$
95   $DHAP(c) + H^+ + NADH(c) \rightarrow GLYC3P(c) + NAD(c)$
96   $GLYC3P(m) + FAD \rightarrow DHAP(m) + H^+ + FADH2$
97   $DHAP(m) \rightarrow DHAP(c)$
98   $GLYC3P(c) \rightarrow GLYC3P(m)$

Metabolic engineering
99   $PYR(c) + CO_2 + H_2O + ATP \rightarrow OXA(c) + ADP + P_i + 2H^+$
100 $GLU(c) + ATP + NH_3 \rightarrow GLN(C) + ADP + Pi$

Abbreviations: (c)=cytosolic; (m)=mitochondrial; AcAc, acetoacetate; AcCoA, acetyl coenzyme A; Ad, adenosine; ADP, adenosine diphosphate; AKG, α-ketoglutarate; ALA, alanine; AMP, adenosine monophosphate; ARG, arginine; ASN, asparagine; ASP, aspartate; (d)ATP, (deoxy)adenosine triphosphate; ARGSUC, Argino-succinate; CaP, carbamyl-phosphate; CHOL, cholesterol; CIT, citrate; CoA, coenzyme A; CO_2, carbondioxide; (d)CTP, (deoxy)cytosine triphosphate; CYS, cysteine; CITR, Citruline; DNA, deoxyribonucleic acid; dTTP, deoxy-thymidine triphosphate; DHAP, Dihydroxy-cetonephosphate; E4P, Erythrose-4-phosphate; FAD(H_2), flavine adenosine dinucleotide (reduced); FORM, Formic acid; FUM, fumarate; F6P, fructose-6-phosphate; F16P, fructose 1,6 biphosphate; GAP, glyceraldehyde 3 phosphate; GDP, guanosine diphosphate; GLC, glucose; GLN, glutamine; GLU, glutamate; GLY, glycine; GLYC3P, glycerol 3 phosphate; (d)GTP, (deoxy)guanosine triphosphate; G5S, L-glutamate 5-semialdehyde; G6P, glucose-6-phosphate; HGS, homogentisate; HIS, histidine; H^+, hydrogen ion; 2Hlipoamid, dihydrolipoamide; H_2O, water; ILE, isoleucine; LAC, lactate; LEU, leucine; LYS, lysine; Mab, monoclonal antibody; MAL, malate; MET, methionine; METHF, N5,N10-Methylene-tetra hydrofolate; MethylX, methylated compound; ML, membrane lipids; MTHF, methyl tetra hydrofolate; MYLTHF, N5,N10-methenyl THF; NAD(H), nicotinamide adenine dinucleotide (reduced); NADP(H), nicotinamide adenine dinucleotide phosphate (reduced); NH_4^+, ammonium; N10FTHF, N10-formyl tetrahydrofolate; OLE, oleic acid; ORN, ornithine; OXA, oxaloacetate; O_2, oxygen; PEP, phospho-enol-pyruvate; PHE, phenyl alanine; P_i, orthophosphate; PPi, pyrophosphate; PRO, proline; PROT, protein; PYR, pyruvate; RNA, ribonucleic acid; R5P, ribose-5-phosphate; S, sulphur; SER, serine; SUC, succinate; SucCoA, succinyl coenzymeA; S7P, Septulose-7-phosphate; TC, total carbohydrates; THF, tetra hydrofolate; THIR(H_2), thioredoxin (reduced); THR, threonine; TRP, tryptophane; TYR, tyrosine; UTP, uracil triphosphate; VAL, valine; X5P, Xylulose 5-phosphate

CHAPTER 10

METABOLIC ENGINEERING

EFFENDI LEONARD*, ZACHARY L. FOWLER* AND MATTHEOS KOFFAS
State University of New York at Buffalo
*These authors contributed equally to the preparation of this document.

Abstract: Metabolic engineering involves the modification of biosynthetic pathways to improve or generate novel cellular phenotypes or to elucidate intracellular biochemical properties. The underlying concept of metabolic engineering originates from the traditional genetic engineering paradigm, yet places its emphasis on integrated, genome-wide biochemical processes. As such, metabolic engineering is one of the pioneering fields to use systems biology for biochemical and biomedical applications. In this chapter, important milestones of implementing experimental and computational metabolic engineering concepts are reviewed. Key examples include the production of high-value chemicals derived from primary and secondary metabolism and elucidation of intracellular metabolic controls through flux balance analysis.

Keywords: metabolic and cellular engineering, protein engineering, metabolic flux analysis, metabolic control analysis

1.   INTRODUCTION

In the broadest sense, the field of metabolic engineering centers on cellular manipulations, aimed at the generation of desirable cellular phenotypes usually for the efficient synthesis of native or nonnative important molecules. Advances in molecular biology and bioinformatics, coupled with the wealth of available sequence information and extensive cataloging allow researchers to choose and utilize 'genetics tools' to construct, remove, and fine tune intracellular genetic machineries. Furthermore, the elucidation of the majority of biosynthetic pathways has provided reaction maps as the guide for engineering novel metabolic phenotypes or predicting the overall

M. Al-Rubeai and M. Fussenegger (eds.), Systems Biology, 301–359.
© 2007 *Springer.*

metabolic effects of a genetic perturbation. This chapter provides an overview of the various strategies for cellular engineering, especially those geared toward exploiting living cells as factories of important chemicals and enzymes. Frontiers of cellular engineering which incorporate signal transduction pathway principals, cell-to-cell communication, and regulations by small RNA species to create regulated engineered systems are also discussed. The chapter closes with a presentation of various aspects of mathematical/computational modeling which has become an integral part of metabolic engineering.

2.   ADDING AND DELETING GENETIC ELEMENTS

Phenotypic variance is a direct result of survival ability throughout the course of selection pressures. The genetic blueprint of the various organisms can be considered as the primary reason behind the observed phenotypic diversity, without underestimating the intrinsic interactions among molecular components that can also influence the overall cellular physiology. In parallel with genetic controls on the expression level of individual genes, gene copy number is an important factor for shaping cellular traits[36,107,195]. Moreover, the plasticity of genetic materials to rearrange and delete increases the possibility of phenotypic variance. Taking lessons from nature, laboratory development of improved cellular functions or novel properties can be conducted by gene over-expression or deletion. These two main genetic perturbations most commonly applied to cellular engineering will be further presented.

2.1   Gene Insertion

Directing a reaction toward the formation of a product can be achieved by the addition of one or more genes that encode identified rate limiting enzymes that mediate the pathway. Similar approaches can also be utilized to introduce an entirely new pathway resulting in the synthesis of nonnative enzymes or metabolites. Advances in molecular cloning allow gene insertions into a cellular host, in which a DNA fragment can be integrated into the chromosome or maintained outside the chromosome through the autonomously replicating circular DNA, a plasmid (see Figure 1). The selections of suitable promoter sequences is used for the facilitation of controlled induction and over expression to optimize enzyme yields and activities.

One of the earliest applications of gene insertion was to address protein isolation and purification problems. A particular protein may exist in low quantity in the native host, making purification and biochemical characterization difficult to achieve. Increasing protein yields can be difficult because methods for inducing expression are often complicated and inefficient because of the existence of complicated regulatory networks in the host. Through molecular cloning, addition of a gene into a cellular host increase the amount of the protein product. Furthermore, the use of surrogate hosts (usually microorganisms) for protein expression is beneficial for increasing protein purity due to the lack of enzymes with similar activities.

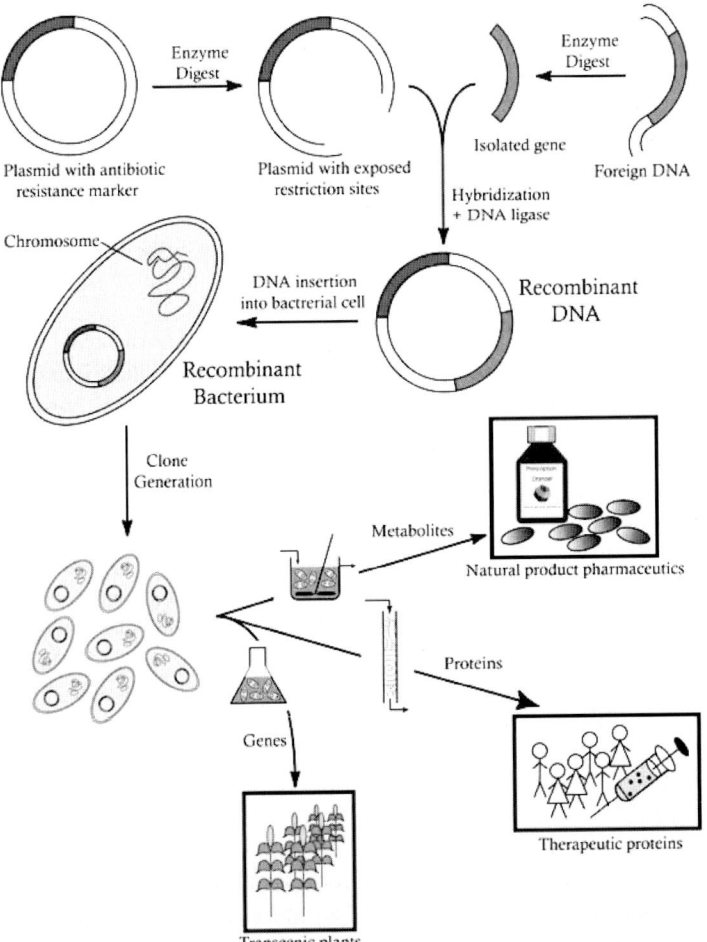

Figure 1. Principle of molecular cloning. Gene of interest (blue) is isolated and inserted into a bacterial carrier DNA (plasmid) through available enzyme restriction sites. Because the carrier plasmid contains a unique selection marker, for example an antibiotic resistance gene (red), successful insertions can be isolated from viable bacterial colonies cultivated in media containing the antibiotic corresponding to the selection marker. Foreign gene amplified through the recombinant bacteria can be further inserted into other organisms, e.g. plants, to create transgenic organisms. Over expression of the foreign gene in recombinant bacteria for biotechnological purposes leads to the production of novel metabolites, therapeutic proteins

2.1.1   Single gene insertion

Initial applications of gene insertion technologies focused on the heterologous expression of important proteins in bacteria. One of the earliest applications of recombinant DNA technology was the engineering of *Escherichia coli* for the synthesis of human insulin[80,227]. In this work, the A and B subunits of insulin were cloned separately into an *E. coli* expression plasmid to obtain a native human protein after chemically adjoined *in vitro*. Similarly, *E. coli* has also been used as a cellular factory for the production of restriction enzymes which are indispensable tools of molecular biology. This is because many bacteria that possess restriction enzymes of interest are pathogenic or difficult to culture. For example, *Haemophilus influenzae*, a human pathogen, is the source of the restriction enzyme Hind III. Metabolic engineering of Hind III synthesis for mass production in a non-pathogenic bacterium was possible by inserting the *Hind III* DNA fragment derived from *H. influenzae* into a non-pathogenic *E. coli* strain, followed by induction of protein expression[147].

In the area of bulk chemicals, the biosynthesis of amino acids such as lysine offers a characteristic example of a single gene insertion for production improvement. The current six hundred thousand tons per year market volume of lysine is derived from industrial fermentations of prokaryote *Corynebacterium glutamicum*. Lysine biosynthesis requires a high amount of NADPH, which is mainly derived from primary metabolism, such as the pentose phosphate pathway (PPP). Therefore, it was speculated that increasing carbon flux through the PPP could lead to lysine production increase[239]. Fructose 1,6-bisphosphatase (FBPase) is an enzyme involved in gluconeogenesis, which is essential for non-carbohydrates carbon source assimilation[169]. FBPase was found to mediate flux through the PPP pathway[119] therefore it could be used to boost the synthesis of NADPH, thus improving lysine synthesis. To engineer a *C. glutamicum* strain for improved lysine production, an extra copy of the gene encoding FBPase was inserted into the bacterium, under the control of the strong promoter for elongation factor TU[14]. The resulting over expression of the FBPase, which is only indirectly involved in the lysine biosynthetic pathway, led to an approximately 2-fold production increase of lysine per cell mass when glucose was used as the carbon source.

In the area of natural products and plant metabolic engineering, a characteristic example of single gene manipulation relates to the biosynthesis of isoflavonoids in tobacco plants. Isoflavonoids, a subgroup of flavonoids, are plant secondary metabolites that play key roles in plant physiology, acting as signal molecules in plant-bacteria interactions (plant nodulation)[56]. They are also high-value human pharmaceutics for potential treatment of hormone-dependent cancers, osteoporosis, menopausal symptoms, and diabetes (for reviews, see Cronwall[45] and Dixon[55]). Isoflavonoids are predominantly synthesized in leguminous plants. Therefore, engineering of isoflavonoid biosynthesis in more commonly consumed crop plants would be expected to increase the availability of these health-promoting metabolites in human diets. Isoflavone synthase (IFS) is the enzyme which catalyzes the conversion of flavanones, which are ubiquitous molecules in plants, into isoflavones. Tobacco, a nonleguminous plant that does not synthesize isoflavones, was used as a case study for metabolic engineering isoflavonoid biosynthesis. In this case, soybean *IFS1* was cloned into a binary vector consisting of the cauliflower mosaic virus 35S promoter. *Agrobacterium tumefaciens* was then transformed with the IFS carrying plasmid, and subsequently used to transfect tobacco plants[246]. Even though the transgenic tobacco plants successfully synthesized isoflavonoids, the amounts were low, because of competing pathways leading to the biosynthesis of another flavonoid class, the anthocyanins.

Mammalian and insect cells have also been utilized as heterologous cellular factories, usually for protein production, or in the case of mammalian cells, for gene therapy. Insertion of a gene into these eukaryotic hosts was usually governed by viral transduction, in which the gene of interest is packaged in viral vector[170]. In general, the efficiency of gene delivery and expression in eukaryotic cells depends on many factors, including the selection of the viral based plasmid, the virus carrier, and the choice of cells for infection. A brief review of gene therapy can be found at Selkirk[188].

2.1.2   Multiple gene insertion

The work of David Hopwood on the biosynthesis of novel isochromanequinone metabolites is considered as the definitive beginning of the era of metabolic engineering for the production of high-value compounds[95]. An excellent review of this work, as well as the extensive work in the area of polyketide biosynthesis has been presented by William Stohl[202]. In this section, we focus on some other representative, examples of multiple gene insertion for the biosynthesis of drugs, nutraceuticals and natural products.

Taxadiene is the key precursor for taxol, the anticancer drug derived from the Pacific yew (*Taxus brevifolia*) plant. However, the inefficient plant extraction (only 1 mg of taxadiene can be isolated by extraction from 750 kg of dry Pacific yew bark) has led to the depletion of natural resources and hindered the production of inexpensive drugs. Although alternative approaches through chemical synthesis are available, the processes are cumbersome, some requiring as many as 25 steps. Therefore, biocatalysis using recombinant microorganisms could offer an environmentally friendly and competitive alternative for the production of this valuable chemical. Taxadiene is synthesized from the isopentenyl diphosphate (IDP) precursor through a 3-step metabolic pathway. The first reaction is the isomerization of IDP to form dimethylallyl diphosphate (DMADP) by IDP isomerase. In the second step, geranyl-geranyl diphosphate (GGDP) is formed from the condensation of three molecules of IDP with one molecule of DMAPP by the enzyme GGDP synthase. Finally, taxadiene synthase catalyzes the cyclization of GGDP to form taxadiene[134]. To achieve substantial biosynthesis of taxadiene in *E. coli*, IDP isomerase, GGDP synthase, and a truncated taxadiene synthase were over expressed in conjunction with deoxyxylulose-5-phosphate (DXP) synthase from *E. coli*, in order to increase the availability of IDP[97]. The genes of heterologous origins were cloned separately into multiple co-replicable *E. coli* expression plasmids, and their expressions were regulated by the strong *T*7 phage promoter. The metabolic engineering strategy resulted in the production of up to 1.3 mg taxadiene per liter from the recombinant *E. coli* strain in batch fermentations.

Artemisinin is another plant-derived drug which belongs to the class of terpenoids. It is currently the anti-malarial drug of choice because the malaria-causing *Plas-*

modium strains have already developed resistance to alternative anti-malarial compounds, such as chloroquine, cycloguanile, and sulfadoxin[164]. Current methods of extraction from the sweet wormwood *Artemisia annua*, are inefficient and result in inadequate production levels that do not accommodate the global need for inexpensive anti-malarial drugs, especially among countries of the developing world where malaria infections are the most frequent. For that reason, *E. coli* has been utilized as a microbial factory to produce the artemisinin precursor, amorphadiene. Specifically, mevalonate kinase, phosphomevalonate kinase, and mevalonate pyrophosphate decarboxylase from *S. cerevisiae*, together with IDP isomerase, farnesyl pyrophosphate synthase from *E. coli* were sewn together to form a synthetic operon in one plasmid, in which the transcription was regulated by the *E. coli lac* promoter. To generate an amorphadiene producing *E. coli* strain, the synthetic operon was co-expressed with a codon modified plant amorphadiene synthase[139]. The resulting recombinant strain generated amorphadiene up to 24 μg caryophyllene equivalent per milliliter.

Remaining in the area of plant secondary metabolites, many of the 8,000 plant polyphenols, known collectively as flavonoids, are currently used as dietary supplements and are being intensively investigated for use as treatments for many human diseases from cancers to diabetes[6,34,96,159–161,248]. Substantial characterization of plant flavonoid pathways has recently allowed the engineering of flavonoid biosynthesis in *E. coli*. The five main classes of flavonoids, namely flavones, flavonols, isoflavones, flavanols, and anthocyanins, are derived from the common flavanone precursors[236]. Flavanones are synthesized from phenylalanine through a five-step enzymatic process. The first step is the conversion of phenylalanine into cinnamic acid by the enzyme phenylalanine ammonia lyase (PAL). Subsequently, cinnamic acid is hydroxylated into *p*-coumaric acid by the enzyme cinnamate 4-hydroxylase (C4H). *p*-Coumaric acid, as well as other propanoic acids are then converted into propanoyl-CoA esters by 4-coumaroyl:CoA ligase (4CL). Next, the chalcone synthase (CHS) performs a condensation reaction of the CoA esters with three molecules of malonyl-CoA to form one molecule of chalcone. Flavanones are formed from chalcones following an isomerization performed by chalcone isomerase (CHI). Subsequent down-stream flavonoid enzymes catalyze the formation of the various flavonoids molecules[236]. The important pharmacological properties of flavonoids and the limited quantities of some purified forms from plants inspired the production of these secondary metabolites from recombinant hosts. *E. coli* strains have been engineered to harbor flavonoid biosynthetic pathways and synthesize flavanones[98,116], flavones[130], flavonols[129], and anthocyanins[241]. In all cases, the biosynthetic genes were episomally inserted into *E. coli* using one or multiple co-replicable plasmids. In order to increase flavonoid production from the recombinant strains, each gene in the pathway was individually regulated by the *T7* phage promoter, instead of following the operon concept.

In higher eukaryotes, the biosynthesis of some novel metabolites requires the action of membrane-bound proteins, such as type II cytochrome dependent P450 monoxygenases. However, functional expression of P450 proteins in *E. coli* is challenging because of the lack of endogeneous electron transport and a membrane anchor system necessary for P450 reactions. Therefore, to functionally express P450 enzymes in *E. coli*, extensive protein engineering is needed. Such methodologies for engineering P450 functional expression in *E. coli* will be discussed in the protein engineering section. As an alternative to *E. coli*, yeast offers the advantage of being able to support the functional expression of such membrane-bound enzymes because it is a eukaryotic species. For that reason, *S. cerevisiae* has also been used as a production platform for flavonoid molecules. For example, the synthesis of the flavonoid naringenin from cinnamic acid, which requires the conversion of cinnamic acid into *p*-coumaric acid was afforded by a recombinant *S. cerevisiae* expressing 4CL, CHS, CHI, and the plant P450, C4H[242]. Flavones, the plant flavonoids with potent medicinal properties[34], are synthesized from flavanones by two distinct enzymes: the soluble flavone synthase I (FSI), and the membrane-bound flavone synthase II (FSII). Because the yeast system can harbor both cytoplasmic and membrane bound enzymes, metabolically engineered *S. cerevisiae* strains were used to investigate flavone biosynthesis by the two distinct enzymes. In the study, consecutive reactions from phenylpropanoic acids to form various flavone molecules were successfully performed by *S. cerevisiae* implanted with the flavanone biosynthetic genes, together with FSI or FSII, and a plant P450 reductase[130]. The yeast model concluded that FSI possessed greater catalytic activities than its P450 analog, resulting in a greater generation of flavones. However, the greater production of the flavone apigenin resulted in an inhibitory effect of the flavanone biosynthetic pathway.

Erythromycin is a potent antibiotic synthesized by the soil bacterium *Saccharopolyspora erythrea*. The macrolytic antibiotic core is synthesized by the large modular polyketide synthases (PKS), capable of subsequent condensation of simple building blocks, such as acetyl-CoA, propionyl-CoA, malonyl-CoA, and methylmalonyl-CoA. For example, 6-deoxyerythronolide B (6dEB) is formed from one propionyl-CoA unit and subsequent elongation of six (2*S*)-methylmalonyl-CoA by the enzyme deoxyerythronolide B synthase (DEBS). The large DEBS (2 megadaltons) is a three-subunit enzyme ($\alpha_2\beta_2\gamma_2$), compose of 2 sets of 28 distinct active sites, seven of which are pantetheinylated upon translation. Because the complex chemical structure hampers the total chemical synthesis of the antibiotic, large-scale production relies on fermentation technologies. Moreover, because of the challenges in developing a scalable fermentation process of actinomyces, researchers seek an alternative solution, that is, production from recombinant *E. coli*. For that reason, the three subunits of DEBS were cloned individually into the *E. coli* expression vector pET21c. In order to facilitate pantetheinylation of the recombinant DEBS and propionyl-CoA synthesis, a phosphopantetheinyl transferase gene (*sfp*) and *prpE* were also inserted into *E. coli* by integration into the *prpRBCD* operon in the genome. Disruption of the *prp* operon, which is responsible for propionate metabolism, was intended to allow optimum conversion of exogeneously supplemented propionate into propionyl-CoA by the *prpE* gene product. Furthermore, the two-subunit propionyl-CoA carboxylase (*pcc*) and the biotin ligase carrier protein (*birA*) were also introduced into the recombinant strain mediated by a co-replicable plasmid. Introduction of the carboxylase gene allowed the conversion of propionate into (2*S*)-methylmalonyl-CoA, which served as an extender unit of the recombinant DEBS. Fermentation of the recombinant *E. coli* strain with propionate supplementation yielded 0.1 mmol of 6dEB per gram cellular protein per day, which was superior to wild-type *S. erythraea*, and compatible to a modified strain for industrial 6dEB production, which generated 0.2 mmol of erythromycin per gram cellular protein per day[156].

2.1.3   Genome insertion

Cellular engineering for a certain phenotypic display is commonly achieved by adopting foreign pathways or over expressing native enzymes, which requires *a priori* knowledge of the genetic make up. However, although genetic insertion strategy has proven useful, it is limited to the availability of genome sequences and gene function information. Moreover, isolating, selecting and culturing the enormous array of unknown organisms for the discovery of novel biosynthetic pathways are cumbersome and require more knowledge of the physiology of the organisms. Therefore, relying on the availability of characterized pathways creates an obstacle to discover and recombinantly synthesize novel enzymes or metabolites.

When engineering cellular functions without the limitation of available data, insertion of the entire genome, or random fragments of the genome, followed by creative selection of desirable traits can be pursued. The reversed engineering technique can then be used to characterize unknown pathways of an end product. For example, the gram-negative bacterium *Ralstonia eutropha* was not known to synthesize pigments. However, the study of the 2-methylcitric acid cycle of *R. eutropha* performed by cloning the entire genome in *E. coli* led to the discovery of an open reading frame of biosynthetic enzymes for indigoids, the blue pigments of plant and bacterial origins which have been used as dyes and pharmaceutics. In this study, the genomic library of *R. eutropha* was inserted into cosmid pHC79, capable of carrying large DNA fragments. Upon transformation of *E. coli*, blue color transformants were identified. Because *E. coli* does not normally produce blue pigments, the pigment synthesis in the recombinant cells must be derived from the action of foreign enzymes from *R. eutropha*. Further isolation and subcloning experiments led to the discovery of an open reading frame of 1251 base-pairs which was responsible for the blue color formation. The open reading frame was identified to encode for a dehydrogenase by comparing sequence similarity with known proteins[57].

A more sophisticated process of isolating new pathways and bioactive molecules is the isolation of microbial genomic DNA from a mixed population from their native habitats, an approach known as metagenomics[84,168,171]. The obvious challenge of this strategy is to clone a large DNA fragment, while removing non-DNA debris that inhibits the cloning processes. Cloning of the metagenome has been successfully achieved by using the 'bacterial artifical chromosome' (BAC) as the cloning vehicle. The BAC vector was derived from the well-studied *E. coli F* factor, which is maintained at low copy number in the cell. This carrier vector is suitable for metagenome applications because of its ability to spontaneously harbor large DNA fragments (as large as 1 megabase pair) while the low copy number reduces the frequency of recombination among the inserts[191]. Even though metagenome cloning for the discovery of new pathways or novel metabolites has proven to be a robust strategy, it is limited by the availability of high throughput screening methods. Furthermore, because *E. coli* was the most widely used host for heterologous expression of the metagenome, many novel pathways will not be captured simply due to the a lack of functional expression of some foreign genes in *E. coli*.

Horizontal gene transfers have been acknowledged to occur between species, especially in bacteria. In many cases, not only small, but large DNA segments were recombined to introduce various genomic variations[91,228]. Recently, advances in molecular biology have allowed insertion of a genome derived from an unrelated species into a bacterial species. The genome of *Bacillus subtilis*, termed BGM vector, can serve as a stable cloning vesicle for large DNA fragments[103]. Based on this finding, the 3.5-Mb genome of the photosynthetic bacterium *Synechocystis* PCC6803 was inserted into the 4.2-Mb genome of the *B. subtilis*[103]. DNA fragments of the *Synechocystis* genome with artificial small DNA sequences flanking each genome fragment were amplified and cloned into the *E. coli* base vector, pBR322. Subsequently, the BGM vector, containing the *Synechocystis* fragment served as the template for the delivery of the next *Synechocystis* fragment through homologous recombination. This process continued until the entire genome of the *Synechocystis* was moved into the BGM vector. The resulting *Bacillus* chimera, which contained the entire *Synechocystis* genome, did not grow well in *Synechocystis* culture media. The authors speculated that the phenomenon was most likely due to the fact that wild-type *Synechocystis* possesses 12 copies of the genome, while only one copy was inserted into the chimera[103]. Furthermore, the 397 *Synechocystis* genes of unknown functions which were not inserted into the *Bacillus* chromosome might contribute to the lack of *Synechocystis* phenotypes of the mutant strain.

2.2 Gene Disruption

Genetic deletions (knockouts) are common methods for the development of mutant strains. Deletions of structural genes have been used to examine the role of the gene products and to search for other unknown indigenous or foreign genes that perform similar functions. Metabolic engineering endeavors for the improvement of a phenotype or metabolite production have often involved deletion of metabolic components, which in the most general sense, is intended to reroute metabolic flux towards a desired reaction pathway. Similar to gene insertion, a phenotypic improvement can involve one or simultaneous deletions of multiple genes[39,86,212,250]. Moreover, deterministic or stochastic methods can be employed to fulfill this goal.

2.2.1   Deterministic approach

Available information on reaction networks and genetic functions allows for rational approaches for selective genetic eliminations to confer a desired phenotype. Microbial catalysts have been used as commercial production platforms for acetate production. The currently used method involves a two-step biocatalytic process, in which yeast is used to catalyze ethanol formation from simple sugars, followed by aerobic oxidation of ethanol to produce acetate by *Acetobacter*[42]. In order to simplify acetate production, and to utilize the robust catalysis, *E. coli* was extensively modified to serve as a sole efficient acetate producer by various genetic deletions[40].

Inspection of glucose metabolism in *E. coli* indicated that the carbon source is converted into various fermentative metabolites, such as acetate, formate, lactate, succinate, and ethanol. Channeling carbon towards the formation of acetate required the elimination of the formation of lactate and ethanol, which directly compete with the acetate pathway. For that reason, lactate dehydrogenase, pyruvate formatelyase, alcohol dehydrogenase, and fumarate reductase were deleted by means of site-specific transposon insertion into the structural genes to halt the production of lactate, formate, alcohol, and succinate. Moreover, to prevent the loss of carbon source through the formation of carbon dioxide, a subunit of a gene in the tricarboxylic acid cycle encoding 2-ketoglutarate dehydrogenase (*sucAB*) was deleted. The result of this disruption was succinate auxothropy, which required succinate supplementation for cell growth. To search and develop an autothroph mutant, the recombinant strain was continuously cultured in media containing succinate in decreasing amount. Succinate autothroph mutant was isolated upon survival in the absence of succinate. Further increasing the carbon flux towards acetate formation was performed by disrupting a portion of $(F1F0)H^+$-ATP synthase in order to eliminate excessive carbon utilization for cell membrane formation under oxidative conditions. Mutations of the ATP synthase were designed to abrogate the oxidative phosphorylation activities, while preserving the crucial catalysis to perform ATP hydrolysis. The extensive metabolic engineering of a single *E. coli* strain resulted in the 2-fold higher acetate production rate from glucose, than from ethanol oxidation, which was competitive for industrial production of acetate. Using multiple deletion strategies, the same group also successfully engineered a new *E. coli* strain that efficiently convert glucose into another high value metabolite, pyruvate[39].

2.2.2   Stochastic approach

Deletion of known gene functions is without a doubt a powerful method for obtaining a desired phenotype by rerouting metabolic fluxes. However, similar to many deterministic methods, it suffers from the disadvantage of relying on available information on gene sequences and functions. Despite the wealth of genome sequences, many gene functions are poorly studied and approximately 30–50% of genome sequences are uncharacterized. With respect to metabolic engineering applications, modifying unknown pathways or gene functions may be useful for the development of a new phenotype. In conjunction with available screening and selection methods, combinatorial strategies can identify a superior phenotype that otherwise would not be discovered by using deterministic approaches. A stochastic deletion of gene targets has been applied for metabolic engineering purposes, for the identification of lycopene overproducing *E. coli* strains[7]. Lycopene is a natural metabolite of high nutritional importance, thus its microbial production could lead to inexpensive productions. Construction of the lycopene producing strain was based on the insertion of *dxs*, *isp*, *idi* into the *E. coli* chromosome, along with the introduction of *crtEB1* through a carrier plasmid.

Random gene deletions were performed by transposon-based mutagenesis, in which a plasmid containing a transposase[5] gene, was introduced into the lycopene producer. The plasmid carried a mutated *tnp* allele, which encoded the broad-range mutant Tn10 transposase to prevent mutational insertion only at certain spots in the genome. Moreover, the plasmid also contained the suicide R6K mutant origin of replication to prevent episomal replication once inside *E. coli* cells. The random genome disruption of the lycopene producing strain resulted in three gene deletions which improved lycopene production, identified by an increased red color intensity of the recombinant strains. The genes were identified to encode for rssB and two hypothetical proteins yjfP, yjiD. To search for further improvement in lycopene production, the authors combined the single mutations found from the transposon library with gene deletions found from computer prediction based on stoichiometric based flux balance analysis (this strategy is discussed in Section 4.3). The combinatorial deletions resulted in lycopene production up to $11,000 \mu g/g$ dry cell weight.

A similar approach that led to carotenoid production alteration from *E. coli* through random chromosomal mutations has also recently been described[206].

2.3   Gene Silencing

It is well established that both prokaryotic and eukaryotic cells use small, non-coding RNA sequences that can bind to RNA transcripts and prevent their translation. Such RNA molecules act as regulators in various signal transduction mechanisms[83] and in plasmid replication and copy number[126,127,210]. Since metabolic engineering aims at examining cellular phenotypes under different genetic perturbations, RNA silencing using artificial antisense RNA could be a powerful tool towards that goal. Indeed, such examples have recently been described in the literature.

Acetate is a spontaneous by-product of *E. coli* fermentation that inhibits protein synthesis. However, acetate biosynthesis plays an important role in *E. coli* as a source of ATP formation under aerobic and anaerobic conditions. Therefore, a simple gene knockout for eliminating acetate buildup would probably not be a wise strategy for increasing foreign protein production. Instead, antisense RNA strategies were employed to down-regulate acetate biosynthetic enzymes[120]. The synthesis of the two major acetate producing enzymes, the phosphotransacetylase (PTA) and acetate kinase (ACK) were blocked by antisense RNA fragments designed specifically for the down regulation of the sense gene expression. Specifically, three antisense RNA constructs were created based on the combinations of 141-bp and 130-bp antisense sequences to *ack* and *pta* respectively. In all cases, to synchronize the expression of the antisense RNA construct with that of the native genes, the expression of all antisense RNA was regulated by the intrinsic *ack* promoter sequence and the *pta* termination sequence. In order to test the effect of silencing *ack* and *pta*, the small RNA generating fragments were introduced into *E. coli*, along with *gfp*, to produce the test protein. The metabolic engineering strategy successfully enhanced the production of GFP by 1.6- to 2.1-fold when *ack* and *pta* were silenced through the artificial synthesis of small RNA fragments.

The gram-positive bacterium *Clostridium acetobutylicum* has been exploited extensively in the past for the commercial production of acetone and butanol. In order to improve such solvent production, the expression of enzymes of the competing pathway leading to butyrate formation was reduced by RNA silencing. Butyrate synthesis was mediated by two enzymes, the phosphotransbutyrylase (PTB) and butyrate kinase (BK). PTB converts butyryl-CoA into butyryl phosphate, and BK catalyzes the subsequent reaction to synthesize butyrate from butyryl phosphate. Aiming to reduce the expression of BK, a synthetic oligonucleotide fragment was developed against *bk*. Specifically; the DNA fragment included 10 codons of *bk*, and its native putative ribosome binding site sequence. To start and end the transcription of the antisense bk, the ptb promoter sequence and a rho-dependent termination sequence were also included in the construct.

Following a similar approach, the antisense fragment for *ptb* was constructed by including a 567-bp *ptb* fragment, its putative ribosome binding site sequence, and an *adc* terminator. Each of the antisense construct was separately inserted into *C. acetobutylicum* through a carrier plasmid and the effectiveness was checked by monitoring protein synthesis and solvent production. The *C. acetobutylicum* strain which expressed antisense butyrate kinase exhibited up to 90% lower BK synthesis and resulted in 50% and 35% higher final concentration of acetone and butanol, respectively. The strain which expressed the *ptb* antisense synthesized 70% lower PTB; however, acetone and butanol concentration were 96% and 75% lower, respectively, compared with that of the native strain. Interestingly, in both recombinant strains, the level of butyrate did not change upon reducing the biosynthetic enzymes. The authors concluded that the levels of the butyrate biosynthetic enzymes did not affect butyrate production[54].

3.   TINKERING THE GENETIC MECHINERY: ENGINEERING ENZYMES

The realization of the diverse functions and versatility of enzymes has motivated the advancement of metabolic engineering strategies through tailoring of these metabolic workhorses. In many cases, the engineering of proteins took advantage of nature-provided blueprints for fine-tuning enzymes for gaining different functions or improving stability by introducing advantageous single or multiple point mutations, and by insertion of amino acid blocks. Some of the most commonly used experimental techniques and their applications will be highlighted in the next few pages.

3.1   Site-Specific and Random Point Mutagenesis

The availability of protein sequence and structure data combined with biochemical characterizations can be used to guide the introduction of specific mutations leading to catalytic activity or stability improvement of an enzyme. Reversely, site-specific mutations can be applied to a particular protein to examine catalytic mechanisms and protein folding. When the desired amino acid exchanges are present in the sequence termini, a simple PCR (Polymerase Chain Reaction) based mutagenesis can be performed by introducing the mutations in the primer sequences used for amplification of the gene. This strategy can be extended to create multiple point mutations throughout the sequence of a gene by step-wise mutations, in which a mutant clone generated from a PCR is used as the template for the next PCR to direct mutations at different sites[19,59,141,211].

Even though in many cases, multiple mutations need to be introduced to fully obtain new enzyme characteristics, a few single mutations can impose a great effect on the catalytic properties of some enzymes. A characteristic example is the sensitivity towards mutations of the plant polyketide synthases type III, the key enzymes in the biosynthesis of structurally diverse valuable natural products (extensively reviewed in Austin[11]). Such enzymes include benzalacetone synthase (BAS) that catalyzes a condensation reaction of 4-coumaroyl-CoA with one malonyl-CoA to form benzalacetone, the key precursor of the anti-inflammatory lindleyin in rhubarb, gingerol and curcumin in ginger plants, and raspberry ketone, the characteristic chemical conferring the raspberry aroma. The amino acid sequences of various plant type III polyketide synthases, for example, CHS, stilbene synthase (STS), 2-pyrone synthase (2-PS), and acridone synthase (ACS) are available. Upon alignment of these sequences, it was shown that the conserved amino acid residue Phe-215, thought to be a crucial integral of the catalytic activities of CHS, is not present in the BAS gene. In order to convert BAS catalysis to perform CHS activities, the amino acids Leu-215, together with its adjacent Ile-214 were replaced with Phe-215 and Leu-214, respectively. The mutations of the BAS sequence resulted in chalcone-forming activities, in which the chalcone naringenin, along with other byproducts were generated from incubation with the substrates[1].

Introducing selective mutations for altered genetic functions suffers from several disadvantages. These include the necessity of sequence data and/or molecular data a priori, careful analysis of the available information, and more importantly, the possibility of excluding useful mutations that can not be predicted from sequence data. Similar to the stochastic approach of genome disruption, a single genetic sequence can be disrupted by randomly introducing point mutations. Clearly, this method does not require the availability of sequence information or molecular structures. However, high through-put methods for isolating and selecting useful mutations are necessary. One example of randomly introducing amino acid exchanges of a biocatalyst for metabolic engineering purposes is the development of deacetoxycephalosporin C synthase (DAOCS) for enhancing the conversion of penicillin G into phenylacetyl-7-aminodeacetoxycephalosporanic acid, a precursor of 7-aminodeacetoxycephalosporanic acid. Cephalosporins are β-lactam antibiotics widely used clinically for treatment of various bacterial infections. The three cephalosporins with medical importance, cephradine, cephalexin, and cephadroxil, are derived from 7-aminodeacetoxycephalosporanic acid (7-ADCA). Industrially, the ring expansion of the thiazolidine ring of penicillin G into phenylacetyl-7-ADCA (G-7-ADCA) is performed chemically, followed by enzymatic side chain cleavage of G-7-ADCA. In order to eliminate the necessity of chemical conversion for the synthesis of the core structure of cephalosporin, a lead enzyme, DAOCS, the Fe(II)- and α-ketoglutarate-dependent dioxygenase that catalyzes the expansion of penicillin N into deacetoxycephalosporin C (DAOC) has been chosen to serve as a biocatalyst for the industrial production of 7-ADCA. This is because when the DAOCS *Streptomyces clavuligerus* gene was introduced into *Penicillium chrysogenum*, the recombinant fungus was able to expand penicillin G into G-7-ADCA.

Despite the availability of the atomic structure, and biochemical characterization of *S. clavuligerus* DAOCS catalysis, rational designs only resulted in mutant enzymes that modestly convert penicillin G. In order to expand the search for other amino acid changes important for shifting the substrate specificity of DAOCS, random mutagenesis of the gene was performed by treating the cDNA with hydroxylamine. The mutated library was then used to transform *E. coli*, and subsequently, colonies were chosen and analyzed to collect three mutants DAOCS that exhibited strong affinity towards penicillin G. Site-specific mutagenesis strategy was also pursued in order to enrich the mutant library. In this case, because the structural data of DAOCS were not available, the structural data of a closely related enzyme, the isopenicillin N synthase (IPNS) were used as a guide for choosing amino acid mutations. To further increase the potential of isolating mutants with high catalytic efficiency for penicillin G conversion, combinatorial mutations were performed among the three mutants derived from random mutagenesis and the three mutants derived from the site-specific mutagenesis. The strategy successfully isolated a triple mutant DAOCS with a 13-fold increase in relative activity toward penicillin G[225].

Besides protein engineering, PCR-based approaches have been used in order to optimize promoter activity. The widely used method for controlling gene expression in metabolic engineering applications is through the use of inducible promoters which respond to inducer concentrations. However, some inducers are expensive, toxic[203], and can result in transcriptional heterogeneity[194]. Therefore, an emerging interest in metabolic engineering is the ability to adjust the level of transcription through fine-tuning promoter sequences[7,105,106]. Generation of the bacteriophage PL- gpromoter with different strength was performed by error-prone PCR, which introduced random mutations during amplification. The promoter library was then attached in front of the GFP (green fluorescent protein) reporter system contained in an *E. coli* plasmid. Upon introduction of the plasmids containing the promoter library, the mutated promoters with increased strength could be identified by observing the fluorescence intensity of the *E. coli* colonies[7]. Similarly, this strategy was also used to breed varying strengths of the TEF1 yeast promoter.

Using the yeast system, the selection process was aided by the yECitrine fluorescence in which stronger/weaker promoters could be identified by monitoring the fluorescence intensity of the yECitrine protein[7].

3.2   Directed Evolution

Directed evolution strategies are inspired by the routes of evolution in nature, in which survival of the fittest is mediated by the accumulation of beneficial genetic mutations that confer phenotypical fitness to withstand selection pressures[123]. The availability of structural studies is normally not a prerequisite to successful laboratory enzyme evolution, because directed evolution employs stochastic methods to generate mutant libraries. However, recently deterministic methods that incorporate structural information have been described.

3.2.1   DNA shuffling

DNA Shuffling methods are based on the iterative processes of random mutation generation, and selection for improved phenotypes in nature[198]. In these methods, random point mutations are generated in the parental DNA sequences to introduce sequence diversity. Subsequently, in order to redistribute the locations of the mutations, the parental DNA pool is randomly fragmented and reassembled (recombined) to regain full-length protein sequences. The genetic library is then inserted into a suitable host, such as *E. coli*. Upon screening, for example by applying a selection pressure, the chimeric genes derived from the clones with improved phenotypes will be used consecutively as the parental sequences for the next generation of DNA shuffling. In some strategies, a pool of homologous parental sequences derived from different species is used as a template of the recombination effort, to increase sequence diversity, which in turn, increases sequence space search for protein evolution purposes[49].

One example of the incorporation of DNA shuffling for metabolic engineering purposes is the generation of various novel carotenoids from *E. coli*[184]. Carotenoids are synthesized from the general terpenoid pathway and involve GGDP synthase (*crtE*) and phytoene synthase (*crtB*) for the production of the carotenoid phytoene. Subsequent desaturation by phytoene desaturase (*crtI*), and further enzymatic modifications generate different carotenoid molecules. To generate a carotene producing strain, *crtE*, *crtB*, *crtI*, and *crtY*, were introduced in *E. coli* through carrier plasmids, thus generating recombinant strains with orange-red color. The generation of novel acyclic carotenoids was then achieved by gene shuffling of the *crtI* genes derived from *Erwinia herbicola* and *Erwinia uredovora*. The mutated library was introduced into *E. coli* harboring *crtE* and *crtB* for subsequent selection of clones that confer to carotenoid colorations. One clone with yellow coloration (I25) and one with pink coloration (I14) were isolated. Further sequence analysis of the mutated *crtE* gene isolated from I14 revealed two amino acid mutations and a replacement of the 39 N-terminus amino acids of *crtE* from *E. uredovora* with that of *E. herbicola*. Sequence analysis of *crtE* isolated from I25 showed 2 amino acid changes. To extend the breeding of novel cyclic carotenoids, *crtY* from *E. uredova* and *E. herbicola* were introduced separately into the carotenoid pathway containing *crtE* from clone I14. When the wild-type *crtE* was introduced into the recombinant *E. coli* expressing the carotenoid pathway, a bright yellow-orange coloration was produced. However, replacing the wild type *crtE* with the desaturase from I14 resulted in bright yellow coloration. A library of *crtY* was also created with shuffling the *crtY* genes from the two origins, and introduced into the *E. coli* carrying the I14 desaturase pathway. Out of 4500 clones screened, 25 colonies with different colorations were selected. Sequencing of the cyclase isolated from a bright red clone revealed the generation of two amino acid changes of the *E. uredova* cyclase, without recombination. A reaction product of this colony was identified to be torulene, a compound nonnative to the artificial reaction pathway introduced.

Recently, the concept of shuffling DNA was extended to shuffling the entire genome of *Streptomyces fradiae*, the bacterium used for the commercial production of tylosin, a complex polyketide antibiotic[249]. In order to generate a new tylosin over-producer, the wild type *S. fradiae* strain was subjected to one round of random mutagenesis using nitrosoguanidine mutagen. When screening 22,000 individual mutants, 11 strains that produced more tylosin than the wild-type were isolated. To generate a genome-shuffled library, protoplasts of the 11 strains were mixed in equal proportion and recursively fused. 1000 clones were screened from the first round of genome shuffling, and 7 identified superior strains were used as the parental strains for the next shuffling cycle. Similarly, 1000 colonies were screened and 7 strains with further improvement of tylosin production were isolated. Analysis of two over-producer strains indicated that tylosin titer was 9-fold higher compare to the wild-type *S. fradiae*. It is compelling that the development of a similar over-producer strain using various mutagens took place in 20 years, requiring 1 million assays. The genome shuffling method achieved the creation of an over-producer strain in the course of 1 year, with only 24,000 assays.

3.2.2   *Nonhomologous in vitro recombination*

Gene shuffling methods require high sequence homology among the parental gene pool for successful reassembly of the fragmented genes. Because cross-over points mostly occur in DNA regions of high sequence identity, DNA shuffling poses limitation in exploring novel cross-over points in regions of low identity that could accelerate enzyme in vitro evolution[20]. A method termed incremental truncation for the creation of hybrid enzyme (ITCHY)[149] allows such exploration by generating all possible fusions between two non-homologous genes[150]. As a case study, ITCHY was applied to create a hybrid *E. coli* and human glycinamide ribonucleotide (GAR) formyltransferase, which only share 50% DNA sequence identity, however with similar structure and active sites. To generate the ITCHY library, the 5′-terminal of the *E. coli* sequence and the 3′-terminal of the human sequence were unidirectionally truncated by enzyme digestion. The two sequences were randomly fused, adjoining two truncated sites, to create a full-length DNA library consisting of the *E. coli* and human fragments. Plasmids containing the library ware used to transform an *E. coli* mutant deficient of GAR transformylase activities. Identification of the functional hybrid enzymes was performed by complementation of the *E. coli* mutant for growth in the absence of purine. When sequencing the hybrid proteins, it was discovered that the sequences generated from ITCHY exhibited wider cross-over distributions and as such scanned a larger protein sequence space than a library generated from DNA shuffling.

It can be argued that even though ITCHY discovered hybrid proteins with non-bias cross-overs, the library contains many nonfunctional enzymes because the adjoining of two gene fragments with various lengths resulted in many chimeras that are not structurally conserved. To address this issue, the method SHIPREC, which stands for sequence homology-independent protein recombination, was introduced[193]. SHIPREC library was created to solubilize a human membrane associated P450 (1A2), by recombination with the heme domain of the soluble P450 from *Bacillus megaterium* (BM3). The library construction was performed by first adjoining the C-terminus of 1A2 with the N terminus of BM3 through a linker that contains a unique restriction site. Both termini of the fusion sequences were randomly truncated by digesting with *DNase I* enzyme. After treating the termini with *S1* nuclease to produce blunt ends, fragments with the same length as the parental sequence were isolated through gel fractation. Subsequently, both termini were ligated to form gene fusions at various cross-overs. To recover head to tail fragments, the circular DNAs were cut using the endonuclease of the unique restriction site present in the linker sequence. It was anticipated that fusing the two gene fragments resulted in many frame shift mutations. Therefore, to select for hybrid sequences that do not contain frame-shift mutations, the library was translationally fused to the *CAT* gene that confer to resistance to the antibiotic chloramphenicol. Upon transformation of the plasmid containing the CAT-fused library into *E. coli*, the library that contained proteins without frame-shift mutations were discovered in the cells that were resistant to chloramphenicol. After removal of sequences containing frame-shift mutations, a second stage screening was perform to isolate chimeras with P450 activities by monitoring absorbance at 450 nm upon CO binding and testing for exclusive 1A2 activities which performs the deethylation of 7-ethoxyresorufin (development of fluorescence). Out of 2000 selected clones, 2 clones exhibited characteristic of P450 absorption and fluorescense. The two clones contained the hybrid sequences fused at two different points.

3.2.3   Custom designing proteins

Methods of generating mutant enzymes by blindly recombining homologous and nonhomologous sequences have discovered chimeras with novel activities. However, the occurance of mutants with highly improved functions is scarce, and not to mention, good screening systems are prerequisite. Proteins can be evolved more efficiently based on their structural data by swapping domains with structural similarity[148,163,166]. The key for successful structure-based recombination is to correctly identify interchangeable modules and the locations of cross-over points. A computational algorithm called SCHEMA was developed to calculate interactions between amino acid residues and determine the level of disruption resulting from replacing a subset of amino acids[221]. The locations of amino acid that correspond to minimum level of disruption will be the potential cross-over points for swapping protein modules. The information obtained from SCHEMA was used to construct a hybrid protein derived from β-lactamases (TEM-1) and PSE-4 that share only 40% amino acid sequence identity but structurally similar. Domains of the two proteins were interexchanged and resulting hybrid sequences were inserted into *E. coli*. The functionality and activities of the hybrid β-lactamases were selected by culturing *E. coli* in different antibiotic concentrations. Using this approach, several hybrid proteins with distinct modular combinations that confer to an increase antibiotic resistance in the host *E. coli* were identified.

Amino acid and structural information can also be used to rationally construct functional proteins with unique properties that are composed of different domains derived from completely unrelated enzymes. One example is the generation of a P450 chimera that was active in *E. coli*. The cytochrome-dependent P450 eukaryotic enzymes mediate the catalysis of myriad reactions, including hormone synthesis and drug metabolisms. Plant P450 enzymes belong to the class of heme-dependent P450s that require a reductase protein to shuttle electrons (not self sufficient). Along with electrons, molecular oxygen is required to perform a stereospecific hydroxylation to the substrate moiety. *E. coli* is perhaps the most favored heterologous host for protein expression because it is well characterized and multiplies rapidly. Therefore high protein yield can be achieved through a relatively short fermentation period. However, functional expression of P450s is hindered in *E. coli*, because of the lack the endogenous electron transfer proteins and endoplasmic reticulum. Extending the engineering of this robust organism to synthesize hydroxylated molecules that can only be catalyzed by P450 enzymes is necessary, since the pharmaceutical potency of some natural metabolites is directly dependent on the number and site of hydroxylation[28,109]. The synthesis of hydroxylated flavonoids in *E. coli* was achieved by incorporating an artificial flavonoid biosynthesis pathway and functional expression of a P450 enzyme, the flavonoid 3′5′-hydroxylase (F3′5′H)[130]. In order to accommodate the lack of electron donor proteins in *E. coli*, the F3′5′H protein was translationally fused with the catalytic domain of a P450-reductase. To prevent the formation of secondary structures that could interfere with the P450 activity, an amino acid sequence glycine-serine-threonine was chosen as a linker of the F3′5′H with the reductase. This design was inspired by the the most catalytically active P450 hydroxylase from *B. megaterium* which contains a heme domain linked to a P450-reductase like domain. Furthermore, because the native plant F3′5′H contains an N-terminal hydrophobic sequence for membrane anchor, which was not suitable for expression in *E. coli*, the first 4 amino acid codons were removed. Upon removal of a portion of the N-terminus, the first and second codons were changed into methionine and alanine, respectively, to allow for transcription initiation and efficiency. Co-expression of the chimeric protein with an artifical flavonoid pathway leading to flavonol synthesis resulted in the synthesis of various hydroxylated flavonoids from the non-hydroxylated substrate.

Recent construction of a designer protein allowed *E. coli* to secrete a black compound as respond to light, functioning as a bacterial photograph. Phytochromes found in plants and some bacteria are two-component systems, which consists of photoreceptor and response-regulator domains. In order to make use of the phytochrome system to generate a response-regulator mechanism in *E. coli*, the photoreceptor domain of phytochrome Cph1 derived from *Synechocystis* was fused with the EnvZ histidine kinase domain from *E. coli*. EnvZ is a part of the well-studied two-component system EnvZ-OmpR involved in the signal transduction mechanism in response to osmotic stress. The bacterial photograph was created by introducing the chimera into *E. coli* containing a chromosomal insertion of *lacZ* reporter gene under the control of the OmpR-dependent *ompC* promoter, and two phycocyanobilin biosynthesis genes, *ho1* and *pcyA* from *Synechocystis*. Without light excitation, phosphorylated histidine kinase acted as an activator for the *lacZ* transcription, which gene product catalyses the formation of a black precipitate from the supplemented substrate, S-gal. In the presence of light, phycocyanobilin response inactivated the phosphorylation of the histidine kinase, hence the expression of lacZ was inactivated, producing a contrasting replica of the image on a lawn of *E. coli*[131].

4.   NEW TREND IN METABOLIC ENGINEERING

The current focus of metabolic engineering for biotechnology applications is the development of efficient cellular factories for the synthesis of novel molecules. In recent years, much attention has been devoted to implement artificial regulation systems in the cells. Concepts of artificial regulatory systems are inspired by natural genetic circuits, mostly involved in signal transduction pathways. The simplest modules of a gene regulatory network consist of a promoter, an output protein, and regulatory elements that dictate the expression of the output protein. Regulatory proteins can promote or halt the expression of the output protein by either the recruitment of RNA polymerase to the promoter or by interfering with transcription initiation. Synthetic circuits were normally built on well-characterized genetic components that exist in nature. For example, the *Lac* operon in *E. coli*, which is composed of *lacZ*, *lacY*, and *lacA* is one of the best characterized regulons. In the absence of lactose, the Lac repressor (LacI) binds to *lacO* sequence adjacent to the *lac* promoter (P_{lac}), to halt the transcription of the Lac operon by the RNA polymerase. The presence of lactose decreases the binding of the LacI repressor to the *lacO* sequence, and allows transcription of the Lac operon[104]. Applications of artificial circuits are diverse, including the study of natural oscillation and 'noise' of gene expression, autoregulated microbial machineries, and riboregulators.

4.1   Natural Circuit Mimicry

Artificial circuits can be used to study the uniformity of protein expression in identical cells as a coordinated function of transcription, RNA degradation, and translation. In order to show that higher stability could be achieved by negative feedback mechanism, an artificial autoregulated circuit was constructed in *E. coli*. The circuit was composed of the $P_L$tet01 promoter to regulate the expression of TetR-EGFP, a fusion of tetracycline repressor (TetR) and the enhanced green fluorescent protein (EGFP), which was designed so that the TetR-EGFP repressed its own transcription. Measuring the fluorescent intensity as a function of cell count, it was shown that the synthetic negative feedback mechanism resulted in improved stability compared to an unregulated circuit, in which the feedback mechanism was eliminated[17]. Similarly, in order to study the effect of a positive feedback regulation on protein expression stability, a synthetic circuit was constructed that contained a tetracycline-responsive transactivator gene fused with GFP in which expression activates (positively regulate) its own promoter. Measuring GFP signals in the yeast *S. cerevisiae* containing the artificial circuit resulted in two distinct yeast populations, in which the protein was expressed in different levels, creating a bistability condition[16]. The construction of more complex artificial circuits to study system bifurcation can be reviewed in Hasty[87], Sprinzak[197], and Kaern[113].

Recently, an artificial circuit was constructed to signal the intracellular state of sugar metabolism in *E. coli*[78]. In *E. coli*, sugars are converted into the intracellular metabolite acetyl-CoA. Subsequently, acetyl-CoA is converted into acetyl phosphate by phosphate acetyltransferase (PTA), and then into acetate by acetate kinase (ACK). The enzyme acetyl-CoA synthetase (ACS) catalyzes the conversion of acetate directly into acetyl-CoA irreversibly. In order to utilize acetyl phosphate as a signaling molecule, *acs* gene was overexpressed in *E. coli* under glnAp2 promoter, which also controlled the expression of Lac repressor (*LacI*). The glnAp2 promoter was used because it is activated through phosphorylation of the nitrogen regulator, which responds to the level of acetyl phosphate. In parallel, *pta* gene was also over expressed in the same cell under the lacO-1 promoter. Therefore, when acetyl phosphate accumulates in high level, it activated the glnAp2 promoter to synthesize ACS. At the same time high acetyl phosphate also repressed PTA over expression. To detect the behavior of the synthetic circuit, GFP was also included in the cell, and its expression was dependent on LacI repressible *tac* promoter. By monitoring the GFP signals emitted from the cells, it could be demonstrated that an oscillation behavior occurred which depicted the fluctuation of the intracellular acetyl phosphate accumulation.

4.2   Artificial Cell-to-Cell Communication

Communication among bacteria (quorum-sensing) by chemical sensing has also been applied to the construction of synthetic circuits. For example, by introducing an autonomously regulated circuit, the population of *E. coli* can be maintained at low level. The construction of the circuit was based on the well-characterized LuxI/LuxR communication system employed by *Vibrio* species reviewed in Withers[237]. The *lux* operon, whose expression was regulated by an inducible *E. coli* promoter was introduced into *E. coli*, along with a killer gene, *ccdB*, expressed under the *lux* promoter P_{luxI}. Upon expression of the *lux* operon, LuxI proteins synthesized small diffusible acyl-homoserine lactone (AHL) molecules. Generated by a high cell density, high concentration of AHL bound to luxR proteins, which subsequently acted as transcription activators of the P_{luxI} promoter to synthesize killer proteins. High levels of killer proteins generated from high cell density caused cell death and lowered the population until the system reached a steady-state[245].

A more elaborate quorum-sensing based circuit was recently created to engineer *E. coli* to form ring-like patterns that depended on the distance from one another[12]. Construction of the 'sender' cells was achieved by introducing $P_{LtetO-1}$ regulated LuxI, to allow the synthesis of AHL upon induction of the tetracycline promoter. The cells receiving the signals generated from the senders were created by introducing three regulators, the response GFP based read-out, and *LuxR*. The expression of LuxR was under the control of $P_{Lux(L)}$ to enable the gene products to serve as AHL sensors. Moreover, the P_{lac} promoter mediated the GFP expression. $LacI_{M1}$, which was expressed under the induction of the $P_{Lux(R)}$ promoter served as the first regulator. The lambda repressor (CI), which expression was under the $P_{Lux(R)}$ promoter was used as the second regulator. The expression of LacI under the control of the λP(R-O12) promoter served as the third regulator in the receiver cells. Induction of the senders generated a chemical gradient of AHL surrounding the cells. Placing a receiver cell within a close proximity to the sender cell inactivated the GFP expression because of the high concentration of AHL. The expression of LuxR was activated by the abundant AHL. The LuxR proteins served as activators of the LacIM1 repressor, whose expression inactivated the GFP expression. In parallel, LuxR proteins also activated the expression of CI, which in turn served as a repressor of LacI. At distance far from the sender cell, where AHL concentration was low, the GFP expression was also off because the inactivation of the CI and $LacI_{M1}$ proteins allowed the expression of the LacI protein to turn off the GFP expression. However, at an intermediate distance, the production of $LacI_{M1}$ was not sufficient to shut off GFP, while the high activities of the CI proteins effectively repressed LacI expression to allow the expression of GFP.

4.3   Synthetic Riboregulator

Another natural genetic component that has been used as a model to design artificial machineries is the small antisense RNAs. Silencing transcription of a specific mRNA sequence is traditionally achieved by designing a small nucleotide fragment that can bind to the transcription signal of the gene. A creative design of a synthetic riboregulator was constructed by formatting a *cis*-repressor and *trans*-activator[102] to allow both gene silencing and translation activation by RNA. The artificial *cis*-repressor was created by attaching a small nucleotide sequence (cr) down stream of the promoter (P_{cr}) of a reporter gene (*GFP*). During RNA synthesis because of P_{cr} induction, the complementary sequence of cr with the ribosome binding site (RBS) allowed the formation of a hair-pin loop with the RBS; hence translation into active proteins was abolished. To reverse the effect of the *cis*-repressor activity, a trans-acting (ta) small RNA was expressed under a second promoter (P_{ta}). The ta RNA was designed to form a low energy complex with cr, forming a stable linear RNA duplex to free RBS, and allow GFP translation.

The construction of a synthetic riboregulator whose activity is dependent on specific binding to small molecules (anti-switch) has also been demonstrated[13]. The design principle of a small *trans*-acting RNA to interfere with transcription of a specific gene only in the presence of a specific small molecule was based on combining an antisense RNA domain and an aptamer. Aptamers are small nucleic acid fragments synthetically synthesized *in vitro* to fold and bind to specific ligands. The synthetic biological polymers are derived from randomly threading nucleotides, followed by in vitro selection for fragments with high affinity binding to a specific ligand[65,66]. To create the antiswitch, a small antisense RNA fragment complementary to a 15-nucleotide stretch around the start codon of a target mRNA encoding GFP was fused with a theophylline aptamer. The stem of the antisense portion which adjoined the folded aptamer was modified so that it also incorporated a reversed complementary sequence of the antisense RNA (antisense stem). Therefore, in the absence of theophylline, the antisense portion formed a double-stranded RNA with the antisense stem, unable to bind to the target gene RNA. Furthermore, the stem of the aptamer portion was also designed to include a short fragment (stem) to base-pair with the antisense stem. Upon binding to theophylline, the aptamer stem bound to the antisense stem, and forced the antisense sequence to linearize. The linearized RNA sequence could bind to the target RNA sequence, halting translation of the GFP.

5.   METABOLIC FLUX ANALYSIS

Metabolic fluxes are defined as the rate at which material is processed through metabolic pathways. Because most reactions affecting a cell at its living state are captured, fluxes can be regarded as the functional determinants of cellular physiology. Perhaps the most essential requirement to quantify *in vivo* molecular fluxes is an accurate, stoichiometric representation of the cellular metabolic network[132]. There are two main strategies that can be applied to elucidate metabolic flux distributions. The first relies on the combination of ^{13}C labeling experiments with metabolite balancing around each intracellular metabolite for an assumed metabolic network. Generally this is referred to as ^{13}C-Metabolic Flux Analysis (MFA). The second, referred to as Flux Balance Analysis (FBA), is an approach that is used to predict cellular fluxes based solely on the metabolic stoichiometric matrix and some constraining assumptions. Both approaches have been proven successful in improving microbial strains used as industrial biocatalysts and in identifying the topology of active reactions and pathways. The following pages highlight some of the aspects and analytical methods (experimental and theoretical) employed by both strategies.

5.1   Metabolite Balancing Essentials

As most industrial processes are operated under quasi-stationary conditions, it is usually assumed that stationary conditions will emerge in all intracellular metabolite pools of the cell. This is a hypothesis that obviously overlooks oscillations that exist in biological systems as well as the different flux patterns of individual cell subpopulations[232]. However, such an assumption can be expected to hold true in a continuous culture or in batch-fed culture with slow variations in input feed[231]. Under these conditions, a set of constraints is developed according to which the sum of fluxes into a metabolite intermediate is equal to the sum of fluxes leaving the intermediate. Application of such balances to cofactors (NADH, NADPH) and ATP, can create links among pathways that are not necessarily related to each other. Such links have been established among reduction-oxidation reactions, ATP regeneration, biomass generation reactions, central carbon metabolism and secondary metabolite pathways and have led to significant improvements in product formation[24,77,145,224,247]. More details of this procedure will be presented in section 3. The procedure of metabolite balance analysis was pioneered in the 1980s and early 1990s and led to the first metabolic flux maps of several microorganisms, including *Clostridium acetobutylicum*[151], *Corynebacterium glutamicum*[213–215], *Bacillus subtilis*[81], *Penicillium chrysogenum*[108] and *Saccharomyces cerevisiae*[146] as well as hybridoma cells[22].

5.2   Isotopic Tracer Methods

It is unfortunate that the application of Metabolite Balance Analysis to large, complex metabolic networks is hindered by reaction dependencies that result in metabolic singularities. Such dependencies exist in the case of parallel metabolic pathways[196], futile cycles or bidirectional reaction steps[233–235] and cause observability problems that can only be resolved through theoretical assumptions about the intracellular biochemistry[23]. It is worth noting that the introduction of additional constraints, particularly in the form of lower value limits (bounds), for irreversible reactions, while limiting the solution space, still do not help overcome the observability problems. Because of these shortcomings, approaches that make use of stable isotopic tracers were introduced in order to generate (over)determined metabolic systems.

5.2.1   Isotope tracer experiment concepts

The basic idea behind isotope tracer experiments is to feed a cell culture that has reached steady-state with ^{13}C-labeled substrates, such as ^{13}C-labeled glucose, and then allow the fractional labeling in all carbon atoms of intracellular metabolites to equilibrate (*isotopic steady state*). Measurement of the ^{13}C enrichment of the various metabolite carbons in combination with isotopomer balancing can then provide additional information about both the structure of the metabolic network and its metabolic flux distribution. The stability of ^{13}C-labeled molecules, the availability of lower cost tools to use for their analysis and their non-radioactive nature are the reasons behind their preference over ^{14}C-labeled compounds.

But what exactly is an isotopomer? Chemical elements containing a different number of neutrons are called *isotopes*. All stable isotopes of an element occur in nature at constant ratios, called natural isotopic abundance. *Isotopic enrichment* of a metabolite atom is the fraction of the molecules in the metabolite pool in which the atom appears in the particular isotopic form. *Isotopomers* are thus considered the different positional configurations of the isotope element in the molecule, since each carbon atom within the molecule can be labeled (^{13}C) or unlabeled (^{12}C). For example, a molecule with n carbon atoms will have 2^n different isotopomers. The percentage (or fraction) of each isotopomer within the overall metabolite pool is referred to as *isotopomer distribution*. There are alternative choices for labeling element used to improve the validity of the network analysis and include labeling of such elements as O, H, N, S and Si. The choice of element used for labeling depends on the network, particular metabolic pathway of interest, or even functional groups present. By using a combination of different labeled metabolites, it is possible to significantly increase the redundancy of the analysis, thereby reducing the error.

The isotopomer distribution of the intracellular and extracellular metabolites will thus depend on both the isotopomer distribution(s) of the feeding precursor(s) and the structure of the overall metabolic network but only provide information about the relative fluxes through the metabolic network. Figure 2 illustrates how the substrate labeling can dictate the information received from this procedure. The current approach for obtaining absolute values of carbon fluxes is through a tight integration of tracer experiments with metabolite balancing measurements, a strategy that requires only few modeling assumptions for the living system. There are two basic schemes to perform isotope balancing analysis: *isotope enrichment analysis* and *isotopomer distribution analysis*.

For the use of isotopes to be an advantage the carbon movement through the metabolic network and the network architecture need to be related. This is done by connecting carbon atoms rather than metabolites with arrows to form a carbon flow network, or just a carbon network. The major distinction between a metabolic and a carbon network is that within a carbon network there are no reactions taking place. This is because the carbon network represents how the carbon atoms simply pass from one metabolite to another and the possible rearrangements for each atom. The carbon network is then used to formulate carbon atom mass balances, following the same principles in those used for metabolites. A very concise way of representing such carbon networks is based on the construction of *atomic mapping matrices*[251] for every reactant-product pair of a metabolite reaction. This uses a matrix AMM_{kl}^n where the $[k, l]$ element is the probability of the carbon transferring from the lth carbon pool of the reactant to the kth carbon pool of the product through reaction n.

5.2.2   Isotope enrichment analysis

In isotope enrichment analysis, the focus is placed on the carbon atoms rather than the metabolites. The fractional enrichments of the network carbon atoms are used to estimate fluxes. The basic idea of this method is to exploit additional constraints, in the form of isotope enrichment balances, which can be obtained when doing

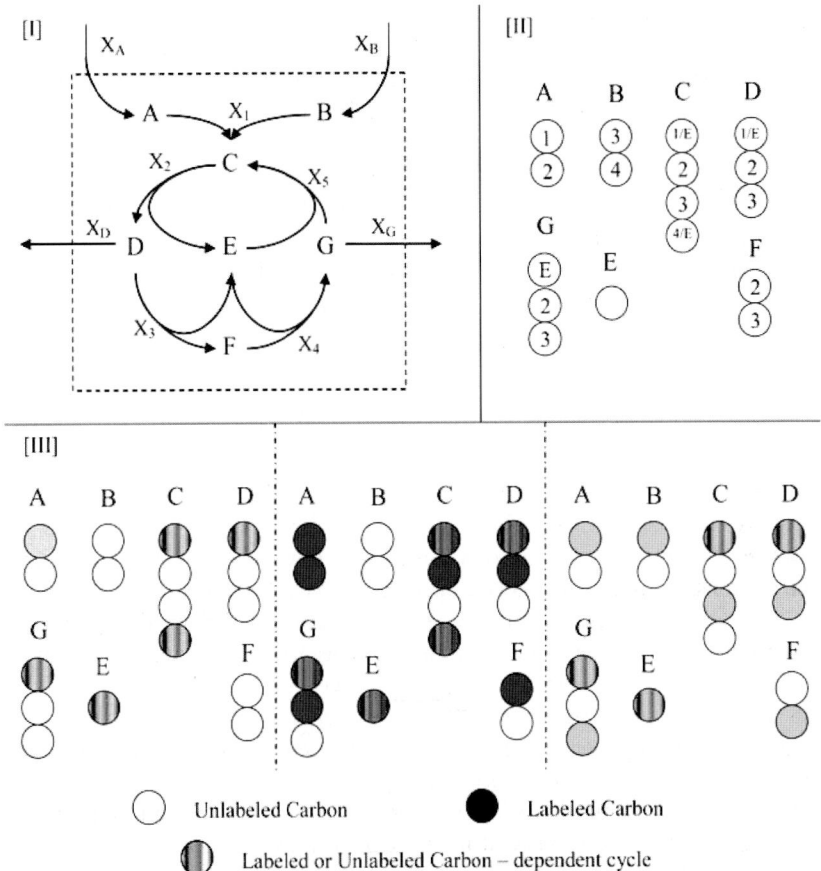

Figure 2. Isotope labeling example with cycle present. [I] The metabolite reaction network for the example system where the metabolites (A-F) are carbon chain molecules. Metabolite A is a two carbon molecule (carbons 1 and 2) and metabolite C is a four carbon molecule. [II] Carbon distribution key for understanding carbon flow through the system in panel [I]; this is not a carbon network map. Metabolite C is made of two carbons from metabolite A and two from B initially. After one loop, the fourth carbon atom in metabolite C can come from two sources; reaction of metabolites A and B or from metabolites E and G reacting. Notice the fourth carbon in metabolite C can be from metabolite B or E. [III] A series of possible labeling outcomes is shown for the example system

a carbon labeling experiment. Since a carbon can contain two different flows, a labeled flow and an unlabeled one, we can set up mole balances around carbon atoms to extract these equations.

Carbon is added to a pool through the uptake of a substrate or through the reaction of metabolites and is removed from a pool through secretion or reaction consumptions. Multiplying the elements of atomic mapping matrices by the carbon enrichment flows over all internal reactions and adding the external feed and secretion enrichment flows, the mole balances for each carbon pool can be formed. The mole balances, with other basic relations result in the isotope enrichment balances. The isotope enrichment balance for carbon pool i is:

$$r_i^{in} CE_i^{in} - r_i^{out} CE_i + \sum_{l=1}^{n} \left(\underline{\underline{A}}_{ij}^l \cdot v_l \cdot CE_l \right) = \frac{dc_i}{dt} \text{ for } i = 1, \ldots, m \qquad (1)$$

where:

r_i are the uptake and excretion rates for the carbon pool i,

v_i is the i element of the vector of unknown intracellular fluxes,

$\underline{\underline{A}}_{ij}^l$ is the piecewise function containing the atomic mapping for reaction l,

CE is the label enrichments of pool i or label enrichment for flux l,

c_i is the carbon of pool i,

n is the number of fluxes in the network,

m is the number of metabolite pools in the network.

To resolve the fluxes via the Isotope Enrichment (IE) Analysis, a least squares method is applied to relate the isotope enrichment measurements to the metabolic fluxes. The computational procedure behind this is extensive and thus will not be presented here. The reader is referred to the literature for a more detail explanation of the mathematical formalism[122,153].

One other important issue to note is that, depending on the structure of the metabolic network, isotopomer enrichment measurements do not always lead to an increase of the number of fluxes that can be quantified. For this reason, it is crucial to have an *a priori* knowledge of the type of measurements that are required in order to increase the observability of the *fluxome* of the cell.

5.2.3   *Isotopomer distribution analysis*

The second approach to isotope balancing places the focus on metabolites by applying isotopomer mass balances for each metabolite isotopomer in each metabolite pool. As in isotopomer enrichment analysis, a method is needed for relating the way isotopomers proceed from every reactant to a product for each reactant-product pair. In a similar fashion to the atomic mapping matrices, *isotopomer mapping matrices*, constructed for every reactant-product pair, can be used for a concise representation of isotopomer conversions[252]. The isotopomer mapping matrices are formed from a simple matrix multiplication. The first matrix contains all possible isotopomer configurations of the reactant in columns while the second matrix is made up of the nonzero rows of the atomic mapping matrix. Therefore the only requirements for assembling the isotopomer mapping matrices are the atomic mapping matrices and a detailed carbon network.

By again forming mole balances, this time around isotopomers, one can establish relations for the isotopomer populations, which can be found experimentally using NMR or MS. More information about these methods will be provided in the following section. If measurements of the relative populations of some or all isotopomers of the network are available, then information about the intracellular flux distribution can be obtained. The balance around isotopomer i for its metabolite pool can be written as:

$$r_i^{in} X_{IS_i}^{in} - r_i^{out} X_{IS_i} + \sum_{p=1}^{m} \left(\underline{\underline{D}}_{pi}^j \cdot v_j \cdot X_{IS_p} \right)_j$$

$$+ \sum_{p=1}^{m} \sum_{u=1}^{m} \left(\underline{\underline{W}}_{pui}^j \cdot v_j \cdot X_{IS_p} \cdot X_{IS_u} \right)_j = \frac{dc_i}{dt}$$

for $j = 1, \ldots, n$ \hfill (2)

where:

m τ is the number of isotopomers for all metabolites in the metabolic network,

n τ the number of unknown intracellular fluxes in the metabolic network,

X_{IS_j} τ is the vector of the relative populations of network isotopomers,

c_i τ is a size in moles of the isotopomer i in the metabolite pool,

$\underline{\underline{D}}_{pi}^j$ τ is a function of the isotopomer mapping matrix for *non-condensation* reactions,

$\underline{\underline{W}}_{pui}^j$ τ is a function of the isotopomer mapping matrix for *condensation* reactions,

r_i τ are the uptake and excretion rates for the isotopomer i due to metabolite flows,

v_i τ is the i element of the vector of unknown intracellular fluxes.

Condensation reactions are those reactions of the type $2A \rightarrow B$ or $A + B \rightarrow C$ where the number of moles is reduced. Non-condensation reactions refer to those where the moles are conserved, such as $C \rightarrow E$.

The combination of metabolite balances with the isotopomer mass balances for all isotopomers in all metabolite pools results in a complicated mathematical framework when applied to realistic networks. To solve this framework, a least squares method is used to fit the experimental data to the developed equations. This method is analogous to the method used in the Isotopomer Enrichment Analysis. While this picture is complicated enough, it becomes even more complicated when taking into consideration the reversibility of the biochemical reactions.

[13]C labeling approaches in general, and isotopomer distribution analysis in particular can exploit data from isotopic enrichment measurements (using ^{13}C-NMR) and the multiple-peak, or *multipet*, structure of ^{13}C-NMR spectra as well as the mass isotopomer fraction measurements from Mass Spectrometry. In that respect, labeling greatly enhances the observability and degree of resolution of the system and even leads to redundancies that allow testing of the experimental measurements and the accuracy of the assumed metabolic network.

5.2.4   Analytical tools for isotopic balance analysis

Perhaps the most pivotal element for advancing the field of metabolic flux analysis using ^{13}C labeling experiments was the development of analytical tools able to detect differences between isotopomers. There are two main types of such techniques currently used: Nuclear Magnetic Resonance (NMR) and Mass Spectrometry (MS) coupled with Liquid or Gas Chromatography (LC-MS or GC-MS). We will describe some of the basic features of both methods, as detailed description goes beyond the scope of the present publication (more extensive information can be found in Sxyperski[205]).

NMR technologies are based on the different nuclear spin of the ^{12}C and ^{13}C isotopes where measurements create peaks depending on the location of atoms and their relationship to nearby atoms. The very first technique applied in metabolite flux balancing was the carbon-proton correlation NMR spectroscopy tailored to link chemical shifts with the resonances of directly attached protons[140]. Because ^1H-NMR spectroscopy is the most sensitive of NMR techniques, it provides an attractive approach for smaller, non-enriched metabolites. It also allows for the observation of each carbon in the metabolite molecule separately from all other positions[231].

On the other hand, ^{13}C-NMR spectra provide a better resolution of the isotopomer distribution through the measurement of the total peak areas as well as through the multiplet pattern of peaks produced in the spectra[185]. The line splitting is due to ^{13}C-^{13}C coupling between adjacent carbon atoms and as a result the various sub-peak areas in a particular multiplet are linearly related to the relative populations of the metabolite isotopomers that generate these peaks[41,138].

In addition to the standard peaks, there are also long-range coupling effects that need to be considered when analyzing a NMR spectrum. These coupling events are usually observed only for a few molecules and are typical of aromatic compounds. Other molecular configurations can lead to difficulties in reading NMR spectra, particularly branch points and symmetric molecules. Most of these issues today however are seen quite regularly and can be compensated for. In all cases, the evaluation of NMR spectra requires care, specially adapted software tools and in some cases specialized measurement procedures[52]. Finally, the relatively low sensitivity of NMR results in experimental inaccuracies and limits its application to extracellular metabolites or metabolites that are produced in high abundance.

Mass spectrometry has recently been introduced as a complimentary and in some cases as an alternative analytical tool for ^{13}C-tracer experiments[121]. In fact, several studies recently underlined the great potential of MS for the labeling measurements in metabolic network characterization[43,50,238]. Among the most attractive properties of MS are the (i) high informational content of data, (ii) high sensitivity, (iii) high accuracy, (iv) versatility, (v) robustness and (vi) rapid rate of data accumulation. While the resolution of specific molecule peaks is high for MS, it can be impeded by more effects such as strong isotope effects changing the retention time of the molecules in the GC column, detector overloads or the natural abundance of isotopes leading to large disturbances in the mass isotopomer spectrum.

Mass spectroscopy consists of five major steps: sample introduction, ion formation, mass separation, detection and data handling. The samples (*analytes*) are ^{13}C labeled metabolites that can exist within complex matrices (media or cell crude extracts) and as a result may require some degree of separation before injection to the MS, thus MS is usually couple with some form of chromatography. In addition to small molecules, biomass constituents such as biopolymers (proteins and polysaccharides) can also be analyzed in order to extract valuable information about metabolic fluxes. Depending on the application, MS can be used to assess molar enrichment (IR MS) and mass and positional isotopomer distributions (GC/MS and MALDI MS).

The recent description of microtiter-plate cultivations coupled with MS analysis opens the possibility of high-throughput metabolic flux analysis that can be applied for mutant screening, drug development and general systems biology[240]. The low cultivation volume also offers the possibility of using tracer substrates that are usually available in small quantities and high cost.

The development of *Corynebacterium glutamicum* strains for the industrial production of lysine was achieved with the application of ^{13}C methods[154,167]. One approach for increasing lysine yield was to increase the flux through the carboxylation of phosphoenolpyruvate (PEP). PEP carboxylase was originally thought to be the main enzyme for the formation of oxaloacetate, but by using labeled carbon, it was revealed that pyruvate carboxylase catalyzes 90% of the carboxylation flux *in vivo*[152,154]. It was also found that the PEP carboxylation flux which seems rather strong *in vitro* is actually highly inhibited under *in vivo* conditions[154]. In addition, Petersen et al. also showed that PEP carboxykinase catalyzes large fluxes of inter-conversion of oxaloacetate and PEP in lysine overproducing *Corynebacterium*, constituting a futile cycle in the system[154,167].

5.3   Flux Balance Analysis: An Optimization Method

To eliminate the dependence of ^{13}C-Metabolic Flux Analysis on values from experimental work prior to predicting engineered solutions, alternative methods relying strictly on the network stoichiometry have been developed having the ability to create goal oriented genetic modification experiments from *in silico* predictions. Of particular interest and growing in popularity is the method of Flux Balance Analysis (FBA) which predicts intracellular fluxes through an optimization procedure of a *cellular flux model*. Before we highlight the advantages and disadvantages of this method, it is important to understand how this method translates biological system into a mathematical framework for analyzing cellular responses to network perturbations.

5.3.1   History of flux balance analysis

The modeling effort began with work done by Amit Varma and Bernhard Palsson in the early 1990's. They set out to define the metabolic capabilities of *Escherichia coli* based on the stoichiometric network alone; firstly for the production of biosynthetic precursors, and secondly making a balanced mix of cellular constituents[216]. At the time, there was no global database containing a compiled list of active metabolic pathways and therefore only an introductory model consisting of only 30 metabolites involved in 53 fluxes was used. It is important to note that this analysis strictly contained aerobic metabolic pathways for the bacterial cell[218]. The scope of the model covers mainly the glycolysis pathway, the citrate cycle (TCA cycle), and the pentose phosphate pathway. Later Varma increased *E. coli* model in size and scope to include some of the biosynthetic reactions to form a 95 reaction model using 107 metabolites. The metabolic capabilities of systems predicted using these small systems have been thoroughly discussed in the referenced material[216–219]. Jeremy Edwards and Bernhard Palsson later expanded the metabolic network of *E. coli* to 627 unique reactions and 438 metabolites[62]. In addition to updating the major pathways in Varma's models, the Edwards' model included pathways for alternate carbon sources, more detailed biosynthetic reactions, and a detailed growth function of cellular constituents[62]. This model was again updated in 2003 by Jennifer Reed and colleagues to include not only a greater number of unique reactions (931) and metabolites (761), but also all protons and water molecules to both elementally and charge balance all but six reactions[165].

To avoid confusion between models, a labeling system has been developed that includes the developer as well as the number of genes characterized in the model and is prefixed by *i* to denote an in silico model. The former *E. coli* model is the iJE660 (developed by J. Edwards with 660 genes) while the later is iJR904 GSM-GPR (developed by J. Reed with 904 genes) where GSM stands for genome-scale model and GPR denotes gene-protein-reaction[165,207]. While this labeling convention is convenient, not all cellular flux models have been identified using this moniker. A variety of other in silico organisms are listed in Table 1 along with the number of reported genes and reactions. The organisms represented are varied from *Saccharomyces cerevisiae*[58], *Helicobacter pylori*[181,207], *Staphylococcus aureus*[15], *Mannheimia succiniciproducens*[94], *Aspergillus niger*[51], and hybridoma cell metabolism in *Mus musculus*[189]. With the advancements in genome sequencing technology and functional genomics, new models for additional organisms will be soon on their way.

Table 1. Sampling of current flux models for a variety of organisms

Organism	Genes	Reactions	Reference
Aspergillus niger	295+	335	51
Escherichia coli	660	739	62
	904	931	165
Helicobacter pylori	291	388	181
	341	476	207
Human cardiac mitochondria	115	189	220
Mannheimia succiniciproducens	320+	373	94
Hybridoma cell of *Mus musculus*	473	1120	189
Saccharomyces cerevisiae	708	842	76
	750	1149	58
Staphylococcus aureus	619	640	15

+Indicates the exact number was not reported and that the above number is an estimate.

5.3.2   Extending metabolic flux analysis

Due to the enormous variation in metabolite composition and stereochemistry, the isolation and measurement of a particular species or a sampling of chemical species from a laboratory experiment is a daunting task. Most metabolic studies are only able to develop a 'metabolic profile' for subsets of existing chemical classes in a sample[118]. In contrast, cellular flux models generally rely on the development of constraints formed by the existing knowledge of biochemical reactions and other processes taking place within the metabolic system. For the purposes of our discussion here, we will consider the biochemical reactions within an individual cell; however this method may be adapted to additional biological systems of interest such as signaling cascades.

The construction of cellular flux models is a multi-step process. It begins with the identification of the metabolic pathways defining the cellular functions within the organism. By comparing the genomic sequence to functional genes already annotated, the metabolic enzymes existing within the genome can be recognized. The enzymatic reactions catalyzed by these enzymes can then be identified to describe the network of metabolic reactions[35]. Database-enabled searches, in particular BLAST type searches, are essential for timely identification of genes and eventually pathways. However, it is important not to over look traditional literature searches and a little intuitive nature. Such is the case when a number of enzymes for a particular pathway exist but there is a missing reaction that with its inclusion would complete the entire metabolic pathway. This is where an insightful modeler would notice the network gap and include it in the model without a functional gene assignment. Thus, based on the physiochemical and biological data known to the researcher, these gaps in the metabolic pathway can be closed to create a fully defined system. The metabolic maps for a number of sequenced organisms can be found on the Kyoto Encyclopedia of Genes and Genomes[115].

The physio-chemical constraints in the form of mass balances of metabolites, balances on energy molecules and known flux limitations, such as ATP maintenance requirements, for particular pathways make up the constraints used to describe and potentially predict cellular functioning[63,128,177,178,217,218]. In addition to enzymatic functional assignments, transport reactions for metabolites entering and leaving the cell must be included. This includes active transport of metabolites or passive transport such as diffusion through pores in the cell membrane or through the membrane itself[219]. Each biochemical reaction, transport flux and other constraints identified are characterized as individual vectors making up an entire convex set of vectors. The vectors are the variables of the mathematical model whose value will be resolved during the optimization iterations. The enzymatic reactions and transport processes that serve to create intracellular metabolites, dissipate those metabolites, and/or convert them to final metabolites, are thus collected to form the constraints for the metabolic system.

With the fluxes defined, taking each metabolite as a node in the network and considering the reactions leading into and out of that node, the mass balances for each species can be formed. The mass balances are the system constraints having the form of equalities with the fluxes forming and consuming each metabolite on the left hand side and the accumulation terms for those metabolites on the right hand side. It is important to note that at this point in the formulation we are still dealing with a dynamic system represented by a set of coupled ordinary differential equations. A matrix of stoichiometric coefficients, $\mathbf{A}$, of size m by n and a vector of the fluxes, $\mathbf{x}$, of size n is typically used to represent this set of equations[22,62,133,217,218].

$$
\begin{bmatrix} \frac{d(\text{metabolite}1)}{dt} \\ \vdots \\ \frac{d(\text{metabolite } m)}{dt} \end{bmatrix} = \begin{bmatrix} \text{fluxes for metabolite1} \\ \vdots \\ \text{fluxes for metabolite } m \end{bmatrix} \Leftrightarrow \begin{bmatrix} b_1 \\ \vdots \\ b_m \end{bmatrix} = [\mathbf{A}] \cdot \begin{bmatrix} x_1 \\ \vdots \\ x_n \end{bmatrix} \quad (3)
$$

For metabolic systems, the number of reactions out numbers the number of metabolites, and therefore an underdetermined system of equations is represented by equations (3)[60]. Unique solutions for the ill defined system can not be solved for unless additional constraints are imposed such that the number of linearly independent equations is equal to the number of independent variables.

5.3.3   Restriction of the solution space

The stoichiometry of the biochemical reactions form the primary constraints for Flux Balance Analysis, where the underdetermined system resulting from the constraints forms a solutions space in the shape of a polyhedral cone[182]. While the solutions space is rather large, by defining maximum and minimum limitations on fluxes, and even the specification of other fluxes from experimental data previously collected, the solution space can be restricted further.

The specification of a flux is a more definitive way to restrict the solution space, but is not commonly applied due to the need for additional experimentation. Typically applied is the inclusion of another class of constraints in the form of bounds where an upper (β) and lower (α) bound is applied to each flux within the network. Two typical fluxes that are bound are the substrate uptake rate and the oxygen uptake rate (for aerobic growth). The substrate of interest is usually limited from -10 to zero while oxygen is typically restricted from -20 to zero. Note the negative lower limit indicating that those fluxes are going into the cell.

In addition to limiting the variables to a particular range, a pseudo steady-state approximation is made because we can assume, with moderate certainty, that the cellular pathways are occurring on time scales where internal metabolites are at a steady-state composition within the system, thus satisfying the mass balances as applied to any given environmental conditions[117]. At this point in the model construction process the following mathematical equations are found where a_{ij} is the coefficient of metabolite i in flux j and the right hand side b_i is zero:

$$\sum_j a_{ij} x_j = b_i$$

$$\alpha_j \leq x_j \leq \beta_j \notin sub, O_2 \qquad (4)$$

$$-10 \leq x_{sub} \leq 0$$

$$-20 \leq x_{O_2} \leq 0$$

The performance capabilities of the metabolic systems are defined by primary constraints and can be refined with the inclusion of more experimental data[179,180,217–219]. In particular, ^{13}C-labeling experiments can be used to gather experimental flux levels which can then be applied in silico to restrict the bounds on fluxes, even fixing them at a prescribed value. However in order to specify a flux, one must be certain of its value from numerous experimental measurements found from isotope labeling[21,151,215]. The number of measurements required depends on the stoichiometric matrix size and is equal to the dimension of the null space[183], thus for large metabolic systems it may not be feasible to assign many values. The intracellular fluxes can be computed from isotope labeling[229] as previously described. Because regulatory networks direct the activation and/or repression of a set of genes in response to environmental stimulus, the application of gene regulatory data from genetic chips can be a useful tool for placing additional limitations on the metabolic network. These limitations are placed by setting fluxes related to those genes either to zero or forced to a minimum level of activity through bounds[48].

5.3.4   Cellular objectives for optimization

To identify the flux distribution resulting from the metabolic constraints, the system of equations above (equations (4)) are optimized with respect to the maximization or minimization of a particular objective function[186]. In order for the cellular flux model to be of any value, the objective must be defined in such a way that the organism's biological goals are translated into a mathematical equation using previously established variables. In other words, for an objective function to comprise some biological relevance and possibly a certain industrial significance[117], it must fully quantify the scope of cellular functioning in the mathematical framework. Any bacterial cell, in theory, has the objective of growth as its sole function, however when modeling more complex cells (i.e. mammalian) this may not be the case. For our purposes we will be restricting our discussion to the simpler, bacterial cell objectives.

For unicellular organisms, the accepted cellular objective is cell growth, and from an evolutionary standpoint this makes sense. To quantify the cellular growth, an additional flux is created that takes in the essential amino acids and other building blocks required for cell growth and leads to a fictitious metabolite, plus remaining metabolic components. While this may seem odd, the fictitious metabolite does have some physical significance as it represents the cell's biomass and is thus typically referred to as biomass in the metabolite listing. The coefficients, and thus the quantities, of the essential amino acids and building blocks required for one mole of biomass for *E. coli* are based on work from Ingraham *et al.* in the 1970's and has been widely accepted for this application[128]. We now have a problem of the form in equations (5).

$$\text{maximize} : Z = x_{growth}$$

$$\text{subject to} : \sum_j a_{ij} x_j = b_i$$

$$\alpha_j \leq x_j \leq \beta_j \notin sub, O_2 \tag{5}$$

$$-10 \leq x_{sub} \leq 0$$

$$-20 \leq x_{O_2} \leq 0$$

This is a standard formulation for a Linear Programming problem. The type of objective function may vary, as we will see in a moment, and thus defines the type of optimization as either Linear Programming or Quadratic Programming. FBA uses a linear objective defined so as to maximize the growth flux within our system of equations. Simulations to calculate the internal fluxes of an underdetermined network conducted with the growth objective function have been shown to be consistent with experimental data[61,99,186].

By optimizing the steady-state model with the objective function defined, simulation results are found[186]. Optimal solutions lie along the edge of the polyhedral cone formed by the stoichiometric limitations of the network and any bounds applied to the fluxes. For a more detailed understanding of Linear Programming, refer to Chvatal's text on the topic[44]. Figure 3 illustrates the steps we have taken so far in Flux Balance Analysis.

Optimal solutions come in two different forms. They may lie at a corner of the polyhedral cone, thereby involving only one possible optimum. Or they may occur as a set of optimal solutions falling along the same path as the objective function creating a situation of multiple optima. This occurs when the line of optimality coincides with a primary linear constraint. For example in Figure 3, the objective function would lie along the same plane as one of the posed constraints. In these situations, while resolving all the optimal distributions is time consuming, using a dual case set of optimizations, the bounds for all fluxes given an objective value can be found[136]. Essentially, a series of maximization and minimizations for every flux is performed with the linear constraints shown in equations (4) along with the specification of the growth flux from solving equations (5).

While the common objective functions generally include some form of maximization related to a biomass, alternative forms have also been considered. As mention earlier, quadratic objective functions can be used to consider perturbations or knockouts in the network from the originally developed system[47]. The Minimization of Metabolic Adjustment (MOMA)[186,187] is one such method using a quadratic objective function that has been used to study a number of different regulatory systems as they are effected by network perturbations resulting in loss of network reactions[8,9,25,79,125]. MOMA uses the solution for a primary analysis with the growth as the objective function by attempting to minimize the difference between the new flux profile created by a perturbation in the flux space and the original flux space. The solution to the metabolic system found by MOMA typically results in a suboptimal growth prediction for a perturbation of positive interest.

The concept behind the MOMA analysis is to minimize the error between the wild-type flux profile and the new knockout flux profile. This is expressed as a simple minimization of squared error or:

$$D = \sqrt{\sum_i (x_i - w_i)^2} \tag{6}$$

where x is the new flux and w is the wild-type flux for reaction i respectively. This can be reformed in vector notation leading us to the formulation of the Quadratic

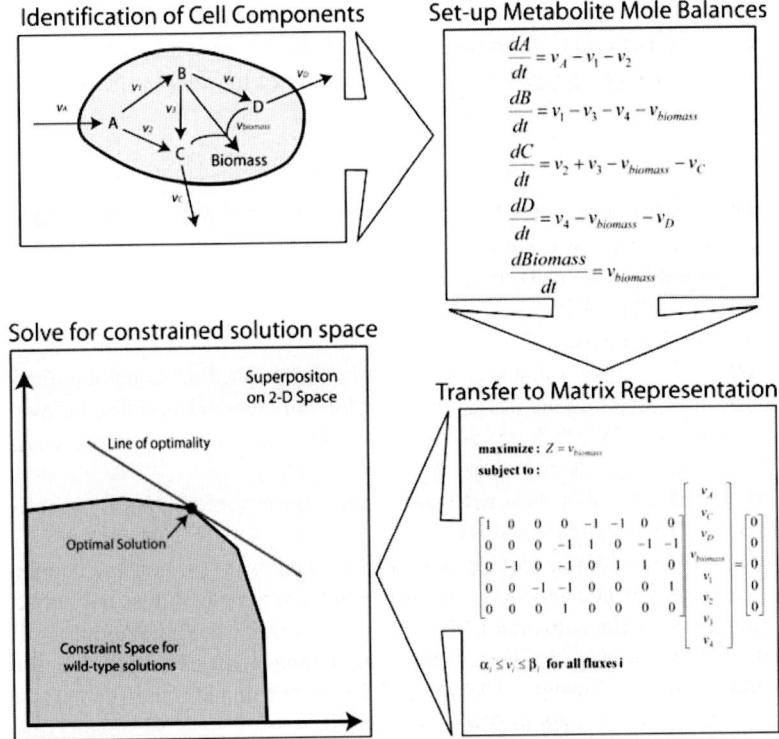

Figure 3. Critical steps in the flux balance analysis optimization procedure. (I) Identify the intracellular conversions and transport processes within the cellular environment. (II) Mole balances for each molecular species in the environment neglecting the reaction kinetics for any process (the v is flux). (III) Reformulation in matrix notation with the inclusion of bounds and cellular objective function. (IV) Superposition of the polyhedral solution space into a two dimensional plane

Programming (QP) problem. The QP problem is then expressed as:

$$\text{maximize}: Z = \sum_{j} \left(x_j - w_j\right)^T \left(x_j - w_j\right)$$

$$\text{subject to}: \sum_{j} a_{ij} x_j = b_i$$

$$\alpha_j \leq x_j \leq \beta_j \notin sub, O_2, k \qquad (7)$$

$$-10 \leq x_{sub} \leq 0$$

$$-20 \leq x_{O_2} \leq 0$$

$$x_k = 0$$

Here, we see the fluxes are again restricted by lower (α) and upper (β) bounds as well as the defining of the substrate and oxygen uptake rates. Additionally some flux, k, has been set to zero to simulate an in silico knockout. For genes controlling multiple cellular reactions, there may be more than one flux to set to zero. By understanding the performance limitations inherent to the system under perturbed metabolic states, one can begin to understand and utilize the cellular performance capabilities associated with the organism.

5.3.5   Limitations of flux balance analysis

As mentioned earlier, the method of FBA utilizes the fact that a steady state or quasi steady state exists for the intracellular pool. This assumption certainly holds true in a continuous culture and, under certain conditions, in the exponential growth phase of a batch culture or in a fed-batch culture with slow input feeds[231]. However limitations of FBA do exist:

1. For two parallel pathways that achieve the same end result, FBA can not indicate which pathway is being used or any flux split that might be taking place between the two pathways. While there are few examples, one such case is the lysine biosynthesis in Corynebacterium glutamicum[231].
2. Any metabolic cycle that does not have a direct measurable output or coupled to an output flux, cannot be resolved.
3. While flux balance analysis requires a prior understanding of the network connectivity, i.e. the enzymatic and spontaneous reactions occurring, it does not require any knowledge of the networks kinetic parameters. With this limitation, it still has allowed for the development of quantitative simulators of cellular function that bring together the results of bioinformatics, metabolomics, and genomics in a systems biology approach to pathway analysis.

By looking at particular metabolite generations as objective functions, it is possible to identify the system's robustness to changes[64,100]. The robustness of the network is explored by varying the maximum flux through a particular pathway and observing the resulting changes in the flux profile and the value of the cellular growth flux[46]. In this way one can investigate how an organism is resistant to changes in specific enzyme activity or overall pathway activities[63]. Additionally, metabolite producibility can be studied, which defines the ability of the network to produce a certain metabolite given prescribed constraints based on the media and genotype[101].

A major disadvantage is the large solution space created by modeling organisms through constraint-based models. However a number of methods have been developed to search such large solution spaces[29,30,187]. The most promising search has been the use of a bi-level mixed-integer linear programming method, termed OptKnock, with the introduction of a cardinality constraint[31,32]. While this method still requires one to select the cardinality of the system, K (i.e. the number of perturbations of interest), it still has provided accurate results[9,25,71–75,157,158,192]. Conceptually, this method uses a bioengineering objective of industrial importance, such as the maximization of a metabolite, as well as the establish LP or QP problem. The mathematical form is shown in equations (8) with a MOMA objective function applied.

$$\text{maximize} : x_{chemical}$$

subject to :

$$
\left[
\begin{array}{l}
\text{minimize} : Z = \sum_j \left(x_j - w_j\right)^T \left(x_j - w_j\right) \\[2mm]
\text{subject to} : \sum A_{ij} x_j = b_i \\[2mm]
\qquad \alpha_j \cdot y_j \le x_j \le \beta_j \cdot y_j \notin sub, O_2 \\[2mm]
\qquad -10 \le x_{sub} \le 0 \\[2mm]
\qquad -20 \le x_{O_2} \le 0
\end{array}
\right] \qquad (8)
$$

$$y_j = \{0, 1\},$$

$$\sum_j \left(1 - y_j\right) \le K$$

Constraint-based analysis has recently been applied for the improvement of secondary metabolite production through the use of in silico predicted genetic knockouts. Alper et al. used a MOMA analysis of the Edwards model with the addition of 15 heterologous reactions that lead to the formation of lycopene to find single, double and triple genetic knockout designs resulting in increased production levels of lycopene in E. coli[8]. E. coli, in the form of the iJR904 GSM-GPR model, has also been optimized for the production of lactic acid using the OptKnock algorithm to predict gene knockout designs that increase production and maintain growth[71]. Again knockout schemes containing single, double and triple knockouts were performed as well as knockouts of 4 and 5 genes.

However by recognizing the limitations of Metabolic Flux Analysis and other modeling methods, it is possible to develop innovative and collaborative, synergistic approaches to combine experimental and modeling methods to order to elucidate to a greater detail the system's flux distributions. Additionally, the large abundance of high-throughput data now available has created a growing need for constraint-based models to incorporate this type of data for cellular function prediction and analysis[230].

6.   METABOLIC CONTROL ANALYSIS

When modeling biological systems, understanding of how system perturbations are naturally developed or formed solely from any *ad hoc* assumptions researchers make is vital. Within this context, molecular mechanisms such as feedback inhibition, post-translational modifications and protein-protein interactions can all potentially shift the carbon fluxes in metabolic pathways from one state to another. Researchers are interested in the sensitivity of metabolic fluxes to changes in particular parameters that are indicative of these cellular mechanisms. The parameters of interest can vary from the kinetic rate constants to an enzyme concentration or activity, or to some other external modifier acting on the system.

Two independent approaches, one by Kacser and Burns[111] and the other by Heinrich and Rapoport[89,90], were developed in the 1970's that began to answer such questions as *'how does the metabolic flux vary?'* or more importantly *'how much does it vary?'*. Their ideas stemmed from the pioneering work of Higgins in the early 1960's[92]. In 1984, the term Metabolic Control Analysis was introduced in order to unify the terminology for defining the same parameters of the metabolic systems[33].

In Metabolic Control Analysis, the relative importance of different enzymes for the control of metabolic fluxes is explored by considering the so-called *flux control coefficients* and the *elasticity coefficients*. It is important to note that Metabolic Control Analysis is not intended to be a complete approach to modeling metabolism; its principle concern is with the sensitivity of a metabolic network to its inherent parameters. One of the more attractive aspects of using Metabolic Control Analysis is that it does not require every system component to be characterized *a priori*[37].

Metabolic Control Analysis actually applies three classes of coefficients and two theorems, in order to determine the systems response to changes in parameters. These coefficients have been termed elasticity, response and control coefficients[26,132]. The control coefficients signify the effects of reaction rates or enzyme concentrations on the steady-state response. Elasticity coefficients govern metabolite effects on enzyme activity while response coefficients are a combination of the two previous[93]. A majority of our discussion here will be in defining these coefficients so that they can be applied to a variety of systems for study.

Other methods proposed to quantify fluxes have been just as successful, most importantly Biochemical System Theory, which was introduced in the late 1960's[173–175]. This method is an elegant way to solve complex systems with the aid of model parameters and experimental data. The approach is founded on the general power-law expression for reaction rates as described by the equation:

$$v_i = \beta_i \prod_j X_j^{h_{ij}} \tag{9}$$

where the system variables (metabolite concentrations, enzyme activities, etc.) are represented by X_j and the parameters β_i and h_{ij} are termed apparent rate constants and apparent kinetics orders[199]. A model constructed from the mass balances around the metabolites and using the above power-law rate expressions has been termed an *S-system*[176]. For a more detailed description of Biochemical Systems Theory and its applications, the reader is referred to the book by Voit[222]. While some have claimed the Metabolic Control Analysis is a special case of Biochemical Systems Theory, it is often a result of the misconception of the objectives and assumptions made by Metabolic Control Analysis[199]. As such, we will take great care in defining the assumptions of Metabolic Control Analysis so one can easily differentiate between the two.

6.1   Assumptions for Control Analysis

In most analysis methods, particular assumptions are made to simplify the analysis. In Metabolic Control Analysis, the assumptions concern the connectivity and stability of the system and the properties of enzymes within the system. These assumptions are:

1. A single connected unit makes up the entirety of the biological system. While this is not a necessary condition for the analysis, it applies to most metabolic systems studied using Metabolic Control Analysis. An alternative view of this was put forth by Kahn and Westerhoff in which the system can be broken into blocks of closely related reactions, separated by key connections representing regulatory or catalytic interactions[114].

   In this sense, there is an overall flow of material through the metabolic system from a set of metabolites acting as the system's *source* leading to another set of metabolites acting as the system's *sink*. These metabolite sources and sinks are boundaries between which the transportation of metabolites, enzyme-catalyzed reactions, and spontaneous reactions transpire. It is between these two boundaries that Metabolic Control Analysis is applied[67].

2. Results from mathematical formulations of metabolic systems operating under stable steady states hold true when applied to the living system. While the concept of a singular steady-state is a mathematical abstract, the idea is of physical relevance, especially as applied to a chemostat. To utilize this assumption, the network reactions require classification into three groups based on their respective time scales. (*i*) Rapid fluxes, i.e. instantaneous reactions and isomerizations, in which the time scale is much shorter than the observation time scale of concern. (*ii*) Fluxes of the same time scale as the observation interval and finally (*iii*) fluxes effecting concentrations at time scales of length much greater than the observation interval[67]. A linear sequence of several reactions can be grouped together as proposed by Brand and colleagues[26,27].

   The first group of reactions are considered to be at steady-state, thus these reactions are not generally included in the metabolic network. Typical examples include ionic equilibria and binding reactions of substrates and enzyme effectors.

The third group of slower reactions, typically the metabolic degradation reactions, either can be treated as constants or simply grouped into the parameters of the analysis. Only the second group of reactions, which also includes transport fluxes, is considered within the framework of Metabolic Control Analysis.

3. Metabolic Control Analysis does not deal with spatial distributions. Thus positional information typically required for reaction-diffusion systems is not required for the description of the system's state at any time point[67]. The metabolites are thought to be evenly and homogenously distributed over the cellular compartment and any enzyme existing in such compartment may act on the metabolite.

4. While this is not the case in all cellular systems, one enzyme effects only one reaction thus the rates of enzyme action are directly proportional to the concentration of the enzyme. Ways for modification of Metabolic Control Analysis have been developed to cope with enzyme action that is disproportionate to enzyme concentration[112,172], enzyme activation of more than one flux[38], and protein-protein interactions[226].

5. Enzymes can not be both variables and parameters unless the system is broken into blocks as proposed by Kahn and Westerhoff[114]. In such cases where blocks exist, enzymes may be defined in one block as a variable with no controllability over the functioning of that block while in another block the enzyme might be acting as regulatory or catalytic entity exhibiting a great deal of control over the blocks functioning.

6. Finally, Metabolic Control Analysis considers all metabolites to be of the free form and not involved in enzyme-metabolite complexes. While this makes measurement of substrate and metabolite quantities difficult, the development of NMR techniques, as discussed earlier, has made this final assumption less of a problem.

A limitation does result from dividing the stoichiometric model into blocks, namely that Metabolic Control Analysis can then only provide quick assessments of the whole system, which are not necessarily accurate ones. The two modes we will discuss for application of Metabolic Control Analysis are direct and indirect. In the direct approach the analysis starts from the broadest portions of the intact system and works towards the details while the opposite is true for the indirect approach[26]. More information about these modes of application will become apparent later through examples.

6.2   Coefficients of Metabolic Control

In general, control coefficients are used to describe how a unique variable of the system, or a property of the system, responds to some minimal variation of a tuning-type parameter. In this sense Metabolic Control Analysis is a methodology similar to process control tuning. The tuning parameters, i.e. the enzyme concentration and activity, can be *tuned* to improve cellular function however we must first unravel an enzyme's ability to induce change or modulate the fluxes. By identifying the activation, inhibition and other regulation taking place within the cell, the adjustment of the cell via genetic modification can be directed for cellular improvement. In order to separate the types of parameters available for tuning, global and local coefficients have been established indicating the scale of their effects[33].

Global coefficients are found through measurements of the entire system's response to a genetic or environmental perturbation and are not isolated to a particular branch in the metabolic network. By definition we define global coefficients as:

$$C_P^V = \lim_{\partial P \to 0} \frac{\partial V/V}{\partial P/P} = \frac{\partial V}{\partial P} \cdot \frac{P}{V} = \frac{\partial \ln |V|}{\partial \ln P} \tag{10}$$

Where V stands for any variable in the system, typically a flux or metabolite pool concentration, and P is any parameter whose modification induces a change in V[33]. *Flux control coefficients* are used to describe the effects an enzyme has on a pathway flux. As with all control coefficients, the partial deferential notation is used to denote that each coefficient is one in a set of coefficients, where the derivative takes place while all other parameters are fixed. For flux control coefficient in particular, the V is usually replaced with J_{ydh} indicating the Jth flux catalyzed by the *ydh* enzyme and P is replaced by E_{xase} to denote the concentration of the *xase* enzyme in the system block. Thus a control coefficient near unity indicates an enzyme with directly proportional control over the flux while less than unity suggests some level of control over the flux.

Alternative definitions have been proposed using similar nomenclature for the definition of the control coefficients with a very general one being introduced by Heinrich, Rapoport and Rapoport[88]. They gave the definition in terms of any parameter p that acts exclusively on an enzyme such that:

$$C_{xase}^{J_{ydh}} = \frac{v_{xase}}{J_{ydh}} \cdot \frac{\partial J_{ydh}/\partial p}{\partial v_{xase}/\partial p} \tag{11}$$

v_{xase} is the reaction velocity in terms of Michaelis-Menten kinetics. While this form looks dramatically different due to the introduction of some parameter p, it simplifies to equation (10) by selecting p as E_{xase} with the provision that the proportionality constant between the enzyme concentration and enzyme activity is unity under normal enzyme kinetics[88]. Importantly, the flux control coefficient of any enzyme is not a constant, but rather is a property of the system and can change with environmental conditions or due to alterations in the flux control coefficients of neighboring enzymes[143].

The naming convention used for the coefficients, while possibly misleading, reflects the fact that the action of genes, DNA manipulation, diet and hormones on metabolism can be mediated through the modulation of the enzyme's activity or concentration and thus making these effects significant when considering control over metabolism[70]. The naming convention however does not allow for the differentiation between a control enzyme and a regulatory enzyme[93]. Not withstanding this, by considering the coefficients as a means to identify 'rate-limiting' ability, one can imply the meaning of control as the enzyme's aptitude for regulation of flux capacity. Extending the concept of flux control to an enzyme's ability to effect metabolite pools, the coefficients are redefined as *concentration control coefficients* where the enzyme acts on metabolite pool levels, rather than on fluxes.

Expanding on the concept of control coefficients originally put forth, Stepanopoulos and Simpson defined a *group flux/concentration control coefficient*[200]. Group control coefficients are defined when one considers an entire reaction group as a single step in the reaction network. These group control coefficients are the sum of the flux control coefficients over all reactions in the grouped pathway[200].

In metabolic networks where the individual enzyme-substrate kinetics are known, Metabolic Control Analysis can derive the individual control coefficients that predict the response of the system to perturbations[68]. For poorly defined systems of unknown kinetics, control coefficients can be experimentally determined by varying the parameter of interest and measuring the flux changes that result[37]. For example, assume that enzyme E is known to exhibit some degree of control over the biosynthesis of the metabolic pathway's end product. By administering an inhibitor to tune enzyme activity in cell cultures and measuring the concurrent enzyme activity and change in metabolite production, the control coefficients can be found.

Figure 4 illustrates two possible situations for enzyme control where enzyme E has a high (Δ) and a low (□) control coefficient. An enzyme is considered to have a high control coefficient when the changes in its activity are reflected in the flux of interest. In the hypothetical case of sole control, the coefficient would be unity. When a change in activity has a small or negligible effect in the flux of interest, the enzyme is said to exhibit low control and the control coefficient has a value near zero.

Metabolic Production as Controlled by Enzyme Activity

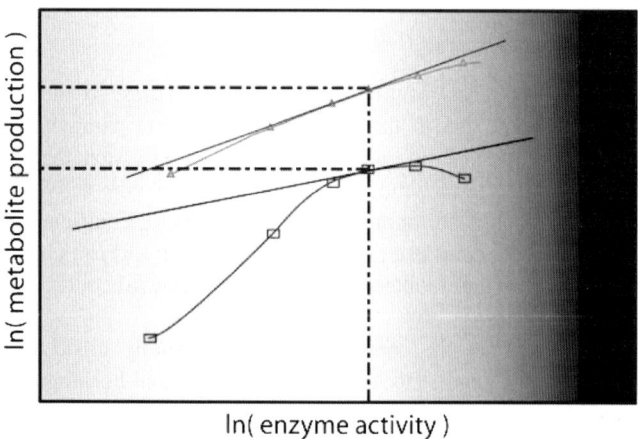

Figure 4. The flux control coefficient for an enzyme E on the metabolite production in two different cases. △ Enzyme has a control coefficient of 0.80 on metabolite production; □ Enzyme has control coefficient of 0.50 on metabolite production

In the same mathematical framework as global coefficients, local coefficients exist to describe the measurements from isolated system pathways. They are used to describe the functional entities of the system, not the major metabolic pathways themselves, and are used to identify a property of an individual enzyme. The primary local coefficient used is the *elasticity coefficient*, or just *elasticity*, and is representative of the effect of a metabolite on the velocity of the enzyme. Using an analogous form to equation (10) we find the form:

$$\varepsilon_S^{xase} = \frac{\partial v_{xase}}{\partial S} \cdot \frac{S}{v_{xase}} = \frac{\partial \ln |v_{xase}|}{\partial \ln S} \tag{12}$$

where v_{xase} stands for the rate of any functional entity isolated from the system, such as an enzyme or a permease, and S represents any molecular species (substrates, products, inhibitors, etc.) which affects the function of such entity directly. When more complex functional entities are used, the term *overall elasticity coefficient* is applied[33]. Elasticities can have either a positive value, for those metabolites stimulating the enzyme activity, or a negative value, for those that slow the reaction. In Metabolic Control Analysis, elasticity is used to characterize the vague concepts of enzyme responsiveness to metabolite changes. In other words, in the statement 'the rate of enzyme is relatively fast when there are variations in concentration of metabolite S' elasticity introduces a quantitative value to the word 'relatively'.

To this point we have considered the effects of parameters internal to the system, but what happens when a metabolite that is external to the system has an effect on a flux internal to the system? The term *response coefficient* has been established to define for the effect of metabolite X on the flux J_{ydh} where X acts as a regulator of the activity of enzyme *xase*. Kascer and Burns[111] identified that the response coefficients are nothing more than the product of the flux control coefficient and the elasticity, thus introducing the definition:

$$R_X^{J_{ydh}} = C_{xase}^{J_{ydh}} \cdot \varepsilon_X^{xase} \tag{13}$$

If X acts on more than one enzyme, the total response is simply the sum of all the responses for each enzyme that X acts on[93]. It is clear from the above equation that any external effectors can have no effect on a metabolic pathway unless the enzyme being acted on by the metabolite has a non-zero flux control coefficient[67]. In order to understand the derivation of these coefficients and their interrelationships in more detail, we need to present a few theorems of Metabolic Control Analysis, from where we will be able to move on to the experimental methodologies.

6.3 Control Theorems

There are a number of constraints and interrelationships, termed theorems of Metabolic Control Analysis, which control coefficients and elasticities of a metabolic system are subjected to. While we will not go into the derivation of these theorems, a majority of them use the assumption of a slowly disappearing, small perturbation of a pathway to outline the correlation between the mathematical and physical worlds. The two major theorems are the summation theorem and the connectivity theorem.

The *summation theorem* for Metabolic Control Analysis states that the sum of all the control coefficients for any one chosen flux J for all the n enzymes in a metabolic system effecting J is one[111]. In mathematical formalism, the theorem says:

$$\sum_{i=1}^{n} C_i^J = 1 \tag{14}$$

While equation (14) demonstrates how the theorem is applied to flux control coefficients, we will later demonstrate its extension for the concentration coefficients. This theorem is linked to the idea that all enzymes that belong to a pathway can share control of flux through the pathway. Can these control coefficients have a negative sign associated with them? In particular, for a linear pathway in which the enzymes follow standard kinetics, the coefficients are either zero or positive. Since the maximum affect any enzyme could have on a flux is a pure proportional increase, the coefficient would have a value of one for that enzyme and the rest would be zero. In this case, the enzyme with a flux control coefficient of one is called the 'rate-limiting' enzyme. In the context of a more general case, the concept of shared pathway control is possible. As such, the possibility exists for control coefficients with unknown kinetics to hold control coefficient values less than zero or greater than one. Even with standard kinetics, there exist the possibility that negative flux control coefficients will arise in pathways that include branches and cycles. It is also possible to have values greater than one for such cases[110].

The summation theorem is not the most meaningful theorem, for it only applies to specialized pathways, unlike the *connectivity theorem*. Because of this, the connectivity theorem is typically regarded as the most useful theorem in Metabolic Control Analysis. It states that if n enzymes respond to the concentration of metabolite S, their flux control coefficients and elasticities are related as follows:

$$\sum_{i=1}^{n} C_i^J \varepsilon_S^i = 0 \tag{15}$$

The theorem provides a means to understanding how the kinetics of the enzymes (as expressed by the elasticities) affect the value of the flux control coefficients[110].

Due to the large number of possible terms in the connectivity theorem, the relationships between two enzymes and the fluxes they control is not always straightforward[67]. The existence of metabolic cycles, such as the tricarboxylic acid cycle, requires that the connectivity theorem be modified for each of the metabolites contained in the conserved group of reactions[69,172]. In such cases the right hand side of equation (15) is no longer zero unless reactions are combined to yield a less complex system which also reduces the number of independent equations.

For linear pathways, the summation and connectivity theorems provide the exact number of equations needed to obtain the flux control coefficients of all the enzymes; however, this is not applicable to branched pathways or pathways containing cycles. Fell and Sauro[69] have proposed additional theorems for relating the flux control coefficients and relative fluxes through different parts of the system.

When applying the summation theorem to the concentration control coefficients, the form is the same as expressed in equation (14) except that flux J is replaced by a variable metabolite S. For the connectivity theorem, there are two possible forms. The first is used when the subject of the control coefficient is the same as the elasticity (equation (16)). The second form is used when the subject for both control coefficient and elasticity are the same (equation (17)).

$$\sum_{i=1}^{n} C_i^A \varepsilon_B^i = 0 \tag{16}$$

$$\sum_{i=1}^{n} C_i^A \varepsilon_A^i = -1 \tag{17}$$

There is generally less attention paid to the concentration control coefficients than flux control coefficients, though in the original derivation[90] it was the concentration forms of control coefficients that led to the formation of the flux control coefficients. We will not present the derivation here but it was made possible through the *elements of the control matrix*. The relationships are shown below in equations (18) and (19).

$$C_i^{J_j} = 1 + \sum_{j=1}^{m} C_{S_j}^i \varepsilon_i^{S_j} \tag{18}$$

$$C_i^{J_k} = \sum_{j=1}^{m} C_{S_j}^k \varepsilon_i^{S_j} \tag{19}$$

As mentioned before, the lumping of reactions based on their time scales helps to determine the control coefficients through experimentation and/or modeling[35,53]. However, without accurate methods to elucidate the time scales of intracellular reactions under *in vivo* fermentation conditions, the results are highly subjective and can only divulge information about network effects due to variable changes, such as enzyme activity and concentration[35]. This is a significant drawback for Metabolic Control Analysis as metabolic engineering often attempts to change the native pathways of an organism (unicellular or multi-cellular), through the insertion of nonnative gene(s) or the deletion of native gene(s). The deletion of gene(s) is not as critical an issue with Metabolic Control Analysis, as the definitions for coefficients are set to reflect this change. In the case of a newly introduced pathway to the metabolic system, there are additional coefficients generated that have not been taken into account in the original model, and must be considered for an accurate analysis.

Metabolic Control Analysis has been applied to a number of different biological systems of interest, including light stimulated lipid biosynthesis[85], glycolysis of potato tube tissue[208,209] and transgenic plants[143], the control of photosynthesis by Rubisco in tobacco plants[155,162,201], the control of lactate dehydrogenase on the formation of formate in *Lactococcus lactis*[10] and has been proposed as a framework for drug discovery and design[37].

7. HEALTH APPLICATIONS OF METABOLIC ENGINEERING

At the heart of metabolic engineering lie the primary goals of (a) understanding and modeling cellular metabolism in a quantitative manner and (b) generating and/or improving metabolic functioning using genetic modification through molecular biology techniques. The major modeling efforts have been previously outlined as well as the experimental techniques used to implement the genetic changes. Although less widely appreciated, metabolic engineering techniques can also be applied to study physiological systems and isolated whole organs *in vivo* to elucidate the metabolic patterns that occur in different physiological states, such as fed, fasted, and in disease[244]. One of the major motivations for metabolic engineering has been the ability to apply it in drug discovery and disease control.

7.1 Metabolism in Organs

One organ easily adaptable to metabolic engineering methods, in particular isotopomer methods, is the heart. The earliest example resulting in significant findings was using ^{13}C-NMR to study the tricarboxylic acid cycle in a perfused heart[137]. In the work by Malloy and colleagues, the isotopomer distribution of glutamate in tissue extracts was used to determine the steady state fluxes in the TCA cycle. Ischemia-reperfusion injury causes an increase in the use of anaplerotic reactions resulting in the deamination of amino acids, specifically aspartate and glutamate, and build up of TCA intermediates; however, exogenously added aspartate and glutamate are not significantly metabolized[137]. This is an interesting finding since it was reported that aspartate and glutamate have a protective effect on myocardial ischemia[244]. Understanding coronary metabolism in altered physiological states has become the subject of more recent extensive analysis. A reduction of blood flow leads to the build up of glycolytic products, especially lactate, and contractile function decrease that in time can lead to cell death[243]. By analyzing the causes of reduced blood flow on a metabolic level, researchers are hoping to provide novel drug targets and/or therapeutic methods.

Thompson and colleagues also looked at the utilization of different substrates within a perfused heart using 1, 2-^{13}C-acetate and 3-^{13}C-lactate thus allowing for a simple analysis of the resulting labeled glutamate[190]. Since the method requires no isotopic and/or metabolic steady states to exist it is highly advantageous for studying highly evolving metabolic systems, although the anaplerotic fluxes can not be predicted. The ischemia induced switch from lactate to acetate utilization was shown to be prevented, at least to a large degree, by lipoamide[204]. More metabolic studies using a wider variety of substrates show that the appreciation of metabolic engineering has the potential to treat angina and myocardial infarction based on drugs that lead to utilization of proper metabolites in the heart for metabolic energy.

While the liver contains a more complex network of metabolism than the heart, it has been one of the more widely studied organs in the body. The major cell type in the liver is hepatocytes and are accountable for a large portion of the liver's metabolic undertaking. One set of recent studies have utilized Metabolic Control Analysis to understand the energy metabolism in primary rat hepatocytes[2,3]. The analysis contained several reaction blocks which were acted on by multiple inhibitors to determine the metabolic control of ATP. It was found that control is shared by reactions consuming ATP and reactions producing ATP, specifically glycolysis and mitochondrial phosphorylation respectively[3]. Hepatocytes were exposed to the hormones adrenaline and glucagon in a second part of the study that found glucagon to be a more homeostatic hormone since intermediate pool concentrations were less affected by glucagon than by adrenaline and glycolysis was affected by glucagon but not adrenaline[2].

7.2   Understanding the Effects of Disease

Type 1, insulin-dependent diabetes is an autoimmune disorder effecting millions of people world wide and is caused by the destruction of insulin producing β cells of the pancreas. While the exact cause of the disorder is still unknown, insulin treatments and careful blood monitoring is a life saving treatment method. Today, researchers are looking to apply a variety of techniques, including protecting existing β cells, regenerating new β cells and even replacement with functional β cells as treatment methods of type 1 diabetes. To better understand type 1 diabetes, metabolic engineering has been employed to quantify the fluxes through the pathways responsible for β cell glucose-stimulated insulin secretion (GSIS) and to engineering GSIS into non-β cells[243].

Due to the multitude of steps in the glycolytic pathway before pyruvate is processed in the tricarboxylic acid cycle, the rate-limiting step for GSIS has been much debated. Using Metabolic Control Analysis to investigate an insulinoma β cell line transfected with an inducible glucokinase gene, the flux control coefficients for the pathway were found. The analysis found a flux control coefficient for glucokinase near unity, indicating the phosphorylation of glucose is indeed the rate limiting step for GSIS[18,223]. Furthermore, the work by Lu and coworkers using ^{13}C-NMR has shown the cycling of pyruvate through the anaplerotic pyruvate carboxylase reaction is proportional to the rate of insulin release[135]. Additionally, the stimulation of the malic enzyme flux leading to pyruvate in the TCA cycle resulted in increased insulin secretion. The malic enzyme produces NADPH, which is also thought to be a signaling molecule within the insulin secretion network[144].

Glycogen phosphorylase is considered a prime target of inhibition for controlling hyperglycemia in type 2 diabetes patients. This has led researchers to use

Metabolic Control Analysis for investigating the control of glycogen phosphorylase on glycogen synthesis. Using cultured rat hepatocytes and transduction of recombinant glycogen phosphorylase, a high degree of control over glycogen synthesis in the liver was induced, therefore identifying a strong candidate for therapeutic intervention for the control of type 2 diabetes[4].

Another type of disorder to which Metabolic Flux Analysis was applied is hereditary fructose intolerance (HFI), an inborn deficiency in the ability of aldolase B to metabolize fructose-1-phosphate, and is caused by the continuous exposure to parental fructose during infancy causing liver cirrhosis, mental retardation and even death[124]. Gopher et al[82] proposed the use of ^{13}C-MFA as a method of noninvasive in vivo diagnosis of HFI using isotopomer mass balances. Traditionally, a final diagnosis for HFI was found through liver biopsy specimens checking for aldolase B deficiency. Plasma isotopomer distributions were analyzed for both control and HFI populations after introduction of labeled fructose. Results showed that for HFI children the conversion of fructose to glucose was significantly lower as compared to control children and thus supporting the validity of doing such a test. Additional evidence also suggested that the direct conversion is mainly supported by phosphofructokinase and not the accepted pathway which only makes up for approximately half of the total conversion[82].

8.   REFERENCES

1. Abe, I., Y. Sano, Y. Takahashi, and H. Noguchi. 2003. Site-directed mutagenesis of benzalacetone synthase. The role of the Phe215 in plant type III polyketide synthases. J Biol Chem **278**:25218–26.
2. Ainscow, E. K., and M. D. Brand. 1999. The responses of rat hepatocytes to glucagon and adrenaline – Application of quantified elasticity analysis. European Journal of Biochemistry **265**:1043–1055.
3. Ainscow, E. K., and M. D. Brand. 1999. Top-down control analysis of ATP turnover, glycolysis and oxidative phosphorylation in rat hepatocytes. European Journal of Biochemistry **263**:671–685.
4. Aiston, S., L. Hampson, A. M. Gomez-Foix, J. J. Guinovart, and L. Agius. 2001. Hepatic glycogen synthesis is highly sensitive to phosphorylase activity – Evidence from metabolic control analysis. Journal of Biological Chemistry **276**:23858–23866.
5. Alexeyev, M. F., and I. N. Shokolenko. 1995. Mini-Tn10 transposon derivatives for insertion mutagenesis and gene delivery into the chromosome of gram-negative bacteria. Gene **160**:59–62.
6. Allister, E. M., N. M. Borradaile, J. Y. Edwards, and M. W. Huff. 2005. Inhibition of microsomal triglyceride transfer protein expression and apolipoprotein B100 secretion by the citrus flavonoid naringenin and by insulin involves activation of the mitogen-activated protein kinase pathway in hepatocytes. Diabetes **54**:1676–1683.
7. Alper, H., C. Fischer, E. Nevoigt, and G. Stephanopoulos. 2005. Tuning genetic control through promoter engineering. Proc Natl Acad Sci U S A **102**:12678–83.
8. Alper, H., Y. S. Jin, J. F. Moxley, and G. Stephanopoulos. 2005. Identifying gene targets for the metabolic engineering of lycopene biosynthesis in Escherichia coli. Metab Eng **7**:155–64.
9. Alper, H., K. Miyaoku, and G. Stephanopoulos. 2005. Construction of lycopene-overproducing E. coli strains by combining systematic and combinatorial gene knockout targets. Nat Biotechnol **23**:612–6.
10. Andersen, H. W., M. B. Pedersen, K. Hammer, and P. R. Jensen. 2001. Lactate dehydrogenase has no control on lactate production but has a strong negative control on formate production in Lactococcus lactis. European Journal Of Biochemistry **268**:6379–6389.

11. Austin, M. B., and A. J. P. Noel. 2003. The chalcone synthase superfamily of type III polyketide synthases. Natural Product Reports **20**:79–110.

12. Basu, S., Y. Gerchman, C. H. Collins, F. H. Arnold, and R. Weiss. 2005. A synthetic multicellular system for programmed pattern formation. Nature **434**:1130–4.

13. Bayer, T. S., L. N. Booth, S. M. Knudsen, and A. D. Ellington. 2005. Arginine-rich motifs present multiple interfaces for specific binding by RNA. Rna **11**:1848–57.

14. Becker, J., C. Klopprogge, O. Zelder, E. Heinzle, and C. Wittmann. 2005. Amplified expression of fructose 1,6-bisphosphatase in Corynebacterium glutamicum increases in vivo flux through the pentose phosphate pathway and lysine production on different carbon sources. Appl Environ Microbiol **71**:8587–96.

15. Becker, S. A., and B. O. Palsson. 2005. Genome-scale reconstruction of the metabolic network in Staphylococcus aureus N315: an initial draft to the two-dimensional annotation. BMC Microbiol **5**:8.

16. Becskei, A., B. Seraphin, and L. Serrano. 2001. Positive feedback in eukaryotic gene networks: cell differentiation by graded to binary response conversion. Embo J **20**:2528–35.

17. Becskei, A., and L. Serrano. 2000. Engineering stability in gene networks by autoregulation. Nature **405**:590–3.

18. Berman, H. K., and C. B. Newgard. 1998. Fundamental metabolic differences between hepatocytes and islet beta-cells revealed by glucokinase overexpression. Biochemistry **37**:4543–52.

19. Bi, W., and P. J. Stambrook. 1998. Site-directed mutagenesis by combined chain reaction. Anal Biochem **256**:137–40.

20. Bogarad, L. D., and M. W. Deem. 1999. A hierarchical approach to protein molecular evolution. Proc Natl Acad Sci U S A **96**:2591–5.

21. Bonarius, H. P., B. Timmerarends, C. D. de Gooijer, and J. Tramper. 1998. Metabolite-balancing techniques vs. 13C tracer experiments to determine metabolic fluxes in hybridoma cells. Biotechnol Bioeng **58**:258–62.

22. Bonarius, H. P. J., V. Hatzimanikatis, K. P. H. Meesters, C. D. deGooijer, G. Schmid, and J. Tramper. 1996. Metabolic flux analysis of hybridoma cells in different culture media using mass balances. Biotechnology and Bioengineering **50**:299–318.

23. Bonarius, H. P. J., G. Schmid, and J. Tramper. 1997. Flux analysis of underdetermined metabolic networks: The quest for the missing constraints. Trends in Biotechnology **15**:308–314.

24. Boonstra, B., D. A. Rathbone, C. E. French, E. H. Walker, and N. C. Bruce. 2000. Cofactor regeneration by a soluble pyridine nucleotide transhydrogenase for biological production of hydromorphone. Appl Environ Microbiol **66**:5161–6.

25. Borodina, I., and J. Nielsen. 2005. From genomes to in silico cells via metabolic networks. Current Opinion in Biotechnology **16**:355–350.

26. Brand, M. D. 1996. Top down metabolic control analysis. Journal Of Theoretical Biology **182**:351–360.

27. Brown, G. C., R. P. Hafner, and M. D. Brand. 1990. A 'top-down' approach to the determination of control coefficients in metabolic control theory. Eur J Biochem **188**:321–5.

28. Brown, J., J. O'Prey, and P. R. Harrison. 2003. Enhanced sensitivity of human oral tumours to the flavonol, morin, during cancer progression: involvement of the Akt and stress kinase pathways. Carcinogenesis **24**:171–177.

29. Burgard, A. P., and C. D. Maranas. 2003. Optimization-based framework for inferring and testing hypothesized metabolic objective functions. Biotechnol Bioeng **82**:670–7.

30. Burgard, A. P., and C. D. Maranas. 2001. Probing the performance limits of the Escherichia coli metabolic network subject to gene additions or deletions. Biotechnology and Bioengineering **74**:364–375.

31. Burgard, A. P., E. V. Nikolaev, C. H. Schilling, and C. D. Maranas. 2004. Flux coupling analysis of genome-scale metabolic network reconstructions. Genome Res **14**:301–12.

32. Burgard, A. P., P. Pharkya, and C. D. Maranas. 2003. Optknock: a bilevel programming framework for identifying gene knockout strategies for microbial strain optimization. Biotechnol Bioeng **84**:647–57.

33. Burns, J. A., A. Cornishbowden, A. K. Groen, R. Heinrich, H. Kacser, J. W. Porteous, S. M. Rapoport, T. A. Rapoport, J. W. Stucki, J. M. Tager, R. J. A. Wanders, and H. V. Westerhoff. 1985. Control Analysis Of Metabolic Systems. Trends In Biochemical Sciences **10**:16–16.

34. Caltagirone, S., C. Rossi, A. Poggi, F. O. Ranelletti, P. G. Natali, M. Brunetti, F. B. Aiello, and M. Piantelli. 2000. Flavonoids apigenin and quercetin inhibit melanoma growth and metastatic potential. International Journal of Cancer **87**:595–600.

35. Cameron, D. C., and F. W. Chaplen. 1997. Developments in metabolic engineering. Curr Opin Biotechnol **8**:175–80.

36. Carrier, T., K. L. Jones, and J. D. Keasling. 1998. mRNA stability and plasmid copy number effects on gene expression from an inducible promoter system. Biotechnol Bioeng **59**:666–72.

37. Cascante, M., L. G. Boros, B. Comin-Anduix, P. de Atauri, J. J. Centelles, and P. W. Lee. 2002. Metabolic control analysis in drug discovery and disease. Nat Biotechnol **20**:243–9.

38. Cascante, M., E. I. Canela, and R. Franco. 1990. Control analysis of systems having two steps catalyzed by the same protein molecule in unbranched chains. Eur J Biochem **192**:369–71.

39. Causey, T. B., K. T. Shanmugam, L. P. Yomano, and L. O. Ingram. 2004. Engineering Escherichia coli for efficient conversion of glucose to pyruvate. Proc Natl Acad Sci U S A **101**:2235–40.

40. Causey, T. B., S. Zhou, K. T. Shanmugam, and L. O. Ingram. 2003. Engineering the metabolism of Escherichia coli W3110 for the conversion of sugar to redox-neutral and oxidized products: homoacetate production. Proc Natl Acad Sci U S A **100**:825–32.

41. Chance, E. M., S. H. Seeholzer, K. Kobayashi, and J. R. Williamson. 1983. Mathematical-Analysis of Isotope Labeling in the Citric-Acid Cycle with Applications to C-13 Nmr-Studies in Perfused Rat Hearts. Journal of Biological Chemistry **258**:3785–3794.

42. Cheryan, M., S. Parekh, M. Shah, and K. Witjitra. 1997. Production of acetic acid by Clostridium thermoaceticum. Adv Appl Microbiol **43**:1–33.

43. Christensen, B., and J. Nielsen. 1999. Isotopomer analysis using GC-MS. Metab Eng **1**:282–90.

44. Chvatal, V. 1983. Linear Programming, Sixteeth printing ed. W. H. Freeman and Company.

45. Cornwell, T., W. Cohick, and I. Raskin. 2004. Dietary phytoestrogens and health. Phytochemistry **65**:995–1016.

46. Covert, M. W., and B. O. Palsson. 2003. Constraints-based models: regulation of gene expression reduces the steady-state solution space. J Theor Biol **221**:309–25.

47. Covert, M. W., and B. O. Palsson. 2002. Transcriptional regulation in constraints-based metabolic models of Escherichia coli. J Biol Chem **277**:28058–64.

48. Cox, S. J., S. Shalel Levanon, G. N. Bennett, and K. Y. San. 2005. Genetically constrained metabolic flux analysis. Metab Eng **7**:445–56.

49. Crameri, A., S. A. Raillard, E. Bermudez, and W. P. Stemmer. 1998. DNA shuffling of a family of genes from diverse species accelerates directed evolution. Nature **391**:288–91.

50. Dauner, M., and U. Sauer. 2000. GC-MS analysis of amino acids rapidly provides rich information for isotopomer balancing. Biotechnol Prog **16**:642–9.

51. David, H., M. Akesson, and J. Nielsen. 2003. Reconstruction of the central carbon metabolism of Aspergillus niger. Eur J Biochem **270**:4243–53.

52. de Graaf, A. A., M. Mahle, M. Mollney, W. Wiechert, P. Stahmann, and H. Sahm. 2000. Determination of full C-13 isotopomer distributions for metabolic flux analysis using heteronuclear spin echo difference NMR spectroscopy. Journal of Biotechnology **77**:25–35.

53. Delgado, J., and J. C. Liao. 1995. Control of Metabolic Pathways by Time-Scale Separation. Biosystems **36**:55–70.

54. Desai, R. P., and E. T. Papoutsakis. 1999. Antisense RNA strategies for metabolic engineering of Clostridium acetobutylicum. Appl Environ Microbiol **65**:936–45.

55. Dixon, R. A. 2004. Phytoestrogens. Annu Rev Plant Biol **55**:225–61.

56. Dixon, R. A., and D. Ferreira. 2002. Genistein. Phytochemistry **60**:205–11.

57. Drewlo, S., C. O. Bramer, M. Madkour, F. Mayer, and A. Steinbuchel. 2001. Cloning and expression of a Ralstonia eutropha HF39 gene mediating indigo formation in Escherichia coli. Appl Environ Microbiol **67**:1964–9.

58. Duarte, N. C., M. J. Herrgard, and B. O. Palsson. 2004. Reconstruction and validation of Saccharomyces cerevisiae iND750, a fully compartmentalized genome-scale metabolic model. Genome Res **14**:1298–309.

59. Dwivedi, U. N., N. Shiraishi, and W. H. Campbell. 1994. Generation of multiple mutations in the same sequence via the polymerase chain reaction using a single selection primer. Anal Biochem **221**:425–8.

60. Edwards, J. S. 1999. Functional Genomics and the Computational Analysis of Bacterial Metabolism. Dissertation. University of California San Diego, San Diego.

61. Edwards, J. S., R. U. Ibarra, and B. O. Palsson. 2001. In silico predictions of Escherichia coli metabolic capabilities are consistent with experimental data. Nat Biotechnol **19**:125–30.

62. Edwards, J. S., and B. O. Palsson. 2000. The Escherichia coli MG1655 in silico metabolic genotype: its definition, characteristics, and capabilities. Proc Natl Acad Sci U S A **97**:5528–33.

63. Edwards, J. S., and B. O. Palsson. 2000. Robustness analysis of the Escherichia coli metabolic network. Biotechnol Prog **16**:927–39.

64. Edwards, J. S., R. Ramakrishna, and B. O. Palsson. 2002. Characterizing the metabolic phenotype: a phenotype phase plane analysis. Biotechnol Bioeng **77**:27–36.

65. Ellington, A. D., and J. W. Szostak. 1990. In vitro selection of RNA molecules that bind specific ligands. Nature **346**:818–22.

66. Ellington, A. D., and J. W. Szostak. 1992. Selection in vitro of single-stranded DNA molecules that fold into specific ligand-binding structures. Nature **355**:850–2.

67. Fell, D. A. 1992. Metabolic control analysis: a survey of its theoretical and experimental development. Biochem J **286 (Pt 2)**:313–30.

68. Fell, D. A. 1997. Understanding the Control of Metabolism, 1 ed. Portlan Press, London.

69. Fell, D. A., and H. M. Sauro. 1985. Metabolic control and its analysis. Additional relationships between elasticities and control coefficients. Eur J Biochem **148**:555–61.

70. Fell, D. A., and J. R. Small. 1986. Fat Synthesis In Adipose-Tissue – An Examination Of Stoichiometric Constraints. Biochemical Journal **238**:781–786.

71. Fong, S. S., A. P. Burgard, C. D. Herring, E. M. Knight, F. R. Blattner, C. D. Maranas, and B. O. Palsson. 2005. In silico design and adaptive evolution of Escherichia coli for production of lactic acid. Biotechnol Bioeng **91**:643–8.

72. Fong, S. S., A. R. Joyce, and B. O. Palsson. 2005. Parallel adaptive evolution cultures of Escherichia coli lead to convergent growth phenotypes with different gene expression states. Genome Res **15**:1365–72.

73. Fong, S. S., J. Y. Marciniak, and B. O. Palsson. 2003. Description and interpretation of adaptive evolution of Escherichia coli K-12 MG1655 by using a genome-scale in silico metabolic model. J Bacteriol **185**:6400–8.

74. Fong, S. S., A. Nanchen, B. O. Palsson, and U. Sauer. 2006. Latent pathway activation and increased pathway capacity enable Escherichia coli adaptation to loss of key metabolic enzymes. J Biol Chem **281**:8024–33.

75. Fong, S. S., and B. O. Palsson. 2004. Metabolic gene-deletion strains of Escherichia coli evolve to computationally predicted growth phenotypes. Nat Genet **36**:1056–8.

76. Forster, J., I. Famili, P. Fu, B. O. Palsson, and J. Nielsen. 2003. Genome-scale reconstruction of the *Saccharomyces cerevisiae* metabolic network. Genome Res **13**:244–253.

77. Fujio, T., and A. Maruyama. 1997. Enzymatic production of pyrimidine nucleotides using Corynebacterium ammoniagenes cells and recombinant Escherichia coli cells: enzymatic production of CDP-choline from orotic acid and choline chloride (Part I). Biosci Biotechnol Biochem **61**:956–9.

78. Fung, E., W. W. Wong, J. K. Suen, T. Bulter, S. G. Lee, and J. C. Liao. 2005. A synthetic gene-metabolic oscillator. Nature **435**:118–22.

79. Gadkar, K. G., F. J. Doyle Iii, J. S. Edwards, and R. Mahadevan. 2005. Estimating optimal profiles of genetic alterations using constraint-based models. Biotechnol Bioeng **89**:243–51.

80. Goeddel, D. V., D. G. Kleid, F. Bolivar, H. L. Heyneker, D. G. Yansura, R. Crea, T. Hirose, A. Kraszewski, K. Itakura, and A. D. Riggs. 1979. Expression in Escherichia-Coli of Chemically

Synthesized Genes for Human Insulin. Proceedings of the National Academy of Sciences of the United States of America **76**:106–110.

81. Goel, A., J. Ferrance, and M. M. Ataai. 1993. Analysis of metabolic fluxes in batch and continuous cultures of *Bacillus subtilis*. Biotechnol Bioeng **42**:686–696.

82. Gopher, A., N. Vaisman, H. Mandel, and A. Lapidot. 1990. Determination of fructose metabolic pathways in normal and fructose-intolerant children: a 13C NMR study using [U-13C]fructose. Proc Natl Acad Sci U S A **87**:5449–53.

83. Gottesman, S. 2005. Micros for microbes: non-coding regulatory RNAs in bacteria. Trends Genet **21**:399–404.

84. Handelsman, J., M. R. Rondon, S. F. Brady, J. Clardy, and R. M. Goodman. 1998. Molecular biological access to the chemistry of unknown soil microbes: a new frontier for natural products. Chem Biol **5**:R245–9.

85. Harwood, J. L., U. S. Ramli, R. A. Page, and P. A. Quant. 1999. Modelling lipid metabolism in plants: a slippery problem? Biochemical Society Transactions **27**:285–289.

86. Hasona, A., S. W. York, L. P. Yomano, L. O. Ingram, and K. T. Shanmugam. 2002. Decreasing the level of ethyl acetate in ethanolic fermentation broths of Escherichia coli KO11 by expression of Pseudomonas putida estZ esterase. Appl Environ Microbiol **68**:2651–9.

87. Hasty, J., D. McMillen, and J. J. Collins. 2002. Engineering Gene circuits. Nature **420**:224–230.

88. Heinrich, R., S. M. Rapoport, and T. A. Rapoport. 1977. Metabolic-Regulation and Mathematical-Models. Progress in Biophysics & Molecular Biology **32**:1–82.

89. Heinrich, R., and T. A. Rapoport. 1974. A linear steady-state treatment of enzymatic chains. Critique of the crossover theorem and a general procedure to identify interaction sites with an effector. Eur J Biochem **42**:97–105.

90. Heinrich, R., and T. A. Rapoport. 1974. A linear steady-state treatment of enzymatic chains. General properties, control and effector strength. Eur J Biochem **42**:89–95.

91. Henne, A., H. Bruggemann, C. Raasch, A. Wiezer, T. Hartsch, H. Liesegang, A. Johann, T. Lienard, O. Gohl, R. Martinez-Arias, C. Jacobi, V. Starkuviene, S. Schlenczeck, S. Dencker, R. Huber, H. P. Klenk, W. Kramer, R. Merkl, G. Gottschalk, and H. J. Fritz. 2004. The genome sequence of the extreme thermophile Thermus thermophilus. Nat Biotechnol **22**:547–53.

92. Higgins, J. 1963. Analysis of sequential reactions. Ann N Y Acad Sci **108**:305–21.

93. Hofmeyr, J. H., and A. Cornish-Bowden. 1991. Quantitative assesment of regulation in metabolic systems. Eur J Biochem **200**:223–236.

94. Hong, S. H., J. S. Kim, S. Y. Lee, Y. H. In, S. S. Choi, J. K. Rih, C. H. Kim, H. Jeong, C. G. Hur, and J. J. Kim. 2004. The genome sequence of the capnophilic rumen bacterium Mannheimia succiniciproducens. Nat Biotechnol **22**:1275–81.

95. Hopwood, D. A., F. Malpartida, H. M. Kieser, H. Ikeda, J. Duncan, I. Fujii, B. A. Rudd, H. G. Floss, and S. Omura. 1985. Production of 'hybrid' antibiotics by genetic engineering. Nature **314**:642–4.

96. Hou, D. X., M. Fujii, N. Terahara, and M. Yoshimoto. 2004. Molecular mechanisms behind the chemopreventive effects of anthocyanidins. Journal of Biomedicine and Biotechnology:321–325.

97. Huang, Q., C. A. Roessner, R. Croteau, and A. I. Scott. 2001. Engineering Escherichia coli for the synthesis of taxadiene, a key intermediate in the biosynthesis of taxol. Bioorg Med Chem **9**:2237–42.

98. Hwang, E. I., M. Kaneko, Y. Ohnishi, and S. Horinouchi. 2003. Production of plant-specific flavanones by Escherichia coli containing an artificial gene cluster. Applied And Environmental Microbiology **69**:2699–2706.

99. Ibarra, R. U., J. S. Edwards, and B. O. Palsson. 2002. Escherichia coli K-12 undergoes adaptive evolution to achieve in silico predicted optimal growth. Nature **420**:186–9.

100. Ibarra, R. U., P. Fu, B. O. Palsson, J. R. DiTonno, and J. S. Edwards. 2003. Quantitative analysis of Escherichia coli metabolic phenotypes within the context of phenotypic phase planes. J Mol Microbiol Biotechnol **6**:101–8.

101. Imielinski, M., C. Belta, A. Halasz, and H. Rubin. 2005. Investigating metabolite essentiality through genome-scale analysis of Escherichia coli production capabilities. Bioinformatics **21**:2008–16.

102. Isaacs, F. J., D. J. Dwyer, C. Ding, D. D. Pervouchine, C. R. Cantor, and J. J. Collins. 2004. Engineered riboregulators enable post-transcriptional control of gene expression. Nat Biotechnol 22:841–7.

103. Itaya, M., K. Tsuge, M. Koizumi, and K. Fujita. 2005. Combining two genomes in one cell: Stable cloning of the Synechocystis PCC6803 genome in the Bacillus subtilis 168 genome. Proceedings of the National Academy of Sciences of the United States of America 102:15971–15976.

104. Jacob, F., and J. Monod. 1961. Genetic regulatory mechanisms in the synthesis of proteins. J Mol Biol 3:318–56.

105. Jensen, P. R., and K. Hammer. 1998. Artificial promoters for metabolic optimization. Biotechnol Bioeng 58:191–5.

106. Jensen, P. R., and K. Hammer. 1998. The sequence of spacers between the consensus sequences modulates the strength of prokaryotic promoters. Appl Environ Microbiol 64:82–7.

107. Jones, K. L., S. W. Kim, and J. D. Keasling. 2000. Low-copy plasmids can perform as well as or better than high-copy plasmids for metabolic engineering of bacteria. Metab Eng 2:328–38.

108. Jorgensen, H., J. Nielsen, J. Villadsen, and H. Mollgaard. 1995. Metabolic flux distributions in *Penicillium chrysogenum* during fed-batch cultivations. Biotechnology and Bioengineering 46:117–131.

109. Jorgensen, L. V., and L. H. Skibsted. 1998. Flavonoid deactivation of ferrylmyoglobin in relation to ease of oxidation as determined by cyclic voltammetry. Free Radic Res 28:335–51.

110. Kacser, H. 1983. The control of enzyme systems in vivo: elasticity analysis of the steady state. Biochem Soc Trans 11:35–40.

111. Kacser, H., and J. A. Burns. 1973. The Control of Flux. Symp Soc Exp Biol 27:65–104.

112. Kacser, H., H. M. Sauro, and L. Acerenza. 1990. Enzyme-enzyme interactions and control analysis. 1. The case of non-additivity: monomer-oligomer associations. Eur J Biochem 187:481–91.

113. Kaern, M., W. J. Blake, and J. J. Collins. 2003. The engineering of gene regulatory networks. Annu Rev Biomed Eng 5:179–206.

114. Kahn, D., and H. V. Westerhoff. 1991. Control-Theory of Regulatory Cascades. Journal of Theoretical Biology 153:255–285.

115. Kanehisa, M. 1997. A database for post-genome analysis. Trends Genet 13:375–6.

116. Kaneko, M., E. I. Hwang, Y. Ohnishi, and S. Horinouchi. 2003. Heterologous production of flavanones in Escherichia coli: potential for combinatorial biosynthesis of flavonoids in bacteria. Journal of Industrial Microbiology & Biotechnology 30:456–461.

117. Kauffman, K. J., P. Prakash, and J. S. Edwards. 2003. Advances in flux balance analysis. Current Opinion in Biotechnology 14:491–496.

118. Kell, D. B. 2004. Metabolomics and systems biology: making sense of the soup. Curr Opin Microbiology 7:296–307.

119. Kiefer, P., E. Heinzle, O. Zelder, and C. Wittmann. 2004. Comparative metabolic flux analysis of lysine-producing Corynebacterium glutamicum cultured on glucose or fructose. Appl Environ Microbiol 70:229–39.

120. Kim, J. Y., and H. J. Cha. 2003. Down-regulation of acetate pathway through antisense strategy in Escherichia coli: improved foreign protein production. Biotechnol Bioeng 83:841–53.

121. Klapa, M. I., J. C. Aon, and G. Stephanopoulos. 2003. Systematic quantification of complex metabolic flux networks using stable isotopes and mass spectrometry. Eur J Biochem 270:3525–42.

122. Klapa, M. I., S. M. Park, A. J. Sinskey, and G. Stephanopoulos. 1999. Metabolite and isotopomer balancing in the analysis of metabolic cycles: I. Theory. Biotechnology and Bioengineering 62:375–391.

123. Koffas, M., and S. D. Cardayre. 2005. Evolutionary metabolic engineering. Metab Eng 7:1–3.

124. Koffas, M., C. Roberge, K. Lee, and G. Stephanopoulos. 1999. Metabolic engineering. Annu Rev Biomed Eng 1:535–57.

125. Koffas, M., and G. Stephanopoulos. 2005. Strain improvement by metabolic engineering: lysine production as a case study for systems biology. Curr Opin Biotechnol 16:361–6.

126. Lacatena, R. M., and G. Cesareni. 1981. Base pairing of RNA I with its complementary sequence in the primer precursor inhibits ColE1 replication. Nature 294:623–6.

127. Lacatena, R. M., and G. Cesareni. 1983. Interaction between RNA1 and the primer precursor in the regulation of ColE1 replication. J Mol Biol **170**:635–50.

128. Lee, S., C. Phalakornkule, M. M. Domach, and I. E. Grossmann. 2000. Recursive MILP model for finding all the alternate optima in LP models for metabolic networks. Computers & Chemical Engineering **24**:711–716.

129. Leonard, E., Y. Yan, and M. A. Koffas. 2006. Functional expression of a P450 flavonoid hydroxylase for the biosynthesis of plant-specific hydroxylated flavonols in Escherichia coli. Metab Eng **8**:172–81.

130. Leonard, E., Y. Yan, K. H. Lim, and M. A. Koffas. 2005. Investigation of two distinct flavone synthases for plant-specific flavone biosynthesis in Saccharomyces cerevisiae. Appl Environ Microbiol **71**:8241–8.

131. Levskaya, A., A. A. Chevalier, J. J. Tabor, Z. B. Simpson, L. A. Lavery, M. Levy, E. A. Davidson, A. Scouras, A. D. Ellington, E. M. Marcotte, and C. A. Voigt. 2005. Synthetic biology: engineering Escherichia coli to see light. Nature **438**:441–2.

132. Liao, J. C., and J. Delgado. 1993. Advances in Metabolic Control Analysis. Biotechnology Progress **9**:221–233.

133. Liao, J. C., S. Y. Hou, and Y. P. Chao. 1996. Pathway analysis, engineering, and physiological considerations for redirecting central metabolism. Biotechnology And Bioengineering **52**:129–140.

134. Lin, X., M. Hezari, A. E. Koepp, H. G. Floss, and R. Croteau. 1996. Mechanism of taxadiene synthase, a diterpene cyclase that catalyzes the first step of taxol biosynthesis in Pacific yew. Biochemistry **35**:2968–77.

135. Lu, D., H. Mulder, P. Zhao, S. C. Burgess, M. V. Jensen, S. Kamzolova, C. B. Newgard, and A. D. Sherry. 2002. 13C NMR isotopomer analysis reveals a connection between pyruvate cycling and glucose-stimulated insulin secretion (GSIS). Proc Natl Acad Sci U S A **99**:2708–13.

136. Mahadevan, R., and C. H. Schilling. 2003. The effects of alternate optimal solutions in constraint-based genome-scale metabolic models. Metab Eng **5**:264–76.

137. Malloy, C. R., A. D. Sherry, and F. M. Jeffrey. 1990. Analysis of tricarboxylic acid cycle of the heart using 13C isotope isomers. Am J Physiol **259**:H987–95.

138. Malloy, C. R., A. D. Sherry, and F. M. H. Jeffrey. 1988. Evaluation of Carbon Flux and Substrate Selection through Alternate Pathways Involving the Citric-Acid Cycle of the Heart by C-13 Nmr-Spectroscopy. Journal of Biological Chemistry **263**:6964–6971.

139. Martin, V. J., D. J. Pitera, S. T. Withers, J. D. Newman, and J. D. Keasling. 2003. Engineering a mevalonate pathway in *Escherichia coli* for production of terpenoids. Nat Biotechnol **21**:796–802.

140. Marx, A., A. A. deGraaf, W. Wiechert, L. Eggeling, and H. Sahm. 1996. Determination of the fluxes in the central metabolism of Corynebacterium glutamicum by nuclear magnetic resonance spectroscopy combined with metabolite balancing. Biotechnology And Bioengineering **49**:111–129.

141. Meetei, A. R., and M. R. Rao. 1998. Generation of multiple site-specific mutations in a single polymerase chain reaction product. Anal Biochem **264**:288–91.

142. Miyahisa, I., M. Kaneko, N. Funa, H. Kawasaki, H. Kojima, Y. Ohnishi, and S. Horinouchi. 2005. Efficient production of (2S)-flavanones by Escherichia coli containing an artificial biosynthetic gene cluster. Appl Microbiol Biotechnol.

143. Morgan, J. A., and D. Rhodes. 2002. Mathematical modeling of plant metabolic pathways. Metab Eng **4**:80–9.

144. Newgard, C. B., D. Lu, M. V. Jensen, J. Schissler, A. Boucher, S. Burgess, and A. D. Sherry. 2002. Stimulus/secretion coupling factors in glucose-stimulated insulin secretion: insights gained from a multidisciplinary approach. Diabetes **51 Suppl 3**:S389–93.

145. Nissen, T. L., M. C. Kielland-Brandt, J. Nielsen, and J. Villadsen. 2000. Optimization of ethanol production in Saccharomyces cerevisiae by metabolic engineering of the ammonium assimilation. Metab Eng **2**:69–77.

146. Nissen, T. L., U. Schulze, J. Nielsen, and J. Villadsen. 1997. Flux distributions in anaerobic, glucose-limited continuous cultures of Saccharomyces cerevisiae. Microbiology **143 (Pt 1)**:203–18.

147. Nwankwo, D. O., L. S. Moran, B. E. Slatko, P. A. Waite-Rees, L. F. Dorner, J. S. Benner, and G. G. Wilson. 1994. Cloning, analysis and expression of the HindIII R-M-encoding genes. Gene **150**:75–80.

148. Ostermeier, M., and S. J. Benkovic. 2000. Evolution of protein function by domain swapping. Adv Protein Chem **55**:29–77.
149. Ostermeier, M., A. E. Nixon, J. H. Shim, and S. J. Benkovic. 1999. Combinatorial protein engineering by incremental truncation. Proc Natl Acad Sci U S A **96**:3562–3567.
150. Ostermeier, M., J. H. Shim, and S. J. Benkovic. 1999. A combinatorial approach to hybrid enzymes independent of DNA homology. Nature Biotechnology **17**:1205–1209.
151. Papoutsakis, E. T. 1984. Equations and Calculations for Fermentations of Butyric-Acid Bacteria. Biotechnology and Bioengineering **26**:174–187.
152. Park, S. M. 1996. Investigation of Carbon Fluxes in Central Metabolic Pathways of Corynebacterium glutamicum. PhD. Massachusetts Institute of Technology, Cambridge.
153. Park, S. M., M. I. Klapa, A. J. Sinskey, and G. Stephanopoulos. 1999. Metabolite and isotopomer balancing in the analysis of metabolic cycles: II. Applications. Biotechnol Bioeng **62**:392–401.
154. Petersen, S., A. A. de Graaf, L. Eggeling, M. Mollney, W. Wiechert, and H. Sahm. 2000. In vivo quantification of parallel and bidirectional fluxes in the anaplerosis of Corynebacterium glutamicum. Journal of Biological Chemistry **275**:35932–35941.
155. Pettersson, G. 1997. Control properties of the Calvin photosynthesis cycle at physiological carbon dioxide concentrations. Biochimica Et Biophysica Acta-Bioenergetics **1322**:173–182.
156. Pfeifer, B. A., S. J. Admiraal, H. Gramajo, D. E. Cane, and C. Khosla. 2001. Biosynthesis of complex polyketides in a metabolically engineered strain of E-coli. Science **291**:1790–1792.
157. Pharkya, P., A. P. Burgard, and C. D. Maranas. 2003. Exploring the overproduction of amino acids using the bilevel optimization framework OptKnock. Biotechnol Bioeng **84**:887–99.
158. Pharkya, P., and C. D. Maranas. 2006. An optimization framework for identifying reaction activation/inhibition or elimination candidates for overproduction in microbial systems. Metab Eng **8**:1–13.
159. Popiolkiewicz, J., K. Polkowski, J. S. Skierski, and A. P. Mazurek. 2005. In vitro toxicity evaluation in the development of new anticancer drugs – genistein glycosides. Cancer Letters **229**:67–75.
160. Potter, S. M., J. A. Baum, H. Y. Teng, R. J. Stillman, N. F. Shay, and J. W. Erdman. 1998. Soy protein and isoflavones: their effects on blood lipids and bone density in postmenopausal women. American Journal of Clinical Nutrition **68**:1375S–1379S.
161. Pouget, C., F. Lauthier, A. Simon, C. Fagnere, J. P. Basly, C. Delage, and A. J. Chulia. 2001. Flavonoids: Structural requirements for antiproliferative activity on breast cancer cells. Bioorganic & Medicinal Chemistry Letters **11**:3095–3097.
162. Quick, W. P., U. Schurr, R. Scheibe, E. D. Schulze, S. R. Rodermel, L. Bogorad, and M. Stitt. 1991. Decreased Ribulose-1,5-Bisphosphate Carboxylase-Oxygenase In Transgenic Tobacco Transformed With Antisense Rbcs .1. Impact On Photosynthesis In Ambient Growth-Conditions. Planta **183**:542–554.
163. Ranganathan, A., M. Timoney, M. Bycroft, J. Cortes, I. P. Thomas, B. Wilkinson, L. Kellenberger, U. Hanefeld, I. S. Galloway, J. Staunton, and P. F. Leadlay. 1999. Knowledge-based design of bimodular and trimodular polyketide synthases based on domain and module swaps: a route to simple statin analogues. Chem Biol **6**:731–41.
164. Rathod, P. K., T. McErlean, and P. C. Lee. 1997. Variations in frequencies of drug resistance in Plasmodium falciparum. Proc Natl Acad Sci U S A **94**:9389–93.
165. Reed, J. L., T. D. Vo, C. H. Schilling, and B. O. Palsson. 2003. An expanded genome-scale model of Escherichia coli K-12 (iJR904 GSM/GPR). Genome Biol **4**:R54.
166. Riechmann, L., and G. Winter. 2000. Novel folded protein domains generated by combinatorial shuffling of polypeptide segments. Proc Natl Acad Sci U S A **97**:10068–73.
167. Riedel, C., D. Rittmann, P. Dangel, B. Mockel, S. Petersen, H. Sahm, and B. J. Eikmanns. 2001. Characterization of the phosphoenolpyruvate carboxykinase gene from Corynebacterium glutamicum and significance of the enzyme for growth and amino acid production. Journal of Molecular Microbiology and Biotechnology **3**:573–583.
168. Riesenfeld, C. S., R. M. Goodman, and J. Handelsman. 2004. Uncultured soil bacteria are a reservoir of new antibiotic resistance genes. Environ Microbiol **6**:981–9.

169. Rittmann, D., S. Schaffer, V. F. Wendisch, and H. Sahm. 2003. Fructose-1,6-bisphosphatase from Corynebacterium glutamicum: expression and deletion of the fbp gene and biochemical characterization of the enzyme. Arch Microbiol **180**:285–92.

170. Rogers, S., and P. Pfuderer. 1968. Use of viruses as carriers of added genetic information. Nature **219**:749–51.

171. Rondon, M. R., P. R. August, A. D. Bettermann, S. F. Brady, T. H. Grossman, M. R. Liles, K. A. Loiacono, B. A. Lynch, I. A. MacNeil, C. Minor, C. L. Tiong, M. Gilman, M. S. Osburne, J. Clardy, J. Handelsman, and R. M. Goodman. 2000. Cloning the soil metagenome: a strategy for accessing the genetic and functional diversity of uncultured microorganisms. Appl Environ Microbiol **66**:2541–7.

172. Sauro, H. M., and H. Kacser. 1990. Enzyme-enzyme interactions and control analysis. 2. The case of non-independence: heterologous associations. Eur J Biochem **187**:493–500.

173. Savageau, M. A. 1970. Biochemical systems analysis. 3. Dynamic solutions using a power-law approximation. J Theor Biol **26**:215–26.

174. Savageau, M. A. 1969. Biochemical systems analysis. I. Some mathematical properties of the rate law for the component enzymatic reactions. J Theor Biol **25**:365–9.

175. Savageau, M. A. 1969. Biochemical systems analysis. II. The steady-state solutions for an n-pool system using a power-law approximation. J Theor Biol **25**:370–9.

176. Savageau, M. A. 1985. A theory of alternative designs for biochemical control systems. Biomed Biochim Acta **44**:875–80.

177. Savinell, J. M., and B. O. Palsson. 1992. Network analysis of intermediary metabolism using linear optimization. I. Development of mathematical formalism. J Theor Biol **154**:421–54.

178. Savinell, J. M., and B. O. Palsson. 1992. Network analysis of intermediary metabolism using linear optimization. II. Interpretation of hybridoma cell metabolism. J Theor Biol **154**:455–73.

179. Savinell, J. M., and B. O. Palsson. 1992. Optimal Selection Of Metabolic Fluxes For Invivo Measurement .1. Development Of Mathematical-Methods. Journal Of Theoretical Biology **155**:201–214.

180. Savinell, J. M., and B. O. Palsson. 1992. Optimal Selection Of Metabolic Fluxes For Invivo Measurement .2. Application To Escherichia-Coli And Hybridoma Cell-Metabolism. Journal Of Theoretical Biology **155**:215–242.

181. Schilling, C. H., M. W. Covert, I. Famili, G. M. Church, J. S. Edwards, and B. O. Palsson. 2002. Genome-scale metabolic model of Helicobacter pylori 26695. J Bacteriol **184**:4582–93.

182. Schilling, C. H., J. S. Edwards, D. Letscher, and B. O. Palsson. 2000. Combining pathway analysis with flux balance analysis for the comprehensive study of metabolic systems. Biotechnol Bioeng **71**:286–306.

183. Schilling, C. H., D. Letscher, and B. O. Palsson. 2000. Theory for the systemic definition of metabolic pathways and their use in interpreting metabolic function from a pathway-oriented perspective. J Theor Biol **203**:229–48.

184. Schmidt-Dannert, C., D. Umeno, and F. H. Arnold. 2000. Molecular breeding of carotenoid biosynthetic pathways. Nat Biotechnol **18**:750–3.

185. Schmidt, K., J. Nielsen, and J. Villadsen. 1999. Quantitative analysis of metabolic fluxes in Escherichia coli, using two-dimensional NMR spectroscopy and complete isotopomer models. Journal of Biotechnology **71**:175–189.

186. Segre, D., D. Vitkup, and G. M. Church. 2002. Analysis of optimality in natural and perturbed metabolic networks. Proc Natl Acad Sci U S A **99**:15112–7.

187. Segre, D., J. Zucker, J. Katz, X. Lin, P. D'Haeseleer, W. P. Rindone, P. Kharchenko, D. H. Nguyen, M. A. Wright, and G. M. Church. 2003. From annotated genomes to metabolic flux models and kinetic parameter fitting. Omics **7**:301–16.

188. Selkirk, S. M. 2004. Gene therapy in clinical medicine. Postgrad Med J **80**:560–70.

189. Sheikh, K., J. Forster, and L. K. Nielsen. 2005. Modeling hybridoma cell metabolism using a generic genome-scale metabolic model of Mus musculus. Biotechnol Prog **21**:112–21.

190. Sherry, A. D., C. R. Malloy, P. Zhao, and J. R. Thompson. 1992. Alterations in substrate utilization in the reperfused myocardium: a direct analysis by 13C NMR. Biochemistry **31**:4833–7.

191. Shizuya, H., B. Birren, U. J. Kim, V. Mancino, T. Slepak, Y. Tachiiri, and M. Simon. 1992. Cloning and stable maintenance of 300-kilobase-pair fragments of human DNA in Escherichia coli using an F-factor-based vector. Proc Natl Acad Sci U S A 89:8794–7.

192. Shlomi, T., O. Berkman, and E. Ruppin. 2005. Regulatory on/off minimization of metabolic flux changes after genetic perturbations. Proc Natl Acad Sci U S A 102:7695–700.

193. Sieber, V., C. A. Martinez, and F. H. Arnold. 2001. Libraries of hybrid proteins from distantly related sequences. Nat Biotechnol 19:456–60.

194. Siegele, D. A., and J. C. Hu. 1997. Gene expression from plasmids containing the araBAD promoter at subsaturating inducer concentrations represents mixed populations. Proc Natl Acad Sci U S A 94:8168–72.

195. Smolke, C. D., and J. D. Keasling. 2002. Effect of copy number and mRNA processing and stabilization on transcript and protein levels from an engineered dual-gene operon. Biotechnol Bioeng 78:412–24.

196. Sonntag, K., L. Eggeling, A. A. De Graaf, and H. Sahm. 1993. Flux partitioning in the split pathway of lysine synthesis in Corynebacterium glutamicum. Quantification by 13C- and 1H-NMR spectroscopy. Eur J Biochem 213:1325–31.

197. Sprinzak, D., and M. B. Elowitz. 2005. Reconstruction of genetic circuits. Nature 438:443–8.

198. Stemmer, W. P. 1994. DNA shuffling by random fragmentation and reassembly: in vitro recombination for molecular evolution. Proc Natl Acad Sci U S A 91:10747–51.

199. Stephanopoulos, G., A. A. Aristidou, and J. Nielsen. 1998. Metabolic Engineering: Principles and Methodologies, 1 ed. Academic Press, San Diego.

200. Stephanopoulos, G., and T. W. Simpson. 1997. Flux amplification in complex metabolic networks. Chemical Engineering Science 52:2607–2627.

201. Stitt, M., W. P. Quick, U. Schurr, E. D. Schulze, S. R. Rodermel, and L. Bogorad. 1991. Decreased Ribulose-1,5-Bisphosphate Carboxylase-Oxygenase In Transgenic Tobacco Transformed With Antisense Rbcs .2. Flux-Control Coefficients For Photosynthesis In Varying Light, Co2, And Air Humidity. Planta 183:555–566.

202. Strohl, W. R. 2001. Biochemical engineering of natural product biosynthesis pathways. Metab Eng 3:4–14.

203. Su, T. Z., H. Schweizer, and D. L. Oxender. 1990. A novel phosphate-regulated expression vector in Escherichia coli. Gene 90:129–33.

204. Sumegi, B., N. B. Butwell, C. R. Malloy, and A. D. Sherry. 1994. Lipoamide influences substrate selection in post-ischaemic perfused rat hearts. Biochem J 297 (Pt 1):109–13.

205. Szyperski, T. 1998. C-13-NMR, MS and metabolic flux balancing in biotechnology research. Quarterly Reviews of Biophysics 31:41–106.

206. Tao, L., R. E. Jackson, and Q. Cheng. 2005. Directed evolution of copy number of a broad host range plasmid for metabolic engineering. Metab Eng 7:10–7.

207. Thiele, I., T. D. Vo, N. D. Price, and B. O. Palsson. 2005. Expanded metabolic reconstruction of Helicobacter pylori (iIT341 GSM/GPR): an in silico genome-scale characterization of single- and double-deletion mutants. J Bacteriol 187:5818–30.

208. Thomas, S., P. F. J. Mooney, M. M. Burrell, and D. A. Fell. 1997. Metabolic Control Analysis of glycolysis in tuber tissue of potato (Solanum tuberosum): Explanation for the low control coefficient of phosphofructokinase over respiratory flux. Biochemical Journal 322:119–127.

209. Thomas, S., P. J. F. Mooney, M. M. Burrell, and D. A. Fell. 1997. Finite change analysis of glycolytic intermediates in tuber tissue of lines of transgenic potato (Solanum tuberosum) overexpressing phosphofructokinase. Biochemical Journal 322:111–117.

210. Tomizawa, J., and T. Itoh. 1981. Plasmid ColE1 incompatibility determined by interaction of RNA I with primer transcript. Proc Natl Acad Sci U S A 78:6096–100.

211. Tu, H. M., and S. S. Sun. 1996. Generation of a combination of mutations by use of multiple mutagenic oligonucleotides. Biotechniques 20:352–4.

212. Underwood, S. A., M. L. Buszko, K. T. Shanmugam, and L. O. Ingram. 2002. Flux through citrate synthase limits the growth of ethanologenic Escherichia coli KO11 during xylose fermentation. Appl Environ Microbiol 68:1071–81.

213. Vallino, J. J., and G. Stephanopoulos. 1994. Carbon Flux Distributions at the Glucose-6-Phosphate Branch Point in Corynebacterium-Glutamicum during Lysine Overproduction. Biotechnology Progress **10**:327–334.

214. Vallino, J. J., and G. Stephanopoulos. 1994. Carbon Flux Distributions at the Pyruvate Branch Point in Corynebacterium-Glutamicum during Lysine Overproduction. Biotechnology Progress **10**:320–326.

215. Vallino, J. J., and G. Stephanopoulos. 1993. Metabolic Flux Distributions in Corynebacterium-Glutamicum during Growth and Lysine Overproduction. Biotechnology and Bioengineering **41**:633–646.

216. Varma, A., B. W. Boesch, and B. O. Palsson. 1993. Stoichiometric interpretation of Escherichia coli glucose catabolism under various oxygenation rates. Appl Environ Microbiol **59**:2465–73.

217. Varma, A., and B. O. Palsson. 1993. Metabolic Capabilities of Escherichia-Coli .1. Synthesis of Biosynthetic Precursors and Cofactors. Journal of Theoretical Biology **165**:477–502.

218. Varma, A., and B. O. Palsson. 1993. Metabolic Capabilities of Escherichia-Coli .2. Optimal-Growth Patterns. Journal of Theoretical Biology **165**:503–522.

219. Varma, A., and B. O. Palsson. 1994. Stoichiometric flux balance models quantitatively predict growth and metabolic by-product secretion in wild-type Escherichia coli W3110. Appl Environ Microbiol **60**:3724–31.

220. Vo, T. D., H. J. Greenberg, and B. O. Palsson. 2004. Reconstruction and functional characterization of the human mitochondrial metabolic network based on proteomic and biochemical data. J Biol Chem **279**:39532–40.

221. Voigt, C. A., C. Martinez, Z. G. Wang, S. L. Mayo, and F. H. Arnold. 2002. Protein building blocks preserved by recombination. Nat Struct Biol **9**:553–8.

222. Voit, E. O. 2000. Computational Analysis of Biochemical Systems: A practical guide for Biochemists and Molecular Biologists, 1st ed. Cambridge University Press, Cambridge, UK.

223. Wang, H., and P. B. Iynedjian. 1997. Modulation of glucose responsiveness of insulinoma beta-cells by graded overexpression of glucokinase. Proc Natl Acad Sci U S A **94**:4372–7.

224. Weckbecker, A., and W. Hummel. 2004. Improved synthesis of chiral alcohols with Escherichia coli cells co-expressing pyridine nucleotide transhydrogenase, NADP+-dependent alcohol dehydrogenase and NAD+-dependent formate dehydrogenase. Biotechnol Lett **26**:1739–44.

225. Wei, C. L., Y. B. Yang, W. C. Wang, W. C. Liu, J. S. Hsu, and Y. C. Tsai. 2003. Engineering Streptomyces clavuligerus deacetoxycephalosporin C synthase for optimal ring expansion activity toward penicillin G. Appl Environ Microbiol **69**:2306–12.

226. Welch, G. R., T. Keleti, and B. Vertessy. 1988. The control of cell metabolism for homogeneous vs. heterogeneous enzyme systems. J Theor Biol **130**:407–22.

227. Wetzel, R., D. G. Kleid, R. Crea, H. L. Heyneker, D. G. Yansura, T. Hirose, A. Kraszewski, A. D. Riggs, K. Itakura, and D. V. Goeddel. 1981. Expression in Escherichia-Coli of a Chemically Synthesized Gene for a Mini-C Analog of Human Proinsulin. Gene **16**:63–71.

228. White, O., J. A. Eisen, J. F. Heidelberg, E. K. Hickey, J. D. Peterson, R. J. Dodson, D. H. Haft, M. L. Gwinn, W. C. Nelson, D. L. Richardson, K. S. Moffat, H. Qin, L. Jiang, W. Pamphile, M. Crosby, M. Shen, J. J. Vamathevan, P. Lam, L. McDonald, T. Utterback, C. Zalewski, K. S. Makarova, L. Aravind, M. J. Daly, K. W. Minton, R. D. Fleischmann, K. A. Ketchum, K. E. Nelson, S. Salzberg, H. O. Smith, J. C. Venter, and C. M. Fraser. 1999. Genome sequence of the radioresistant bacterium Deinococcus radiodurans R1. Science **286**:1571–7.

229. Wiback, S. J., R. Mahadevan, and B. O. Palsson. 2003. Reconstructing metabolic flux vectors from extreme pathways: defining the alpha-spectrum. Journal Of Theoretical Biology **224**:313–324.

230. Wiback, S. J., R. Mahadevan, and B. O. Palsson. 2004. Using metabolic flux data to further constrain the metabolic solution space and predict internal flux patterns: The Escherichia coli spectrum. Biotechnology And Bioengineering **86**:317–331.

231. Wiechert, W. 2001. 13C metabolic flux analysis. Metab Eng **3**:195–206.

232. Wiechert, W. 2002. Modeling and simulation: tools for metabolic engineering. J Biotechnol **94**:37–63.

233. Wiechert, W., and A. A. De Graaf. 1997. Bidirectional reaction steps in Metabolic Networks: I. Modeling and simulation of carbon isotope labeling experiments. Biotechnology and Bioengineering 55:101–117.

234. Wiechert, W., and A. A. De Graaf. 1997. Bidirectional reaction steps in metabolic networks: II. Flux estimation and statistical analysis. Biotechnology and Bioengineering 55:118–135.

235. Wiechert, W., M. Mollney, N. Isermann, M. Wurzel, and A. A. de Graaf. 1999. Bidirectional reaction steps in metabolic networks: III. Explicit solution and analysis of isotopomer labeling systems. Biotechnol Bioeng 66:69–85.

236. Winkel-Shirley, B. 2001. Flavonoid biosynthesis. A colorful model for genetics, biochemistry, cell biology, and biotechnology. Plant Physiol 126:485–93.

237. Withers, H., S. Swift, and P. Williams. 2001. Quorum sensing as an integral component of gene regulatory networks in Gram-negative bacteria. Curr Opin Microbiol 4:186–93.

238. Wittmann, C., and E. Heinzle. 1999. Mass spectrometry for metabolic flux analysis. Biotechnol Bioeng 62:739–750.

239. Wittmann, C., and E. Heinzle. 2001. Modeling and experimental design for metabolic flux analysis of lysine-producing Corynebacteria by mass spectrometry. Metab Eng 3:173–91.

240. Wittmann, C., H. M. Kim, and E. Heinzle. 2004. Metabolic network analysis of lysine producing Corynebacterium glutamicum at a miniaturized scale. Biotechnol Bioeng 87:1–6.

241. Yan, Y. J., J. Chemler, L. X. Huang, S. Martens, and M. A. G. Koffas. 2005. Metabolic engineering of anthocyanin biosynthesis in Escherichia coli. Applied and Environmental Microbiology 71:3617–3623.

242. Yan, Y. J., A. Kohli, and M. A. G. Koffas. 2005. Biosynthesis of natural flavanones in Saccharomyces cerevisiae. Applied and Environmental Microbiology 71:5610–5613.

243. Yarmush, M. L., and S. Banta. 2003. Metabolic engineering: advances in modeling and intervention in health and disease. Annu Rev Biomed Eng 5:349–81.

244. Yarmush, M. L., and F. Berthiaume. 1997. Metabolic engineering and human disease. Nat Biotechnol 15:525–8.

245. You, J., J. P. Yu, X. B. Ren, C. L. Wang, P. Zhang, and X. Z. Zhang. 2004. [Generation of T cell-mediated antitumor response in vitro by autologous dendritic cells pulsed with tumor lysates in patients with non-small cell lung cancer]. Zhonghua Zhong Liu Za Zhi 26:333–6.

246. Yu, O., W. Jung, J. Shi, R. A. Croes, G. M. Fader, B. McGonigle, and J. T. Odell. 2000. Production of the isoflavones genistein and daidzein in non-legume dicot and monocot tissues. Plant Physiol 124:781–94.

247. Yun, H., H. L. Choi, N. W. Fadnavis, and B. G. Kim. 2005. Stereospecific synthesis of (R)-2-hydroxy carboxylic acids using recombinant E. coli BL21 overexpressing YiaE from Escherichia coli K12 and glucose dehydrogenase from Bacillus subtilis. Biotechnol Prog 21:366–71.

248. Zava, D. T., and G. Duwe. 1997. Estrogenic and antiproliferative properties of genistein and other flavonoids in human breast cancer cells in vitro. Nutrition and Cancer-an International Journal 27:31–40.

249. Zhang, Y. X., K. Perry, V. A. Vinci, K. Powell, W. P. Stemmer, and S. B. del Cardayre. 2002. Genome shuffling leads to rapid phenotypic improvement in bacteria. Nature 415:644–6.

250. Zhou, S., K. T. Shanmugam, and L. O. Ingram. 2003. Functional replacement of the Escherichia coli D-(−)-lactate dehydrogenase gene (ldhA) with the L-(+)-lactate dehydrogenase gene (ldhL) from Pediococcus acidilactici. Appl Environ Microbiol 69:2237–44.

251. Zupke, C., and G. Stephanopoulos. 1994. Modeling of Isotope Distributions and Intracellular Fluxes in Metabolic Networks Using Atom Mapping Matrices. Biotechnology Progress 10:489–498.

252. Zupke, C., R. Tompkins, D. Yarmush, and M. Yarmush. 1997. Numerical isotopomer analysis: Estimation of metabolic activity. Analytical Biochemistry 247:287–293.

CHAPTER 11

CHEMICAL ORGANISATION THEORY

PETER DITTRICH AND PIETRO SPERONI DI FENIZIO*

Bio Systems Analysis Group, Jena Centre for Bioinformatics and Department of Mathematics and Computer Science, Friedrich Schiller University Jena, D-07743 Jena, Germany

Abstract: Complex dynamical reaction networks consisting of many molecular species are difficult to understand, especially, when new species may appear and present species may vanish completely. This chapter outlines a technique to deal with such systems. The first part introduces the concept of a chemical organisation as a closed and self-maintaining set of molecular species. This concept allows to map a complex (reaction) network to its set of organisations, providing a new view on the system's structure. The second part connects dynamics with the set of organisations, which allows to map a movement of the system in state space to a movement in the set of organisations. The relevancy of this approach is underlined by a theorem that says that given a differential equation describing the chemical dynamics of the network, then every stationary state is an instance of an organisation. Finally, the relation between pathways and chemical organisations is sketched

Keywords: reaction networks, constraint based network analysis, hierarchical decomposition, constructive dynamical systems

1.   INTRODUCTION

The rapidly increasing size and complexity of reaction system models requires novel mathematical and computational techniques in order to cope with their complexity[9]. This chapter describes a technique that allows to identify for a given reaction network important sub-structures, called chemical organisations[38,10]. These

*Both authors contributed equally. May 10, 2006

M. Al-Rubeai and M. Fussenegger (eds.), Systems Biology, 361–393.
© 2007 *Springer.*

organisations allow to explain the (potential) behaviour of the reaction system from a global and more abstract perspective.

The theory aims at those systems where different combinations of molecular species (compounds) are present at different points in time. These systems are characterised by the fact that they are changing not only *quantitatively*, that is, by a change in the concentration of a molecular species, but also *qualitatively* when new molecular species appear or a present species completely vanish. Fontana and Buss[20] called systems that display the production of novelty *constructive (dynamical) systems*.

Classical approaches describe the dynamics of a reaction system as a "quantitative" movement in a fixed state space[22], where a state is usually described by a concentration vector[16]. Here, we will operate on a higher level of abstraction and consider *qualitative* movements from a set of molecular species to another set of molecular species. We can interpret this qualitative change as a movement that goes from state space to state space, as new molecular species appear and old species disappear.

The lack of a theory for such constructive dynamical systems has been presented, identified, and discussed in detail by Fontana and Buss[20] in the context of a theory for biological organisation. As a partial solution, they suggest the important concept of a (biological) organisation as a set of molecules that are algebraically closed and dynamically self-maintaining.

Closure means that no new molecular species can be generated by reactions among molecules inside the organisation (Section 3.1). As such no novelty can spontaneously appear. Note that closure, as a property of a set of molecules, should not be confused with the thermodynamical closure of a system, which are two different and separated concepts.

Self-maintaining roughly means that every consumed molecule of the organisation has a way to be generated within the organisation such that it does not disappear from the system (Section 3.3).

Although closure and self-maintenance do not assure that a set of species will remain unchanged in time, the lack of them does imply that the system will eventually *qualitatively* move to a different set of molecules (Section 4.2). In a vast class of systems, which we call *consistent* (Section 3.5), it is possible to define a generator operator such that for any set of molecules an organisation is uniquely defined. The organisation generated by a set *A* represents the largest possible set of molecules that can stably exists when starting with *A*.

This implies that organisations partition the set of all possible sets of molecules, where a partition consists of all combinations of molecular species that generate the same organisation. Thus, as the system qualitatively progress from one set to another we can follow it on the more tractable set of all possible organisations (Section 4.3, Figure 4). The study of this movement together with a theorem relating fixed points to organisations will be the core concepts of the dynamical part of chemical organisation theory (Section 4).

Before we enter into the theory, some prerequisites are introduced in the following section.

2. REACTION SYSTEMS

The theory described herein aims at understanding *reaction systems*. A reaction system consists of molecules, and interaction rules among molecules that lead to the appearance or disappearance of other molecules.

Note that we have to distinguish between a reaction system as an abstract description of all possible molecular species (and their reactions), and an actual reaction vessel, which contains some concrete instances of molecules from the set of all possible molecules. In order to refer to an element of a reaction system, we will use a series of terms equivalently: *molecule, compound, molecular species*, or simply *species*; keeping in mind that the term "molecule" is somehow imprecise, since it can also refer to a concrete physical instance. Similarly, we call an interaction rule among molecules shortly a *reaction*.

In passing we note that reaction systems are not only used to model chemical phenomena. Their applications range from ecology[35], protobiology[36], systems biology[33] to computer science[2] and reach even the study of language and social systems[11].

The description of a reaction system can be subdivided into three parts[12]: (1) the set of all possible molecules $\mathcal{M}$, (2) the set of all possible reactions among all the possible molecules $\mathcal{R}$, and (3) the dynamics (e.g. kinetic laws), which describes how the reactions are applied to a collection of molecules inside a reaction vessel[16,23].

2.1 The Molecules $\mathcal{M}$

Step (1) requires that we identify all players, that is, the set of all molecular species that can appear in the model. The easiest way to specify this set is to enumerate explicitly all molecules. For example: $\mathcal{M} = \{H_2, O_2, H_2O, H_2O_2\}$. Alternatively, $\mathcal{M}$ can be defined implicitly, e.g., $\mathcal{M} = \{$all polymers that can be made from two monomers$\}$. In this case, the set of all possible molecules can even become infinite, or at least quite large. For simplicity, but without loss of generality, we will consider here only small, explicit sets of molecular species. And we will refer to molecular species just by an index $i \in \mathcal{M}$, neglecting their structure.

2.2 The Reaction Rules $\mathcal{R}$

A reaction rule like $2H_2 + O_2 \rightarrow 2H_2O$ can be interpreted as a transformation of molecules, e.g., the transformation of hydrogen and oxygen molecules into water molecules. We will represent reaction rules by two matrixes $(l_{i,\rho})$ and $(r_{i,\rho})$, where $l_{i,\rho}$ and $r_{i,\rho}$ is the stoichiometric coefficient of molecule $i \in \mathcal{M}$ in reaction $\rho \in \mathcal{R}$ on the lefthand side and on the righthand side, respectively. The set of all molecules $\mathcal{M}$ and a set of reaction rules $\mathcal{R}$ define a reaction network:

Definition 1 (reaction network)[1] *Given a set $\mathcal{M}$ of elements, called molecules or molecular species, and a set of reaction rules $\mathcal{R}$ given by the lefthand side and righthand side stoichiometric matrices $(l_{i,\rho})$ and $(r_{i,\rho})$, respectively, with $i \in \mathcal{M}$ and $\rho \in \mathcal{R}$. We call the pair $\langle \mathcal{M}, \mathcal{R} \rangle$ a reaction network (or algebraic chemistry, as in Ref.[10]).*

In the following $\mathrm{LHS}(\rho)$ denotes the set of molecules that appear on the left-hand side of reaction $\rho \in \mathcal{R}$. And $\mathrm{RHS}(\rho)$ the molecules on the righthand side. Furthermore we define $\mathcal{R}_A \subseteq \mathcal{R}$ as the subset of reactions that can "fire" when the molecules of the set A are present; formally $\mathcal{R}_A = \{\rho \in \mathcal{R} | \mathrm{LHS}(\rho) \subseteq A\}$.

The *stoichiometric matrix* $\mathbf{S}$ is defined as

$$\mathbf{S} = (s_{i,\rho}) = (r_{i,\rho} - l_{i,\rho}). \tag{1}$$

An entry $s_{i,\rho}$ of the stoichiometric matrix denotes the net amount of molecules of type i produced in reaction ρ.

Example 1 (reaction network, three species) *The reaction network consists of $m = 4$ molecular species $\mathcal{M} = \{H_2, O_2, H_2O, H_2O_2\}$ and $r = 2$ reaction rules $\mathcal{R} = \{\rho_1 : 2H_2 + O_2 \rightarrow 2H_2O, \rho_2 : 2H_2O_2 \rightarrow 2H_2O + O_2\}$.*

$$\mathrm{LHS}(\rho_1) = \{H_2, O_2\}, \quad \mathrm{RHS}(\rho_1) = \{H_2O\} \tag{2}$$

$$\mathrm{LHS}(\rho_2) = \{H_2O_2\}, \quad \mathrm{RHS}(\rho_2) = \{H_2O, O_2\}. \tag{3}$$

$$For\ A = \{H_2\}: \quad \mathcal{R}_A = \{\} \tag{4}$$

$$For\ A = \{H_2O_2\}: \quad \mathcal{R}_A = \{\rho_2 : 2H_2O_2 \rightarrow 2H_2O + O_2\}. \tag{5}$$

$$For\ A = \{H_2, O_2\}: \quad \mathcal{R}_A = \{\rho_1 : 2H_2 + O_2 \rightarrow 2H_2O\}. \tag{6}$$

$$(l_{i,\rho}) = \begin{pmatrix} 2 & 0 \\ 1 & 0 \\ 0 & 0 \\ 0 & 2 \end{pmatrix}, \quad (r_{i,\rho}) = \begin{pmatrix} 0 & 0 \\ 0 & 1 \\ 2 & 2 \\ 0 & 0 \end{pmatrix} \begin{matrix} H_2 \\ O_2 \\ H_2O \\ H_2O_2 \end{matrix}, \tag{7}$$

$$\mathbf{S} = \begin{pmatrix} -2 & 0 \\ -1 & 1 \\ 2 & 2 \\ 0 & -2 \end{pmatrix} \begin{matrix} H_2 \\ O_2 \\ H_2O \\ H_2O_2 \end{matrix}. \tag{8}$$

---

[1] From a theoretical point of view, a reaction network is a directed bipartite graph whose nodes represent molecules and reaction rules, respectively, and whose edges are weighted by stoichiometric coefficients. Further note that a reaction network as defined here with whole-numbered stoichiometric coefficients is equivalent to a Petri net[31].

In the following, we will denote molecules with lowercase characters like a, b, c, d and sets of molecules by uppercase characters like A, B, C, O, S. A character like a stands for the name of a molecular species like H_2O. The following abstract example will be used throughout this chapter to illustrate the various concepts.

Example 2 (reaction network, five species) *There are five molecular species* $M = \{a, b, c, d, s\}$, *which react according to the following reaction rules*

$$\mathcal{R} = \left\{ \begin{array}{ll} \rho_1 : a+s \rightarrow 2a, & \rho_6 : \emptyset \rightarrow s, \\ \rho_2 : b+s \rightarrow 2b, & \rho_7 : a \rightarrow \emptyset, \\ \rho_3 : a+b \rightarrow 2a, & \rho_8 : b \rightarrow \emptyset, \\ \rho_4 : b+c \rightarrow 2b, & \rho_9 : c \rightarrow \emptyset, \\ \rho_5 : a+d \rightarrow c+d, & \rho_{10} : s \rightarrow \emptyset \end{array} \right\}.$$

A graphical representation of this reaction network can be found in Figure 1. The lefthand side and righthand side stoichiometric matrixes read:

$$(l_{i,\rho}) = \begin{pmatrix} 1 & 0 & 1 & 0 & 1 & 0 & 1 & 0 & 0 & 0 \\ 0 & 1 & 1 & 1 & 0 & 0 & 0 & 1 & 0 & 0 \\ 0 & 0 & 0 & 1 & 1 & 0 & 0 & 0 & 1 & 0 \\ 0 & 0 & 0 & 0 & 0 & 0 & 0 & 0 & 0 & 0 \\ 1 & 1 & 0 & 0 & 0 & 0 & 0 & 0 & 0 & 1 \end{pmatrix} \begin{array}{l} a \\ b \\ c \\ d \\ s \end{array} \tag{9}$$

and

$$(r_{i,\rho}) = \begin{pmatrix} 2 & 0 & 2 & 0 & 0 & 0 & 0 & 0 & 0 & 0 \\ 0 & 2 & 0 & 2 & 0 & 0 & 0 & 0 & 0 & 0 \\ 0 & 0 & 0 & 0 & 1 & 0 & 0 & 0 & 0 & 0 \\ 0 & 0 & 0 & 0 & 0 & 0 & 0 & 0 & 0 & 0 \\ 0 & 0 & 0 & 0 & 0 & 1 & 0 & 0 & 0 & 0 \end{pmatrix} \begin{array}{l} a \\ b \\ c. \\ d \\ s \end{array} \tag{10}$$

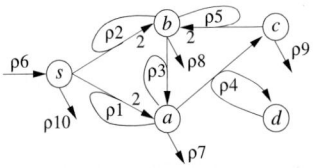

Figure 1. Reaction network of Example 2. A "2" refers to the stoichiometry, e.g. in the reaction rule $\rho_1 : a+s \longrightarrow 2a$

Subtracting $(l_{i,\rho})$ *from* $(r_{i,\rho})$ *leads to the stoichiometric matrix* $\mathbf{S} = (s_{i,\rho}) = (r_{i,\rho}) - (l_{i,\rho})$:

$$
\mathbf{S} = \begin{pmatrix}
1 & 0 & 1 & 0 & -1 & 0 & -1 & 0 & 0 & 0 \\
0 & 1 & -1 & 1 & 0 & 0 & 0 & -1 & 0 & 0 \\
0 & 0 & 0 & -1 & 1 & 0 & 0 & 0 & -1 & 0 \\
0 & 0 & 0 & 0 & 0 & 0 & 0 & 0 & 0 & 0 \\
-1 & -1 & 0 & 0 & 0 & 1 & 0 & 0 & 0 & -1
\end{pmatrix}
\begin{matrix} a \\ b \\ c. \\ d \\ s \end{matrix}
\tag{11}
$$

2.3   Dynamics

A reaction network $\langle \mathcal{M}, \mathcal{R} \rangle$ specifies the structure of a reaction system, but does not contain any notion of time. A common way to specify the dynamics of the reaction system is by using a system of ordinary differential equations of the following form:

$$
\dot{\mathbf{x}}(t) = \mathbf{S}\mathbf{v}(\mathbf{x}(t))
\tag{12}
$$

where $\mathbf{x} = (x_1, \ldots, x_m)^T \in \mathbb{R}^m$ is a concentration vector depending on time t, $\mathbf{S}$ a stoichiometric matrix, and $\mathbf{v} = (v_1, \ldots, v_r)^T \in \mathbb{R}^r$ a flux vector depending on the current concentration vector. A flux $v_\rho \geq 0$ describes the velocity or turnover rate of reaction $\rho \in \mathcal{R}$. The actual value of v_ρ depends usually on the concentration of the species participating in the reaction ρ (i.e., LHS(ρ)). In order to avoid unwanted mathematical effects, we demand that $\mathbf{v}$ is differentiable, meaning intuitively that it depends "smoothly" on the concentration vector $\mathbf{x}$. Beside this mathematical assumption, there are (at least) two further assumptions that are due to the nature of reaction systems. These assumptions relate the function $\mathbf{v}$ to the reaction rules $\mathcal{R}$:

Assumptions 1: If a species i is necessary for a reaction ρ to take place, it must appear on the lefthand side of that reaction (i.e., $i \in \text{LHS}(\rho)$). This implies that for all molecules $i \in \mathcal{M}$ and reactions $\rho \in \mathcal{R}$ with $i \in \text{LHS}(\rho)$, if $x_i = 0$ then $v_\rho = 0$. The flux v_ρ must be zero, if the concentration x_i of a molecule appearing on the lefthand side of this reaction is zero. This assumption meets the obvious fact that a molecule has to be present to react.

Assumptions 2: If all species LHS(ρ) of a reaction $\rho \in \mathcal{R}$ are present in the reactor (e.g. for all $i \in \text{LHS}(\rho)$, $x_i > 0$) the flux of that reaction is positive, (i.e., $v_\rho > 0$). In other words, the flux v_ρ must be positive, if all molecules required for that reaction are present, even in small quantities (cf. Definition 14 (instance) and Definition 15 (abstraction)).

There is a large amount of kinetic laws fulfilling these assumptions, including all laws that are usually applied in practice. The most fundamental of such kinetic laws is mass-action kinetics, which is just the product of the concentrations of the interacting species:

$$
v_\rho = \prod_{i \in \mathcal{M}} x^{l_{i,\rho}}.
\tag{13}
$$

It should be noted that more complicated laws like Michaelis-Menten kinetics are derived from mass-action kinetics. This is especially true for many laws describing inhibition[23]. So, we may interpret these more complicated laws as "syntactic sugar" that allows to describe large reaction systems in a more compact way, that is, with a smaller number of species.

2.4 Modifiers

When applying chemical organisation theory, we can neglect modifiers that inhibit a reaction, because they are not necessary for the reaction to occur (Assumption 2) and they usually do not switch off the reaction completely, which follows from the fact that laws of inhibition are derived from Michaelis-Menten kinetics. However, in case an inhibitory effect is so strong that it practically switches off a reaction completely, it has to be considered (not shown here).

For the theory, we can also neglect modifiers that enhance a reaction, as long as they are not necessary for that reaction to take place. Note that a modifier i that is necessary for a reaction ρ has to appear as a catalyst on the lefthand side of that reaction, that is, $i \in \text{LHS}(\rho)$ (Assumption 2) and on its righthand side (i.e., $i \in \text{RHS}(\rho)$).

2.5 Input and Output

There are many processes that give rise to an inflow and outflow, such as, incident sunlight, decaying molecules, or a general dilution flow. In this chapter we interpret the reaction rule $\emptyset \to a$ as an input of a, and $a \to \emptyset$ as an output of a ($\emptyset$ denoting the empty set, which is, from a mathematical point of view, assumed to be always present). In Example 2 (Figure 1) there is an inflow of s whereas a, b, c, and s decay spontaneously.

2.6 Unbalanced Reaction Systems

As sometimes otherwise stated, in chemical reaction system models the masses on the left hand side and right hand side can differ. That means that, formally, a reaction can produce or consume mass. This might appear unrealistic, since it is generally assumed that in a real chemical reaction mass is conserved.

However in a reaction system *model* it makes sense to consider unbalanced reaction rules, too, which can lead to more elegant and simpler models. We have already encountered two examples in the previous section, namely, inflow and outflow reaction rules. Further examples are models of exponential growth (e.g., $a \to 2a$) and other models assuming implicitly an unlimited substrate. This substrate is removed to obtain a simpler model, compare Example 9 (hypercycle without an explicit substrate, Section 3.5.1) with Example 10 (hypercycle with an explicit substrate, Section 3.5.2).

Chemical organisation theory can also deal with unbalanced reaction systems, including those where some molecular species cannot operate at steady state and can exhibit unlimited growth.

3.   CHEMICAL ORGANISATION THEORY: STATIC PART

The first part of the theory deals with the static structure of a reaction system, that is, the molecules $\mathcal{M}$ and the reactions $\mathcal{R}$. Instead of considering a state, e.g. a concentration vector, we limit ourself to the analysis of the set of molecules present in that state.

In classical analysis, we study the movement of the system in state space. Instead, here, we consider the movement from one set of molecules to another. As in the classical analysis of the dynamics of the system, where fixed points and attractors are considered more important than other states, some sets of molecules are more important than others.

In order to find those sets, we introduce some properties that define them, namely: closure, semi-self-maintenance, semi-organisation, self-maintenance, and finally being an organisation. All definitions herein refer to a reaction network $\langle \mathcal{M}, \mathcal{R} \rangle$.

3.1   Closed Sets

The first property of a set of molecules, called closure, assures that no new molecular species can be generated by the reactions inside the set or equivalently that all molecules that can be generated by reactions inside the set are already inside that set.

Definition 2 (closed set[20]) *A set* $C \subseteq \mathcal{M}$ *is closed, if for all reactions* $\rho \in \mathcal{R}_C$, $\mathrm{RHS}(\rho) \subseteq C$.

Given a set $A \subseteq \mathcal{M}$, we can always *generate* its closure $G_{CL}(A)$ according to the following definition:

Definition 3 (generate closed set[20]) *Given a set of molecules* $A \subseteq \mathcal{M}$, *we define* $G_{CL}(A)$ *as the smallest closed set* C *containing* A. *We say that* A *generates the closed set* $C = G_{CL}(A)$ *and we call* C *the closure of* A.

In passing we note that this definition is unambiguous: Let us suppose, ad absurdum, that we can find two smallest closed sets $C_1 \neq C_2$ both containing C. Against our assumption, their intersection would be an even smaller closed set containing C, because the intersection of closed sets is obviously closed, too.

We can generate the closed set for a given set A efficiently by the following algorithm: add all reaction products among molecules of A and insert them into A and repeat this procedure until no new molecule can be inserted anymore (for infinite systems, a limit has to be taken).

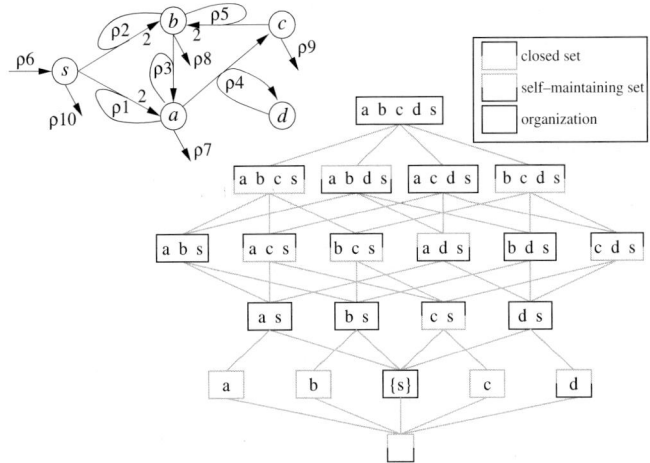

Figure 2. Fraction of the lattice of all sets of molecules from Example 2. Shown are all sets containing molecule *s* and small sets like {*a*} and {}. Note that an organisation is a closed *and* self-maintaining set

Example 3 (closed sets) *In our Example 2, Figure 2, there are 14 closed sets:* $\mathcal{O}_{CL} = \{\{s\}, \{a, s\}, \{b, s\}, \{c, s\}, \{d, s\}, \{a, b, s\}, \{a, c, s\}, \{b, d, s\}, \{c, d, s\}, \{a, b, c, s\}, \{a, c, d, s\}, \{b, c, d, s\}, \{a, b, c, d, s\}\}.$ *The empty set is not closed, because there is an inflow of molecule s. The set* $A = \{a, d, s\}$ *is not closed, because c can be produced by the reaction* $a + d \rightarrow d + c$*. The closed set generated by A is:* $C = G_{CL}(A) = \{a, c, d, s\}.$

A common algebraic concept that we shall use very often, from now on, is the lattice. A lattice, is a partially ordered set (poset) in which any two elements have a greatest lower bound (here, their intersection) and a least upper bound (here, their union).

Given the generate closed set operator we can define two basic operations, a union operation $(U \sqcup_{CL} V)$ and an intersection operation $(U \sqcap_{CL} V)$ on closed sets:

$$U \sqcup_{CL} V \equiv G_{CL}(U \cup V), \quad \text{and} \tag{14}$$

$$U \sqcap_{CL} V \equiv G_{CL}(U \cap V). \tag{15}$$

Trivially, closed sets, with the operations $\sqcup_{CL}$ and $\sqcap_{CL}$, form a lattice $\langle \mathcal{O}_{CL}, \sqcup_{CL}, \sqcap_{CL} \rangle$.

Closure is important because the closed set generated by a set (its closure) represents the largest possible set that can be reached from a given set of molecules. Furthermore a set that is closed cannot generate new molecules and is in that sense more stable with respect to novelty. As such the concept of closure alone can already give valuable insight into the structure and organisation of complex chemical networks as shown by Ebenhöh et al.[13].

3.2   Semi-Self-Maintaining Sets

Before we introduce the second important property called "self-maintenance", an intermediate step will be taken by defining a property that is necessary for a set to be self-maintaining but not sufficient. This property, called *semi-self-maintenance*, is easier to check then the property self-maintenance. Furthermore, in reaction systems like *catalytic flow system* (Section 3.5.1), every semi-self-maintaining set is also self-maintaining, and thus it is sufficient to check in those systems just the semi-self-maintenance property.

The property *semi-self-maintaining* assures that every molecule that is consumed within a set, is produced within that set.

We say that a molecule $i \in \mathcal{M}$ is *produced* within a set $A \subseteq \mathcal{M}$, if there exists a reaction $\rho \in \mathcal{R}_A$ with $s_{i,\rho} > 0$. In the same way, we say that a molecule $i \in A$ is *consumed* within the set A, if there is a reaction $\rho \in \mathcal{R}_A$ with $s_{i,\rho} < 0$.

Definition 4 (semi-self-maintaining set[10]) *A set of molecules $S \subseteq \mathcal{M}$ is called semi-self-maintaining, if all molecules $i \in S$ that are consumed within S are also produced within that set S.*

Example 4 (semi-self-maintaining sets) *In our Example 2 there are 13 semi-self-maintaining sets:* $\mathcal{O}_{SSM} = \{\{\}, \{s\}, \{d\}, \{a, s\}, \{b, s\}, \{d, s\}, \{a, b, s\}, \{a, d, s\}, \{b, d, s\}, \{a, b, c, d\}, \{a, b, d, s\}, \{a, c, d, s\}, \{a, b, c, d, s\}\}.$

Note that the concept of (semi-) self-maintenance is closely related to the concept of an autocatalytic set[34,14,27]. An autocatalytic set is usually defined as a set of molecules such that each molecule is produced by at least one catalytic reaction within that set[25].

3.3   Self-Maintaining Sets

In a semi-self-maintaining set, all molecules that are consumed are produced; yet this does not guarantee that the total amount of mass can be maintained.

Example 5 (reversible reaction) *A simple counterexample is the following reversible reaction in a flow reactor:* $\mathcal{M} = \{a, b\}$, $\mathcal{R} = \{a \rightarrow b, b \rightarrow a, a \rightarrow \emptyset, b \rightarrow \emptyset\}$. *Both molecules, a and b, decay.* $S = \{a, b\}$ *is a semi-self-maintaining set, because a is produced by the reaction $b \rightarrow a$, and b is produced by the reaction* *a → b. But, obviously, the system {a, b} is not stable, in the sense that there cannot be a stationary state in which the two molecules a and b have positive concentrations: both molecules decay and cannot be sufficiently reproduced, and thus they will finally vanish.*

The solution to this problem is to consider the overall ability of a set to maintain its total mass. We call such sets simply *self-maintaining*.

Definition 5 (self-maintaining[10]) *Given an algebraic chemistry $\langle \mathcal{M}, \mathcal{R} \rangle$ with $m = |\mathcal{M}|$ molecules and $r = |\mathcal{R}|$ reactions, and let $\mathbf{S} = (s_{i,j})$ be the $(m \times r)$ stoichiometric matrix implied by the reaction rules $\mathcal{R}$, where $s_{i,\rho}$ denotes the number of molecules of type i produced in reaction ρ. A set of molecules $S \subseteq \mathcal{M}$ is called self-maintaining, if there exists a flux vector $\mathbf{v} \in \mathbb{R}^r$ such that the following three conditions apply: (1) for all reactions $\rho \in \mathcal{R}_S$ the flux $v_\rho > 0$; (2) for all remaining reactions $\rho \notin \mathcal{R}_S$, the flux $v_\rho = 0$; and (3) for all molecules $i \in S$, the production rate $(\mathbf{Sv})_i \geq 0$.*

v_ρ denotes the element of $\mathbf{v}$ describing the flux (i.e., velocity) of reaction ρ. $(\mathbf{Sv})_i$ is the production rate of molecule i given flux vector $\mathbf{v}$. It is practically the sum of fluxes producing i minus the fluxes consuming i.

Example 6 (reversible reaction (cont.)) *For the reversible reaction, Example 5, the stoichiometric matrix becomes $\mathbf{S} = ((-1, 1), (1, -1), (-1, 0), (0, -1))$, and we can see that there is no positive flux vector $\mathbf{v} \in \mathbb{R}^4$, such that $\mathbf{Sv} \geq \mathbf{0}$. In other words, no matter how we chose the velocity of the reactions, it is not possible to keep a and b in a reaction vessel. In fact, in that example, only the empty set {} is self-maintaining. In case a and b would not decay, $\mathcal{R} = \{a \to b, b \to a\}$, the set {a, b} would be (as desired) self-maintaining, because there is a flux vector, e.g., $\mathbf{v} = (1.0, 1.0)$, such that $\mathbf{Sv} = \mathbf{0} \geq \mathbf{0}$ with $\mathbf{S} = ((-1, 1), (1, -1))$.*

Example 7 (self-maintaining) *In our five species network (Example 2, Figure 2) all semi-self-maintaining sets except {a, b, c, d} are also self-maintaining. The criterion for self-maintenance will be illustrated by looking at the self-maintaining set $S = \{s, a, d\}$ in more detail. First we have to find all the reactions active within S, this means to find all reactions $\rho \in \mathcal{R}_S \subseteq \mathcal{R}$ where the left hand side consists of molecules from S (i.e., $\mathrm{LHS}(\rho) \subseteq S$). There are five such reactions: $\mathcal{R}_S = \{\rho_1 : a + s \to 2a, \rho_5 : a + d \to d + c, \rho_6 : \emptyset \to s, \rho_7 : a \to \emptyset, \rho_{10} : s \to \emptyset\}$ These reactions correspond to column 1, 5, 6, 7, and 10 of the stoichiometric matrix $\mathbf{S}$, Eq. (11), and to the fluxes v_1, v_5, v_6, v_7, and v_{10}. According to the definition of self-maintenance, these five fluxes must be positive while the remaining fluxes must be zero. Now, in order to show that $S = \{s, a, d\}$ is self-maintaining, we have to find positive values for v_1, v_5, v_6, v_7, and v_{10} such that a, c, and s are produced at a non-negative rate.*

Here, this can be achieved by setting $v_1 = 3$, $v_5 = 1$, $v_6 = 10$, $v_7 = 1$, $v_{10} = 1$. When multiplying this flux vector with the stoichiometric matrix,

$$
\mathbf{Sv} =
\begin{pmatrix}
1 & 0 & 1 & 0 & -1 & 0 & -1 & 0 & 0 & 0 \\
0 & 1 & -1 & 1 & 0 & 0 & 0 & -1 & 0 & 0 \\
0 & 0 & 0 & -1 & 1 & 0 & 0 & 0 & -1 & 0 \\
0 & 0 & 0 & 0 & 0 & 0 & 0 & 0 & 0 & 0 \\
-1 & -1 & 0 & 0 & 0 & 1 & 0 & 0 & 0 & -1
\end{pmatrix}
\begin{pmatrix}
3 \\ 0 \\ 0 \\ 0 \\ 1 \\ 10 \\ 1 \\ 0 \\ 0 \\ 1
\end{pmatrix}
=
\begin{pmatrix}
1 \\ 0 \\ 1 \\ 0 \\ 6
\end{pmatrix}
\begin{matrix} a \\ b \\ c, \\ d \\ s \end{matrix}
$$

we can see that all molecules of $S = \{a, d, s\}$ are produced at a non-negative rate. For example, d is produced at rate 0 and s is produced at rate 6. So, we can conclude that $\{a, d, s\}$ is self-maintaining. Note further that, in Example 2, molecule d will always be produced at zero rate $((\mathbf{Sv})_4 = 0)$, independently on how we chose $\mathbf{v}$.

In a self-maintaining set S every molecule that is consumed by a reaction $\rho \in \mathcal{R}_s$ must be also produced by a reaction within that set, in order to achieve a non-negative production of that molecule. Therfore we can conclude, as mentioned previously:

Lemma 1 [10] *Every self-maintaining set is semi-self-maintaining (proof in Ref.* [10]*).*

In other words, if a set is not semi-self-maintaining – a property easy to check – the set cannot be self-maintaining. The opposite is not generally true; there are sets that are semi-self-maintaining but not self-maintaining, such as $\{a, b, c, d\}$ in Example 2.

Note that the empty set is always self-maintaining (as it has nothing to maintain) and the set of all possible molecules is always closed (as there is nothing that can be added to it). On the other hand the empty set is not necessarily closed, and the set of all possible molecules is not necessarily self-maintaining.

3.4   Organisations

Together, closure and self-maintenance lead to the central definition of this approach[2]:

Definition 6 (organisation[20,10]) *A set of molecules $O \subseteq \mathcal{M}$ that is both closed and self-maintaining is called an organisation.*

---

[2] Note that our definition of an organisation intentionally reads like the definition by Fontana and Buss[20]. However, we are using a more general definition of self-maintenance, which includes the definition by Fontana and Buss and thus ensures "compatibility" with their approach.

An organisation represents an important combination of molecular species, which are likely to be observed in a large reaction vessel on the long run (cf. Theorem 1, Section 4.2). A set of molecules that is not closed would not exist for a long time, because new molecules will appear, changing that set. A set of molecules that is not self-maintaining will also inevitably change, since molecules not-maintained will vanish.

In the same way as we defined an organisation, we can define a *semi-organisation* as a closed and semi-self-maintaining set of molecules. From Lemma 1 trivially follows that every organisation is a semi-organisation.

Example 8 (organisations) *In our example (Example 2) there are 14 closed sets, 13 semi-self-maintaining sets, 12 self-maintaining sets, and 8 semi-organisations. All 8 semi-organisations are also organisations:* $O = \{\{s\}, \{a, s\}, \{b, s\}, \{d, s\}, \{a, b, s\}, \{b, d, s\}, \{a, c, d, s\}, \{a, b, c, d, s\}\}$. *Although the reaction system is small, its organisational structure is already difficult to see when looking at the rules or their graphical representation. In Figure 2, a Hasse diagram with all 16 possible sets of molecules containing the substrate s together with some smaller sets is shown.*

Finding all organisations of a general reaction system appears to be computationally difficult (Section 6.2). One approach is to find the semi-organisations first, and then check, which of them are also self-maintaining.

3.5   Consistent Reaction Systems

The property of the set of organisations and semi-organisations depends strongly on the type of system studied. In this section we discuss a class of systems, called consistent reaction systems, where the set of organisations always forms an algebraic lattice, that is, there is a unique union and intersection of organisations and there is a unique largest organisation (if there is a finite number of organisations).

Definition 7 (consistent[10]) *A reaction network is called* consistent, *if (1) given any two (semi-)self-maintaining sets A and B then their set-union $A \cup B$ is also (semi-)self-maintaining; and (2) the closure $G_{CL}(A)$ of any (semi-)self-maintaining set A is (semi-)self-maintaining.*

Now, in Sections 3.5.1–3.5.3, we will present three types of consistent systems, which can be easily (i.e., in linear time) identified by looking at the reaction rules only. In practice it makes sense to check first, whether the system to be analysed falls into one of these classes.

3.5.1   Catalytic flow system

In a *catalytic flow system* all molecules $i \in \mathcal{M}$ are consumed by first-order reactions of the form $i \to \emptyset$ (dilution) and there is no molecule consumed by any other reaction. So, each molecule i decays spontaneously, or equivalently, is removed by a dilution flow. Apart from this, each molecule can appear only as a catalyst (without being consumed).

Definition 8 (catalytic flow system[10]) *A reaction network $\langle \mathcal{M}, \mathcal{R} \rangle$ is called a catalytic flow system, if for all molecules $i \in \mathcal{M}$: (1) there exists a first-order decay reaction $\rho \in \mathcal{R}$ with $\mathrm{LHS}(\rho) = \{i\}$ and $s_{i,\rho} = -1$; (2) for all reaction $\rho \in \mathcal{R}$ with $s_{i,\rho} < 0$, $s_{i,\rho} = -1$ and $\mathrm{LHS}(\rho) = \{i\}$ (reactions that consume i must be first-order decay reactions).*

Examples of *catalytic flow system* are the replicator equation[35], the hyper-cycle[14,15], the more general *catalytic network equation*[40], various models of auto-catalytic sets[24,3], and AlChemy[19]. Furthermore some models of genetic regulatory networks and social system[11] are *catalytic flow systems*.

Example 9 (catalytic flow system) *The three-membered elementary hyper-cycle[14,15] under flow condition can be represented by three molecular species $\mathcal{M} = \{a, b, c\}$ and six reaction rules:*

$$\mathcal{R} = \{a + b \to a + 2b, \qquad a \to \emptyset,$$
$$b + c \to b + 2c, \qquad c \to \emptyset$$
$$c + a \to c + 2a, \qquad b \to \emptyset\}.$$

We can see, that all three molecules decay by first-order reactions (representing the dilution flow), and that no molecule is consumed by any of the three remaining reactions. There are two organisations: $\{\}$ and $\{a, b, c\}$. The catalytic reaction rules are not balanced, since a substrate available at a constant concentration is implicitly assumed to be consumed.

In a *catalytic flow system* we can easily check a set for being self-maintaining, because:

Lemma 2 [10] *In a catalytic flow system, all semi-self-maintaining sets are self-maintaining (Proof in Ref.[10]).*

From which follows immediately:

Lemma 3 [10] *In a catalytic flow system, every semi-organisation is an organisation.*

So, in a *catalytic flow system* we can easily check, whether a set O is an organisation by just checking whether it is closed and whether each molecule in that set is produced by that set. Furthermore, given a set A, we can always generate an organisation by adding all molecules produced by A until A is closed and then removing molecules that are not produced until A is (semi-) self-maintaining. With respect to the intersection and union of (semi-) organisations the set of all (semi-) organisations of a *catalytic flow system* forms an algebraic lattice (see below). A result which has already been noted by Fontana and Buss[20].

3.5.2 Reactive flow system

In a *reactive flow system* all molecules are consumed by first-order reactions of the form $\{i\} \rightarrow \emptyset$ (dilution). But as opposed to the previous system, we allow arbitrary additional reactions in $\mathcal{R}$ and do not restrict these reactions to be catalytic. Thus note how a *catalytic flow system* is a particular kind of *reactive flow system*.

Definition 9 (reactive flow system[10]) *A reaction network* $\langle \mathcal{M}, \mathcal{R} \rangle$ *is called a reactive flow system, if for all molecules* $i \in \mathcal{M}$, *there exists a first-order decay reaction* $\rho \in \mathcal{R}$ *with* $\mathrm{LHS}(\rho) = \{i\}$ *and* $s_{i,\rho} = -1$.

This is a typical situation for chemical flow reactors or bacteria that grow and divide[33]. Note that in a cell that grows and divides, every molecule including the genome is subject to a dilution flow.

In a *reactive flow system*, semi-organisations are not necessarily organisations. Nevertheless, both the semi-organisations and the organisations form a lattice $\langle \mathcal{O}, \sqcup, \sqcap \rangle$. Moreover, the union ($\sqcup$) and intersection ($\sqcap$) of any two organisations is an organisation (see below).

Example 10 (reactive flow system) *As an example, we take the three-membered elementary hypercycle as before, but add an explicit substrate s. There are four molecules* $\mathcal{M} = \{a, b, c, s\}$ *and seven reaction rules:*

$$\mathcal{R} = \{a+b+s \rightarrow a+2b, \quad a \rightarrow \emptyset,$$
$$b+c+s \rightarrow b+2c, \quad b \rightarrow \emptyset,$$
$$c+a+s \rightarrow c+2a, \quad c \rightarrow \emptyset,$$
$$s \rightarrow \emptyset\}.$$

We can see, that again all molecules decay by first-order reactions (representing the dilution flow), but now a molecule, s, is consumed by other reactions. Note that in this example there is only one organisation (the empty organisation), and even no semi-organisation that contains one or more molecules, because there is no inflow of the substrate. If we would add a reaction equation $\emptyset \rightarrow s$ *representing an inflow of the substrate, we would obtain two organisations:* $\{s\}$ *and the "hypercycle"* $\{a, b, c, s\}$.

3.5.3   Reactive flow system with persistent molecules

In a *reactive flow system with persistent molecules* there are two types of molecules: persistent molecules and non-persistent molecules. All non-persistent molecules are consumed (as in the two systems before) by first-order decay reactions, whereas a persistent molecule is not consumed by any reaction at all.

Definition 10 (reactive flow system with persistent molecules[10]) *A   reaction network $\langle \mathcal{M}, \mathcal{R} \rangle$ is called a reactive flow system with persistent molecules, if we can partition the set of molecules in persistent P and non-persistent molecules $\bar{P}(\mathcal{M} = P \cup \bar{P}, P \cap \bar{P} = \emptyset)$ such that: (i) for all non-persistent molecules $i \in \bar{P}$: there exists a first-order decay reaction $\rho \in \mathcal{R}$ with $\mathrm{LHS}(\rho) = \{i\}$ and $s_{i,\rho} = -1$; and (ii) for all persistent molecules $i \in P$: there does not exist a reaction $\rho \in \mathcal{R}$ with $s_{i,\rho} < 0$.*

An example of a *reactive flow system with persistent molecules* is Example 2, where d is a persistent molecule. The *reactive flow system with persistent molecules* is the most general of the three systems where the semi-organisations and organisations always form a lattice, and where the generate organisation operator can properly be defined (see below). As in a *reactive flow system*, not all semi-organisations are organisations.

Lemma 4 [10] *A reactive flow system with persistent molecules is consistent (Proof in Ref.[10]).*

Remember that the intersection of two closed sets is again always closed. In a similar way, in consistent reaction systems, the union of self-maintaining sets is again self-maintaining. However, Section 3.7 will demonstrate that the latter is not necessarily true for general reaction systems.

3.6   Common Properties of Consistent Reaction Systems

Consistent reaction systems (including those discussed in Sections 3.5.1–3.5.3) possess some comfortable properties that allow us to present a series of useful definitions and lemmas.

In a consistent reaction system, given a set of molecules A, we can uniquely *generate* a semi-self-maintaining set, a semi-organisation, a self-maintaining set, and an organisation in a similar way as we have generated a closed set. And like for closed sets, we can define the union and intersection on semi-self-maintaining sets, semi-organisations, self-maintaining sets, and organisations, respectively. Furthermore, each, the semi-self-maintaining sets, semi-organisations, self-maintaining sets, and organisations form a lattice together with their respective union and intersection operators. This does not generalise to general reaction systems (see Section 3.7), because for a general reaction system we cannot uniquely generate a semi self-maintaining set, a semi-organisation, a self-maintaining set, nor an organisation as is the case with consistent reaction systems.

3.6.1 Generate semi-self-maintaining set

Definition 11 (generate semi-self-maintaining set[10]) *Given a set of molecules $A \subseteq \mathcal{M}$ of a consistent reaction system, we define $G_{SSM}(C)$ as the biggest semi-self-maintaining set S contained in A. We say that A generates the semi-self-maintaining set $S = G_{SSM}(A)$.*

In order to calculate the semi-self-maintaining set generated by **A**, we remove those molecules that are consumed and not produced within **A**, until all molecules consumed are also produced, and thus reaching a semi-self-maintaining set. The operator G_{SSM} (generate semi-self-maintaining set) implies the union $\sqcup_{SSM}$ and intersection $\sqcap_{SSM}$ on semi-self-maintaining sets: Given two semi-self-maintaining sets S_1 and S_2, the semi-self-maintaining sets generated by their union $(S_1 \sqcup_{CL} S_2)$ and intersection $(S_1 \sqcap_{CL} S_2)$ are defined as: $S_1 \sqcup_{SSM} S_2 \equiv G_{SSM}(S_1 \cup S_2)$, and $S_1 \sqcap_{SSM} S_2 \equiv G_{SSM}(S_1 \cap S_2)$, respectively.

3.6.2 Generate self-maintaining set

Definition 12 (generate self-maintaining set[10]) *Given a set of molecules $C \subseteq \mathcal{M}$ of a consistent reaction system, we define $G_{SM}(C)$ as the biggest self-maintaining set S contained in C. We say that C generates the self-maintaining set $S = G_{SM}(C)$.*

For consistent reaction systems, $G_{SM}(C)$ is always defined, because the union ($\cup$) of two self-maintaining sets is self-maintaining; and further, every set is either self-maintaining, or it contains a unique biggest self-maintaining set. Thus from every set we can generate a self-maintaining set. Note that self-maintaining sets are also semi-self-maintaining, $G_{SM}(G_{SSM}(S)) \equiv G_{SSM}(S)$, which is a useful property, because $G_{SSM}(S)$ is easier to compute. As usual, the union $\sqcup_{SM}$ and intersection $\sqcap_{SM}$ of self-maintaining sets S_1, S_2 are defined as $S_1 \sqcup_{SM} S_2 \equiv G_{SM}(S_1 \cup S_2)$, $S_1 \sqcap_{SM} S_2 \equiv G_{SM}(S_1 \cap S_2)$, respectively. Thus also the set of all self-maintaining sets $\mathcal{O}_{SM}$ forms a lattice $\langle \mathcal{O}_{SM}, \sqcup_{SM}, \sqcap_{SM} \rangle$. If S is self-maintaining, its closure $G_{CL}(S)$ is self-maintaining, too (again, not valid for *general reaction systems*).

3.6.3 Generate organisation

There are many ways in which we can generate an organisation from a set. We will present here the simplest one, which implicitly assumes that molecules are produced quickly and vanish slowly. This assumption leads to the largest possible organisation generated by a set:

Definition 13 (generate organisation[20,10]) *Given a set of molecules $A \subseteq \mathcal{M}$ of a consistent reaction system, we define*

$$G(A) = G_{SM}(G_{CL}(A)). \tag{16}$$

We say that A generates the organisation $O = G(A)$.

Equivalently we can also generate a semi-organisation $O = G_{SO}(A) = G_{SSM}(G_{CL}(A))$.

Following the same scheme as before, the union $\sqcup$ and intersection $\sqcap$ of two organisations O_1 and O_2 is defined as the organisation generated by their set-union and set-intersection:

$$O_1 \sqcup O_2 \equiv G(O_1 \cup O_2), \tag{17}$$

$$O_1 \sqcap O_2 \equiv G(O_1 \cap O_2). \tag{18}$$

Equivalently $G(A) = G_{SM}(G_{SSM}(G_{CL}(A)))$, which allows to compute the organisation generated by a set more easily in three steps.

Thus, for consistent reaction systems, also the set of all organisations $\mathcal{O}$ forms a lattice $\langle \mathcal{O}, \sqcup, \sqcap \rangle$. This important fact should be emphasised by the following lemma:

Lemma 5 [10] *Given a reaction network* $\langle \mathcal{M}, \mathcal{R} \rangle$ *of a consistent reaction system and all its organisations* $\mathcal{O}$, *then* $\langle \mathcal{O}, \sqcup, \sqcap \rangle$ *is a lattice.*

Knowing that the semi-organisations and organisations form a lattice, and that we can uniquely generate an organisation for every set, is a useful information. Then we know, for example, that there is a largest organisation. Furthermore we can map every state of a reaction vessel uniquely to an organisation (Section 4.1), which allows to characterise states and to partition the state space, uniquely.

In order to find the whole set of organisations, it is impractical just to check all the possible sets of molecules. Instead, we can start by computing the lattice of semi-organisations, and then test only those sets for self-maintenance. Furthermore, if the semi-organisations form a lattice, we can start with small sets of molecules and generate their semi-organisations, while the $\sqcup_{SO}$ operator can lead us to the more complex semi-organisations.

3.7   General Reaction Systems

When we consider general reaction systems, that is, reaction networks without any constraints, we cannot always generate a self-maintaining set uniquely. This implies that in a general reaction system neither the set of organisation nor the set of semi-organisations necessarily form a lattice. Examples of this can be found in planetary atmosphere chemistries[42,8].

Example 11 (Reaction system without a lattice of organisation) *In this example we present a simple reaction network where the set of organisations does not form a lattice, and where we can not always generate an organisation for any given set of molecules:*

$$\mathcal{M} = \{a, b, c\}, \quad \mathcal{R} = \{a + b \rightarrow c\}. \tag{19}$$

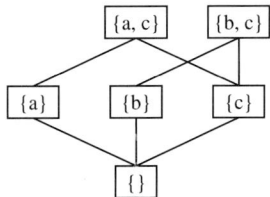

Figure 3. Example of a general reaction system that does not have a lattice of organisations. There are three molecules $\mathcal{M} = \{a, b, c\}$ and just one reaction rule $\mathcal{R} = \{a + b \rightarrow c\}$

This example can be interpreted as an isolated system, where there is no inflow nor outflow. a and b simply react to form c. Obviously, every set that does not contain a together with b is an organisation. So, there are 6 organisations: {}, {a}, {b}, {c}, {a, c}, and {b, c}. As illustrated by Figure 3, there is no unique largest organisation and therefor the set of organisations does not form a lattice. Furthermore, given the set {a, b, c}, we can not generate an organisation uniquely, because there does not exist a unique largest self-maintaining set contained in {a, b, c}. There are two self-maintaining sets of equal size: {a, c} and {b, c}. Why can this not happen in a reactive flow system with persistent molecules ? In a reactive flow system with persistent molecules, the set-union of two self-maintaining sets is again self-maintaining. Therefore there can not exist two largest self-maintaining sets within a set A, because their union would be a larger self-maintaining set within A (see Lemma 5). Note that each organisation makes sense, because each organisation represents a combination of molecules that can stably exists in a reaction vessel, which does not allow an outflow of any of the molecules according to the rules $\mathcal{R}$.

4. CHEMICAL ORGANISATION THEORY: DYNAMICAL PART

The previous section deals with molecules $\mathcal{M}$ and their reaction rules $\mathcal{R}$, but not with the evolution of the system in time. However, the structures that we identified (i.e., the organisations) possess a strong relation to the potential dynamical behaviour of the reaction system. We will now get back the dynamics into our consideration.

To add dynamics to the theory, we have to formalise the dynamics of a system. In a very general approach, the *dynamics* is given by a *state space* X and a formal definition (mathematical or algorithmic) that describes all possible movements in X for any possible initial state $\mathbf{x}_0 \in X$. For simplicity, we assume a deterministic dynamical process described by a system of ordinary differential equations (Eq. (12), Section 2.3).

4.1 Connecting to the Static Theory

Let us assume now that $\mathbf{x} \in X$ represents the state of a reaction vessel, which contains molecules from $\mathcal{M}$. In the static part of the theory we consider just the set of molecular species present in the reaction vessel, but not their concentrations, spatial distributions, velocities, and so on.

Now, given the state $\mathbf{x}$ of the reaction vessel, we need a function that maps uniquely this state to the set of molecules present. Vice versa, given a set of molecules $A \subseteq \mathcal{M}$, we need to know, which states from X correspond to this set of molecules. For this reason we introduce a mapping ϕ called *abstraction*, from X to $\mathcal{P}(\mathcal{M})$, which maps a state of the system to the set of molecules that are present in the system being in that state. The exact mapping can be defined precisely later, depending on the state space, on the dynamics, and on the actual application.

The concept of *instance* is the opposite of the concept of abstraction. While $\phi(\mathbf{x})$ denotes the molecules represented by the state $\mathbf{x}$, an instance $\mathbf{x}$ of a set A is a state where exactly the molecules from A are present according to the function ϕ.

Definition 14 (instance[10]) *Given a function $\phi : X \to \mathcal{P}(\mathcal{M})$ (called abstraction), which maps a state to a set of molecules, we say that a state $\mathbf{x} \in X$ is an instance of $A \subseteq \mathcal{M}$, if and only if $\phi(\mathbf{x}) = A$.*

In particular, we can define an instance of an organisation O (if $\phi(\mathbf{x}) = O$) and an instance of a generator of O (if $G(\phi(\mathbf{x})) = O$). Loosely speaking we can say that $\mathbf{x}$ *generates organisation* O.

Note that a state $\mathbf{x}$ of a consistent reaction system (Section 3.5) is *always* an instance of a generator of one and only one organisation O. This leads to the important observation that a lattice of organisations partitions the state space X, where a partition $X_O \subseteq X$ implied by organisation O is defined as the set of all instance of all generators of O:

$$X_O = \{\mathbf{x} \in X | G(\phi(\mathbf{x})) = O\} \quad \text{(states generating organisation } O\text{).} \quad (20)$$

Note that as the system state evolves over time, the organisation $G(\phi(\mathbf{x}(t)))$ generated by $\mathbf{x}(t)$ might change (see below, Figure 4).

4.2   Fixed Points are Instances of Organisations.

Now we will describe a theorem that relates fixed points to organisations, and by doing so, underlines the relevancy of organisations. We will show that, given an ordinary differential equation (ODE) of a form that is commonly used to describe the dynamics of reaction systems, every fixed point of this ODE is an instance of an organisation. We therefore assume in this section that $\mathbf{x}$ is a concentration vector $\mathbf{x} = (x_1, x_2, \ldots, x_{|\mathcal{M}|})$, $X = \mathbb{R}^{|\mathcal{M}|}$, $x_i \geq 0$ where x_i denotes the concentration of molecular species i in the reaction vessel, and $\mathcal{M}$ is finite.

The dynamics is given by an ODE of the form $\dot{\mathbf{x}} = \mathbf{S}\mathbf{v}(\mathbf{x})$ where $\mathbf{S}$ is the stoichiometric matrix implied by the reaction network $\langle \mathcal{M}, \mathcal{R} \rangle$ (see Section 2.2). $\mathbf{v}(\mathbf{x}) = (v_1(\mathbf{x}), \ldots, v_r(\mathbf{x})) \in \mathbb{R}^r$ is a flux vector depending on the current concentration $\mathbf{x}$, where r denotes the number of reaction rules. A flux $v_\rho(\mathbf{x}) \geq 0$ describes the rate of a particular reaction ρ. For the function v_ρ we require only that $v_\rho(\mathbf{x})$ is positive,

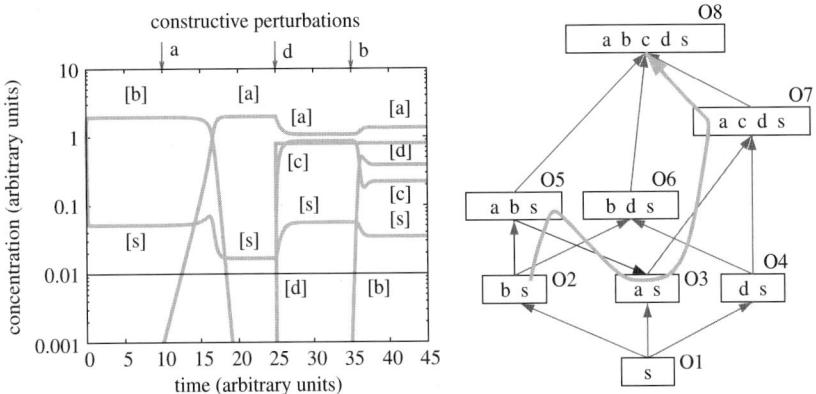

Figure 4. Example of a constructive dynamics in state space (left) and its visualisation in the lattice of organisations (right). Reaction network: Example 2 (five species) as shown in Figure 1. The following constructive perturbations are performed: At $t = 10$ we add a tiny amount (i.e., 0.001 units) of a; at $t = 25$ we add 0.8 units of d; and at $t = 35$ we add 0.001 units of b. Simulation parameters: $v_1 = 30[a]^2[s]$, $v_2 = 10[b]^2[s]$, $v_3 = [a][b]$, $v_4 = 10[b][c]$, $v_5 = [a][d]$, $v_6 = 2$, $v_7 = v_8 = v_9 = 1$. Initial state: $[a] = 0$, $[b] = 1$, $[c] = 0$, $[d] = 0$, $[s] = 1$. ODE: Eq. (12) with Eq. (11). Threshold $\Theta = 0.01$ (cf. Eq. (21))

if and only if the molecules on the left hand side of the reaction ρ are present in the state $\mathbf{x}$, and otherwise it must be zero (see Assumption 2, Section 2.3). Often it is also assumed that $v_\rho(\mathbf{x})$ increases monotonously, but this is not required here.

Given the dynamical system as $\dot{\mathbf{x}} = \mathbf{Sv}(\mathbf{x})$, we can define the abstraction of a state $\mathbf{x}$ formally by using a (small) threshold $\Theta \geq 0$ such that all fixed points have positive coordinates greater than Θ.

Definition 15 (abstraction[10]) *Given a dynamical system $\dot{\mathbf{x}} = f(\mathbf{x})$ and let $\mathbf{x}$ be a state in X, then the abstraction $\phi(\mathbf{x})$ is defined by*

$$\phi(\mathbf{x}) = \{i \mid x_i > \Theta, i \in \mathcal{M}\}, \quad \phi : X \to \mathcal{P}(\mathcal{M}), \quad \Theta \geq 0 \qquad (21)$$

where x_i is the concentration of molecular species i in state $\mathbf{x}$, and Θ is a threshold chosen such that it is smaller than any positive coordinate of any fixed point of $\dot{\mathbf{x}} = f(\mathbf{x})$, $x_i \geq 0$.

Setting $\Theta = 0$ is a safe choice, because in this case ϕ always meets the definition above. But for practical reasons, it makes often sense to apply a positive threshold greater zero, e.g., when we take into consideration that the number of molecules in a reaction vessel is finite.

Theorem 1 *Let us consider a general reaction system whose reaction network is given by the reaction network $\langle \mathcal{M}, \mathcal{R} \rangle$ and whose dynamics is given by $\dot{\mathbf{x}} = \mathbf{Sv}(\mathbf{x}) = f(\mathbf{x})$ as defined before. Let $\mathbf{x}' \in X$ be a fixed point, that is, $f(\mathbf{x}') = \mathbf{0}$, and let us consider an abstraction ϕ as given by Definition 15, which assigns a set of molecules to each state $\mathbf{x}$; then $\phi(\mathbf{x}')$ is an organisation (Proof in Ref.[10]).*

In other words, each fixed point $\mathbf{x}'$ is an instance of an organisation. Intuitively this has to be true, for the following reasons: Assume, ad absurdum, that there is a fixed point formed by a set of species that is not closed. These molecules would inevitably produce new species, which is in contradiction with the definition of a fixed point. Now assume, ad absurdum, that there is a fixed point formed by a set of species that is not self-maintaining. In this case, there is no flux vector such that all species can be produced at a non-negative rate, which again contradicts the definition of a fixed point.

From this theorem it follows immediately that a fixed point is an instance of a closed set, a semi-self-maintaining set, and of a semi-organisation. Let us finally mention that even if each fixed point is an instance of an organisation, an organisation does not necessarily possess a fixed point. A well known example is exponential growth: $\mathcal{M} = \{a\}$, $\mathcal{R} = \{a \rightarrow 2a\}$, $\dot{x} = \mathbf{S}v(x)$ with $\mathbf{S} = 1$ and $\mathbf{v}(x) = x$. There are two organisation the empty organisation $\{\}$ and the organisation $\{a\}$, which represents an exponentially growing population. Obviously there is no fixed point with $x > 0$.

4.3   Movement from Organisation to Organisation

As opposed to an ODE, molecular systems are inherently discrete. The amount of molecules present in a reaction vessel is countable and can be represented by a natural number. In finite time, a molecular species can vanish completely, so that its concentration becomes exactly zero. Therefore, the composition of molecular species in a reaction vessel can change while new species enter or present species vanish.

This change of the composition of molecular species can be interpreted as a movement in the set of all possible sets of molecules $\mathcal{P}(\mathcal{M})$. In that case we can track the dynamics in the lattice of all possible sets of molecules. Note that $\mathcal{P}(\mathcal{M})$ is usually much smaller than X, however it can still be quite large, since it grows exponentially with the number of molecules.

As we can see in Figure 2, the lattice of all sets can already by quite complicated. A solution to this problem is to track a constructive dynamics in the lattice (or set) of organisations (Figure 4). This level of abstraction will filter out those changes that do not lead to a new organisation. Thus providing a quite high-level view.

4.3.1   ODEs and movement in the set of organisations

In an ODE like Eq. (12) molecules cannot vanish completely in finite time. They can only *tend to* zero as time *tends to* infinity. So, even if in reality a molecule disappears, in an ODE model it might still be present in a tiny quantity. A molecule whose concentration tends to zero in an ODE can be interpreted as a molecule that would vanish completely in reality, which would change the set of molecules present.

A common approach to overcome this problem is to introduce a concentration threshold Θ, below which a molecular species is considered not to be present. We use this threshold in order to define the abstraction ϕ, which just returns the set of molecules present in a certain state. Additionally, we might use the threshold to manipulate the numerical integration of an ODE by setting a concentration to zero, when it falls below the threshold. In this case, a constructive perturbation (i.e., a perturbation that causes a new molecular species to appear) has to be greater than this threshold.

4.3.2   Downward movement

Not all organisations are stable. The fact that there exits a flux vector, such that no molecule of that organisation vanishes, does not imply that this flux vector can be realized when taking dynamics into account. As a result a molecular species can disappear. Each molecular species that disappears simplifies the system. Some molecules can be generated back. But eventually the system can move from a state that generates organisation O_1 into a state that generates organisation O_2, with O_2 always below $O_1 (O_2 \subset O_1)$. We call this spontaneous movement a *downward movement*.

   Figure 4 illustrates this downward movement using the five-species example (Example 2). Starting with high concentration of the molecular species $\{a, b, s\}$ (at time $t = 15$ in Figure 4) the system moves spontaneously down to organisation $\{a, s\}$.

4.3.3   Upward movement

Moving up to an organisation above requires that a new molecular species appears in the system. This new molecular species cannot be produced by a reaction among present molecules (condition of closure). Thus moving to an organisation above is more complicated then the movement down and requires a couple of specifications that describe how new molecular species enter the system. Here we assume that new molecular species appear by some sort of random perturbations or purposeful interference, called *constructive perturbation*. We assume that a small quantity of molecules of that new molecular species (or a set of molecular species) suddenly appears.

Definition 16 (constructive perturbation) *A perturbation that moves a state* **x** *to a perturbed state* **x**′ *where the molecular species present in* **x** *are different from those present in* **x**′ *is called a constructive perturbation.*

   Often, in practice, a constructive perturbation (appearance of new molecular species) has a much slower time scale than the internal dynamics (e.g., chemical reaction kinetics) of the system.

4.3.4   Visualising movements in the set of organisations

In order to display potential movements in the lattice or set of organisations, we can draw links between organisations. As exemplified in Figure 4, these links can indicate possible downward movements (down-link, blue) or upward movements (up-link, red). A neutral link (black line) denotes that neither the system can move spontaneously down, nor can a constructive perturbation move the system up. Whether the latter is true depends on the definition of "constructive perturbation" applied. For the example in Figure 4 we defined a constructive perturbation as inserting a small quantity of *one* new molecular species.

The dynamics in between organisations is more complex than this intuitive presentation might suggest, for example in some cases it is possible to move from one organisation O_1 to an organisation O_2, with O_2 above (or below) O_1 without passing through the organisations in between O_1 and O_2.

5. ORGANISATIONS IN REAL SYSTEMS

Speroni et al.[39] have shown that artificial chemical reaction networks that are based on a structure-to-function mapping[20,3] possess a more complex lattice of organisation than networks created randomly. From this observation we can already expect that natural networks possess non-trivial organisation structures. Investigation of models of planetary photo-chemistries[42,8] and bacterial metabolism[33,7], revealed lattices of organisations that vanish when the networks are randomised, indicating a non-trivial structure.

Here two brief examples from ongoing research[29,7] are presented. The first example, studied by Matsumaru et al.[29], is a model of HIV-immune system dynamics comprising four species. The simplicity of that model allows to validate the approach analytically. The second example, taken from Ref.[7], is based on a model of the central sugar metabolism of *E.coli* by Puchalka and Kierzek[33] comprising 92 species. It shows that non-trivial network complexity can be tackled and that sub-structures in the model can be identified not known before.

5.1 Example: HIV-Immune System Dynamics

Wodarz and Nowak[41] developed a model of immunological control of HIV in order to explain the effect of various drug treatment strategies. Especially the model shows, why a specific drug treatment strategy does not try to remove the virus, but aims at stimulating the immune defence, such that the immune system controls the virus at low but positive quantities.

In the model, there are four molecular species: $\mathcal{M} = \{x, y, w, z\}$: uninfected CD4$^+$ T cells x, infected CD4$^+$ T cells y, cytotoxic T Lymphocyte (CTL) precursors w, and CTL effectors z. The concentration of each species is specified by x, y, w, and z, respectively. The dynamics is given by an ordinary differential equation (ODE) with kinetic parameters a, b, c, d, h, p, q, β, and λ:

$$\dot{x} = \lambda - dx - \beta xy,$$
$$\dot{y} = \beta xy - ay - pyz,$$
$$\dot{w} = cxyw - cqyw - bw, \tag{22}$$
$$\dot{z} = cqyw - hz.$$

From the given deterministic ODE model we derive chemical reaction rules, which form a reaction network (Fig. 5, right top):

$$\emptyset \rightarrow x,$$

$$\mathcal{R} = \{ \begin{array}{ll} x + y \rightarrow 2y, & x \rightarrow \emptyset, \\ y + z \rightarrow z, & y \rightarrow \emptyset, \\ x + y + w \rightarrow x + y + 2w, & w \rightarrow \emptyset, \\ y + w \rightarrow y + z, & z \rightarrow \emptyset \}. \end{array}$$

The ODE model includes a decay term for each species. Therefore, for each species, we have a reaction rule transforming that molecular species into the empty set: $x \rightarrow \emptyset$, $y \rightarrow \emptyset$, $w \rightarrow \emptyset$, and $z \rightarrow \emptyset$. We observe in passing that in this particular case, since all species decay, the system is a reactive flow system (Definition 3.5.2), thus consistent (Lemma 4), and therefore the set of organisations must be a lattice (Lemma 5), with one well defined largest and one well defined smallest organisation.

A graphical representation of the network is shown in Fig. 5, upper right corner. The corresponding 4×9 stoichiometric matrix $\mathbf{S}$ reads:

$$\mathbf{S} = \begin{array}{c} x \\ y \\ w \\ z \end{array} \begin{pmatrix} 1 & -1 & 0 & 0 & 0 & -1 & 0 & 0 & 0 \\ 0 & 0 & -1 & 0 & 0 & 1 & -1 & 0 & 0 \\ 0 & 0 & 0 & -1 & 0 & 0 & 0 & 1 & -1 \\ 0 & 0 & 0 & 0 & -1 & 0 & 0 & 0 & 1 \end{pmatrix}$$

where each row corresponds to a molecular species x, y, w, z (from the top) and each column corresponds to a reaction. As mentioned previously, the stoichiometric matrix does not contain all information of the reaction network. For example, the reaction rule $y + z \rightarrow z$ appears only as the column vector $(0, -1, 0, 0)^T$.

5.1.1 Lattice of organisations

For applying the theory we check every possible set of species (i.e., 16 sets) whether it is closed and self-maintaining. As a result three organisations are found. The Hasse diagram is depicted in Figure 5, middle right. The smallest organisation consists only of the "healthy cells" x (uninfected CD4$^+$ T cells). There can not be a smaller organisation (e.g., the empty set), because x is an input species and therefore the empty set is not closed. Since x is an input species, the set $\{x\}$ is obviously self-maintaining. Looking at the reaction rules we can see that x alone can not produce anything else, thus the set $\{x\}$ is closed, too. Formally, we can show that $\{x\}$ is an organisation.

The second organisation, $\{x, y\}$ contains "healthy cells" x together with "ill cells" (infected CD4$^+$ T cells). Looking at the reaction network, we can see that $\{x, y\}$ is closed, because there is no reaction rule that allows to produce w or z just using x and y alone. With the flux vector $\mathbf{v} = (10, 1, 1, 0, 0, 1, 0, 0, 0)^T$ we can show that, according to Definition 5, $\{x, y\}$ is self-maintaining, e.g., $\mathbf{Sv} = (8, 0, 0, 0)^T$.

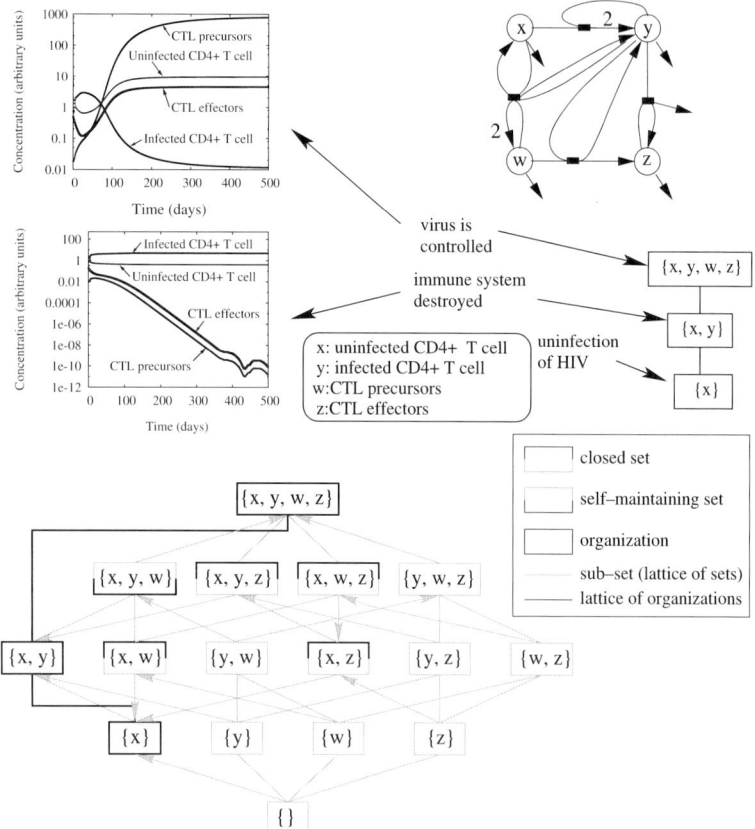

Figure 5. Illustration of the analysis of the HIV immunological response model by Wodarz and Nowak [41]. The ODE model given in Eq. 22 is transformed to a chemical reaction network (right top). The resulting hierarchy of organisations is shown as a Hasse diagram (right middle). Two of the organisations represent the attractors: virus under control (top organisation) and immune system destruction (middle organisation). Dynamic simulations [28] leading to both attractors are shown on the left. Parameters were taken from Ref. [41] as follows: $\lambda = 1$; $d = 0.1$; $\beta = 0.5$; $a = 0.2$; $p = 1$; $c = 0.1$; $b = 0.01$; $q = 0.5$; $h = 0.1$. Initial concentrations for left, top plot: $x = 0.74$; $y = 0.75$; $w = 0.018$; $z = 0.49$. Initial concentrations for left, bottom plot: $x = 0.75$; $y = 0.14$; $w = 0.0095$; $z = 0.17$. At the bottom, the full lattice of sets is shown including closed and self-maintaining sets. See Matsumaru et al.[29] for details. Figure reproduced from Ref.[10]

The largest organisation contains all species and is thus obviously closed. Looking at the reaction rules, we can see that since x can be produced at an arbitrarily high rate, we can also produce y, z, and w at arbitrarily high rates, because we can freely choose the flux vector **v** according to the definition of self-maintenence (Definition 5). Actually, the production rate **Sv** of all four species $\{x, y, w, z\}$ can be positive, when we chose for example a flux vector like $\mathbf{v} = (100, 1, 1, 1, 50, 1, 50, 10)^T$.

No further organisation exists, which implies according to Theorem 1 (Section 4.2) that there is no other combination of species that can form a stationary state.

5.1.2   Connecting with dynamics and explaining a drug treatment strategy

From a mathematical analysis[41] and simulation studies[28] it is known that the model has two modes of behaviour belonging to two asymptotically stable fixed points: One of the attractors is characterised by high virus load and no CTL precursors and effectors present. This state is interpreted as the complete destruction of the immune defence. The organisation $\{x, y\}$ represents this attractor. When the HIV virus is controlled by the immune defence, all four molecular species are present in the system, constituting the other attractor. This state is reflected in the largest organisation $\{x, y, w, z\}$. The smallest organisation $\{x\}$ can be interpreted as the condition where no $CD4^+$ T cell is infected by the HIV virus.

After identifying the lattice of organisations, we can use it to explain the strategy of a drug therapy: Looking at the lattice of organisations, we can describe two strategies for a drug therapy: The first one tries to move the system into the smallest organisations $\{x\}$, where no virus is present at all. An alternative strategy may move the system into the largest organisation, where the virus is present, but also an immune system response controlling the virus.

There are drugs available that can bring down the virus load by several orders of magnitude. If by this procedure the virus could be completely removed, the system would move into the smallest organisation, because the set $\{x, w, z\}$ generates[3] organisation $\{x\}$. However, it has been observed that although the virus load can be decreased below detection limit, the virus can not be fully removed so that the virus appears again after stopping the treatment. Therefore, the actual strategy of a drug therapy is not to move the system into the lowest organisation, but into the highest organisation. In practice, this is achieved by applying the drug periodically allowing the immune defence to increase[41].

We can see that the strategy of a drug treatment can be explained on a relatively high (i.e., less detailed) level of abstraction using the lattice of organisations, namely as a movement from an organisation representing an ill state to an organisation representing a healthy state. It is important to note that choosing the right level of abstraction depends on what should be explained. The lattice of organisations is a suitable level of abstraction for describing the overall strategy, i.e., the quality of a drug treatment. However, *how* an actual drug treat should look like quantitatively in order to move the system into the largest organisation can not be answered by our theory. For this we have to chose a more detailed level of abstraction, e.g., the ODE model, which provides information on how the system can move from one organisation to another.

---

[3] Note that we use the word "generate" as a precisely defined technical term (Eq. (16), Section 3.6.3).

5.2   Example: Central Sugar Metabolism of E. Coli

In order to demonstrate that chemical organisation theory can reveal structures in networks of non-trivial size, Centler et al.[7] applied the theory to a large stochastic network model of the central sugar metabolism of E. Coli as introduced by Puchalka and Kierzek[33].

The model consists of 92 species and 197 reactions, including gene expression, signal transduction, transport, and enzymatic activities. The model was simplified by ignoring inhibitory links. By doing so we assume that an inhibition has only a quantitative effect and does not shut off a reaction pathways completely (cf. Section 2.4).

Two slightly different reaction networks can be studied depending on whether we consider activators as necessary or not in the production of proteins.

When we consider activators to be necessary, a relatively complex lattice of organisations appears as depicted in Figure 6. The smallest organisation, O_1, contains 76 molecules including the glucose metabolism and all input molecules. The input molecules, chosen according to Ref.[33], include the external food set (Glcex, Glyex, Lacex) and all promoters. Two other organisations, O_3 and O_4, contain the Lactose and Glycerol metabolism, respectively. Their union results in the largest organisation O_5 that contains all molecules.

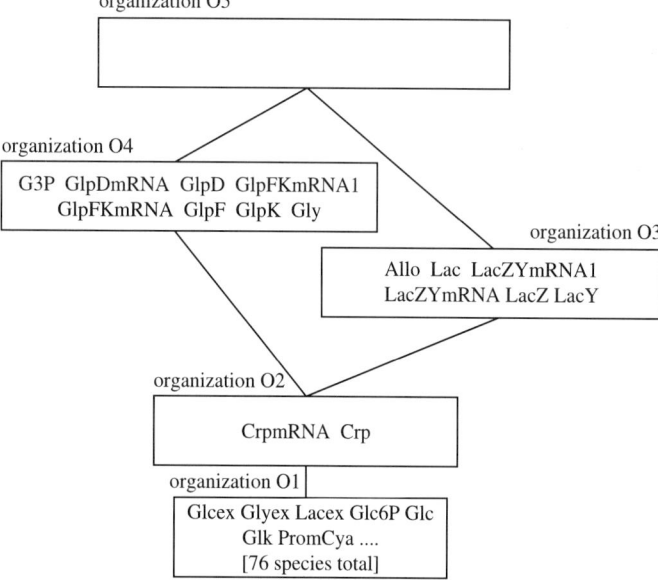

Figure 6. Lattice of organisations of a model of the central sugar metabolism of *E.coli.* [33]. In an organisation, only names of new molecular species are printed that are not present in an organisation below. The vertical position of an organisation correlates with the number of chemical species it contains. Organisation O_5 (top) is the largest one, containing all species from O_4 and O_3. Figure from Refs.[7] and [10]

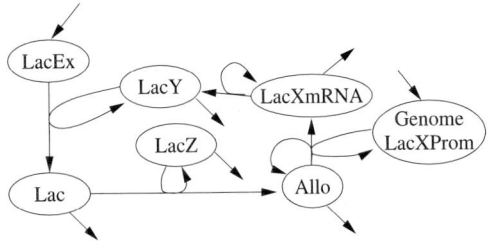

Figure 7. Illustration of the Mechanism leading to Organisation O_3 in Figure 6. *LacEx* represents external lactose, whereas *Lac* represents lactose inside the cell

In order illustrate the mechanism that leads to higher organisations, we take a closer look at organisation O_3, which includes the lactose uptake. Following Puchalka and Kierzek[33], we assume that external lactose (*LacEx*) is alway present, as well as the genetic information, including the promoter (*LacXProm*) to make messenger RNA (*LacXmRNA*). This is considered by defining a fixed inflow for those molecular species: $\emptyset \rightarrow LacEx$ and $\emptyset \rightarrow LacXProm$. Figure 7 shows the fraction of the reaction network, which explains, why, in the model, the lactose uptake can not be generated from the input species: *LacY* is required for lactose uptake, which is expressed by the reaction $LacEx + LacY \rightarrow LacY + Lac$. However, *LacY* cannot be generated given only the input species, such as *LacEx* and *LacXProm*. Therefore, the closure of the input species does not include molecules like *LacY*. If we add *LacY* to the input species, organisation O_3 can be generated.

If instead we consider the more precise version of the model, which does not assume the necessary presence of activators for gene expression, we have a more realistic model, yet a model that focuses at a longer timescale (as the base level expression of a gene without the activator appears to be relatively slow). In this case the resulting lattice collapses into a single huge organisation. In such case what the theory suggests is a general stability of the system against various perturbations, since no matter what kind of initial combination of molecular species we start off, all other molecular species can be generated, given the input flux mentioned above and sufficient time such that a gene depending on an activator can be expressed in the absence of its activators.

6. DISCUSSION AND OUTLOOK

The theory of chemical organisation, as sketched here, creates a first, rough map of the structure and potential dynamical behaviour of a reaction system. The obtained scaffold (i.e., the set of organisations) can guide further more detailed analysis, which may study the dynamics within or in-between organisations using classical tools from dynamical systems theory. The results of more detailed studies can in turn be explained and summarised with respect to the lattice of organisations resulting in a global picture.

6.1   Related Work

There are a number of other approaches that operate purely on the reaction network's topology in order to infer potential dynamical properties.

Classical reaction network theory provides powerful theorems, which can predict for a specific class of reaction networks whether a network possesses positive stationary states or whether positive periodic solutions are possible. For example, the deficiency-zero theorem states roughly that a weakly reversible mass-action reaction system with deficiency zero contains one unique equilibrium point in each positive reaction simplex[18]. This line of research provides probably the strongest mathematical results that link network structure to potential dynamics. However, the research focuses on *positive* solutions, i.e., solutions where all molecular species are present. Here, we are interested in states of the reaction system where only a subset of species are present. Furthermore, our theory is not restricted to systems with deficiency zero or one. We do not require for our theorem that the reaction system is governed by mass-action kinetics; in turn of course, chemical organisation theory can not predict the stability of an organisation. Furthermore, we do not focus on stationary behaviour, but we aim at understanding complex transitive dynamics, such as the movement between organisations.

Another class of methods that operate on the network structure identifies so called *flux modes*[23,30]. A flux mode is a set of reaction rules that can operate at a steady state. Flux modes are similar to T-invariants, a concept from Petri net theory[31]. Obviously, flux modes can be linearly combined and thus form a complete lattice (when we also consider the empty flux mode). A flux mode implies a set of molecules, namely the set of molecules participating in the reactions of that flux mode. Therefore, a flux mode is similar to the concept of self-maintenence. However, the set of participating molecules is not necessarily self-maintaining nor closed[26]. And not all self-maintaining sets are represented by flux modes[26]. For example, a self-replicating molecule whose concentration growths for ever is self-maintaining. But, since it does not reach a steady state, it is not captured by a flux mode.

Boolean networks or logical networks are another approach to handle networks without requiring detailed kinetics. A boolean network consists of a set of boolean functions that take their own result of a previous time step as input. It is not yet clear, how stationary states of a boolean network are related to organisations, which is an interesting aspect for future investigations. Logical networks are a useful tool to model signalling and gene regulatory networks[17]. However, a boolean network does not consider stoichiometry, which is essential in general reaction networks as considered by our theory.

6.2   Computational Complexity

This chapter describes the mathematical base of our theory and does not focus on algorithmic issues. So far some preliminary algorithms are available. One of them computes the set of organisations from the bottom up by starting from the smallest organisation and then, recursively, adding molecules in order to generated all the organisations above. With this algorithm we can currently analyse networks containing up to 200 molecular species. For less than 30 species, a brute force algorithm appears faster. The brute force algorithm simply tests every possible set of molecules whether it is closed and self-maintaining.

6.3   Structure-to-Function Mapping

The aim of the theoretical framework introduced in this chapter is to deal with constructive dynamical systems. However, the examples we presented for illustrating the new concepts where relatively simple: A set of molecules has always been defined as a list of symbols. And also the reaction rules were given as an explicit list. In other words, in these examples, all molecules that could appear were already listed explicitly in the definition of the set of molecules.

When designing the presented theoretical framework we had already more complex reaction systems in mind, namely those where the set of molecules and reaction rules are defined *implicitly*. In these systems, molecules possess a structure, that is, there is a grammar specifying their syntax, and reaction rules are defined implicitly by referring to that structure. A simple example is the prime number chemistry[2,4], where the set of molecules are all natural numbers and the reaction rules are defined by the numerical devision operator. More complex examples from the field of artificial chemistry are AlChemy[19,20], the combinator chemistry[37], or the more realistic toy chemistry by Benkö et al.[5]. But also in biochemistry and systems biology we observe a growing number of models where the reaction network is defined implicitly, which usually leads to a combinatorial explosion in size. Examples are models of DNA assembly[21], DNA computing[1], or combinatorial signalling networks[32,6].

Note that in our approach the set of molecules $\mathcal{M}$ and the set of reactions $\mathcal{R}$ of a reaction network can be defined implicitly. Furthermore, the dynamics that we assumed in Section 4 is quite general, so that we can theoretically apply our framework also to the systems mentioned above. However, computational tools for an automatic analysis of implicitly defined reaction systems have yet to be developed, which is a significant challenge for future research.

7.   ACKNOWLEDGEMENTS

We grateful to Florian Centler, Naoki Matsumaru, and Christoph Kaleta contributing examples and valuable discussions. Financial support by the *Federal Ministry of Education and Research* (BMBF) Grant 0312704A and the *German Research Foundation* (DFG) Grant Di852/4-1 is greatly acknowledged.

8. REFERENCES

1. L. M. Adleman. Molecular computation of solutions to combinatorial problems. *Science*, 266(5187):1021–1024, 1994.
2. J.-P. Banâtre and D. L. Métayer. The GAMMA model and its discipline of programming. *Sci. Comput. Program.*, 15(1):55–77, 1990.
3. W. Banzhaf. Self-replicating sequences of binary numbers – foundations I and II: General and strings of length n = 4. *Biol. Cybern.*, 69:269–281, 1993.
4. W. Banzhaf, P. Dittrich, and H. Rauhe. Emergent computation by catalytic reactions. *Nanotechnology*, 7(1):307–314, 1996.
5. G. Benkö, C. Flamm, and P. F. Stadler. A graph-based toy model of chemistry. *J. Chem. Inf. Comput. Sci.*, 43(4):1085–1093, 2002.
6. N. Borisov, N. Markevich, H. J.B., and K. B.N. Signaling through receptors and scaffolds: independent interactions reduce combinatorial complexity. *Biophys. J*, 89(2):951–966, 2005.
7. F. Centler, P. S. di Fenizio, N. Matsumaru, and P. Dittrich. Chemical organizations in the central sugar metabolism of escherichia coli. In *Modeling and Simulation in Science Engineering and Technology, Post-proceedings of ECMTB 2005*, pages 1–8. Birkhäuser, 2006. (in print).
8. F. Centler and P. Dittrich. Chemical organizations in atmospheric photochemistries: a new method to analyze chemical reaction networks. Submitted.
9. M. E. Csete and J. C. Doyle. Reverse engineering of biological complexity. *Science*, 295:1664–1669, 2002.
10. P. Dittrich and P. S. di Fenizio. Chemical organization theory. *Bull. Math. Biol.*, 2006. In print.
11. P. Dittrich, T. Kron, and W. Banzhaf. On the formation of social order – modeling the problem of double and multi contingency following luhmann. *Journal of Artifical Societies and Social Simulation*, 6(1), 2003.
12. P. Dittrich, J. Ziegler, and W. Banzhaf. Artificial chemistries – a review. *Artificial Life*, 7(3):225–275, 2001.
13. O. Ebenhöh, T. Handorf, and R. Heinrich. Structural analysis of expanding metabolic network. *Genome Informatics*, 15(1):35–45, 2004.
14. M. Eigen. Selforganization of matter and the evolution of biological macromolecules. *Naturwissenschaften*, 58(10):465–523, 1971.
15. M. Eigen and P. Schuster. The hypercycle: a principle of natural self-organisation, part A. *Naturwissenschaften*, 64(11):541–565, 1977.
16. P. Érdi and J. Tóth. *Mathematical Models of Chemical Reactions: Theory and Applications of Deterministic and Stochastic Models*. Pinceton University Press, Princeton, NJ, 1989.
17. C. Espinosa-Soto, P. Padilla-Longoria, and E. R. Alvarez-Buylla. A gene regulatory network model for cell-fate determination during Arabidopsis thaliana flower development that is robust and recovers experimental gene expression profiles. *Plant Cell*, 16(11):2923–2939, 2004.
18. M. Feinberg and F. J. M. Horn. Dynamics of open chemical systems and the algebraic structure of the underlying reaction network. *Chem. Eng. Sci.*, 29(3):775–787, 1973.
19. W. Fontana. Algorithmic chemistry. In C. G. Langton, C. Taylor, J. D. Farmer, and S. Rasmussen, editors, *Artificial Life II*, pages 159–210, Redwood City, CA, 1992. Addison-Wesley.
20. W. Fontana and L. W. Buss 'The arrival of the fittest': Toward a theory of biological organization. *Bull. Math. Biol.*, 56(1):1–64, 1994.
21. K. V. Gothelf and R. S. Brown. A modular approach to DNA-programmed self-assembly of macromolecular nanostructures. *Chemistry*, 11(4):1062–1069, Feb 2005.
22. C. Heij, A. e. C. M. Ran, and F. v. Schagen. *Introduction to Mathematical Systems Theory Linear Systems, Identification and Control*. Birkhäuser, 2006.
23. R. Heinrich and S. Schuster. *The Regulation of Cellular Systems*. Chapman and Hall, New York, NY, 1996.
24. S. Jain and S. Krishna. Autocatalytic sets and the growth of complexity in an evolutionary model. *Phys. Rev. Lett.*, 81(25):5684–5687, 1998.
25. S. Jain and S. Krishna. A model for the emergence of cooperation, interdependence, and structure in evolving networks. *Proc. Natl. Acad. Sci. U. S. A.*, 98(2):543–547, 2001.

26. C. Kaleta, F. Centler, and P. Dittrich. Analyzing molecular reaction:networks: from pathways to chemical organizations. *Molecular Biotechnology*, 2006. In print.

27. S. A. Kauffman. Cellular homeostasis, epigenesis and replication in randomly aggregated macro-molecular systems. *J. Cybernetics*, 1:71–96, 1971.

28. N. Matsumaru, F. Centler, and P. D. Klaus-Peter Zauner. Self-adaptive scouting – autonomous experimentation for systems biology. In G. R. Raidl, S. Cagnoni, J. Branke, D. Corne, R. Drechsler, Y. Jin, C. G. Johnson, P. Machado, E. Marchiori, F. Rothlauf, G. D. Smith, and G. Squillero, editors, *Applications of Evolutionary Computing, EvoWorkshops 2004*, volume 3005 of *LNAI*, pages 52–62. Springer, Berlin, 2004.

29. N. Matsumaru, F. Centler, P. Speroni di Fenizio, and P. Dittrich. Chemical organization theory applied to virus dynamics. *it – Information Technology*, 48(3), 2006. In print.

30. J. A. Papin, J. Stelling, N. D. Price, S. Klamt, S. Schuster, and B. O. Palsson. Comparison of network-based pathway analysis methods. *Trends Biotechnol*, 22(8):400–405, Aug 2004.

31. C. A. Petri. *Kommunikation mit Automaten*. PhD thesis, University of Bonn, Bonn, 1962.

32. R. Pinkas-Kramarski, I. Alroy, and Y. Yarden. Erbb receptors and egf-like ligands: cell lineage deter-mination and oncogenesis through combinatorial signaling. *J. Mammary. Gland. Biol. Neoplasia*, 2(2):97–107, 1997.

33. J. Puchalka and A. Kierzek. Bridging the gap between stochastic and deterministic regimes in the kinetic simulations of the biochemical reaction networks. *Biophys. J.*, 86(3):1357–1372, 2004.

34. O. E. Rössler. A system theoretic model for biogenesis (in German). *Z. Naturforsch. B*, 26(8):741–746, 1971.

35. P. Schuster and K. Sigmund. Replicator dynamics. *J. Theor. Biol.*, 100:533–8, 1983.

36. D. Segré, D. Lancet, O. Kedem, and Y. Pilpel Graded autocatalysis replication domain (GARD): Kinetic analysis of self-replication in mutually catalytic sets. *Orig. Life Evol. Biosph.*, 28(4–6):501–514, 1998.

37. P. Speroni di Fenizio. A less abstract artficial chemistry. In M. A. Bedau, J. S. McCaskill, N. H. Packard, and S. Rasmussen, editors, *Artificial Life VII*, pages 49–53, Cambridge, MA, 2000. MIT Press.

38. P. Speroni Di Fenizio and P. Dittrich. Artificial chemistry's global dynamics. movement in the lattice of organisation. *The Journal of Three Dimensional Images*, 16(4):160–163, 2002.

39. P. Speroni di Fenizio, P. Dittrich, J. Ziegler, and W. Banzhaf. Towards a theory of organizations. In *German Workshop on Artificial Life*, Bayreuth, 5.-7. April, 2000. In print, available online: di.ttri.ch/p/SDZB2001gwal.html.

40. P. F. Stadler, W. Fontana, and J. H. Miller. Random catalytic reaction networks. *Physica D*, 63:378–392, 1993.

41. D. Wodarz and M. A. Nowak. Specific therapy regimes could lead to long-term immunological control of hiv. *Proc. Nat. Acad. Sci. USA*, 96(25):14464–9, 1999.

42. Y. L. Yung and W. B. DeMore. *Photochemistry of Planetary Athmospheres*. Oxford University Press, New York, 1999.

CHAPTER 12

PROKARYOTIC SYSTEMS BIOLOGY

AMY K. SCHMID AND NITIN S. BALIGA

Institute for Systems Biology, 1441 N 3415st Seatlle, WA 98103, USA

Abstract: Prokaryotic systems biology is a holistic biological approach that enables comprehensive understanding of an organism. However, two opposing strategies have been proposed to attain such understanding: the top-down and bottom-up approaches. Here we present a review of the current status of the prokaryotic systems biology field against the backdrop of the top-down vs. bottom-up debate, including such topics as current experimental and computational methods and recent literature findings. We use four prokaryotic model systems as examples, including *E. coli*, *Caulobacter crescentus*, *Halobacterium NRC-1*, and *Helicobacter pylori*. We posit that systems biology programs which pursue an integrated combination of both approaches will be the most successful in attaining comprehensive systems-level understanding of prokaryotic model organisms

Keywords: Prokaryote, systems biology, *Halobacterium*, top-down approach, bottom-up approach, microarray, proteomics, ChIP-chip, networks, circuits

1.   INTRODUCTION

1.1   Definition and Goals of Systems Biology

Systems biology is a holistic scientific approach to understand complex biological processes by monitoring the behavior of all parts of the system simultaneously (genes, proteins, metabolites *etc.*) rather than one individual part at a time. Systems biology approaches reveal the dynamic interrelationship between system components, thus enabling the discovery of novel biology. For example, the system can be interpreted and organized into an intricate web of interacting networks such as transcriptional regulatory circuits (which are comprised of both protein-protein and protein-DNA interaction networks), intra- and intercellular communication, protein trafficking, signal transduction, *etc.* It is a major goal of systems biology to cultivate an understanding of system behavior in the context of these networks through

395

M. Al-Rubeai and M. Fussenegger (eds.), Systems Biology, 395–423.
© 2007 *Springer.*

mathematical modeling of system responses to genomic or environmental perturbations. Such comprehensive models enable biological systems to be re-engineered
for specific, controlled, or novel outputs, as well as correct undesirable system
states, such as those perturbed to an abnormal or diseased state.

A system can be defined at many different levels of complexity, starting from
small circuits of interacting proteins or DNA-protein interactions, and working
up to progressively larger systems such as subcellular structures (e.g. organelles),
single cells, multicelluar organs, or complete multicellular organisms. Reaching a
molecular systems-level understanding for any biological organism or system will
require iterative cycles of global experiments, computational data integration and
system-level interpretation (Fig 1). However, current experimental techniques and
computational modeling capabilities for eukaryotic organisms fall short of allowing
satisfactory description or interpretation of such intricate systems. In light of this, the
relatively simple prokaryotic model systems provide an excellent starting point for
attaining comprehensive systems-level insights into complex biological phenomena.
Some of the major practical advantages of working with prokaryotic organisms
include relatively small, fully sequenced genomes ($\sim$2–6 Mb), and generally less
elaborate information processing mechanisms (including signal transduction, transcriptional and posttranscriptional processes).

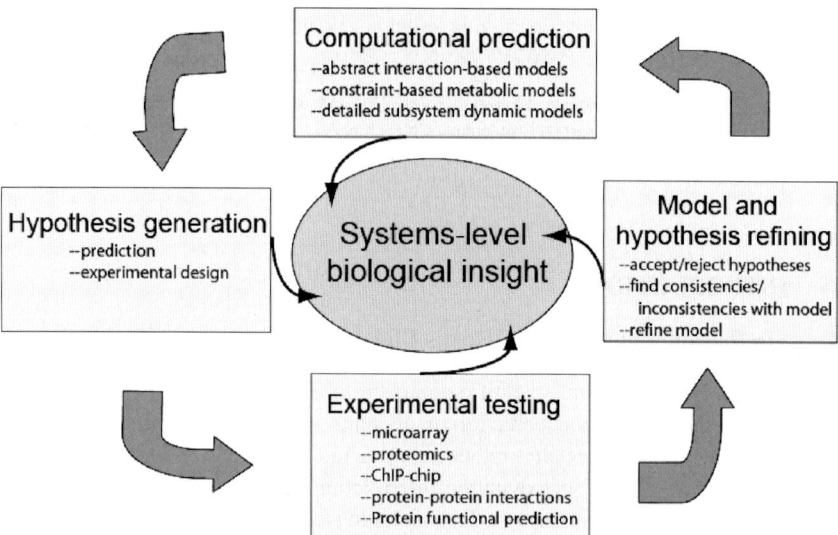

Figure 1. The systems biology inquiry cycle. Several systems biology experimental data types are
analyzed to build quantitative predictive mathematical models. Systems-level hypotheses are generated
from these predictive models, and further experiments are designed to test the hypotheses. Results of
the experiments are compared to the model. Consistencies verify the model and inconsistencies point to
flaws of the model, which is then refined. Systems-level biological insight is gained at every step in the
iterative cycle

Ultimately a systems-level approach towards the study of prokaryotic biology will become a crucial component in continued understanding of the microbial world including infectious bacterial agents and their likely roles in host-pathogen interactions. The depth and detail of our systems-level understanding should eventually lead to mathematical models of corresponding sophistication that will motivate the increasingly intricate re-engineering of microbes for targeted industrial, environmental and medical purposes. Moreover, while applications for the treatment and eradication of infectious disease are likely to grow from the experimental and computational techniques developed and perfected in the prokaryotic forum, these methodologies can likely be transferred and applied successfully for use in studying eukaryotic organisms[2] and human disease.

1.2 Focus and Direction of the Chapter

In this chapter, we will discuss (1) the current status of the field of prokaryotic systems biology; (2) what further developments are required in order to accomplish the ultimate goal of engineering complex biological networks for desired purposes; and (3) what therapeutic, biotechnology, and nanotechnology applications of systems level prokaryotic cell engineering are currently in place. These three themes will be described against the backdrop of a recent debate over so called "top-down" and "bottom-up" approaches to systems biology. Traditional bottom-up approaches build transcriptional regulatory networks in a modular fashion, learning as many mechanistic details as possible about each gene and its function in the regulatory circuit, then linking the circuits together[3,4]. In contrast, top-down approaches compile the regulatory network *de novo* in organisms for which very little information is known by simultaneously monitoring all globally measurable aspects of physiology in response to various stimuli (e.g. growth rate, transcription profiles, DNA-protein interactions, protein-protein interactions, proteomic profiles)[5,6].

Because systems biology is such a new and burgeoning scientific field that produces overwhelming amounts of data, it is imperative that we evaluate the efficacy of each approach in order to: (1) focus and define effective scientific questions for experimental design and data mining; (2) delineate the minimal set of biochemical parameters required for attaining a comprehensive understanding of various biological networks; and (3) determine the level of detail at which biological systems are interrogated. We posit that, because both the top-down and bottom-up approaches have unique advantages and disadvantages, applying a calculated mix of both approaches is the most effective means to overcome the three challenges listed above (See Box 1). This dual method will ultimately navigate the future direction of systems biology efforts, which will enable the comprehensive understanding of biological systems.

Throughout the course of the discussion in this chapter, we will use a series of examples to highlight and evaluate the efficacy of the top-down and bottom-up approaches by focusing on four specific prokaryotic model organisms for which systems-level inquiry is well underway: *Escherichia coli* and *Caulobacter crescentus* will be used as examples in which the bottom-up approach has been implemented, whereas *Helicobacter pylori* and *Halobacterium sp. NRC-1* are examples of the top-down approach.

BOX1. Quick guide to systems biology approaches.

Top-down approach

+ accelerates knowledge discovery in understudied organisms

+ regulatory relationships between biomodules are evident

+ emphasizes the importance of delineating the kinetics of transient global responses

-- Not all biochemical details of the system can be defined using top-down approaches

-- High false positive rates (e.g. in yeast-2-hybrid protein-protien interaction networks) may result in false assumptions or conclusions

Bottom-up approach

+ Provides in-depth biochemical information regarding well-studied model organisms (e.g. E. coli).

+ Literature-based information is detailed and comprehensive, bootstrapped by decades of evidence

+ Allows for the building of accurate, detailed biochemical computational models.

-- Slow and time-intensive

-- Preconceived notions may bias results

--integrating data generated by different experimental methods, time points, and conditions from different labs is quite challenging and often not possible

Integrating bottom-up and top-down approaches may be the key to gaining a comprehensive systems-level understanding of biological complexity.

2.   APPROACHES AND PROCESSES FOR SYSTEMS BIOLOGY IN PROKARYOTES

2.1   The Bottom-up Approach to Prokaryotic Systems Biology: Building the Network in a Modular Fashion, Circuit by Circuit

In this section we will focus on the more traditional bottom-up approach, in which the response of sub-networks or modules to a specific stimulus is emphasized. A module is defined as "groups of proteins that work together to execute a biological function"[7], or more relatively speaking, a module is the functional unit encompassing several regulatory motifs or circuits[8]. A bottom-up approach has been taken to study two well-known bacterial model systems: *Escherichia coli* and *Caulobacter crescentus*.

2.1.1   Cataloging network components: parts lists for C. crescentus and E. coli

The first step to compiling a comprehensive parts lists began with sequencing of the *E. coli* genome in 1997[9] and that of *C. crescentus* in 2001[10]. Subsequently, experimental evidence has detected the functional expression of 25% of the *E. coli*[11] proteome and 33% of the *C. crescentus*[12,13] proteome. The availability of these tools has allowed for advancements in systems-level transcriptional regulatory, metabolic, and dynamic signaling networks for these organisms. Below we include a brief summary of the current systems-level understanding for the purpose of providing an example of the implementation of the bottom-up approach.

2.1.2   Compiling a transcriptional network model for C. crescentus cell cycle

C. crescentus is a free-living aquatic microbe that undergoes a specified developmental program during its cell cycle, which is coordinated with morphogenesis: *C. crescentus* alternates between motile swarmer and sessile stalked cellular morphotypes by undergoing oscillating asymmetric divisions[14] (Fig. 2). Because of this fascinating cellular program and its relevance to the human cell cycle and cancer progression, the elements regulating the morphogenic switch during cell cycle progression in this organism have been the subject of intense experimentation and have resulted in the discovery of several transcriptional regulatory factors. Coupled with non-mathematical models, intriguing transcription regulatory and dynamic signal transduction sub-network topologies for the *C. crescentus* cell cycle have been revealed. Based on the wealth of information that exists for these sub-networks, *C. crescentus* exemplifies results that can be expected from the first stages of a bottom-up approach.

Initial transcriptional analysis of cell cycle progression time courses using microarray analysis of *ctrA* mutants first established CtrA (*c*ell cycle *t*ranscriptional *r*egulator *A*) as a master regulator of the cell cycle in this organism. This study showed that CtrA directly or indirectly activates or represses various cell cycle responsive genes, thus delineating the preliminary cell cycle sub-network topology for *C. crescentus*[14]. Follow-up transcription factor binding localization studies using the ChIP-chip technique demonstrated that CtrA directly regulates 55 genes during cell cycle progression[15]. In ChIP-chip technology, the first "ChIP" refers to *ch*romatin *i*mmuno*p*recipitation, in which specific *in vivo* transcription factor-DNA complexes are stabilized by chemical cross-linking and enriched using antibodies against the transcription factor. The second "chip" refers to localizing the position of these enriched DNA fragments in the genome using a high density

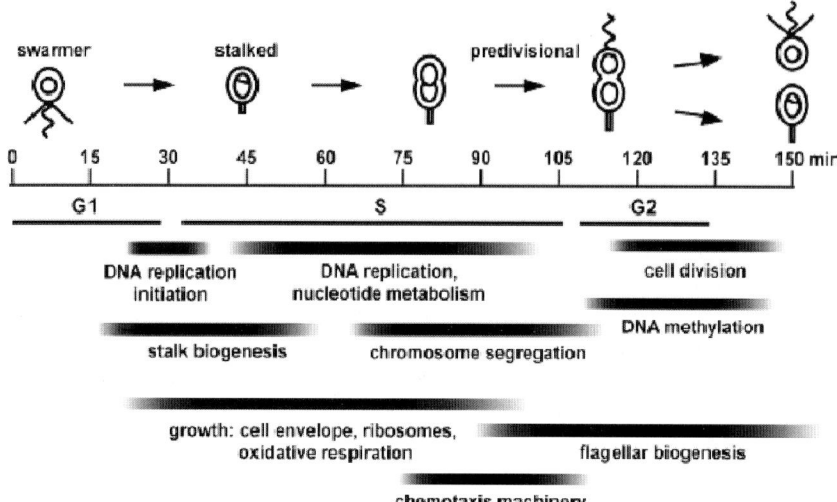

Figure 2. The Caulobacter crescentus *cell cycle.* Swarmer cells, which contain a single polar flagella (curved line) and pili (straight lines), differentiate into stalked cells during the G1 to S phase transition. Only stalked cells can initiate DNA replication. Division is asymmetric, and the timing of various cellular events occurring during division is indicated below the time scale. Reproduced from reference 20 (©2005, Skerker et al.)

microarray *chip*(s) which contain PCR products or synthetic oligonucleotides spanning the whole genome (including intergenic regions). Using this technology, one can monitor genome-wide distribution of cognate *cis*-regulatory DNA binding-sites of transcription factors[15,16].

Another master cell cycle regulator called GcrA was shown to work together with CtrA. Together, the two regulators function as an oscillating switch that temporally controls the global expression of genes important for the cell cycle[17]. Along with detailed single gene experiments, these global studies have enabled the construction of a transcriptional regulatory network model governing cell cycle progression in *C. crescentus*. This model, built gradually module by module, consists of several interconnecting transcriptional regulatory circuits regulating cell cycle progression[18].

Similar transcriptional microarray analyses in genetically perturbed backgrounds of *C. crescentus* have been conducted in cells subjected to fluctuations in oxygen concentrations, which found that the response to oxygen is regulated by a circuit in which four transcription factors (FixL, J, K and T) interact to control electron transport chain genes[19]. Integrating the cell cycle and oxygen transcription regulatory sub-networks would be the first step towards comprehensive systems-level transcriptional regulatory network model for *C. crescentus*. Subsequently, sub-network modules describing the behavior of *C. crescentus* under the influence of other environmental conditions or genetic mutations would also need to be generated and integrated into the existing framework.

2.1.3   Dynamics of the cell cycle sub-network in C. crescentus

Cell cycle progression requires accurate temporal transcription of effectors, appropriate spatial localization of factors within the cell, and rapid signaling between regulators *via* phosphorelays[18]. These tenets of the cell cycle sub-network dynamics have been demonstrated by several recent *C. crescentus* studies. For instance, various histidine kinases and response regulator complexes assemble at cell poles and delocalize to the membrane and the cytosol at different times during the cell cycle. These complexes interact with polar localization factors and flagellar assembly proteins. Polar complexes also enable proper positioning of the chromosome during replication. Biochemical phosphotransfer experiments combined with knockout mutant analysis on all 106 two-component regulatory proteins in *C. crescentus* identified several novel phosphorelay regulatory systems involved in the cell cycle regulatory dynamics, all of which are under spatial and temporal transcriptional control by CtrA[20,21].

Although several non-mathematical descriptive models have been constructed for these sub-networks, as the complexity of the cell cycle module of the *C. crescentus* network unfolds, mathematical models will be required to integrate this information with the transcriptional regulatory wiring and dynamic signaling modules as they are assembled. No mathematical model currently exists for the cell cycle module in *C. crescentus*, so the challenge awaits[18].

2.1.4   The transcriptional regulatory network model and its dynamics in E. coli

Since the advent of the molecular biology in the 1940's, *E. coli* has been the model organism of choice for studying prokaryotes because of its genetic tractability, fast generation time, and medical relevance as a major enteric pathogen in human hosts. Therefore, a great wealth of information is available that has enabled the *E. coli* system to be by far the most advanced with respect to building a systems-level understanding using bottom-up experimental approaches. Systems biologists have made effective use of the wealth of information available in the literature by building a plethora of mathematical models that attempt to accurately capture aspects of physiology at varying levels of detail. Appropriately designed and executed experiments have tested model predictions, which have lead to model refinement[4]. Three main types of mathematical models have been generated for *E. coli* regulatory networks: (1) interaction-based, (2) constraint-based, and (3) mechanism-based models[4]. Interaction-based models are the most global but more generalized, leaving out the details of kinetics, concentration of components, and other biochemical information (*e.g.* protein-protein interaction and transcriptional regulatory networks). Constraint-based models include stoichiometric parameters of biochemical components. Mechanism-based models are narrow in scope, focusing in exquisite detail on a single sub-circuit. Mechanism-based models typically include kinetic and biochemical parameters. Brief examples of each type of model are discussed below.

2.1.4.1   Interaction-based mathematical models: In 1998, Collado-Vides and colleagues compiled the first interaction-based model of the *E. coli* transcription regulatory network from literature sources[22], culminating in the organization of the information into a database called RegulonDB[23], which has been continuously updated[24]. This network, like other interaction-based models, is static, and does not include stoichiometric or kinetic parameters[4]. However, the global nature of the network and high level of abstraction allowed the identification of three highly overrepresented network motifs: (1) feed-forward loops (FF), in which cascades of regulatory genes are transcriptionally induced in temporal order; (2) single input modules (SIM), in which a single autorepressed transcription factor regulates a specific subset of genes; and (3) dense overlapping regulons (DOR), in which multiple transcription factors overlap to regulate a large subset of genes in response to a given environmental condition[25]. The discovery of these motifs elucidated the higher-level hierarchical organization of the transcription regulatory network: it was composed of loosely interconnected small regulatory sub-networks[22,25,26]; and sub-networks with similar architecture tended to perform analogous regulatory and physiological functions[26]. Interestingly, the latter feature of the network raises the possibility of using these regulatory motifs to perform homology searches analogous to sequence comparisons in order to assign regulatory and physiological functions to unknown components of the network, not only in *E. coli*, but in related organisms as well[25]. Another notable feature is that, in spite of the fact that it is estimated that only 25% of the *E. coli* transcription regulatory network has been experimentally defined, it is predicted that much of the remaining network will be arranged into similarly organized motifs (i.e. FF, DOR, SIM), since more than 1/4 of the existing connections in the network could be removed or rearranged with little effect on the existing network motif topology[22,25].

2.1.4.2   Constraint-based mathematical models: Palsson and colleagues (2000) constructed the first global constraint-based metabolic model for *E. coli* using the flux-balance analysis (FBA) method, which includes stoichiometric parameters[27]. FBA predicts the capabilities of cellular metabolism by discerning the energetically most parsimonious use of metabolic flux through pathways predicted by the genome sequence. Thus, FBA identifies which states of the metabolic network will result in optimal growth rates and which will not. Hence, all imaginable fluxes through the system are constrained within the metabolic reaction space of what is feasible for the organism, allowing prediction of nutritional requirements[27–30]. For *E. coli*, metabolic fluxes were calculated globally and the effect of various gene deletions were predicted by setting the flux through the metabolic pathway catalyzed by the gene product of interest to zero. Predictions were compared to the literature, in which phenotypes of the various gene deletions were experimentally determined. Predictions which matched the model served as verifications, whereas inconsistencies identified shortcomings. Through this analysis, the FBA model was refined[27]. Using a similar strategy, metabolic models have been constructed for several other organisms, including *Methylobacterium extorquens AM1*[31], *Haemophilus influenzae*[32], *Geobacter sulfurreducens*[33] and *Helicobacter pylori*[34]. The proliferation of such models has allowed comparative analysis of FBA-derived networks for various organisms, which identified a suite of highly conserved central metabolic core reactions that were non-redundant and utilized under all 30,000 growth conditions tested *in silico* for all organisms considered in the analysis[28]. The amino acid sequences of enzymes catalyzing these reactions were also highly conserved, and their gene deletions lethal[28]. Clearly, the conservation of modular features in transcriptional networks identified by interaction-based models can also be extended to metabolic constraint-based networks, giving further clues to the evolutionary influences acting on networks[7].

Taken together, the global metabolic and transcription regulatory network models for *E. coli* represent the first iteration in the systems-level experimentation/computation analysis cycle (Fig. 1). To progress through to a second iteration, the FBA metabolic model was combined with Boolean logic statements describing global transcription regulatory reactions[30] into a single *in silico* "strain" called iMC1010v1[35,36]. Analysis of this combined network led to the formulation of several hypotheses about phenotypes and gene expression levels in response to various growth conditions. These hypotheses were tested by high-throughput experimental phenotype analysis[37] combined with microarray transcriptional analysis of mutants predicted to be important in central cellular functions[35]. The results of these analyses were used to refine the model, yielding iMC1010v2. Emergent features of this new network include: (1) compared to the number of possible growth conditions ($\sim 15,000$), *E. coli* exhibits few functional metabolic states, and (2) the organism is constrained to these states primarily by the available terminal electron acceptor[36]. This analysis represents the culmination of the bottom-up systems approach in which more than 40 years of research were combined into a single metabolic and transcriptional network.

2.1.4.3 Mechanism-based mathematical models: Despite enormous progress on systems scale modeling of *E. coli*, much work remains to achieve a global-scale mechanism-based model for *E. coli*, which would include stoichiometric and kinetic parameters as well as estimates of the relative concentrations of each molecular component[4,38]. The daunting nature of the task is exemplified by the level of detail included in dynamic mechanism-based mathematical models that have been developed for well-defined sub-networks. These in-depth mathematical examinations of bacterial sub-networks have revealed emergent properties of the sub-systems. For example, circuit motifs such as switches, oscillators, and robust integral feedback loops have been observed in lambda phage lysis/lysogeny decisions[39], *C. crescentus* cell cycle[18], and the chemotaxis two-component sensory/regulatory system, respectively[40,41] (for a complete review of circuit motifs, see[8]). In this section we will focus on the example of the chemotaxis network, since it is one of the best-understood physiological networks in prokaryotes[42].

The signaling cascade of the chemotaxis system in *E. coli* contains sensory, signal transduction, and effector components (Fig. 3). Upon introduction to a chemical gradient (e.g. limiting nutrients), bacteria are propelled by polar flagella up the gradient by a series of runs (forward swimming) and tumbles (direction changing).

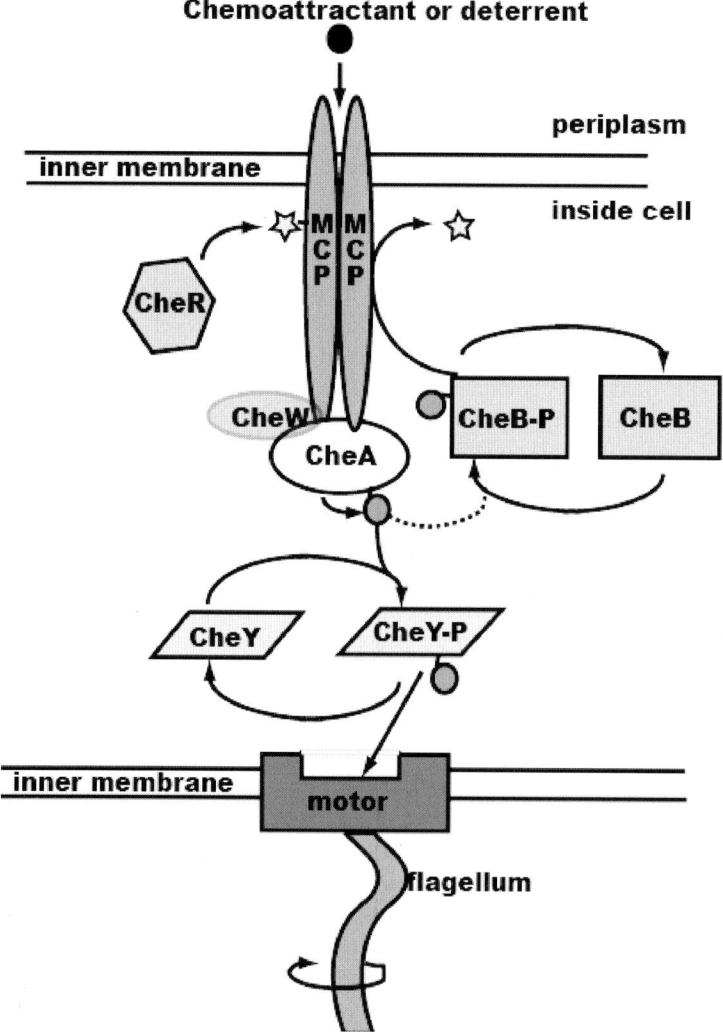

Figure 3. Schematic of the components of the E. coli *chemotaxis signal transduction network.* In the presence of decreased chemoattractants, periplasmic sensor methyl-accepting chemotaxis proteins (MCPs, pink ellipses embedded in inner membrane) cause the autophosphorylation of the associated CheA sensory histitine kinase (white ellipse). CheW (grey ellipse) assists the association of CheA with MCPs. CheA phosphorylates CheY (light blue trapezoids), which binds to the flagellar motor (blue rectangle) and changes the direction of the flagellum (green), enabling the cell to locate higher attractant concentrations. Upon sensing higher concentrations of attractant, CheA autophosphorylation decreases, CheY-P levels are reduced, and the rate of motor direction switching increases, allowing longer unidirectional swimming. Furthermore, the system is modulated by the methylation level (methyl groups are represented by yellow stars) of the MCPs, which is regulated by CheR (purple hexagon, methyltransferase) and CheB (green square, methylesterase which competes with CheY for phosphate). Phosphoryl groups are represented by small orange circles. Figure adapted from references 41 and 42.

One of the key features of the system is precise adaptation: in the presence of constant concentrations of stimulus, the transduction system returns to exactly prestimulus levels[40,42]. Deciphering the mechanism behind adaptation has been the subject of intense study (Fig. 3)[43], which has resulted in detailed knowledge of kinetic parameters and component concentrations of the chemotaxis network. Several groups have used these parameters for precise mathematical modeling of the network. For example, Barkai and Leibler's two-state model (1997) assumes the network is either in an active or inactive state and uses ordinary differential equations to predict that (1) the chemotaxis system responds sensitively to wide dynamic ranges of attractant concentration; (2) reaction rate constants and other biochemical parameters could be varied by several orders of magnitude with little to no effect on the precise adaptation of the system[40].

Direct experimental evidence verified the hypothesis of integral feedback robustness[43], and later experimental/computational iterations refined the model by including additional system components that were not considered in earlier models[41]. The robustness of bacterial chemotaxis has also been observed in other sub-networks, suggesting that robustness may be a general property of biological systems, enabling organisms to maintain homeostasis despite environmental fluctuations[3,43,44]. It has been suggested that not all chemotaxis system components have been defined, and that further experiments are necessary to further refine the model, which would be important in demonstrating the true source of network robustness[45]. In contrast, some argue that, since one of the key features of robust systems is their insensitivity to changes in the biochemical parameters of the network, completely defining sub-network modules at the molecular mechanistic level may not be necessary for discerning most cellular functions[3,40]. This debate raises several important questions as to which of the three model types discussed above is appropriate for systems biology. Should we comprehensively model systems at the more abstract, interaction-based level or should we completely define parameters by intensive experimental measurements of biochemical subsystem components? How should systems biology proceed?

2.2   Advantages and Disadvantages of the Bottom-up Approach

It is clear from the examples of *Caulobacter crescentus* and *E. coli* systems that the bottom-up approach to building network models results in detailed mechanistic insight about important pathways or subsystems gained by years of high quality genetic and biochemical evidence. These insights allow for accurate mathematical modeling and detailed description of sub-network behavior. In addition, the literature-based information regarding single-gene studies is impressive and quite comprehensive. Thus, any conclusions drawn from these studies are bootstrapped by decades of detailed evidence.

However, three disadvantages of the bottom-up approach make systems-level insights more difficult to achieve. First, generating a comprehensive computational systems model requires the integration of several disparate data types (e.g. microarray, proteomics, protein-protein interaction); however, integrating data generated by different experimental methods, time points, and conditions from different labs is quite challenging and often not possible. Often, global experiments need to be repeated under standardized conditions to build accurate and consistent models[35]. Second, important conclusions resulting from years of single-gene studies are difficult to ignore when interpreting large, complex systems-level datasets. In other words, preconceived notions about favorite genes or pathways may bias the results or interpretation of systems-level studies, especially in reference to large datasets[3,44]. Third, bottom-up approaches are slow and time-intensive. The first draft of the transcription regulatory network for *E. coli* took ~40 years to complete[23]. Furthermore, additional connections may exist between modules within the network, but these cross-regulatory interactions and overlaps could continue to elude observation if each module is studied in isolation.

Indeed, although traditional approaches have clearly yielded crucial systems-level insights, in light of the drawbacks described above, it may be not be possible to decipher global-scale dynamic models for prokaryotic organisms comprehensively in a reasonable time frame using a bottom-up approach exclusively[18]. However, integrating mechanism-based models formulated by bottom-up approaches with abstract models from top-down approaches may be the key to gaining a comprehensive systems-level understanding of biological complexity. By combining the two approaches, the resolution at which a system is modeled can then be dictated by the resolution at which the system is observed and the data are collected. This allows experimentation and computation to drive one another and agree closely. Using such a strategy, appropriate interpretation of system dynamics is possible (See Box 1).

Box 2.   Experimental and computational steps to a successful top-down systems biology approach.

1. Generate a parts list: genome sequence, non-coding and coding sequences and translated proteins, function assignment etc.
2. Perturb system (genetically and/or environmentally) and measure dynamic changes of mRNA, proteins, protein-protein and protein-DNA interactions, protein modifications and metabolite concentrations on a genome-wide scale.
3. Integrate and analyze all resultant data types simultaneously using statistical learning tools to generate a regulatory circuit in which the protein-protein and protein-DNA interactions constitute the wiring diagram
4. Evaluate the network for capability to recapitulate previously made observations and predict new behaviors by conducting additional rounds of targeted experimentation.
5. Repeat steps 2, 3 and 4 iteratively to resolve conflicts and refine the circuit.

Below we describe an alternative to bottom-up systems biology approaches: the top-down approach. We evaluate its effectiveness in systems studies by using the example of two prokaryotic organisms for which the top-down method has recently been applied: *Halobacterium NRC-1* and *Helicobacter pylori*.

2.3 The Top-down Approach to Prokaryotic Systems Biology: Simultaneous Measurement of Cellular Outputs

In top-down systems approaches, changes at all information levels (mRNA and protein levels, protein-protein and protein-DNA interactions, protein modifications, etc.) during a cellular response are measured and analyzed simultaneously to formulate predictive models that describe cell behavior in changing environments (see Box 2). The model is refined by testing predictions through additional rounds of experiments[46]. The top-down approach compiles a network *de novo* for an organism for which very little molecular or biochemical information exists.

2.3.1 *Cataloging system components: Parts lists for H. pylori and* Halobacterium

The genome sequences of *Helicobacter pylori* and *Halobacterium NRC-1* were completed in 1997 and 2000, respectively[47,48], enabling the global functional prediction of gene products. However, approximately 31% of the *H. pylori* genes and 60% of *Halobacterium* genes did not match any functions in the databases[47,48]. Therefore, an in-depth annotation pipeline incorporating protein family (PFAM), protein data bank (PDB) and PSI-BLAST searches was devised and applied to the *Halobacterium* genome to improve functional prediction and annotation. When combined with prediction of protein structures *de novo* using the Rosetta algorithm[49], this pipeline reduced the unknown gene count for *Halobacterium* to 10%[49], and resultant functions are cataloged in the annotation database called SBEAMS (http://halo.systemsbiology.net/halobacterium). For both organisms, global whole-cell proteomic approaches have been applied to detect the expression of the predicted protein coding sequences. For *Halobacterium*, expression of up to 35% of the predicted proteome has been detected in three separate studies[50–52], and ~13% has been detected for *H. pylori*[53,54].

2.3.2 *Phenotypic characterization in* Halobacterium

After generating a parts list, the first step in a systems-level investigation of an understudied prokaryotic model organism is to characterize its physiological growth and survival phenotype in response to a variety of environmental stimuli (Box 2) . *Halobacterium NRC-1* is a heterotrophic, obligately halophilic archaeon requiring nearly saturated salt conditions (~4.5 M) to survive and grow. Marine salterns, salt evaporation ponds, and hypersaline lakes comprise typical habitats, where high cell densities and rapid evaporation can lower oxygen tension and deplete nutrient concentrations[55]. Under these harsh environmental conditions, flotation toward air-water interfaces mediated by gas vesicle production allows for greater availability of light and oxygen, and flagellar-mediated chemotaxis enables access to richer nutrient sources. However, little is known regarding the global molecular response to these conditions. What doses of stress cause changes in gene expression but allow survival of the organism? What are the nutritional requirements of the organism? To answer these questions in *Halobacterium*, the growth and survival were tested under conditions the organism would likely encounter in its natural environment, including UV irradiation[56], sunlight[57], high and low oxygen[58], and various concentrations of transition metals[59]. The results of these experiments demonstrated that the organism is highly resistant to UV radiation and arrests its growth at specific concentrations of metals. Moreover, these experiments defined the environment in which to analyze the mechanisms of how cell behavior is regulated *Halobacterium*, and laid the foundation for determining the global transcription regulatory network circuitry.

2.3.3   *Deciphering regulatory interaction networks using the top-down approach*

Several top-down approaches exist for deciphering the global regulatory interaction network of an organism. Two particularly fruitful methods, including global DNA-protein interaction mapping by ChIP-chip (see also section 2.1.2) and global protein-protein interaction mapping will be discussed below.

2.3.3.1   *DNA-protein interactions: ChIP-chip.* Archaeal transcriptional machinery, like in eukaryotes, consists of the basal transcription factors II B (TFB), TATA binding protein (TBP), and an RNA-Pol II-like polymerase. However, unlike in eukaryotes, which require over thirty proteins to initiate transcription, a TBP and TFB are sufficient to direct RNA polymerase to archaeal promoters[60]. In contrast, repressor and activator components of archaea, which serve to modulate transcription from the basal apparatus, appear to most closely resemble those of bacteria[61].

Intriguingly, *Halobacterium* is unique among archaea because its genome encodes thirteen predicted basal transcription factors: six TBP and seven TFB genes, which evokes the hypothesis that these genes could combine in 42 different combinations to regulate Halobacterial genes[62]. To determine whether the organism makes use of this potential regulatory flexibility, DNA-protein interactions between all 7 TFB regulatory factors and promoter targets were investigated using ChIP-chip under standard experimental growth conditions (rich medium, no stressors present)[63]. Each regulator was overexpressed to ensure detectable levels. Surprisingly, the resultant regulatory network was partitioned into discrete functional groupings (or regulons) by the TFB's, with some promoters exclusively regulated by one TFB. In contrast, several overlaps in regulatory groups were also seen, with functionally related promoters bound by several TFB's[63]. Moreover, the TFB/TBP's were often autoregulated or interacted with the promoters of other TFB/TBP's, suggesting that the regulatory network is organized into a distinct hierarchical structure, with TFB/TBP's constituting the highest level, and downstream modules of gene groups with distinct functions regulated by TFB/TBP pairs representing the lower levels[63]. These results enabled the elucidation of key components of the *Halobacterium* basal transcription regulatory network in a relatively short amount of time ($\sim$2 yrs).

However, the network is not complete, since it awaits the investigation of the bacterial-like transcription activators and repressors.

2.3.3.2   Protein-protein interactions.   Three methods have been developed for the top-down determination of global protein-protein interaction networks: (1) yeast two-hybrid screening (2) co-immunoprecipitation, and (3) BIAcore assays. The yeast two-hybrid system exploits the yeast galactose utilization transcription regulatory system: the GAL4 DNA binding domain is translationally fused to "bait" proteins or domains, and "fish" proteins/domains to the GAL4 activation domain. When bound to each other, the two hybrid proteins bind and activate the Gal1 upstream activation sequence (UAS), and transcription of a reporter gene proceeds[64]. With the use of shotgun protein domain libraries, the two-hybrid method is applicable to *in vivo* global protein interaction studies, and has deciphered global interaction networks for several organisms, including yeast, *H. pylori, Plasmodium falciparum*, and human[65–69]. For *H. pylori*, the resultant protein-protein interaction network assigned functions for previously uncharacterized proteins, pinpointed known interactions to specific protein domains, and identified novel complexes[67]. However, the high incidence of false positives inherent in the two-hybrid method precludes global interaction maps generated using this technique from being definitive[70]. Therefore, two-hybrid interaction maps require extensive verification by other methods such as co-immunoprecipitation and surface plasmon resonance (BIAcore).

Co-immunoprecipitation (CoIP) is an *in vivo* method in which binding partners are precipitated from cell extracts using an antibody specific to the protein of interest. Although not yet useful as a global application, this technique verifies global data and delineates protein-protein interactions in sub-networks such as the *Halobacterium* TFB and TBP protein interaction network[63]. When integrated with data from chromosomal localization ChIP-chip studies, the TFB/TBP interactome led to further refinement of the basal regulatory network by revealing a second protein interaction regulatory layer[63]. CoIP coupled to mass spectrometric approaches is also very powerful when applied to identification of novel members of complexes and protein machines involved in important cellular processes. For example, three novel members of the yeast core transcriptional machinery were identified and verified using CoIP coupled to tandem mass spectrometry[71].

The BIAcore system uses sensor chip technology integrated with a flow cell to monitor *in vitro* molecular binding interactions in real time using surface plasmon resonance (SPR)[72] (www.biacore.com). SPR simultaneously determines molecular binding interactions, kinetics (on/off rates) and affinities. The technology is high-throughput and has been applied to measure the binding kinetics of an archaeal TBP to various stress response promoters[73] and to detect the binding of bacterial pathogens to various host cell surface receptors[74,75].

2.3.3.3   Caveats.   It is important to note that each of the methods described above has its own strengths and limitations. BIAcore assays, while high-throughput and precise, are conducted *in vitro* and require purified components and expensive, specialized equipment. CoIP experiments, while reliable, are time-consuming and low-throughput. Two-hybrid methods, as mentioned above, are applicable on a global scale but have a high false positive discovery rate. Moreover, in both BIAcore and two-hybrid systems one can only look at interactions between two molecular components at a time (although BIAcore in principle could be adapted to analyze complexes with more than two members) and in both cases the interactions are measured under non-native conditions. Therefore, combining these methods with the co-IP technique may give a more complete and reliable picture of global protein-protein interaction networks. Another important caveat is that most interaction networks described to date have been deciphered by over-expression of proteins under standard culture conditions that result in optimum growth of the organism (i.e. in rich medium during logarithmic growth). Ideally, interaction experiments should be conducted under conditions of native protein concentrations in contexts where the functions of the binding partners are most likely to be active. Using such conditions, results are more likely to be physiologically relevant and the global interaction circuit can subsequently be correctly interpreted.

2.3.4   Computational regulatory network prediction in Halobacterium NRC-1

The first step toward interpreting the global gene regulatory network in the context of the physiology particular to a given prokaryotic organism is determining how cell behavior changes in response to various stress conditions. DNA microarrays are a useful tool to assess global transcriptional responses to environmental and genetic perturbations, and have been used extensively in *Halobacterium*, for which nearly 500 microarray experiments were conducted in response to UV and gamma irradiation[56,57], metal toxicity[59], and mutations in regulators of arginine fermentation and phototrophy[51]. These data were used to drive statistically-based algorithms which predict gene regulatory networks developed for *Halobacterium*. These computational methods are described below.

2.3.4.1   Regulatory network Inference models: interaction-based predictions.

A three-step procedure for inferring regulatory networks from diverse systems biology data has been developed for *Halobacterium*[76,77] (Fig. 4A). This algorithm includes (1) "cMonkey", a clustering program which identifies groups of genes called "biclusters" that are co-expressed under subsets of conditions (Fig. 4B), and (2) the "Inferelator", a statistical-learning algorithm for inferring the regulatory influences of putative transcription factors and environmental cues on these biclusters (Fig. 4C), and (3) "Evolvolator", an algorithm that evaluates dynamics and predictive power of the inferred regulatory network. To identify putatively co-regulated sets of genes, cMonkey integrates RNA expression data with additional sources of evidence such as transcription factor binding sites (detected during the clustering process) and functional, physical, or experimentally-derived associations (including DNA-protein interactions) to constrain the clustering. Inferelator then scans the complete sets of putative transcription regulators (identified through the compendium of genome annotation procedures) and environmental factors to

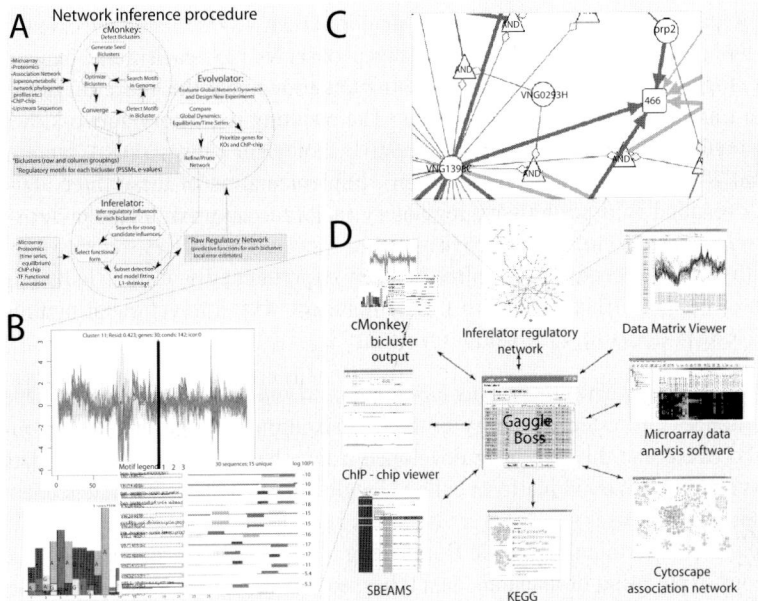

Figure 4. Computational tools for the top-down deciphering of the Halobacterium *gene regulatory network.*
A. Network inference procedure. Step 1: cMonkey detects genes co-regulated over a subset of conditions
(biclusters). Step 2: Inferelator scans all possible transcription regulators and environmental factors to isolate
the minimal set of regulatory influences that sufficiently explain the behavior of each biclusters. This yields
a raw regulatory network. Step 3: Evolvolator assesses robustness of this network for dynamics and potential
conflicts requiring further experimental refinement. **B.** cMonkey bicluster output. The cMonkey algorithm
selects groups of genes with correlated expression (listed bottom right) over a selected subset of microarray
conditions (top graph). The top graph displays the profiles of clustered genes over all microarray conditions.
The most significantly correlated profiles over a subset of conditions are shown to the left of the thick
black line. cMonkey also finds putative conserved regulatory sequence motifs in the upstream region of
each gene (bottom left). Upstream sequences are represented in the bottom right by the multicolored lines.
Red, green, and blue boxes above and below each sequence indicate the location of the motif displayed
on the bottom left. The cMonkey bicluster output can be visualized and integrated into the Gaggle (D).
C. Inferelator regulatory influence prediction algorithm. A sub-network selected out of the full predicted
regulatory network is shown. Circles represent transcription factors and bold circles represent environmental
influences. Squares indicate biclusters. The length of the bicluster represents the number of conditions
and the height displays the number of genes included in the bicluster. Red arrows are regulatory induction
influences and green are repression influences. Triangles represent logic gates (AND, OR), and blue lines
connect two or more transcription regulators to a logic gate. The Inferelator predictive network is also
incorporated into the Gaggle. **D.** Gaggle: A software tool for seamless integration and exploration of different
types of data such as mRNA profiles (microarray analysis software and Data Matrix Viewer), protein-DNA
interactions (ChIP-chip), gene association networks (Cytoscape), metabolic pathways (KEGG), Inferelator,
cMonkey, and the *Halobacterium* genome database (SBEAMS). The Gaggle functions by user-specified
"broadcasting" (i.e. transmitting information; indicated by double-headed arrows) groups of genes from the
Boss to the various linked applications. Once a broadcast is received by that application (Goose), the user
performs analysis on the given group of genes with the aid of the given application, and the information
is broadcast back through the Gaggle Boss to other applications. The cMonkey and Inferelator outputs are
integrated into the Gaggle, which facilitates exploration and analysis of those datasets in the biological
context of all other systems-level information available for *Halobacterium*. Figure adapted from references
76, 77, and 78.

identify the parsimonious sets of regulators/environmental factors that sufficiently describe the observed transcript level changes observed for each gene in each biclusters (Fig. 4B). The Evolvolator then attempts to recreate the entire transcriptome from transcript levels of the regulators. The resulting regulatory network then describes regulatory influences and the putative biological causes (DNA binding motifs) and effects (changes in transcription). Implementation of these three algorithms has resulted in a preliminary regulatory influence network for behavior of *Halobacterium NRC-1* in diverse environments. Several regulatory relationships predicted by this procedure have already been experimentally tested, including regulation of copper-efflux by an Lrp family regulator with a novel metal-binding domain (TRASH-domain), (see below for detail)[59].

2.3.4.2   Visualizing diverse systems biology data through a single portal.   One of the challenges of systems biology is the visualization of large, heterogeneous datasets. To overcome this, we have developed a Java-based, user-interactive data integration and visualization platform called Gaggle[78]. In this platform, independent applications and databases ("geese") are linked and communicate with each other via a central command terminal ("Gaggle Boss") (Fig. 4D). Such applications allow the simultaneous exploration of disparate data types, and include: Cytoscape for visualization of regulatory networks (e.g. Inferelator output)[79]; KEGG and SBEAMS (the *Halobacterium* genome database) for exploration of functional annotations; plotting tools (Data Matrix Viewer) for the visualization of microarray and proteomics data; and an integrated R statistical analysis programming terminal for data manipulations. Using the Gaggle, a biologist obtains a very rich environment for data mining and manipulation.

*2.3.5   Regulatory network dynamics and experimental verification
            of computational prediction in* Halobacterium

Using the full computational suite of cMonkey, Inferelator, Evolvolator and Gaggle algorithms, systems-level predictions were made for *Halobacterium* and hypotheses were generated and tested experimentally (see below). These experiments, which included microarrays, classical genetic approaches with in-frame deletion strains, ChIP-chip, and protein-protein interactions, verified several predictions of the computational algorithms. These results will be discussed as examples to highlight four important features of the top-down approach: (1) global network attributes are revealed; (2) regulatory relationships between biomodules become evident; (3) accelerated knowledge discovery of understudied organisms is possible; and (4) the importance of delineating the kinetics of transient global response to stress is emphasized

2.3.5.1   Global network attributes are revealed.   *Halobacterium* was exposed to UV irradiation and the global transcriptional response was interrogated during light and dark DNA damage repair 30 and 60 minutes after exposure. A total of 273 genes changed in response to damage, 40 of which were shared between light and dark repair[56]. Surprisingly, 12% of all genes encoded in the genome of

Halobacterium, distributed among 15 functional categories, were down-regulated, in contrast to the up-regulation of only a small number of DNA and protein repair genes. These results suggested that UV radiation may cause a widespread global shut-down of many aspects of metabolism (Fig. 5), a phenomenon that has been hypothesized to prevent further cellular damage and conserve energy[56].

2.3.5.2 *Regulatory relationships between biomodules become evident.* The

metabolism of *Halobacterium* transitions from organic heterotrophy to bacteriorhodopsin-mediated phototrophy under conditions of intense light and anaerobiosis[55]. To investigate the regulatory control of this conversion, proteomic and microarray analyses were conducted for two mutant strains of a putative phototrophy transcription factor, Bat (bacteriorhodopsin activator): (1) *bat+*, a constitutively phototrophic strain, and (2) *bat−*, a phototrophy-deficient strain[51]. Surprisingly, both strains exhibited coordinated yet inverse control of the arginine fermentation and phototrophy biomodules. In *bat+* strains, phototrophy pathway genes were induced, whereas arginine fermentation genes were repressed. Conversely, in *bat−* strains, phototrophy was repressed and arginine fermentation was induced. In addition, pyrimidine biosynthesis and glutamate metabolism pathways were induced in *bat−* strains, suggesting a functional link between these modules and arginine fermentation, and possibly revealing novel functions for the Bat transcription factor. These results not only delineated the role of a previously understudied transcription factor, but also suggested a novel functional link between two anaerobic energy production modules, generating novel systems hypotheses.

2.3.5.3 *Accelerated knowledge discovery in understudied organisms is

possible.* *Halobacterium* and other halophilic archaea regularly encounter high metal concentrations during evaporation cycles in their natural environment[80]. However, very little was previously known regarding the physiological protective response to transition metal toxicity in *Halobacterium.* Therefore, a systems-level analysis was conducted in the organism in response to varying concentrations of transition metals. The study, which took only 18 months to complete, included 66 microarray experiments, phenotypic analysis of wild type and 17 gene knockout strains, and $\sim 6,000$ protein-DNA interaction datapoints[59]. Some of these data were generated to drive the Inferelator prediction algorithm, and some to test its predictions. As a result, a comprehensive systems-level physiological reconstruction model of metal detoxification in *Halobacterium* was constructed, which in large part verified predictions of the Inferelator and led to several novel systems-level insights (Fig. 6)[59]: (1) Preliminary architecture of the metal regulatory network was delineated. Thirteen putative metal-binding transcription factors may differentially regulate 43 other transcription factors in response to various metals. In addition, two novel transcriptional regulators, SirR and VNG1179C were identified and characterized: SirR may function as a Mn(II)-dependent repressor of Mn(II) uptake genes, and VNG1175C as a Cu(II)-dependent TRASH domain-containing activator of Cu(II) efflux genes. (2) Specific mechanisms accounting for the resistance to metals were discovered, including efflux and metal chelating. (3) Each

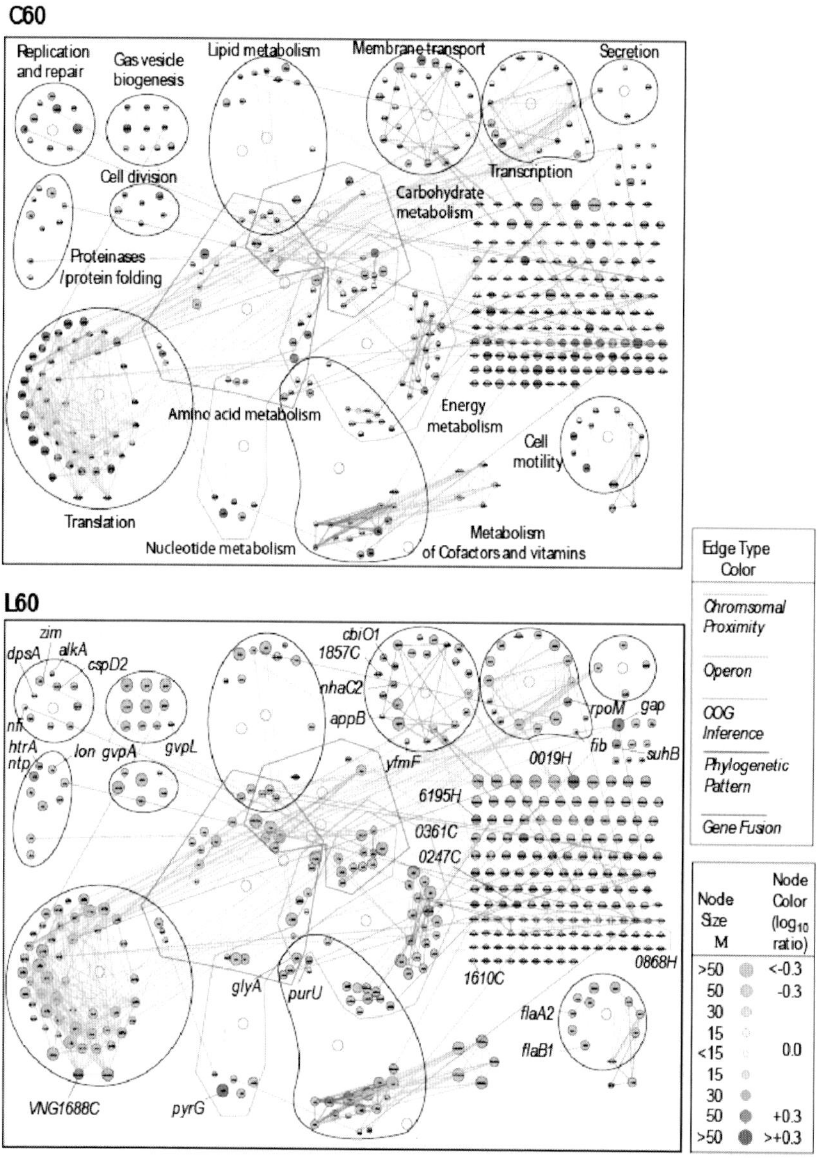

Figure 5. Gene association network layout of the global response to UV radiation in *Halobacterium NRC-1*. C60, control at 60 minutes. L60, irradiated cells after 60 minutes of light recovery. Small circles (nodes) represent genes, the size of the node represents the significance of mRNA change, and the intensity of color the magnitude of change (see inset legend). Lines (edges) connecting genes represent five different associations (see inset for color code). Genes changing in response to radiation are grouped according to functional category (large circles and polygons). The majority of genes are repressed, but the few DNA damage repair genes that are induced are labeled in larger font. Reproduced with permission from reference 56.

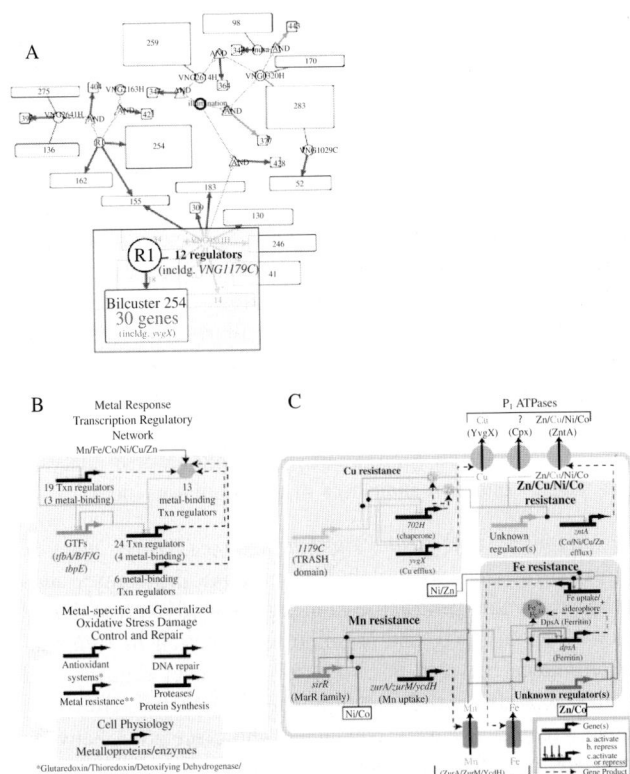

Figure 6. Systems-level verification of Inferelator prediction resulted in a physiological reconstruction of the metal toxicity response in *Halobacterium NRC-1*. **A.** Inferelator sub-network prediction for the regulation of copper efflux systems. R1 represents a given regulator, one of 12 possible factors influencing 30 genes. This hypothesis and others were tested by global assays (see text), resulting in **B.** Metal response transcription regulatory network. Genes or groups of genes are represented by bold black lines and regulatory influences by solid blue or dotted black arrows. Grey, green, and blue watermark arrows indicate putative feedback loops. Grey watermark box represents the highest level in the transcriptinoal regulatory heirarchy, whereas the blue and green boxes illustrate cellular response or output. **C.** Specific metal responses that were predicted by the Inferelator and verified by systems-level experiments (microarrays, DNA-protein interactions, mutant analysis). In the blue (copper resistance) box, VNG1179C positively directly or indirectly regulates *yvgX* and *VNG0702H*, which encode copper efflux systems. In the purple box (manganese resistance), SirR directly or indirectly regulates manganese uptake systems. In the orange (iron resistance) box, DpsA is responsible for iron storage, and in the grey box (zinc, copper, nickel, and cobalt), as yet unknown regulators induce cross-metal protection by increasing the expression of metal efflux pumps (ZntA). It is also important to note that several cross-regulatory influences were found between the various metal-response modules. For example, SirR and VNG1179C, although central to manganese and copper resistance mechanisms, respectively, also played a role in protection against iron, zinc, nickel, and cobalt. B and C reproduced with permission from reference 59 .

metal elicited a response in a unique set of genes as well as shared or overlapping gene sets. For example, at steady state *zntA* transcription changed only in response to zinc, whereas *yvgX* responded to copper and zinc. Both genes encode P1 ATPase efflux pumps. Therefore, this top-down systems-level study facilitated rapid global insight into a phenomenon for which virtually no prior knowledge existed.

2.3.5.4   The importance of delineating the kinetics of transient global response to stress is emphasized.   As another part of the metal response study in *Halobacterium*, a 9-point microarray time-course experiment was conducted in response to iron excess. In contrast to steady-state experiments, which were conducted at 300 minutes after the addition of iron to mid-log phase cultures, the time course experiment showed that 5–25 minutes after iron addition, more than 20% of all genes in *Halobacterium* changed transcriptionally in a transitory manner[59]. This result suggests that single-point steady state experiments can miss the most dramatic physiological changes in the system. Clearly, the responses to several stress conditions are transient, emphasizing the need for systems-level time course analyses rather than single-point or single-condition snapshots of the regulatory network.

2.3.5.5   Integrating proteomics and transcriptomics: uncovering network dynamics.   Integrating microarray data with global quantitative proteomics adds another level of depth to the assessment gene expression regulatory dynamics. Initial global single-point studies of phototrophy and arginine fermentation in *Halobacterium* estimated 65% discordance between RNA and protein level changes, and this RNA/protein discrepancy was also corroborated in other systems[51,81]. However, when extended time-course experiments were conducted, the correlation between the expression profile patterns of RNA and protein was markedly higher, albeit with differences in the timing of peak expression[57,58,82,83]. Such differences could be attributed to the differential dynamics of RNA and protein synthesis and degradation rates, since only steady state levels are measured in current proteomics and microarray procedures[84]. Recent studies have measured RNA and proteomic degradation rates globally and showed that, while RNA decay rates tend to be on the order of minutes[85], protein half-lives vary widely depending on the polypeptide in question[86]. Concurrent measurement of RNA and protein synthesis rates in single *E. coli* cells suggests that bursts in transcription, stochasticity, and noise at the RNA level contribute to variable synthesis rates which are largely reflected at the protein level[82,87,88]; however, the simultaneous measurement of transcription and translation rates in a microbial population awaits advances in proteomics technology (improved detection sensitivity, measurements of degradation rates, etc.). In sum, these studies again emphasize the importance of time course studies in dynamic systems biology experiments.

2.3.6   Advantages and disadvantages of the top-down approach

In summary, the studies in *Halobacterium* have made nearly two full experimental/computational iterations of the systems biology cycle (Fig. 1). Innovative global computational and experimental methods generated top-down systems-level interaction-based models of transcription regulation, protein-protein interaction, and DNA-protein interaction networks that predict or explain global behavior. These models were generated in a short time frame ($\sim$2 years). More importantly, top-down approaches in general provide global insight into biological properties of an organism for which little *a priori* knowledge exists. The unbiased global approach also reveals overlaps and interconnections between subsystems and regulatory mechanisms that might be missed by the bottom-up approach[6]. However, since not all biochemical details of the system are known, many global mathematical models cannot be constructed top-down (e.g. mechanism-based models or metabolic reconstruction by FBA), and those that exist are not necessarily complete. In addition, several types of networks that are constructed top-down tend to be static, generated under only one experimental condition (especially protein-protein interaction maps). This particular condition may or may not be relevant to the experiment used to verify the model, which can lead to false interpretations of global biological phenomena. However, combining results from static interaction networks with dynamic information generated by bottom-up approaches can aid appropriate interpretation.

3.   PERSPECTIVES. BRIDGING THE GAP: HOW TO PROGRESS FROM SUBSYSTEM TO GLOBAL CIRCUIT UNDERSTANDING

A major goal of systems biology is to sufficiently understand an organism at the comprehensive global level to facilitate the re-wiring or engineering of biological circuits for a desired purpose (e.g. therapeutics and treatment of disease). However, current understanding is relatively complete only at the sub-system level (e.g. bacterial cell cycle and chemotaxis), and more global gene regulatory networks (e.g. for *Halobacterium* and *H. pylori*) are incomplete. It is especially important to gain comprehensive systems-level understanding of biological networks when applying systems-level approaches to complex eukaryotic organisms, for example in human therapeutic applications. Therefore, given the strengths and weaknesses of the top-down and bottom-up protocols, it is critical to combine the approaches to attain systems biology goals. In one possible scenario for the combination of the top-down and bottom-up approaches, predictive global models generated for a new model organism for which little is known can be experimentally tested at the local sub-network level with biochemical experiments. Concurrently, sub-network motifs and modules discovered and dissected at the detailed mechanistic level can be used to search global networks for functional homologs. Using this two-pronged approach, systems biology will eventually bridge the gap between sub-system and whole-organism understanding. However, the gap is still wide. What types of data or information are required before we can begin re-engineering cells? In general, how will whole-organism engineering be achieved?

The field of synthetic biology has made the first steps toward whole-cell engineering by generating synthetic gene regulatory circuits in microbes, which has been accomplished by manipulating network architecture with little requirement for recombinant proteins. In synthetic bacterial strains, simplified exogenous circuits control cellular system behavior, the most successful of which have been constructed by closely coupling predictive mathematical models with *in vivo* experimental circuit design[89,90]. For instance, toggle switches, in which two or more repressors regulate each other's transcription, direct cellular behavior toward bistability, exhibiting memory and oscillatory expression of a green fluorescent protein (GFP) reporter[91,92]. In these studies, oscillations were transmitted to daughter cells, and were independent of the cell cycle. However, in contrast to natural oscillators, the amplitude and frequency of the synthetic oscillations varied widely from cell to cell, and daughter cells exhibited phase shifting[92]. Robustness and noise resistance was gained in synthetic oscillatory circuits by (1) integrating the synthetic genetic regulatory circuit with cellular metabolism, allowing for stable period length that was controllable by changing the glycolytic rate of the cell[93]; or (2) integrating an autoregulated activator into the circuit[94]. Building upon these simple artificial circuits, more recent studies have developed more complicated synthetic strains in which the native sensor, regulatory, and effector output modules were re-wired, effectively uncoupling the modules from their natural regulatory connections to produce non-native population behavioral phenotypes from known environmental stimuli. For example, *E. coli* cells were engineered to produce biofilms in response to an irradiation stimulus[95], and in an analogous study, the quorum sensing circuit was re-wired to induce a regulated death signal upon reaching a cell density threshold, thus controlling population size[96].

Design and construction of synthetic circuits are a platform for the study of larger, more complex whole-cell regulatory networks: they reveal design principles (e.g. motif architecture, bistability, memory), and more importantly, sharply delineate shortcomings in the current understanding of regulatory networks. By combining synthetic biology with whole-cell systems-level experimental/computational studies, the minimal set of biochemical parameters required for attaining a comprehensive understanding of genetic regulatory networks will begin to take shape.

4.   APPLICATIONS

The potential positive impact of systems-scale analysis and subsequent cell engineering on therapeutics, biotechnology, and industry is profound, and the viability of many prospective applications has already been demonstrated. In terms of therapeutics, *E. coli* cells have been engineered to invade cancer cells in a targeted manner by expressing invasion-specific surface proteins in anoxic environments[97]. In addition, non-infectious, engineered, live *Salmonella* strains expressing genes whose products metabolize tumor-specific drugs into cytotoxic forms have been used in cancer patient clinical trials[98]. Microbial systems engineering efforts have produced efficient microbial factories, which express exogenous recombinant biochemical pathways, allowing for large-scale production of therapeutic and industrial products in *E. coli* from inexpensive precursors. A selected list of therapeutics includes precursors to anti-malarials[99], antibiotics, and anticancer drugs[100]. Similar engineering strategies in organisms other than *E. coli* produced recombinant bacteria capable of *in situ* degradation of environmentally persistent and cancer-causing organopollutants such as toluene[101] and the insecticide parathion[102]. Systems-level metabolic engineering has also increased the productivity of microbial factories, making additional industrial applications possible, such as the increased microbial production of biodegradable plastic, a side-product of central carbon metabolism in several bacterial species[103]. Other highlights from recent developments in biotechnology applications include bacterial pili as electrical nanowires[104], light-sensing bacteria as photographic film[105], and microbial biosensors for explosives (TNT) and serotonin[106].

5. SUMMARY AND CONCLUSION

The process of systems biology aims to decipher cellular circuits and regulatory networks by generating mathematical models from complex systems biology experimental data. Once regulatory networks are complete, novel emergent behaviors can be engineered into the networks, and desired outputs can be realized. However, discovering new therapeutic, industrial, and biotechnological applications using systems biology requires a broader perspective, gathering information from both the sub-network and full network level; and moving from steady-state snapshots to detailed dynamic spatial and temporal resolution of organism behavior. Accomplishing the transition will require collaboration and cooperation among systems biologists to affect the synthesis of top-down and bottom-up approaches.

6. ACKNOWLEDGEMENTS

We would like to thank Marc Facciotti and Kenia Whitehead for their insightful comments on the manuscript.

7. REFERENCES

1. Zeth, K., Offermann, S., Essen, L. O. & Oesterhelt, D. Iron-oxo clusters biomineralizing on protein surfaces: structural analysis of Halobacterium salinarum DpsA in its low- and high-iron states. Proc Natl Acad Sci U S A 101, 13780–5 (2004).
2. Kramer, B. P., Fischer, M. & Fussenegger, M. Semi-synthetic mammalian gene regulatory networks. Metab Eng 7, 241–50 (2005).
3. Kitano, H. Computational systems biology. Nature 420, 206–10 (2002).
4. Stelling, J. Mathematical models in microbial systems biology. Curr Opin Microbiol 7, 513–8 (2004).
5. Bray, D. Molecular networks: the top-down view. Science 301, 1864–5 (2003).
6. Ideker, T. & Lauffenburger, D. Building with a scaffold: emerging strategies for high- to low-level cellular modeling. Trends Biotechnol 21, 255–62 (2003).
7. McAdams, H. H., Srinivasan, B. & Arkin, A. P. The evolution of genetic regulatory systems in bacteria. Nat Rev Genet 5, 169–78 (2004).

8. Wolf, D. M. & Arkin, A. P. Motifs, modules and games in bacteria. Curr Opin Microbiol 6, 125–34 (2003).

9. Blattner, F. R. et al. The complete genome sequence of Escherichia coli K-12. Science 277, 1453–74 (1997).

10. Nierman, W. C. et al. Complete genome sequence of Caulobacter crescentus. Proc Natl Acad Sci U S A 98, 4136–41 (2001).

11. Corbin, R. W. et al. Toward a protein profile of Escherichia coli: comparison to its transcription profile. Proc Natl Acad Sci U S A 100, 9232–7 (2003).

12. Grunenfelder, B. et al. Proteomic analysis of the bacterial cell cycle. Proc Natl Acad Sci U S A 98, 4681–6 (2001).

13. Vohradsky, J. et al. Proteome of Caulobacter crescentus cell cycle publicly accessible on SWICZ server. Proteomics 3, 1874–82 (2003).

14. Laub, M. T., McAdams, H. H., Feldblyum, T., Fraser, C. M. & Shapiro, L. Global analysis of the genetic network controlling a bacterial cell cycle. Science 290, 2144–8 (2000).

15. Laub, M. T., Chen, S. L., Shapiro, L. & McAdams, H. H. Genes directly controlled by CtrA, a master regulator of the Caulobacter cell cycle. Proc Natl Acad Sci U S A 99, 4632–7 (2002).

16. Ren, B. et al. Genome-wide location and function of DNA binding proteins. Science 290, 2306–9 (2000).

17. Holtzendorff, J. et al. Oscillating global regulators control the genetic circuit driving a bacterial cell cycle. Science 304, 983–7 (2004).

18. McAdams, H. H. & Shapiro, L. A bacterial cell-cycle regulatory network operating in time and space. Science 301, 1874–7 (2003).

19. Crosson, S., McGrath, P. T., Stephens, C., McAdams, H. H. & Shapiro, L. Conserved modular design of an oxygen sensory/signaling network with species-specific output. Proc Natl Acad Sci U S A 102, 8018–23 (2005).

20. Biondi, E. G. et al. A phosphorelay system controls stalk biogenesis during cell cycle progression in Caulobacter crescentus. Mol Microbiol 59, 386–401 (2006).

21. Skerker, J. M., Prasol, M. S., Perchuk, B. S., Biondi, E. G. & Laub, M. T. Two-component signal transduction pathways regulating growth and cell cycle progression in a bacterium: a system-level analysis. PLoS Biol 3, e334 (2005).

22. Thieffry, D., Huerta, A. M., Perez-Rueda, E. & Collado-Vides, J. From specific gene regulation to genomic networks: a global analysis of transcriptional regulation in Escherichia coli. Bioessays 20, 433–40 (1998).

23. Gutierrez-Rios, R. M. et al. Regulatory network of Escherichia coli: consistency between literature knowledge and microarray profiles. Genome Res 13, 2435–43 (2003).

24. Salgado, H. et al. RegulonDB (version 5.0): Escherichia coli K-12 transcriptional regulatory network, operon organization, and growth conditions. Nucleic Acids Res 34, D394–7 (2006).

25. Shen-Orr, S. S., Milo, R., Mangan, S. & Alon, U. Network motifs in the transcriptional regulation network of Escherichia coli. Nat Genet 31, 64–8 (2002).

26. Resendis-Antonio, O. et al. Modular analysis of the transcriptional regulatory network of E. coli. Trends Genet 21, 16–20 (2005).

27. Edwards, J. S. & Palsson, B. O. The Escherichia coli MG1655 in silico metabolic genotype: its definition, characteristics, and capabilities. Proc Natl Acad Sci U S A 97, 5528–33 (2000).

28. Almaas, E., Oltvai, Z. N. & Barabasi, A. L. The Activity Reaction Core and Plasticity of Metabolic Networks. PLoS Comput Biol 1, e68 (2005).

29. Edwards, J. S., Ibarra, R. U. & Palsson, B. O. In silico predictions of Escherichia coli metabolic capabilities are consistent with experimental data. Nat Biotechnol 19, 125–30 (2001).

30. Covert, M. W., Schilling, C. H. & Palsson, B. Regulation of gene expression in flux balance models of metabolism. J Theor Biol 213, 73–88 (2001).

31. Van Dien, S. J. & Lidstrom, M. E. Stoichiometric model for evaluating the metabolic capabilities of the facultative methylotroph Methylobacterium extorquens AM1, with application to reconstruction of C(3) and C(4) metabolism. Biotechnol Bioeng 78, 296–312 (2002).

32. Edwards, J. S. & Palsson, B. O. Systems properties of the Haemophilus influenzae Rd metabolic genotype. J Biol Chem 274, 17410–6 (1999).

33. Mahadevan, R. et al. Characterization of Metabolism in the Fe(III)-Reducing Organism Geobacter sulfurreducens by Constraint-Based Modeling. Appl Environ Microbiol 72, 1558–68 (2006).

34. Schilling, C. H. et al. Genome-scale metabolic model of Helicobacter pylori 26695. J Bacteriol 184, 4582–93 (2002).

35. Covert, M. W., Knight, E. M., Reed, J. L., Herrgard, M. J. & Palsson, B. O. Integrating high-throughput and computational data elucidates bacterial networks. Nature 429, 92–6 (2004).

36. Barrett, C. L., Herring, C. D., Reed, J. L. & Palsson, B. O. The global transcriptional regulatory network for metabolism in Escherichia coli exhibits few dominant functional states. Proc Natl Acad Sci U S A 102, 19103–8 (2005).

37. Bochner, B. R., Gadzinski, P. & Panomitros, E. Phenotype microarrays for high-throughput phenotypic testing and assay of gene function. Genome Res 11, 1246–55 (2001).

38. Bailey, J. E. Complex biology with no parameters. Nat Biotechnol 19, 503–4 (2001).

39. McAdams, H. H. & Shapiro, L. Circuit simulation of genetic networks. Science 269, 650–6 (1995).

40. Barkai, N. & Leibler, S. Robustness in simple biochemical networks. Nature 387, 913–7 (1997).

41. Kollmann, M., Lovdok, L., Bartholome, K., Timmer, J. & Sourjik, V. Design principles of a bacterial signalling network. Nature 438, 504–7 (2005).

42. Wadhams, G. H. & Armitage, J. P. Making sense of it all: bacterial chemotaxis. Nat Rev Mol Cell Biol 5, 1024–37 (2004).

43. Alon, U., Surette, M. G., Barkai, N. & Leibler, S. Robustness in bacterial chemotaxis. Nature 397, 168–71 (1999).

44. Kitano, H. Systems biology: a brief overview. Science 295, 1662–4 (2002).

45. Yi, T. M., Huang, Y., Simon, M. I. & Doyle, J. Robust perfect adaptation in bacterial chemotaxis through integral feedback control. Proc Natl Acad Sci U S A 97, 4649–53 (2000).

46. Facciotti, M. T., Bonneau, R., Hood, L. & Baliga, N. S. Systems biology experimental design–considerations for building predictive gene regulatory network models for prokaryotic systems. Current Genomics 5 527–544 (2004).

47. Tomb, J. F. et al. The complete genome sequence of the gastric pathogen Helicobacter pylori. Nature 388, 539–47 (1997).

48. Ng, W. V. et al. Genome sequence of Halobacterium species NRC-1. Proc Natl Acad Sci U S A 97, 12176–81 (2000).

49. Bonneau, R., Baliga, N. S., Deutsch, E. W., Shannon, P. & Hood, L. Comprehensive de novo structure prediction in a systems-biology context for the archaea Halobacterium sp. NRC-1. Genome Biol 5, R52 (2004).

50. Goo, Y. A. et al. Proteomic analysis of an extreme halophilic archaeon, Halobacterium sp. NRC-1. Mol Cell Proteomics 2, 506–24 (2003).

51. Baliga, N. S. et al. Coordinate regulation of energy transduction modules in Halobacterium sp. analyzed by a global systems approach. Proc Natl Acad Sci U S A 99, 14913–8 (2002).

52. Gan, R. R. et al. Proteome analysis of Halobacterium sp. NRC-1 facilitated by the biomodules analysis tool BMSorter. Mol Cell Proteomics (2006).

53. Pleissner, K. P. et al. Web-accessible proteome databases for microbial research. Proteomics 4, 1305–13 (2004).

54. Bumann, D., Meyer, T. F. & Jungblut, P. R. Proteome analysis of the common human pathogen Helicobacter pylori. Proteomics 1, 473–9 (2001).

55. Robb, F. T. et al. Archaea: A laboratory manual (Cold Spring Harbor Laboratory Press, Cold Spring Harbor, NY, 1995).

56. Baliga, N. S. et al. Systems level insights into the stress response to UV radiation in the halophilic archaeon Halobacterium NRC-1. Genome Res 14, 1025–35 (2004).

57. Whitehead, K., Kish, A., Pan M., Reiss, D.J., King, N., Hohmann, L., DiRuggiero, J. & Baliga, N.S. An integrated system approach for understanding cellular responses to gamma radiation. In press at Nature Molecular Systems Biology.

58. Schmid, A. K. et al. In preparation

59. Kaur, A. et al. Survival strategies of an archaeal halophile to withstand stress from transition metals. Genome Research. 16(7): 841–54 (2006).

60. Reeve, J. N., Sandman, K. & Daniels, C. J. Archaeal histones, nucleosomes, and transcription initiation. Cell 89, 999–1002 (1997).
61. Ouhammouch, M. & Geiduschek, E. P. An expanding family of archaeal transcriptional activators. Proc Natl Acad Sci U S A 102, 15423–8 (2005).
62. Baliga, N. S. et al. Is gene expression in Halobacterium NRC-1 regulated by multiple TBP and TFB transcription factors? Mol Microbiol 36, 1184–5 (2000).
63. Facciotti, M. T. et al. Expansion of a family of TBP and TFIIB orthologs in an archaeal organism echoes the development of the eukaryotic basal transcription apparatus in preparation.
64. Fields, S. & Song, O. A novel genetic system to detect protein-protein interactions. Nature 340, 245–6 (1989).
65. Ito, T. et al. A comprehensive two-hybrid analysis to explore the yeast protein interactome. Proc Natl Acad Sci U S A 98, 4569–74 (2001).
66. LaCount, D. J. et al. A protein interaction network of the malaria parasite Plasmodium falciparum. Nature 438, 103–7 (2005).
67. Rain, J. C. et al. The protein-protein interaction map of Helicobacter pylori. Nature 409, 211–5 (2001).
68. Rual, J. F. et al. Towards a proteome-scale map of the human protein-protein interaction network. Nature 437, 1173–8 (2005).
69. Uetz, P. et al. A comprehensive analysis of protein-protein interactions in Saccharomyces cerevisiae. Nature 403, 623–7 (2000).
70. Van Criekinge, W. & Beyaert, R. Yeast Two-Hybrid: State of the Art. Biol Proced Online 2, 1–38 (1999).
71. Ranish, J. A. et al. The study of macromolecular complexes by quantitative proteomics. Nat Genet 33, 349–55 (2003).
72. Jonsson, U. et al. Real-time biospecific interaction analysis using surface plasmon resonance and a sensor chip technology. Biotechniques 11, 620–7 (1991).
73. Conway de Macario, E., Rudofsky, U. H. & Macario, A. J. Surface plasmon resonance for measuring TBP-promoter interaction. Biochem Biophys Res Commun 298, 625–31 (2002).
74. Clyne, M. et al. Helicobacter pylori interacts with the human single-domain trefoil protein TFF1. Proc Natl Acad Sci U S A 101, 7409–14 (2004).
75. Oli, M. W., McArthur, W. P. & Brady, L. J. A whole cell BIAcore assay to evaluate P1-mediated adherence of Streptococcus mutans to human salivary agglutinin and inhibition by specific antibodies. J Microbiol Methods (2005).
76. Bonneau, R. et al. The Inferelator: an algorithm for learning parsimonious regulatory networks from systems biology data sets. Genome Biology. 7(5) : R36 (2006).
77. Reiss, D. J., Bonneau, R. & Baliga, N. S. in BMC Bioinformatics 7:280 (2006).
78. Shannon, P., Reiss, D. J., Bonneau, R. & Baliga, N. S. BMC Bioinformatics 7:176 (2006).
79. Shannon, P. et al. Cytoscape: a software environment for integrated models of biomolecular interaction networks. Genome Res 13, 2498–504 (2003).
80. Wang, G., Kennedy, S. P., Fasiludeen, S., Rensing, C. & DasSarma, S. Arsenic resistance in Halobacterium sp. strain NRC-1 examined by using an improved gene knockout system. J Bacteriol 186, 3187–94 (2004).
81. Gygi, S. P., Rochon, Y., Franza, B. R. & Aebersold, R. Correlation between protein and mRNA abundance in yeast. Mol Cell Biol 19, 1720–30 (1999).
82. Golding, I., Paulsson, J., Zawilski, S. M. & Cox, E. C. Real-time kinetics of gene activity in individual bacteria. Cell 123, 1025–36 (2005).
83. Kislinger, T. et al. Proteome dynamics during C2C12 myoblast differentiation. Mol Cell Proteomics 4, 887–901 (2005).
84. Schmid, A. K. et al. Global whole-cell FTICR mass spectrometric proteomics analysis of the heat shock response in the radioresistant bacterium Deinococcus radiodurans. J Proteome Res 4, 709–18 (2005).
85. Wang, Y. et al. Precision and functional specificity in mRNA decay. Proc Natl Acad Sci U S A 99, 5860–5 (2002).

86. Pratt, J. M. et al. Dynamics of protein turnover, a missing dimension in proteomics. Mol Cell Proteomics 1, 579–91 (2002).

87. Elowitz, M. B., Levine, A. J., Siggia, E. D. & Swain, P. S. Stochastic gene expression in a single cell. Science 297, 1183–6 (2002).

88. Rosenfeld, N., Young, J. W., Alon, U., Swain, P. S. & Elowitz, M. B. Gene regulation at the single-cell level. Science 307, 1962–5 (2005).

89. Hasty, J., McMillen, D. & Collins, J. J. Engineered gene circuits. Nature 420, 224–30 (2002).

90. McDaniel, R. & Weiss, R. Advances in synthetic biology: on the path from prototypes to applications. Curr Opin Biotechnol 16, 476–83 (2005).

91. Gardner, T. S., Cantor, C. R. & Collins, J. J. Construction of a genetic toggle switch in Escherichia coli. Nature 403, 339–42 (2000).

92. Elowitz, M. B. & Leibler, S. A synthetic oscillatory network of transcriptional regulators. Nature 403, 335–8 (2000).

93. Fung, E. et al. A synthetic gene-metabolic oscillator. Nature 435, 118–22 (2005).

94. Atkinson, M. R., Savageau, M. A., Myers, J. T. & Ninfa, A. J. Development of genetic circuitry exhibiting toggle switch or oscillatory behavior in Escherichia coli. Cell 113, 597–607 (2003).

95. Kobayashi, H. et al. Programmable cells: interfacing natural and engineered gene networks. Proc Natl Acad Sci U S A 101, 8414–9 (2004).

96. You, L., Cox, R. S., 3rd, Weiss, R. & Arnold, F. H. Programmed population control by cell-cell communication and regulated killing. Nature 428, 868–71 (2004).

97. Anderson, J. C., Clarke, E. J., Arkin, A. P. & Voigt, C. A. Environmentally controlled invasion of cancer cells by engineered bacteria. J Mol Biol 355, 619–27 (2006).

98. Nemunaitis, J. et al. Pilot trial of genetically modified, attenuated Salmonella expressing the E. coli cytosine deaminase gene in refractory cancer patients. Cancer Gene Ther 10, 737–44 (2003).

99. Martin, V. J., Pitera, D. J., Withers, S. T., Newman, J. D. & Keasling, J. D. Engineering a mevalonate pathway in Escherichia coli for production of terpenoids. Nat Biotechnol 21, 796–802 (2003).

100. Weissman, K. J. & Leadlay, P. F. Combinatorial biosynthesis of reduced polyketides. Nat Rev Microbiol 3, 925–36 (2005).

101. Lange, C. C., Wackett, L. P., Minton, K. W. & Daly, M. J. Engineering a recombinant Deinococcus radiodurans for organopollutant degradation in radioactive mixed waste environments. Nat Biotechnol 16, 929–33 (1998).

102. Gilbert, E. S., Walker, A. W. & Keasling, J. D. A constructed microbial consortium for biodegradation of the organophosphorus insecticide parathion. Appl Microbiol Biotechnol 61, 77–81 (2003).

103. Aldor, I. S. & Keasling, J. D. Process design for microbial plastic factories: metabolic engineering of polyhydroxyalkanoates. Curr Opin Biotechnol 14, 475–83 (2003).

104. Reguera, G. et al. Extracellular electron transfer via microbial nanowires. Nature 435, 1098–101 (2005).

105. Levskaya, A. et al. Synthetic biology: engineering Escherichia coli to see light. Nature 438, 441–2 (2005).

106. Looger, L. L., Dwyer, M. A., Smith, J. J. & Hellinga, H. W. Computational design of receptor and sensor proteins with novel functions. Nature 423, 185–90 (2003).

Cell Engineering

1. M. Al-Rubeai (ed.): *Cell Engineering*. 1999   ISBN 0-7923-5790-6
2. M. Al-Rubeai (ed.): *Cell Engineering*. Vol. 2: Transient Expression. 2000
                                                                   ISBN 0-7923-6596-8
3. M. Al-Rubeai (ed.): *Cell Engineering*. Vol. 3: Glycosylation. 2002
                                                                   ISBN 1-4020-0733-7
4. M. Al-Rubeai and M. Fussenegger (eds.): *Cell Engineering*. Vol. 4: Apoptosis. 2004
                                                                   ISBN 1-4020-2216-6
5. M. Al-Rubeai and M. Fussenegger (eds.): *Cell Engineering*. Vol. 5: Systems Biology.
   2007   ISBN 978-1-4020-5251-4